PROGRESS IN NONLINEAR ANALYSIS RESEARCH

PROGRESS IN NONLINEAR ANALYSIS RESEARCH

ERIK T. HOFFMANN
EDITOR

Nova Science Publishers, Inc.
New York

Library of Congress Cataloging-in-Publication Data

Available upon request.
ISBN 978-1-60456-359-7

Published by Nova Science Publishers, Inc. ✢ New York

CONTENTS

PREFACE

Nonlinear analysis is a broad, interdisciplinary field characterized by a mixture of analysis, topology, and applications. Its concepts and techniques provide the tools for developing more realistic and accurate models for a variety of phenomena encountered in fields ranging from engineering and chemistry to economics and biology. This new book presents recent and important research in the field.

Chapter 1 – This work is motivated by the recent interest in modeling, control, optimization, and stability analysis of systems involving nonstationary Markov chains in discrete and continuous time. By imposing simple conditions and using a spectrum gap property, we have established convergence and ergodicity of certain classes of such Markov chains.

Chapter 2 – The authors construct an age-dependent mathematical epidemic model of chlamydial infection, which fits the demonstrative data accumulated by the STDs (sexually transmitted diseases) surveillance conducted by the Japanese Government. Performing numerical simulations of the model, the authors assess the present/future dynamic phase of chlamydial infection. It follows from the assessment that the present/future situation in chlamydial infection is very critical in Japan.

Chapter 3 – This chapter describes a new class of nonlinear integral equations, which involve endogenous leads presented by the unknown upper limit of integration. Such equations are crucial for a successful investigation of diverse age-dependent mathematical models of significant phenomena in economics, operations research, management sciences, biology, and other scientific areas. Understanding the dynamics of their solutions enhances a progress in solving some important open applied issues. The chapter offers the qualitative analysis and numeric simulation of the integral equations with leads. It answers the question of solvability and describes qualitative properties of the solution. Real-data examples illustrate and confirm presented theoretical outcomes.

Chapter 4 – In this paper, the authors introduce and study a system of nonlinear set-valued implicit variational inclusions with relaxed cocoercive mappings in real Banach spaces. By using the resolvent operator technique for H-accretive operators, we prove the convergence of a new class of perturbed iterative algorithms for solving this system of set-valued implicit variational inclusions in q-uniformly smooth Banach spaces. Our results generalize and improve the corresponding results of recent works.

Chapter 5 – The homogeneous Monge-Ampère equation (HMAE) $u_{xx}u_{yy} - u_{xy}^2 = 0$ describes the developable surfaces in $\mathbb{R}^3$. An explicit formula for its general solution was found by Vitaly Ushakov using a series of changes of the involved variables deduced by geometric arguments. In the present work the authors obtain the general solution of the HMAE by applying a single contact transformation. Further the authors obtain the general solution of the HMAE in the higher-dimensional case using the same approach. The authors also discuss the Lie point symmetries of the HMAE.

Chapter 6 – The focus of this work is on the solutions to some open problems of the global weak solutions of the Cauchy problems for a general nonlinear dissipative partial differential equation

$$\mathbf{u}_t - \varepsilon\Delta\mathbf{u}_t - \alpha\Delta\mathbf{u} + \mathcal{D}\mathbf{u} + \mathcal{N}(\mathbf{u}, \nabla\mathbf{u}) = \mathbf{0}, \qquad \mathbf{u}(\mathbf{x}, 0) = \mathbf{u}_0(\mathbf{x}),$$

in n-dimensional space, where $n \geq 1$ is an integer, $\alpha > 0$ and $0 \leq \varepsilon \leq 1$ are real constants, and $\Delta = \sum_{k=1}^{n} \frac{\partial^2}{\partial x_k^2}$ denotes the classical Laplace operator. More precisely, suppose that the initial function $\mathbf{u}_0 \in L^1(\mathbb{R}^n) \cap H^2(\mathbb{R}^n)$, let $\mathbf{u} = \mathbf{u}(\mathbf{x},t,\mathbf{u}_0)$ represent the global solutions of the Cauchy problem, we will study the limit

$$\lim_{t\to\infty} \left\{ (1+t)^{2m+\lambda+n/2} \int_{\mathbb{R}^n} \left[|\Delta^m \mathbf{u}(\mathbf{x}, t)|^2 + \varepsilon |\nabla \Delta^m \mathbf{u}(\mathbf{x}, t)|^2 \right] d\mathbf{x} \right\}$$

in terms of the initial function $\mathbf{u}_0$ and the model parameters, such as the dissipation coefficient, where $m \geq 0$ is any integer, $\lambda = 0$ if $\int_{\mathbb{R}^n} \mathbf{u}_0(\mathbf{x})d\mathbf{x} \neq 0$, and $\lambda = 1$ if $\int_{\mathbb{R}^n} \mathbf{u}_0(\mathbf{x})d\mathbf{x} = 0$. The limit problem has been open for a long time. The general model includes the n-dimensional Burgers equation, the n-dimensional Benjamin-Bona-Mahony-Burgers equation, the one-dimensional nonlinear cubic Korteweg-de Vries-Burgers equation, the one-dimensional nonlinear Benjamin-Ono-Burgers equation, the two-dimensional nonlinear nonlocal quasi-geostrophic equation, the n-dimensional incompressible Navier-Stokes equations and the n-dimensional incompressible Magnetohydrodynamics equations as particular examples.

The main ideas in the analysis are Fourier transform, Plancherel's identity, new decomposition of frequency space, lower limit estimate and upper limit estimate.

Chapter 7 – The problems of linear stability of steady axial-symmetric sheared jet flows of non-viscous ideally conducting incompressible fluid with free surface in the magnetic field are being investigated. The sufficient conditions for stability, the necessary and sufficient conditions for stability, or the sufficient conditions for instability of these flows regarding small axial-symmetric long-wave perturbations are gained by Lyapunov's direct method. The a priori upper and lower exponential estimates, which are significative of the possible time growth of the investigated small perturbations, are constructed for those stationary flows at issue which turned out to be unstable. The examples of the steady flows and their small perturbations evolving in time according to the constructed estimates are presented.

Chapter 8 – Adaptive control by means of neural networks for nonlinear plant wide dynamical systems is an open but promising issue. For real world applications, practitioners have to paid attention to external disturbances, parameters uncertainty and measurement noise, as long as these factors will influence the stability and robustness of the closed loop system. This chapter presents some of the most popular control schemes based on behavioural models and adaptive control with neural networks. Stability and robustness are discussed and the main difficulties are mentioned: the initialization and pre training phases, the determination of the networks size, and the arbitrary value of the adaptive rate. Then an indirect adaptive control scheme is detailed. This scheme is based on fully connected neural networks and is inspired from the standard real time recurrent learning. It is characterized by a small number of neurons that depends only on the number of system inputs and outputs and by a permanent updating of all parameters. The stability analysis is concerned by combining Lyapunov approach and linearization around the nominal parameters to establish analytical sufficient conditions for the global robust stability of the closed loop system. The scheme is applied to control the Tennessee Eastman Challenge Process. Performance evaluation such as set point stabilization, processing modes changes and disturbances rejection are pointed out, and results are discussed according to the Down and Vogel control objectives.

Chapter 9 – In this work some mathematical aspects induced by the strongly nonlinearity of diffusion equations with convective terms modeling flows in porous media are investigated. Specifically, the interest lies in studying the properties of the solutions to some types of diffusion equations in which the diffusion coefficient and the convective term are nonlinear discontinuous functions of the solution. Particularly, this kind of equations can arise in soils science, describing the water infiltration in nonhomogeneous saturated-unsaturated soils characterized by strongly nonlinear hydraulic properties, but generally, they may be adequate for modeling the dynamics of various fluids in porous materials, as well as other physical diffusion processes, such as those arising in biology.

The mathematical approach is illustrated in the case of fast diffusion equations with flux and Robin boundary conditions and is developed in the framework of the theory of evolution equations with m-accretive nonlinear multivalued operators in Hilbert spaces. First, the study of the existence of the solution to an appropriate abstract approximating problem involving a quasi m-accretive operator will be done. Next, compactness results and a passing to the limit technique will prove the existence of the solution to the original problem. Additional properties of the solutions to some other models will be discussed. The theoretical results will be illustrated at the end by numerical applications to a real problem of water infiltration in nonlinear soils.

Chapter 10 – This work studies the use of fuzzy sets for handling multicriteria decision making problems. The multicriteria approach is needed to solve:

- problems whose solution consequences cannot be estimated with a single criterion;
- problems that, initially, may require a single criterion, but their unique solutions are unachievable, due to the existence of decision uncertainty regions, which can be contracted using additional criteria.

According to this, two classes of models, <*X*, *M*> and <*X*, *R*>, can be constructed. The analysis of <*X*, *M*> models, based on applying the Bellman-Zadeh approach to decision making in a fuzzy environment, is briefly described. The analysis of <*X*, *R*> models is based on four techniques for fuzzy preference modeling. These techniques permit the evaluation, comparison, selection, and/or ordering of alternatives with the use of quantitative estimates, as well as qualitative estimates, based on knowledge, experience, and intuition of professionals. With the availability of different techniques, the most appropriate one can be chosen, considering the sources of information and its uncertainty. To extend the results associated with analyzing <*X*, *R*> models, two rational consensus schemes are discussed.

They permit one to generalize the analysis of <*X*, *R*> models to multiperson multicriteria decision making. These schemes can also be used for evaluating priority weights for criteria in analyzing <*X*, *M*> models. Finally, the results of the present work are illustrated by using a multiperson multicriteria decision making framework to solve an enterprise strategy planning problem, generated with the use of the Balanced Scorecard methodology.

Chapter 11 – The aim of this paper is in a first time to recall some results existing on Secant-type methods and in a second time to study the Steffensen-type method for solving a variational inclusion in the form $0 \in f(x) + G(x)$ where f is a single function and G is a set-valued map. Under a center-Hölder condition on the first order divided difference and using a well-known fixed point theorem for set-valued maps we prove the existence and the superlinear convergence of a sequence (x_k) satisfying $0 \in f(x_k) + [g_1(x_k), g_2(x_k); f](x_{k+1} - x_k) + G(x_{k+1})$ where g_1 and g_2 are some continuous functions parameter.

Chapter 12 – This paper concerns global classical solutions for a class of quasilinear hyperbolic systems of balance laws in one space dimension. It is shown that the Cauchy problem for a class of quasilinear weakly linearly degenerate hyperbolic systems of balance laws with small and decaying initial data admits a unique global C^1 solution $u=u(t, x)$ on $t \geq 0$. This result is also applied to the flow equations of a model class of fluids with viscosity induced by fading memory.

Chapter 13 – This paper deals with the two-dimensional Navier-Stokes equations in which the source term involves a Dirac delta function and describes the elastic reaction of the immersed boundary. The authors analyze the existence of the approximate solution with Dirac delta function approximated by differentiable function. The authors obtain the result via the Banach Fixed Point Theorem and the properties of the solutions to the Navier-Stokes equations of viscous incompressible fluids with periodic boundary conditions.

Chapter 14 – In this paper the authors use the penalty approach for constrained minimization problems in infinite dimensional spaces. A penalty function is said to have the exact penalty property if there is a penalty coefficient for which a solution of an unconstrained penalized problem is a solution of the corresponding constrained problem. The authors discuss the exact penalty property for several large classes of constrained minimization problems.

Chapter 15 – The authors propose a two-stage approach for estimating Value-at-Risk (VaR) that can simultaneously reflect two stylized facts displayed by most asset return series: volatility clustering and the heavy-tailedness of conditional return distributions over short horizons. The proposed method combines the bias-corrected exponentially weighted moving average (EWMA) model for estimating the conditional volatility and the extreme value theory (EVT) for estimating the tail of the innovation distribution. In particular, for minimizing bias in the estimation procedure, the proposed method makes minimal assumptions about the underlying innovation distribution and concentrates on modeling its tail using the non-parametric Hill estimator and uses the moment-ratio Hill estimator for the shape parameter of the extreme value distribution. To validate the model, the authors conducted an empirical investigation on the daily stock market returns of eight emerging Asian markets: China, India, Indonesia, Malaysia, Philippines, South Korea, Taiwan, and Thailand. In addition, the proposed method was compared with J.P. Morgan's RiskMetrics approach. The empirical results show that the proposed method provides a more accurate forecast of VaR for lower probabilities of VaR violation from 0.1% to 1%. Furthermore, the authors demonstrate that applying the Hill estimator to estimate the tail of the innovation distribution can better capture additional downside risk faced during times of greater fluctuation than the second-order moment-ratio Hill estimator.

Chapter 16 – In this chapter, a system of ordinary differential equations with impulse effect at fixed instants is considered. The system is assumed to have the zero solution. It is shown that the existence of a corresponding Lyapunov function is a necessary and sufficient condition for the uniform asymptotic stability of the zero solution. Restrictions on perturbations of the right-hand sides of differential equations and impulse effect are obtained under which the uniform asymptotic stability of the zero solution of the "unperturbed" system implies the uniform asymptotic stability of the zero solution of the "perturbed" system. In the case of a periodic system with impulse effect, it is shown that if the trivial solution of the system is stable or asymptotically stable, then it is uniformly stable or uniformly asymptotically stable, respectively. By using the method of Lyapunov functions, the criteria of asymptotical stability and instability are obtained.

Chapter 17 - Next we review the development and solution of the Moment Problem. We also provide solutions of the Moment Problem for the examples. In addition, we provide the scaled times using the solution of the moment problem for the examples.

In: Progress in Nonlinear Analysis Research
Editor: Erik T. Hoffmann, pp. 1-4

ISBN: 978-1-60456-359-7

Expert Commentary

IMPORTANCE OF NONLINEAR ANALYSIS IN BIOLOGICAL SCIENCES

Vikram Kumar Yeragani[a,b*], Shravya Yeragani[b], Pratap Chokka[b], Manuel Tancer[a] and Karl J. Bar[c]

[a]Department of Psychiatry and Behavioral Neurosciences, Wayne State University School of Medicine, Detroit, Michigan, USA.
[b]University of Alberta, Edmonton, Canada and [c]Friedrich Schiller University, Germany.

Abstract

Recent literature has focused on the importance of nonlinear techniques of analysis in biological sciences, especially in the field of medicine. Traditional statistical terms that appeared for years in medical literature such as means, standard deviations and 'p' values are now increasingly accompanied by "size of the treatment effect" and "odds ratios". Now many articles in medical literature and other biological sciences include measures of nonlinear methods derived from the theories of entropy and deterministic chaos. These indices include fractal dimension, approximate entropy, Lyapunov Exponents and several others. These techniques are borrowed from the physical sciences and appear to have considerable relevance to study the condition of health as well as disease. The eventual application and adaptation of these nonlinear statistical techniques will depend on how relevant these methods are to different biological sciences. Some of these measures appear to be valuable to understand the pathophysiology and prognosis of different diseases. However, one has to understand some of the limitations and pitfalls in the quantification of these indices. It may take considerable amount of time for the readers, authors, reviewers, and especially, clinicians and their patients to understand the background and the importance of nonlinear analysis in medicine. This commentary is an attempt to illustrate some of the advantages and difficulties in the quantification and interpretation of these techniques in medical fields such as cardiology, neurology and psychiatry.

Key Words: linear, nonlinear, variability, ECG, EEG, medicine, biology, disease, mortality, prognosis

---

* E-mail address: Vikramyershr@yahoo.com. Tel # 780-434-1986; Fax: 011-91-80-23610508. Address correspondence to Dr. V.K. Yeragani, Professor of Psychiatry, #411, 11135-83 Ave, Edmonton, Alberta, Canada

Applied mathematics has played a major role in the understanding of differences in various biological measures whether it is between two groups of subjects or in one group of subjects before and after a particular intervention. This is very important to understand the pathophysiology of any disease or to evaluate the effect of treatment of any particular illness. Traditionally, we have examined the means and standard deviations as the measures of 'gold standard'. There is enough reason to do this as most of the physiological measures fall into Gaussian distribution and can be reasonably quantified by mean and standard deviation. If they don't lend themselves to this distribution, we can use the nonparametric statistics such as the Mann-Whitney U test. One can also transform these measures using common logarithm or natural logarithm of these values to make the distributions somewhat 'normal'. However, many of these measures such as the series of heart rate, blood pressure, signals of electroencephalogram (EEG), fluctuations of mood and hormonal levels are irregular, jagged and may not lend themselves to linear statistics so that the above comparisons can be made between different groups of subjects or to evaluate the effects of treatment.

This article mainly focuses on the recent explosion of studies dealing with nonlinear measures. In simple terms, nonlinear measures include measures of regularity such as entropy, measures of complexity such as fractal dimension (FD), symbolic dynamics and measures of predictability such as the Largest Lyapunov exponent (LLE). It's important that the average clinician understands these techniques/measures so that the ultimate goal, the clinical utility of these measures can be achieved. It is rather ironic that often times, we still use the comparisons of means and standard deviations of these so called nonlinear measures even when they are not normally distributed. Of course the use of nonparametric statistics is perfectly reasonable in these instances. From a practical viewpoint, we have noted that most of the time, the results of either parametric or nonparametric tests are about the same in several studies. Hence, one may not have to be too rigid to dismiss the parametric tests in these instances. This appears to be true also with correlational statistics such as Pearson product-moment and Spearman rank-order tests.

We need to closely examine whether it is really important to apply these techniques to clinical sciences. Even though it may be important, does the understanding of complex mathematics justify the application of these techniques to biological sciences? This is the age of computers and we have seen a technological leap in every sphere of human life. One such example would be the transformation of the speed of personal computers. Is it necessary for the common person to have the currently available computer with the fastest speed? In one sense the answer is 'yes', as it lends itself to several advantages such as the video resolution, computer aided designs and the increasing use of video conferencing.

Now that we made a case for the need and utility of nonlinear measures in clinical medicine, the first question is whether these should replace the existing classical-moment statistics such as the means and standard deviations. The answer is a 'no' as it appears that these nonlinear measures are complimentary to the traditional techniques in understanding different diseases and the prediction of the effectiveness of a particular treatment. Hence these are only additional measures and we have no reason to abandon the traditional 'gold standard' measures.

Now let's look at what is so specific about 'biology' or 'nature' and what does 'fractal physics' mean? Nonlinear dynamics studies systems in which the output is not proportional to the input. Mandelbrot dealt with "fractals" extensively in his work, and he defines fractal as irregular but with a self-similar underlying pattern. This self-similarity is obviously apparent at different levels of magnification. Fractal structures have been described in different biological systems such as bronchi, neural networks, vascular branching, and gastrointestinal folds. Examples of fractal structures abound in nature and include the trees, coastlines, mountain ranges, and clouds. Fractal structures have also been described in different biological systems. We should also familiarize ourselves with another concept, 'stationarity'. Quantification of stationarity has importance in several fields including engineering, physics, mathematics, economics, and medicine. The statistical characterization of many random processes is time invariant. A random process that occurs at a particular time is strictly stationary, if the mean is a constant, for all time shifts. Then the autocorrelation function of a stationary random process depends only on the time difference. One should note that not all wide-sense stationary processes are strictly stationary. Here let me give a simple example: when a person receives a drug that increases heart rate depending up on the dosage, the graphical plot would look like a gradual steep slope and the standard deviations of such series will be very high and can't be compared to the resting conditions meaningfully. Here, one has to understand the concept of trends in a particular time series. These can be both linear and nonlinear. Simple linear detrending may be a useful technique before a time series is subjected to spectral analysis and some may argue that other techniques such as cubic detrending may be more relevant in the analysis of time series such as heart rate. However, many of these techniques yield comparable results in the end in most instances. Hence, there may not be a need to make these analyses too complex to understand.

Now let's examine the usefulness of these nonlinear techniques in a few clinical situations. Nonlinear approaches for the quantification of EEG using the FD and entropy can be effective in certain situations such as the study of psychoses and sleep (1-3). Some of these techniques have also been used to predict epileptic seizures (4). The study of cardiovascular system is another fruitful area for the application of nonlinear techniques. There are several studies in regards to heart rate variability in health and disease showing the effectiveness of these techniques (5-7). Similarly these techniques can also be applied to the study of beat-to-beat QT interval variability (8), blood pressure variability (9) and respiratory variability (10, 11). It's also important to note the usefulness of these measures to study mood fluctuations in humans as the linear statistics may not do complete justice to these data (12-14).

Thus there is no doubt that further evaluation of these techniques is needed, especially in medicine. However, one should note the pitfalls of these techniques as these measure may need long stationary time series and need a good bit of mathematical understanding. Once the software programs become user friendly, these measures can be valuable clinical as well as research tools in biological sciences.

References

[1] Keshavan MS, Cashmere JD, Miewald J, Yeragani VK. Decreased nonlinear complexity and chaos during sleep in first episode schizophrenia: a preliminary report. *Schizophrenia research* 2004;71:263-72.

[2] Abasolo D, Hornero R, Espino P, Alvarez D, Poza J. Entropy analysis of the EEG background activity in Alzheimer's disease patients. *Physiological measurement* 2006;27:241-53.

[3] Acharya UR, Faust O, Kannathal N, Chua T, Laxminarayan S. Non-linear analysis of EEG signals at various sleep stages. *Computer methods and programs in biomedicine* 2005;80:37-45.

[4] Adeli H, Ghosh-Dastidar S, Dadmehr N. A wavelet-chaos methodology for analysis of EEGs and EEG subbands to detect seizure and epilepsy. *IEEE transactions on biomedical engineering* 2007;54:205-11.

[5] Yeragani VK, Rao KA, Smitha MR, Pohl RB, Balon R, Srinivasan K. Diminished chaos of heart rate time series in patients with major depression. *Biological psychiatry* 2002;51:733-44.

[6] Rao RK, Yeragani VK. Decreased chaos and increased nonlinearity of heart rate time series in patients with panic disorder. *Auton Neurosci* 2001;88:99-108.

[7] Wu ZK, Vikman S, Laurikka J, Pehkonen E, Iivainen T, Huikuri HV, Tarkka MR. Nonlinear heart rate variability in CABG patients and the preconditioning effect. *Eur J Cardiothorac Surg* 2005;28:109-13.

[8] Yeragani VK, Rao KA. Nonlinear measures of QT interval series: novel indices of cardiac repolarization lability: MEDqthr and LLEqthr. *Psychiatry research* 2003;117:177-90.

[9] Yeragani VK, Mallavarapu M, Radhakrishna RK, Tancer M, Uhde T. Linear and nonlinear measures of blood pressure variability: increased chaos of blood pressure time series in patients with panic disorder. *Depression and anxiety* 2004;19:85-95.

[10] Yeragani VK, Rao R, Tancer M, Uhde T. Paroxetine decreases respiratory irregularity of linear and nonlinear measures of respiration in patients with panic disorder. A preliminary report. *Neuropsychobiology* 2004;49:53-7.

[11] Burioka N, Cornelissen G, Halberg F, Kaplan DT, Suyama H, Sako T, Shimizu E. Approximate entropy of human respiratory movement during eye-closed waking and different sleep stages. *Chest* 2003;123:80-6.

[12] Pincus SM. Approximate entropy as a measure of irregularity for psychiatric serial metrics. *Bipolar disorders* 2006;8:430-40.

[13] Yeragani VK, Pohl R, Mallavarapu M, Balon R. Approximate entropy of symptoms of mood: an effective technique to quantify regularity of mood. *Bipolar disorders* 2003;5:279-86.

[14] Sree Hari Rao V, Raghvendra Rao C, Yeragani VK. A novel technique to evaluate fluctuations of mood: implications for evaluating course and treatment effects in bipolar/affective disorders. *Bipolar disorders* 2006;8:453-66.

In: Progress in Nonlinear Analysis Research
Editor: Erik T. Hoffmann, pp. 5-19
ISBN 978-1-60456-359-7

Chapter 1

TIME-INHOMOGENEOUS MARKOV CHAINS AND ERGODICITY ARISING FROM NONLINEAR DYNAMIC SYSTEMS AND OPTIMIZATION

G. George Yin[1,*], *Son Luu Nguyen*[1,†], *Le Yi Wang*[2,‡] *and Chengzhong Xu*[2,§]
[1]Department of Mathematics, Wayne State University
Detroit, Michigan 48202
[2]Department of Electrical and Computer Engineering,
Wayne State University, Detroit, Michigan 48202

Abstract

This work is motivated by the recent interest in modeling, control, optimization, and stability analysis of systems involving nonstationary Markov chains in discrete and continuous time. By imposing simple conditions and using a spectrum gap property, we have established convergence and ergodicity of certain classes of such Markov chains.

Key Words. nonlinearity, nonstationary Markov chain, ergodicity.

1. Introduction

Owing to the rapid progress in science and technology, there have been increasing demands for modeling, control, and optimization of nonlinear dynamic systems. Revolutionary advancements in internet and communication networks have provided a great opportunity for

[*]E-mail address: gyin@math.wayne.edu. The research of this author was supported in part by the National Science Foundation under grant DMS-0603287, and in part by the National Security Agency under MSPF-068-029.

[†]E-mail address: luu@math.wayne.edu. Research of this author was supported in part by the National Science Foundation under DMS-0624849.

[‡]E-mail address: lywang@ece.eng.wayne.edu. Research of this author was supported in part by the National Science Foundation under ECS-0329597 and DMS-0624849.

[§]E-mail address: czxu@ece.eng.wayne.edu. Research of this author was supported in part by the National Science Foundation under CCF-0611750 and DMS-0624849.

a wide array of system applications. At the same time, they have also introduced substantial challenges on system analysis and design. In this emerging area, the systems often display hybrid features involving both continuous dynamics and discrete events. Thus hybrid system formulation in control and optimization becomes an important task. One particular class of hybrid systems uses a Markov chain to model the discrete events, regime changes, and/or environment dynamics. As a result, the usual difference, or differential, or stochastic differential equations are, by themselves alone, no longer adequate to describe the overall dynamic systems. As a remedy, much effort has been placed on treating systems of the form

$$\begin{aligned} &x_{n+1} = x_n + \varepsilon f(n, \alpha(n)), \quad \text{or} \\ &x_{n+1} = x_n + \varepsilon f(x_n, \alpha(n)) + \sqrt{\varepsilon}\sigma(x_n, \alpha(n)), \quad \text{or} \\ &\dot{x} = f(x(t), \alpha(t)), \quad \text{or} \\ &dx = f(x(t), \alpha(t))dt + \sigma(x(t), \alpha(t))dw. \end{aligned} \tag{1.1}$$

In this formulation, the nonlinear dynamic systems are represented in either discrete time or continuous time. The $f(\cdot)$ and $\sigma(\cdot)$ are some appropriate functions, $\varepsilon > 0$ is a small parameter, and $\alpha(n)$ and $\alpha(t)$ are discrete-time and continuous-time Markov chains, respectively. (Note that the use of the small parameter $\varepsilon > 0$ as step size is motivated by the approach in stochastic approximation; see [12] for details.) In this work, we focus on the case that the Markov chains are time-inhomogeneous or non-stationary. Thus their transition probability matrices or generators are time varying. It is the time-varying nature that makes the underling problems more difficult to deal with.

Much of our current study is motivated by applications arising from a wide variety of situations. For example, the first study on nowadays known as regime-switching asset price models was given in [2]. Subsequent investigations for options pricing were in [3, 5]. One of the earliest slowly varying discrete-time systems was presented in [4]. Regime-switching models for adaptive filtering type algorithms and their corresponding switching diffusion limits were in [14], whereas two-time-scale models and associated limit systems could be found in [17, 18]. For applications of stochastic hybrid systems to communication systems, the reader is referred to [8] and references therein. A systematic study on nonlinear switching diffusion systems is in [21]; see also [19]. Study on randomly switched nonlinear differential equations was given in [22], in which behaviors different from the usual Hartman-Grobman phenomena were discovered. Numerical methods for approximating invariant measures of switching diffusion processes were given in [15].

Many of the nonlinear dynamic systems mentioned above with switching must operate over a long period of time. Thus long-time behavior of the underlying systems plays an important role in control, optimization, and related problems. Take, for instance, the following average cost per unit time problem:

$$\begin{aligned} &\text{Minimize } \frac{1}{n} E \sum_{k=0}^{n} C(x_n, u_n, \alpha(n)), \\ &\text{subject to } x_{n+1} = x_n + f(x_n, u_n, \alpha(n)), \end{aligned} \tag{1.2}$$

where $\alpha(n)$ is a Markov chain with time-varying transition matrices $P(n)$ and state space $\mathcal{M} = \{1, \dots, m\}$, u_n is the control used at time n, $f(\cdot,\cdot,\cdot) : \mathbb{R}^r \times \mathbb{R}^l \times \mathcal{M} \to \mathbb{R}^r$, and $C(\cdot,\cdot,\cdot) : \mathbb{R}^r \times \mathbb{R}^l \times \mathcal{M} \to \mathbb{R}$ is an appropriate cost function. Because of the time-varying nature of the transition matrices $P(n)$, the calculation using instantaneous measures is far from trivial. However, when n is large, we may approximate the instantaneous measure by that of the stationary measure (if it exists). This will substantially reduce the amount of computational effort. Another problem deals with stability of systems described in (1.1). In any event, an issue crucial to these problems is the ergodicity of the associated Markov chain, which is the main concern of this paper.

In this work, we study finite state Markov chains. There are several reasons for this consideration. First, many applications require handling of Markov chains with a finite state space. Second, for nonlinear systems, we usually cannot obtain closed form solutions. In carrying out approximations, finite state space cases are computationally more trackable. In addition, Markov chains with a countable state space may be approximated by suitable finite state space models. The rationale of our analysis is to use time-homogeneous quantities to approximate the time-inhomogeneous characteristics. We use simple analysis argument to establish the desired results. The main idea lies in utilizing irreducibility in an essential way.

The rest of the paper is arranged as follows. Section 2 begins with the formulation of the problem in discrete time, in which a number of definitions are provided. Section 3 concentrates on ergodicity of nonstationary Markov chains. Section 4 proceeds with studies on continuous-time Markov chains. Section 5 provides some discussions together with supporting examples. Finally, Section 6 concludes the paper with a few more remarks.

2. Time-inhomogeneous Markov Chains in Discrete Time

In this section, we consider time-inhomogeneous or nonstationary Markov chains in discrete time. Let $\alpha(k)$ be a discrete-time Markov chain with a state space $\mathcal{M} = \{1, 2, \ldots, m\}$, and time-varying one-step transition matrices $\{P(k)\} = \{(p_{ij}(k))\}$. Define

$$Y(k) = (P(\alpha(k) = 1), \ldots, P(\alpha(k) = m)) \in \mathbb{R}^{1\times m}.$$

Associated with the transition matrices $P(k)$, the forward equation

$$Y(k+1) = Y(k)P(k),\ Y(0) = Y_0 \tag{2.1}$$

describes the dynamics of the probability distributions, where $Y_0 = (y_{0,1}, \ldots, y_{0,m}) \in \mathbb{R}^{1\times m}$ with $y_{0,i} \geq 0$ and $\sum_{i=1}^{m} y_{0,i} = 1$). Iterating on (2.1), it is easily seen that for any positive integer k_0,

$$Y(k) = Y(k_0)T(k|k_0),$$

where $Y(k_0)$ is a probability vector, and $T(k|k_0)$ is a product of the transition matrices representing the $(k-k_0)$-step transition matrix given by

$$T(k|k_0) = \begin{cases} \prod_{j=k_0}^{k-1} P(j) = P(k-1)\cdots P(k_0), & \text{if } k > k_0, \\ I, & \text{if } k = k_0. \end{cases} \tag{2.2}$$

It has been known that this product is crucial in studying properties of the corresponding stochastic processes in discrete time. There is a well-known result by Kesten and Furstenberg for product of random matrices [10]; see also [6], [11], and references therein. To study long-time properties of probability distributions or transition probabilities, we shall also concentrate on the asymptotics of this product of matrices.

In many dynamic systems, and the associated control and optimization problems, one often wishes to replace the instantaneous probability measures by their ergodic measures so as to substantially ease the computational effort. To facilitate such an effort, we need to have a thorough understanding of the ergodic measures. Crucial to many systems is: Under what conditions, will the systems be ergodic? Even before answering this question, because we are now working with time-inhomogeneous Markov chains, we need to make sure that we have workable definitions of ergodicity. In what follows, we present the definitions of weak and strong ergodicity.

Definition 2.1 For a time-inhomogeneous Markov chain $\alpha(k)$, if for any $k_0 \geq 0$ and any two probability vectors u and $v \in \mathbb{R}^{1\times m}$,

$$\lim_{k\to\infty} |Y^u(k) - Y^v(k)| = 0, \tag{2.3}$$

where $Y^u(k)$ and $Y^v(k)$ denote the solutions of (2.1) with $Y(k_0) = u$ and $Y(k_0) = v$, respectively, then the corresponding Markov chain is said to be weakly ergodic.

Definition 2.2 For the time-inhomogeneous Markov chain $\alpha(k)$, if for any $k_0 \geq 0$ and any probability vector $u \in \mathbb{R}^{1\times m}$, there is a probability vector $\xi \in \mathbb{R}^{1\times m}$ such that

$$\lim_{k\to\infty} Y^u(k) = \xi, \tag{2.4}$$

where $Y^u(k)$ denotes the solution of (2.1) with initial data $Y(k_0) = u$, then the corresponding Markov chain is said to be strongly ergodic.

Remark 2.3. Note that in both Definitions of 2.1 and 2.2, the ergodicity is independent of the initial conditions, which coincides with the usual ergodicity definitions for stationary Markov chains.

3. Ergodicity

We pose the following condition.

(A1) There is a positive integer $n_0 > 0$ such that $P(n_0)$ is irreducible and aperiodic.

Note that for the fixed n_0, $P(n_0)$ is a constant transition matrix. Denote by $\mathbb{1} \in \mathbb{R}^m$ the vector consisting of 1's in all of its entries. The irreducibility and aperiodicity in (A1) imply that there is a stationary distribution $\nu(n_0)$ such that

$$\begin{aligned} &[P(n_0)]^k \to \mathbb{1}\nu(n_0) \text{ as } k \to \infty, \\ &|[P(n_0)]^k - \mathbb{1}\nu(n_0)| \leq K\lambda^k \end{aligned} \tag{3.1}$$

for some $0 < \lambda < 1$ and some $K > 0$. That is, the k-step transition matrix $[P(n_0)]^k$ approaches a matrix with identical rows. Each of its rows consists of the stationary distribution $\nu(n_0)$. Moreover, the difference of $[P(n_0)]^k$ and $\mathbb{1}\nu(n_0)$ goes to zero exponentially fast. Note that the second line of (3.1) is often referred to as a spectrum gap condition. The irreducibility implies the convergence of the n-step transition matrix in the first line of (3.1). It is easily seen that $P(n_0)$ has an eigenvalue 1; the aperiodicity implies that all other eigenvalues are inside the unit disk. This in turn yields the second line of (3.1). Using (A1), we proceed to establish weak ergodicity and strong ergodicity next.

Theorem 3.1 *Suppose that the time-inhomogeneous Markov chain* $\alpha(k)$ *satisfies condition* (A1). *Then* $\alpha(k)$ *is weakly ergodic.*

Proof. Using the variation of constant formula for (2.1), we obtain that

$$\begin{aligned} Y(k+1) &= Y(k)P(n_0) + Y(k)[P(k) - P(n_0)] \\ &= Y(k_0)[P(n_0)]^{k+1-k_0} + \sum_{j=k_0}^{k} Y(j)[P(j) - P(n_0)][P(n_0)]^{k-j}. \end{aligned} \tag{3.2}$$

With different initial data u and v, denote the corresponding solutions by $Y^u(k+1)$ and $Y^v(k+1)$, respectively. Then using (3.2), by adding and subtracting $\mathbb{1}\nu(n_0)$,

$$\begin{aligned} &Y^u(k+1) - Y^v(k+1) \\ &\quad= (u-v)[P(n_0)]^{k+1-k_0} + \sum_{j=k_0}^{k} [Y^u(j) - Y^v(j)](P(j) - P(k_0))[P(n_0)]^{k-j} \\ &\quad= (u-v)[P(n_0)^{k+1-k_0} - \mathbb{1}\nu(n_0)] \\ &\qquad+ \sum_{j=k_0}^{k} [Y^u(j) - Y^v(j)](P(j) - P(n_0))[P(n_0)]^{k-j} \\ &\quad= (u-v)[P(n_0)^{k+1-k_0} - \mathbb{1}\nu(n_0)] \\ &\qquad+ \sum_{j=k_0}^{k} [Y^u(j) - Y^v(j)](P(j) - P(k_0))[P(n_0)^{k-j} - \mathbb{1}\nu(n_0)]. \end{aligned} \tag{3.3}$$

In the third line and the last line of (3.3), we have used

$$(u-v)\mathbb{1} = 0 \text{ and } (P(j) - P(k_0))\mathbb{1} = 0, \text{ respectively.}$$

The well-known Gronwall's inequality yields that

$$\begin{aligned} &|Y^u(k+1) - Y^v(k+1)| \\ &\quad\le K|u-v||[P(n_0)]^{k+1-k_0} - \mathbb{1}\nu(n_0)| \\ &\qquad\times \exp\Big[\sum_{j=k_0}^{k} |(P(j) - P(n_0))||[P(n_0)]^{k-j} - \mathbb{1}\nu(n_0)|\Big] \\ &\quad\le K\lambda^{k+1-k_0} \to 0 \text{ as } k \to \infty. \end{aligned} \tag{3.4}$$

The desired weak ergodicity then follows. □

Note that in the above, only for some n_0, $P(n_0)$ being irreducible and aperiodic is needed. The condition used appears to be weaker than that of [9, 13, 20] and references therein. Roughly, in the aforementioned references, they require that for each k, $P(k)$ is irreducible. In [9, 20], the authors require the associated stationary distributions $\nu(k)$ in fact be slowly varying in the sense that

$$\sum_{k=1}^{\infty} |\nu(k+1) - \nu(k)| < \infty. \tag{3.5}$$

The convergence of the infinite series above implies that $\nu(k+1) - \nu(k) \to 0$ as $k \to \infty$. In view of (3.5), not only are the variations of $\nu(k+1) - \nu(k)$ small, but also they diminish sufficiently fast. Here in our development, we shall require only for some n_0, $P(n_0)$ be irreducible and aperiodic. In lieu of time-varying $P(k)$, we then make use of the constant matrix $P(n_0)$ to obtain the desired convergence property.

Theorem 3.2 *Suppose that the time-inhomogeneous Markov chain $\alpha(k)$ satisfies condition* (A1). *Then $\alpha(k)$ is strongly ergodic.*

Proof. Consider $Y(n_0) = u$. Then in view of (2.1) and (2.2), we have

$$Y(k+1) = uT(k+1|k_0). \tag{3.6}$$

However, using the variation of constant formula (3.2) and using $u\mathbb{1} = 1$, (3.6) may also be written as

$$\begin{aligned}
Y(k+1) &= u\{[P(n_0)]^{k+1-k_0} - \mathbb{1}\nu(n_0)\} + u\mathbb{1}\nu(n_0) \\
&\quad + \sum_{j=k_0}^{k} Y(j)[P(j) - P(n_0)][P(n_0)]^{k-j} \\
&= u\{[P(n_0)]^{k+1-k_0} - \mathbb{1}\nu(n_0)\} + \nu(n_0) \\
&\quad + \sum_{j=k_0}^{k} Y(j)[P(j) - P(n_0)]\{[P(n_0)]^{k-j} - \mathbb{1}\nu(n_0)\}.
\end{aligned} \tag{3.7}$$

We claim that (i) a finite limit $\lim_{k\to\infty} Y(k+1)$ exists, and (ii) the limit is a probability vector. The claims will be verified below.

To verify (i), we note that by virtue of spectrum gap condition–the second inequality in (3.1),

$$u\{[P(n_0)]^{k+1-k_0} - \mathbb{1}\nu(n_0)\} \to 0 \ \text{ as } k \to \infty. \tag{3.8}$$

Moreover, by virtue of the boundedness of $Y(j)$ and $P(j) - P(n_0)$,

$$\begin{aligned}
&\left| \sum_{j=k_0}^{k} Y(j)[P(j) - P(n_0)]\{[P(n_0)]^{k-j} - \mathbb{1}\nu(n_0)\} \right| \\
&\quad \le \sum_{j=k_0}^{k} |Y(j)||P(j) - P(n_0)||[P(n_0)]^{k-j} - \mathbb{1}\nu(n_0)| \\
&\quad \le K \sum_{j=k_0}^{k} \lambda^{k-j} \le K < \infty,
\end{aligned}$$

where K is a generic positive constant independent of k. In the above, we have used the convention $K + K = K$ and $KK = K$. We will also use such a convention henceforth. Therefore, the series

$$\lim_{k\to\infty}\sum_{j=k_0}^{k} Y(j)[P(j) - P(n_0)]\{[P(n_0)]^{k-j} - \mathbb{1}\nu(n_0)\} \tag{3.9}$$

converges absolutely and uniformly. Thus, using (3.8) and (3.9) in (3.7), we obtain (i) as claimed. In fact, the limit has the representation

$$\lim_{k\to\infty} Y(k+1) = \nu(n_0) + \lim_{k\to\infty}\sum_{j=k_0}^{k} Y(j)[P(j) - P(n_0)]\{[P(n_0)]^{k-j} - \mathbb{1}\nu(n_0)\}. \tag{3.10}$$

To prove (ii), we note that (3.6) yields that $Y(k) \geq 0$ for all k. (Here, by $Y(k) \geq 0$, we mean that each component of $Y(k)$ is great than or equal to 0.) Thus the limit vector $\lim_{k\to\infty} Y(k+1) \geq 0$. It thus suffices to verify that

$$\sum_{i=1}^{m}(\lim_{k\to\infty} Y(k+1))_i = 1, \tag{3.11}$$

i.e., the components of the limit vector add up to 1.

Note that

$$\begin{aligned}&\nu(n_0)\mathbb{1} = 1, \text{ and}\\ &\{[P(n_0)]^{k-j} - \mathbb{1}\nu(n_0)\}\mathbb{1} = \mathbb{1} - \mathbb{1} = 0 \ \text{ for each } j \leq k.\end{aligned}$$

To prove (3.11), multiplying from the right of (3.10) by the column vector $\mathbb{1}$, we obtain that

$$\begin{aligned}\lim_{k\to\infty} Y(k+1)\mathbb{1} &= [\nu(n_0)\\ &\quad + \lim_{k\to\infty}\sum_{j=k_0}^{k} Y(j)[P(j) - P(n_0)]\{[P(n_0)]^{k-j} - \mathbb{1}\nu(n_0)\}]\mathbb{1}\\ &= 1 + \lim_{k\to\infty}\sum_{j=k_0}^{k} Y(j)[P(j) - P(n_0)]\{[P(n_0)]^{k-j} - \mathbb{1}\nu(n_0)\}\mathbb{1}\\ &= 1.\end{aligned} \tag{3.12}$$

Thus (ii) is verified. This completes the proof of the theorem. □

4. Continuous-time Markov Chains

In this section, we are concerned ourselves with continuous-time Markov chains that are time-inhomogeneous. Again, we consider finite-state cases.

Suppose that we have a Markov chain $\alpha(t)$ with state space $\mathcal{M} = \{1, \ldots, m\}$ and time-varying generator $Q(t) \in \mathbb{R}^{m\times m}$.

Denote $X(t) = (P(\alpha(t) = 1), \ldots, P(\alpha(t) = m))$. Then $X(t)$ satisfies the forward equation

$$\frac{dX(t)}{dt} = X(t)Q(t). \tag{4.1}$$

The solution of (4.1) with initial data $X(t_0)$ can be represented by

$$X(t) = X(t_0)\Phi(t|t_0), \tag{4.2}$$

where $\Phi(t|t_0)$ is known as the principal matrix solution to (4.1) that satisfies $\Phi(t_0|t_0) = I$; see [7]. Next, we define the ergodicities as follows.

Definition 4.1 For a time-inhomogeneous Markov chain $\alpha(t)$, if for any $t_0 \geq 0$ and any two probability vectors u and $v \in \mathbb{R}^{1\times m}$,

$$\lim_{t\to\infty} |X^u(t) - X^v(t)| = 0, \tag{4.3}$$

where $X^u(t)$ and $X^v(t)$ denote the solutions of (4.1) satisfying $X(t_0) = u$ and $X(t_0) = v$, respectively, then the corresponding Markov chain is said to be weakly ergodic.

Definition 4.2 For the time-inhomogeneous Markov chain $\alpha(t)$, if for any $t_0 \geq 0$ and any probability vector $u \in \mathbb{R}^{1\times m}$, there is a probability vector $\xi \in \mathbb{R}^{1\times m}$ such that

$$\lim_{t\to\infty} X^u(t) = \xi, \tag{4.4}$$

where $X^u(t)$ denotes the solution of (4.1) with initial data $X(t_0) = u$, then the corresponding Markov chain is said to be strongly ergodic.

We pose the following condition. The main requirement is irreducibility of one generator of the Markov chain.

(A2) The $Q(\cdot)$ is Lipschitz continuous with a Lipschitz constant L. There is a constant $\tau > 0$ such that $Q(\tau)$ is irreducible in the sense that the system of equations

$$\begin{cases} fQ(\tau) = 0 \\ f\mathbb{1} = 1 \end{cases}$$

has a unique nonnegative solution.

Remark 4.3. Similar to the case of discrete-time Markov chains. Condition (A2) implies that there is a stationary distribution $\nu(\tau) \in \mathbb{R}^{1\times m}$ such that

$$\begin{aligned} &\exp(Q(\tau)t) \to \mathbb{1}\nu(\tau) \ \text{ as } \ t \to \infty, \\ &|\exp(Q(\tau)t) - \mathbb{1}\nu(\tau)| \leq K\exp(-\kappa t), \end{aligned} \tag{4.5}$$

for some $\kappa > 0$. Again, we refer to the inequality in the second line above as the spectrum gap conditions.

Theorem 4.4 *Suppose that the time-inhomogeneous Markov chain $\alpha(t)$ satisfies condition (A2). Then $\alpha(t)$ is weakly ergodic.*

Proof. Working with (4.1), we write its solution as

$$X(t) = X(t_0)\exp(Q(\tau)(t-t_0)) + \int_{t_0}^{t} X(s)[Q(s)-Q(\tau)]\exp(Q(\tau)(t-s))ds. \quad (4.6)$$

Consider two different initial conditions $X(t_0) = u$ and $X(t_0) = v$. For the initial data u and v, denote the associated solution of (4.1) by $X^u(t)$ and $X^v(t)$, respectively. Note that

$$\begin{aligned}&(u-v)\mathbb{1}\nu(\tau) = \nu(\tau) - \nu(\tau) = 0,\\ &[Q(s)-Q(\tau)]\mathbb{1}\nu(\tau) = 0.\end{aligned}$$

Then

$$\begin{aligned} X^u(t) - X^v(t) &= (u-v)\exp(Q(\tau)(t-t_0)) \\ &\quad + \int_{t_0}^{t} (X^u(s) - X^v(s))[Q(s)-Q(\tau)]\exp(Q(\tau)(t-s))ds \\ &= (u-v)[\exp(Q(\tau)(t-t_0)) - \mathbb{1}\nu(\tau)] \\ &\quad + \int_{t_0}^{t} (X^u(s) - X^v(s))[Q(s)-Q(\tau)][\exp(Q(\tau)(t-s)) - \mathbb{1}\nu(\tau)]ds. \end{aligned} \quad (4.7)$$

In the above, we have used

$$(u-v)\mathbb{1} = 0 \ \text{ and } \ (Q(s)-Q(\tau))\mathbb{1} = 0 \ \text{ for each } s.$$

Taking norm in (4.7) and applying the well-known Gronwall's inequality, we obtain

$$\begin{aligned} |X^u(t) - X^v(t)| &\le K|\exp(Q(\tau)(t-t_0)) - \mathbb{1}\nu(\tau)| \\ &\quad + K\int_{t_0}^{t} |s-\tau||\exp(Q(\tau)(s-t_0)) - \mathbb{1}\nu(\tau)| \\ &\quad \times |\exp(Q(\tau)(t-s)) - \mathbb{1}\nu(\tau)| \\ &\quad \times \int_{s}^{t} |\zeta-\tau||\exp(Q(\tau)(t-\zeta)) - \mathbb{1}\nu(\tau)|d\zeta ds \\ &\le K\exp(-\kappa(t-t_0)) \\ &\quad + K(t-t_0)\exp(-\kappa(t-t_0))\int_{t_0}^{t}\int_{s}^{t} |\zeta-\tau|\exp(-\kappa(t-\zeta))d\zeta \\ &\to 0 \ \text{ as } t\to\infty. \end{aligned} \quad (4.8)$$

Thus the weak ergodicity is established. □

Theorem 4.5 *Suppose that the time-inhomogeneous Markov chain $\alpha(t)$ satisfies condition* (A2). *Then $\alpha(t)$ is strongly ergodic.*

Proof. For the initial data $X(t_0) = u$, using (4.6), we obtain

$$X(t) = u\exp(Q(\tau)(t-t_0)) + \int_{t_0}^{t} X(s)[Q(s)-Q(\tau)]\exp(Q(\tau)(t-s))ds. \quad (4.9)$$

Note that as $t \to \infty$, for the second term on the right-hand side of (4.9), by the spectrum gap condition (4.5) and noting the boundedness of $X(t)$ (since it is a probability vector),

$$\begin{aligned}
&\left|\int_{t_0}^{t} X(s)[Q(s)-Q(\tau)]\exp(Q(\tau)(t-s))ds\right| \\
&\quad \le K\int_{t_0}^{t} |X(s)||s-\tau|\exp(-\kappa(t-s))ds \\
&\quad \le K\int_{t_0}^{t} |s-\tau|\exp(-\kappa(t-s))ds.
\end{aligned}$$

Thus

$$\lim_{t\to\infty}\int_{t_0}^{t} X(s)[Q(s)-Q(\tau)]\exp(Q(\tau)(t-s))ds \text{ converges.} \tag{4.10}$$

As for the first term on the right-hand side of (4.9),

$$u\exp(Q(\tau)(t-t_0)) \to u\mathbb{1}\nu(\tau) = \nu(\tau) \text{ as } t\to\infty. \tag{4.11}$$

Therefore, (4.10) and (4.11) yield that $\lim_{t\to\infty} X(t) = \xi$ exists. We next demonstrate that the limit is a probability vector.

Denote $\xi = (\xi_1, \ldots, \xi_m)$. First, all the components ξ_i satisfy $\xi_i \ge 0$. This can be seen easily from (4.2) and the probability meaning of $X(t)$. Next we verify that $\xi\mathbb{1} = 1$. Note that

$$\begin{aligned}
\exp(Q(\tau)(t-s)) &= I + \frac{Q(\tau)(t-s)}{1!} + \frac{Q^2(\tau)(t-s)^2}{2!} + \frac{Q^3(\tau)(t-s)^3}{3!} + \cdots \\
&= \sum_{i=0}^{\infty} \frac{Q^i(\tau)(t-s)^i}{i!}.
\end{aligned}$$

As a result,

$$\exp(Q(\tau)(t-s))\mathbb{1} = \mathbb{1}.$$

In addition, $\mathbb{1}\nu(\tau)\mathbb{1} = \mathbb{1}$since $\nu(\tau)\mathbb{1} = 1$. Thus, we have

$$\begin{aligned}
\lim_{t\to\infty} X(t)\mathbb{1} &= 1 + \lim_{t\to\infty}\int_{t_0}^{t} X(s)[Q(s)-Q(\tau)]\exp(Q(\tau)(t-s))\mathbb{1}ds \\
&= 1 + \lim_{t\to\infty}\int_{t_0}^{t} X(s)[Q(s)-Q(\tau)]\mathbb{1}ds \\
&= 1.
\end{aligned}$$

Hence the desired condition is verified. The proof of the theorem is concluded. □

5. Examples

5.1. Slowly-varying Chains

One class of Markov chains that are of particular interest is slowly-varying chains in discrete time. Such chains are allowed to have large jump changes. However the changes are infrequent. If the two consecutive transition matrices differ only slightly, we quantify the corresponding chain as slowly-time-varying chain. Under suitable conditions, corresponding to $P(k)$, there is an associated stationary distribution $\nu(k)$. In the literature, [9, 20] defined a slowly Markov chain as one whose stationary distributions satisfy (3.5). Thus, by slow chains, it was meant that the associated stationary distributions change slowly. We say that a time-inhomogeneous Markov chain $\alpha(k)$ is said to be slowly varying, or in short a slow chain, if the one-step transition matrices satisfy

$$P(k+1) - P(k) = O(\varepsilon), \quad \text{for each } k \geq 0 \text{ and for some } \varepsilon > 0. \tag{5.1}$$

As a particular example of slowly-varying Markov chains, we consider a Markov chain whose transition probability matrix is given by

$$P(k) = P^{\varepsilon}(k) = P + \varepsilon Q(k), \tag{5.2}$$

where $\varepsilon > 0$ is a sufficiently small parameter, P is a constant transition matrix that is irreducible and aperiodic, and for each k, $Q(k)$ is a generator of a continuous-time Markov chain. Such slowly-varying chains have been used in wireless communications, discrete optimization, multi-user detection, and many other applications; see [17] and references therein. For this class, we obtain the following ergodicity result.

Proposition 5.1. *Assume that the constant transition matrix P given in* (5.2) *is irreducible and aperiodic and that the function $Q(\cdot) : \mathbb{R} \to \mathbb{R}^{m\times m}$ is bounded. Then the associated Markov chain is both weakly ergodic and strongly ergodic.*

Proof. We proceed to valid the assertions. The proof is divided into two parts covering weak ergodicity and strong ergodicity.

(i) To prove the weak ergodicity, as in the proof of Theorem 3.1,

$$\begin{aligned} Y(k+1) &= Y(k)P + \varepsilon Y(k)[Q(k)] \\ &= Y(k_0)P^{k+1-k_0} + \varepsilon \sum_{j=k_0}^{k} Y(j)[Q(j)]P^{k-j}. \end{aligned} \tag{5.3}$$

Corresponding to P, denote the stationary distribution by ν, Then,

$$\begin{aligned} &Y^u(k+1) - Y^v(k+1) \\ &= (u-v)P^{k+1-k_0} + \varepsilon \sum_{j=k_0}^{k} [Y^u(j) - Y^v(j)]Q(j)P^{k-j} \\ &= (u-v)[P^{k+1-k_0} - \mathbb{1}\nu] + \varepsilon \sum_{j=k_0}^{k} [Y^u(j) - Y^v(j)]Q(j)P^{k-j} \\ &= (u-v)[P^{k+1-k_0} - \mathbb{1}\nu] \\ &\quad + \varepsilon \sum_{j=k_0}^{k} [Y^u(j) - Y^v(j)]Q(j)[P^{k-j} - \mathbb{1}\nu]. \end{aligned} \tag{5.4}$$

Using $Q(j)\mathbb{1} = 0$ for each j, for some $0 < \lambda < 1$, similar as before, the well-known Gronwall's inequality yields that

$$\begin{aligned}
&|Y^u(k+1) - Y^v(k+1)| \\
&\leq K|u-v||P^{k+1-k_0} - \mathbb{1}\nu| \exp\Big[\sum_{j=k_0}^{k} |Q(j)||[P^{k-j} - \mathbb{1}\nu|\Big] \\
&\leq K\lambda^{k+1-k_0} \to 0 \text{ as } k \to \infty.
\end{aligned} \tag{5.5}$$

The desired weak ergodicity then follows.

(ii) To prove the strong ergodicity, we write

$$\begin{aligned}
Y(k+1) &= u\{P^{k+1-k_0} - \mathbb{1}\nu\} + u\mathbb{1}\nu + \varepsilon \sum_{j=k_0}^{k} Y(j)[Q(j)]P^{k-j} \\
&= u\{[P^{k+1-k_0} - \mathbb{1}\nu\} + \nu + \varepsilon \sum_{j=k_0}^{k} Y(j)Q(j)\{P^{k-j} - \mathbb{1}\nu\}.
\end{aligned} \tag{5.6}$$

We then proceed that the limit $\lim_k Y(k)$ exists and the limit is a probability vector. The arguments are similar to the proof of Theorem 3.2. We thus omit the details. □

5.2. Fast-varying Chains in Continuous-time

Consider a continuous-time Markov chain with generator given by

$$Q(t) = \frac{Q}{\varepsilon} + \widehat{Q}(t), \tag{5.7}$$

where both Q and $\widehat{Q}(t)$ are generators of suitable Markov chains. We note that using the variation of constant, we can write the solution of (4.1) as

$$X(t) = X(t_0)\exp(\frac{Q(t-t_0)}{\varepsilon}) + \int_{t_0}^{t} X(s)\widehat{Q}(s)\exp(\frac{Q(t-s)}{\varepsilon})ds. \tag{5.8}$$

We state the following results, but omit the verbatim proof for brevity. The proof is similar to the weak and strong ergodicity for the continuous-time systems.

Proposition 5.2. *Suppose that the generator Q is irreducible, and that $\widehat{Q}(\cdot)$ has polynomial growth of its argument. Then the associated Markov chain $\alpha(t)$ is both weakly ergodic and strongly ergodic.*

5.3. An Example of a Controlled Dynamic System

In the last section, we examined a time-inhomogeneous model with a parameter $\varepsilon > 0$ involved. Our study was concerned with a fixed $\varepsilon > 0$. A related problem in under the framework of the two-time-scale formulation, in which $\varepsilon \to 0$. The problem is closely related to the problem studied in the previous sections, but with somewhat different formulation. This class of Markov chains has been motivated by our recent study in two-time-scale approach for Markovian systems; see [16]. For the many applications in manufacturing, production planning, and queueing theory, we refer the reader to [16].

We illustrate the problem by considering a controlled dynamic system for $t \in [0, T]$ for a $T > 0$ as follows:

$$\begin{aligned} &\text{Minimize } E \int_0^T C(x^\varepsilon(t), u(t), \alpha^\varepsilon(t))dt \\ &\text{subject to } \dot{x}^\varepsilon(t) = f(x^\varepsilon(t), u(t), \alpha^\varepsilon(t)), \end{aligned}$$

where $\alpha^\varepsilon(t)$ is a continuous-time Markov chain with generator $Q/\varepsilon + \widehat{Q}(t)$ and state space $\mathcal{M} = \{1, \ldots, m\}$, $f(\cdot, \cdot, \cdot) : \mathbb{R}^r \times \mathbb{R}^l \times \mathcal{M} \mapsto \mathbb{R}$, $C(\cdot, \cdot, \cdot) : \mathbb{R}^r \times \mathbb{R}^l \times \mathcal{M} \mapsto \mathbb{R}$ is known as the cost rate function, and $u(\cdot)$ is the control used.

Assume that Q is irreducible. Using the singular perturbation techniques developed in [16], we can show that there is a limit problem:

$$\begin{cases} \text{Minimize } \displaystyle\int_0^T \overline{C}(x(t), u(t))dt \\ \text{subject to } \dot{x} = \overline{f}(x(t), u(t)), \end{cases}$$

where

$$\begin{aligned} \overline{C}(x, u) &= \sum_{i=1}^m \nu_i C(x, u, i) \\ \overline{f}(x, u) &= \sum_{i=1}^m \nu_i f(x, u, i), \end{aligned}$$

and ν is the stationary distribution associated with Q. Then we can use comparison control techniques to design controls for the original process leading to near optimality.

6. Concluding Remarks

Originated from nonlinear system analysis, this paper is devoted to inhomogeneous Markov chains. Both discrete-time and continuous-time problems are treated. The motivation of our study stems from a wide range of applications where nonlinear dynamic systems involve both the usual dynamics and discrete events. These discrete events are modeled by Markov chains and they jump change at random times. The switching processes are used to depict random environment, sudden changes in the systems, and uncertain factors that cannot be covered by the usual deterministic dynamic systems. We have focused on Markov chains having finite state space. The rationale is that in the usual computation using digital computers, only finite state cases can be handled. In addition, countable state spaces can also be approximated by finite state spaces with large number of states. Since such systems are often in operation for a long period of time, the large time behavior of the systems is important. In this paper, under fairly simple conditions, we have established the ergodicity of the Markov chains. The results obtained will be useful for many control, optimization, and stability analysis of many nonlinear dynamic systems subject to regime switching driven by Markov chains.

References

[1] Artzrouni, M. (1996). On the dynamics of the linear process $Y(k) = A(k)Y(k-1)$ with irreducible matrices $A(k)$, *SIAM J. Matrix Anal.*, Vol. 17, pp. 822–833.

[2] Barone-Adesi, G. & Whaley, R. (1987). Efficient analytic approximation of American option values, *J. Finance*, Vol. 42, pp. 301–320.

[3] Buffington, J. & Elliott, R.J. (2002). American options with regime switching, *Internat. J. Theoretical Appl. Finance*, Vol. 5, pp. 497–514.

[4] Desor, C.A. (1970). Slowly varying discrete system, *Electronics Lett.*, Vol. 6, pp. 339–340.

[5] Di Masi, G.B., Kabanov, Y.M., & Runggaldier, W.J. (1994). Mean variance hedging of options on stocks with Markov volatility, *Theory of Probability and Applications*, Vol. 39, pp. 173–181.

[6] Guo, L. (1993). *Time-varying Stochastic Systems*, Jilin Sci. Tech. Press.

[7] Hale, J. (1980). *Ordinary Differential Equations*, 2nd Ed., R.E. Krieger Pub. Co., Malabar, FL.

[8] Hespanha, J.P. (2004). *Stochastic Hybrid Systems: Application to Communication Networks*, Springer, Berlin.

[9] Issason, D.L. (1988). Conditions for strong ergodicity using intensity matrices, *J. Appl. Probab.*, Vol. 25, pp. 34–42.

[10] Kesten, H. & Furstenberg, H. (1960). Products of random matrices, *Ann. Math. Statist.* Vol. 31, pp. 457–469.

[11] Khasminskii, R.Z. (1980). *Stochastic Stability of Differential Equations* , Sijthoff and Noordhoff, Alphen aan den Rijn, Netherlands.

[12] Kushner, H.J. & Yin, G. (2003). *Stochastic Approximation and Recursive Algorithms and Applications*, 2nd Edition, Springer-Verlag, New York.

[13] Wu, J.-W. & Chen, K,-C. (2007). On the ergodicity of slow-varying nonstationary Markov chains, preprint.

[14] Yin, G. & Krishnamurthy, V. (2005). Least mean square algorithms with Markov regime switching limit, *IEEE Tran. Automatic Control*, Vol. 50, pp. 577–593.

[15] Yin, G., Mao, X.R., & Yin, K. (2005). Numerical approximation of invariant measures for hybrid diffusion systems, *IEEE Trans. Automat. Control*, Vol. 50, pp. 577–593.

[16] Yin, G. & Zhang, Q. (1998). *Continuous-time Markov Chains and Applications: A Singular Perturbations Approach* , Springer-Verlag, New York, NY.

[17] Yin, G. & Zhang, Q. (2005). *Discrete-time Markov Chains: Two-time-scale Methods and Applications*, Springer, New York.

[18] Yin, G. Zhang, Q., & Badowski, G. (2000). Asymptotic properties of a singularly perturbed Markov chain with inclusion of transient states, *Ann. Appl. Probab.*, Vol. 10, pp. 549–572.

[19] Yin, G. & Zhu, C. (2007). On the notion of weak stability and related issues of hybrid diffusion systems, *Nonlinear Anal.: Hybrid System*, Vol. 1, pp. 173–187.

[20] Zeifman, A.I. & Issacson, D.L. (1994). On strong ergodicity for nonhomogeneous continuous-time Markov chains, *Stochastic Process. Appl.*, Vol. 50, pp. 263–273.

[21] Zhu, C. & Yin, G. (2007). Asymptotic properties of hybrid diffusion systems, *SIAM J. Control Optim.*, Vol. 46, pp. 1155–1179.

[22] Zhu, C., Yin, G., & Song, Q.S., Stability of random-switching systems of differential equations, to appear in *Quarterly Appl. Math.*

In: Progress in Nonlinear Analysis Research
Editor: Erik T. Hoffmann, pp. 21-29

ISBN: 978-1-60456-359-7

Chapter 2

A MATHEMATICAL-MODEL APPROACH TO CHLAMYDIAL INFECTION IN JAPAN

Minoru Tabata[1*], Toshitake Moriyama[2], Satoru Motoyama[2] and Nobuoki Eshima[3]

[1]Department of Mathematical Sciences, Graduate School of Engineering / School of Engineering, Osaka Prefecture University, Sakai, Osaka 599-8531, Japan
[2]Department of Obstetrics and Gynecology, Kobe University, Graduate School of Medicine, Kobe 650-0017, Japan
[3]Department of Statistics, Faculty of Medicine, Oita University, Oita 879-5593 Japan

Abstract

We construct an age-dependent mathematical epidemic model of chlamydial infection, which fits the demonstrative data accumulated by the STDs (sexually transmitted diseases) surveillance conducted by the Japanese Government. Performing numerical simulations of the model, we assess the present/future dynamic phase of chlamydial infection. It follows from the assessment that the present/future situation in chlamydial infection is very critical in Japan.

Keywords: age-dependent mathematical epidemic model, chlamydial infection, numerical simulation, STDs surveillance in Japan, discrete dynamical system

Introduction

Recently in Japan the amount of infected persons with STDs has increased very rapidly because of drastic changes in sexual habits and practices. In order to assess the present dynamic phase of infection of STDs, the Japanese Ministry of Health, Labor and Welfare conducted large-scale surveillance over STDs from 1998 to 2002 (see [1-4]). The recovery

[*] E-mail address: mnrtabata@luck.ocn.ne.jp. Fax: +81-72-254-9916. Correspondence should be addressed to: Minoru Tabata, Department of Mathematical Sciences, Graduate School of Engineering / School of Engineering, Osaka Prefecture University, Gakuen-cho, Sakai, Osaka 599-8531, Japan

rate of the surveillance is 84.4%, and the population in the region where statistics were gathered is 24.7% of the Japanese population. All the Japanese take out the *Japanese National Health Insurance*. Hence, if a male/female is symptomatically infected with STDs, then he/she can consult a doctor easily with the aid of the health insurance system, and the doctor notifies his/her infection to the surveillance. Hence, we can consider that almost all persons detected being infected with STDs are counted by the surveillance. The result of surveillance shows that all STDs infection increase very rapidly. In particular, chlamydial infection increases more rapidly than the other STDs infection and reaches over about 30% of all the STDs infection. Hence, in the present paper we will study chlamydial infection in Japan.

We must note that chlamydial infection has two different forms; one is a *symptomatic* form, and the other is *asymptomatic*. If a person is detected being infected with chlamydia, then he/she is counted by the surveillance. Hence, each symptomatically infected person is counted. However, the number of asymptomatically infected persons is not considered at all in the surveillance, because there are no effective measures to assess the number of asymptomatically infected males/females. Practically, it is impossible to examine an unspecified number of persons for chlamydial infection from the viewpoint of individual privacy protection. Therefore, this surveillance cannot accurately reflect a real-life extensive situation of chlamydial infection in Japan. In this paper we will analyze the real-life extensive situation by taking a mathematical-model approach to the demonstrative date accumulated by the surveillance. Performing numerical simulations of an age-dependent mathematical epidemic model that fits the demonstrative data, we will assess the amount of asymptomatically infected males/females. From the assessment, we find that the present/future situation in chlamydial infection is critical in Japan, e.g., at present about 11% of all 23-year-old females are asymptomatically infected, and after 10 years about 28% of all 23-year-old females will be asymptomatically infected.

The birth rate in Japan has decreased rapidly, and it is expected that the Japanese population will begin to decrease in 2007. Since the most troubling sequela with chlamydial infection is infertility, the spread of chlamydial infection will make the birthrate decrease more rapidly, which will cause extremely serious social problems in Japan.

Chlamydial Epidemiology

For simplicity we assume that the population is constant without regard to age, sex, and year. By rescaling the unit of population, we can impose the following assumption on the model with no loss of generality:

Assumption 1. The population of males/females of j years old is equal to 1 for each age $j \geq 0$ in each year.

Although chlamydia trachomatis is a pathogen that causes disease such as conjunctivitis, pneumonia, and so on, in the present paper we will regard chlamydial infection as a venereal disease. Because chlamydial infection is transmitted from infected persons to susceptible persons mainly through heterosexual intercourse and partly through homosexual intercourse, for simplicity we consider that chlamydia is transmitted only through heterosexual intercourse. Moreover, it follows from the STDs surveillance that there are only a few persons infected with chlamydia under 15 years of age or over 44 years old (see [1-4]). Therefore we impose the following assumption on the model:

Assumption 2. Chlamydial infection is transmitted from infected males/females between the ages of 15 and 44 to susceptible females/males between the ages of 15 and 44 only through heterosexual intercourse.

In general Japanese males (females, respectively) show a tendency to have sexual intercourse with females (males, respectively) who are younger (older, respectively) than themselves. Hence, in addition to Assumption 2, we impose the following assumption on the model:

Assumption 3. For each $j = 15,...,44$, each male (female, respectively) of j years old have sexual intercourse only with females (males, respectively) between the ages of $\max\{j-E,15\}$ and $\min\{j+F,44\}$ ($\max\{j-F,15\}$ and $\min\{j+E,44\}$, respectively), where E and F are nonnegative integers such that $F < E$.

We denote the rate of the number (per year) of newly asymptomatically infected males (females, respectively) in the total number (per year) of newly infected males (females, respectively) by c_m (c_f, respectively), where c_m and c_f are positive constants such that c_m, $c_f <$ 1. If the total number (per year) of newly infected males is equal to a positive constant C_0, then the number (per year) of newly asymptomatically (symptomatically, respectively) infected males is equal to

$$c_m C_0 \quad ((1-c_m)C_0, \text{ respectively}). \tag{1}$$

On the basis of the results in [1-4] and [7], we reasonably assume that

$$c_m = 1/2, \; c_f = 4/5. \tag{2}$$

If a person is infected asymptomatically, then he/she develops few symptoms and does not consult a doctor, but there is the possibility that he/she is cured by chance. For example, if he/she is newly infected symptomatically with other STDs like gonorrhea, then he/she develops the symptoms of those STDs and consults a doctor, who is supposed to detect him/her being infected also with chlamydia. Moreover, pregnant females are supposed to be examined for preventing prenatal chlamydial infection. Furthermore also patients with sterility are supposed to be examined for chlamydial infection for screening. By b_m (b_f, respectively) we denote the rate (per year) of the number (per year) of asymptomatically infected males (females, respectively) thus cured in the total number of asymptomatically infected males (females, respectively), where b_m and b_f are positive constants such that b_m, b_f < 1. Considering how asymptomatically infected males/females are detected by chance, we easily see that asymptomatically infected females can be detected more easily than asymptomatically infected males, i.e., that

$$b_m < b_f. \tag{3}$$

The Mathematical Model

We define a discrete dynamical system that describes our model. In what follows throughout the paper we assume that i is a nonnegative integer denoting the *year variable* and that j denotes the *age variable* such that $j = 15,...,44$. By $I_{ma} = I_{ma}(i,j)$ ($I_{fa} = I_{fa}(i,j)$, respectively) we denote the number of asymptomatically infected males (females, respectively) of j years old in a year i. By $I_{ms} = I_{ms}(i,j)$ ($I_{fs} = I_{fs}(i,j)$, respectively) we denote the number of symptomatically infected males (females, respectively) of j years old in a year i. By $I_m = I_m(i,j)$ ($I_f = I_f(i,j)$, respectively) we denote the number of infected males (females, respectively) of j years old who can infect susceptible persons in a year i. If a male/female is infected asymptomatically, then he/she can infect susceptible persons until he/she is cured by chance. However, if a male/female is infected symptomatically, then he/she develops the symptoms immediately and consults a doctor, i.e., there are only a few days for him/her to infect susceptible persons. Hence we easily see that

$$I_m(i,j) = \delta I_{ma}(i,j) + \varepsilon I_{ms}(i,j),\ \ I_f(i,j) = \delta I_{fa}(i,j) + \varepsilon I_{fs}(i,j), \tag{4}$$

where δ (ε, respectively) is a positive constant such that $0 \leq \delta \leq 1$ ($0 \leq \varepsilon \leq 1$, respectively), and is in proportion to the mean length of period such that asymptomatically (symptomatically, respectively) infected persons can infect susceptible persons. We reasonably consider that δ is very close to 1 and ε is very close to 0. Hence, for simplicity we assume that $\delta = 1$ and $\varepsilon = 0$ in what follows.

Even if a person infected with chlamydia is completely cured, then he/she can be infected again. Hence our model needs to be a *criss-cross SI model* (see, e.g., [6, p. 329]). We denote the number of susceptible females (males, respectively) of j years old in a year i by $S_f = S_f(i,j)$ ($S_m = S_m(i,j)$, respectively). By Assumption 1, we see that

$$S_m(i,j) = 1 - I_m(i,j),\ \ S_f(i,j) = 1 - I_f(i,j). \tag{5}$$

In the same way as [6, pp. 327-331], we assume that the number (per year) of newly infected males is in proportion to the product of the number of susceptible males and the number of infected females who have sexual intercourse with the susceptible males. Hence, making use of Assumption 3, we see that the number (per year) of newly infected males of j years old is in proportion to the product of the number of susceptible males of j years old and the number of infected females who are between the ages of $\max\{j-E,15\}$ and $\min\{j+F,44\}$. Hence we see that the number (per year) of newly infected males of j years old is equal to

$$a_m(i,j)S_m(i,j) \sum_{k=\max\{j-E,15\}}^{\min\{j+F,44\}} I_f(i,k), \tag{6}$$

where $a_m = a_m(i,j)$ denotes the coefficient of proportion, which is a nonnegative-valued function of i and j.

Recalling the definitions of b_m and c_m (see (1)), and making use of (6), we see that

$$I_{ma}(i+1,j+1)-I_{ma}(i,j) = -b_m I_{ma}(i,j)+c_m a_m(i,j)S_m(i,j) \sum_{k=\max\{j-E,15\}}^{\min\{j+F,44\}} I_f(i,k), j=15,...,43. \quad (7)$$

Since each person of j years old becomes j+1 years old one year later, we see that the left-hand side of (7) represents the growth (per year) of the number of asymptomatically infected males (see [6, pp. 361-362]). The first term of the right-hand side represents the decrease (per year) of the number of asymptomatically infected males, which is caused by the fact that asymptomatically infected males are cured by chance. By (6) we see that the last term of the right-hand side represents the number (per year) of males of j years old who become asymptomatically infected in a year i.

Applying (4-5) with $\delta = 1$ and $\varepsilon = 0$ to (7), we obtain

$$I_{ma}(i+1,j+1)-I_{ma}(i,j)=-b_m I_{ma}(i,j)+c_m a_m(i,j)(1-I_{ma}(i,j)) \sum_{k=\max\{j-E,15\}}^{\min\{j+F,44\}} I_{fa}(i,k), j=15,...,43. \quad (8)$$

In the same way, we have

$$I_{fa}(i+1,j+1)-I_{fa}(i,j) = -b_f I_{fa}(i,j)+c_f a_f(i,j)(1-I_{fa}(i,j)) \sum_{k=\max\{j-F,15\}}^{\min\{j+E,44\}} I_{ma}(i,k), j = 15,...,43, \quad (9)$$

where $a_f = a_f(i,j)$ is a nonnegative-valued function of i and j. From Assumption 2, we see that no person of 15 years old is infected asymptomatically. Hence we make the following boundary conditions:

$$I_{ma}(i,15) = 0, I_{fa}(i,15) = 0, \text{ for each } i. \quad (10)$$

Our discrete dynamical system is defined by (8-10).

The STDs surveillance picked up each male/female detected being infected. By $\mathbf{J}_m = \mathbf{J}_m(i,j)$ ($\mathbf{J}_f = \mathbf{J}_f(i,j)$, respectively) we denote the number of infected males (females, respectively) of j years old thus picked up in a year i, where i = 1998,...,2002 and j = 15,....,44. Their graphs are described in Figs. 1-2. By $J_m = J_m(i,j)$ ($J_f = J_f(i,j)$, respectively) we denote the quantity that represents $\mathbf{J}_m = \mathbf{J}_m(i,j)$ ($\mathbf{J}_f = \mathbf{J}_f(i,j)$, respectively) in the model. Recalling the definition of the model, we see that $J_m = J_m(i,j)$ is equal to the sum of the number (per year) of asymptomatically infected males of j years old who are detected by chance in a year i and the number (per year) of males of j years old who are symptomatically infected in a year i. Hence, recalling (1), in the same way as (8-9) we deduce that

$$J_m(i,j) = b_m I_{ma}(i,j) + (1-c_m)a_m(i,j)(1-I_{ma}(i,j)) \sum_{k=\max\{j-E,15\}}^{\min\{j+F,44\}} I_{fa}(i,k), j = 15,...,44. \quad (11)$$

In the same way as (11) we have

$$J_f(i,j) = b_f I_{fa}(i,j) + (1-c_f)a_f(i,j)(1-I_{fa}(i,j)) \sum_{k=max\{j-F,15\}}^{min\{j+E,44\}} I_{ma}(i,k), j = 15,...,44. \quad (12)$$

We note that $\mathbf{J}_m = \mathbf{J}_m(i,j)$ and $\mathbf{J}_f = \mathbf{J}_f(i,j)$ are the demonstrative data accumulated by the surveillance, and that $J_m = J_m(i,j)$ and $J_f = J_f(i,j)$ are their corresponding data obtained from the model. If $J_m = J_m(i,j)$ and $J_f = J_f(i,j)$ are close to $\mathbf{J}_m = \mathbf{J}_m(i,j)$ and $\mathbf{J}_f = \mathbf{J}_f(i,j)$ respectively for $i = 1998,...,2002$ and $j = 15,...,44$, then we can consider that the model fits the demonstrative data of the surveillance.

The Infection Coefficients

We call $a_m = a_m(i,j)$ and $a_f = a_f(i,j)$ the *infection coefficients*. No large-scale surveillance over STDs has been conducted before 1997 in Japan. However, it is well known among a large number of doctors in Japan that the number (per year) of males/females detected being infected with chlamydia was almost constant before 1983. Moreover, it is known that the number of such males/females is about one fifth as many as the mean value of numbers of infected males/females picked up by the STDs surveillance between 1998 and 2002 (see [1-4]). Hence we reasonably impose the following assumption on the model:

Assumption 4. (i) Before 1983 the discrete dynamical system reached an equilibrium.

(ii) For each age j, the mean value of numbers of infected males/females of j years old picked up by the STDs surveillance is quintuple as many as the number (per year) of infected males/females of j years old detected in each year before 1983.

The spread of STDs seemed to be tangible in around 1985 in Japan. For example, the first patient with AIDS in Japan was found in 1986, which had a great impact on the Japanese society. Moreover, many Japanese sociologists consider that sexual practices and habits changed drastically from around 1985 to around 1994 (see, e.g., [5]). Hence, we reasonably consider that the change made the infection coefficients increase rapidly from around 1985 to around 1994. Moreover, we reasonably consider that the infection coefficients in 1995 are several times as large as those before 1983. Therefore we impose the following assumption on the model:

Assumption 5. (i) If $i \leq 1983$, then $a_m = a_m(i,j)$ and $a_f = a_f(i,j)$ are equal to positive constants that are independent of the year variable i and dependent on the age variable j. If $1983 < i \leq 1995$, then $a_m = a_m(i,j)$ and $a_f = a_f(i,j)$ increase rapidly with i for each j. If $i \geq 1995$, then $a_m = a_m(i,j)$ and $a_f = a_f(i,j)$ increase slowly with i for each j.

(ii) $a_m(1995,j)$ and $a_f(1995,j)$ are close to $\gamma a_m(1983,j)$ and $\gamma a_f(1983,j)$ respectively for each age j, where γ is a constant such that $\gamma > 1$.

From Assumptions 4-5, we see that if $i \leq 1983$, then

$$I_{ma}(1983,j) = I_{ma}(i,j),\ I_{fa}(1983,j) = I_{fa}(i,j),\ a_m(1983,j) = a_m(i,j), \text{ for each } j. \quad (13)$$

Applying (13) with $i = 1982$ to (8) with $i = 1982$, we deduce that

$$I_{ma}(1983,j+1) - I_{ma}(1983,j) = -b_m I_{ma}(1983,j) + c_m a_m(1983,j)(1-I_{ma}(1983,j)) \sum_{k=\max\{j-E,15\}}^{\min\{j+F,44\}} I_{fa}(1983,k). \tag{14}$$

Eliminating the summation term of the right-hand side of (14) by making use of (11) with $i = 1983$, we obtain

$$I_{ma}(1983,j+1) - I_{ma}(1983,j) = -b_m I_{ma}(1983,j)/(1-c_m) + c_m J_m(1983,j)/(1-c_m). \tag{15}$$

By Assumption 4, (ii), we reasonably define $J_m = J_m(1983,j)$ as follows:

$$J_m(1983,j) = (1/5)\{\mathbf{J}_m(1998,j)+\cdots+\mathbf{J}_m(2002,j)\}/5, \text{ for each } j. \tag{16}$$

Substituting (16) in (15), and making use of the boundary conditions (10) with $i = 1983$, we can obtain $I_{ma} = I_{ma}(1983,j)$ for each $j = 15,...,44$. In the same way we obtain $I_{fa} = I_{fa}(1983,j)$ for each $j = 15,...,44$. Substituting these functions in (14), we can obtain $a_m = a_m(1983,j)$. In the same way we obtain $a_f = a_f(1983,j)$. By Assumption 5, we reasonably consider that if $i \geq 1995$, then $a_m = a_m(i,j)$ and $a_f = a_f(i,j)$ are in the neighborhood of $\gamma a_m(1983,j)$ and $\gamma a_f(1983,j)$ for each age j.

Numerical Simulations

If we impose an initial condition on (8-10), then we can perform numerical simulations of the model, i.e., we can obtain $I_{ma} = I_{ma}(i,j)$ and $I_{fa} = I_{fa}(i,j)$ numerically. Furthermore, by (11-12) we obtain $J_m = J_m(i,j)$ and $J_f = J_f(i,j)$. Since the surveillance began in 1998, we should make such an initial condition when $i = 1998$. By Assumption 5 we can consider that if $1995 \leq i \leq 2002$, then $a_m = a_m(i,j)$ and $a_f = a_f(i,j)$ in the neighborhood of $\gamma a_m(1983,j)$ and $\gamma a_f(1983,j)$ respectively for each j. For simplicity of numerical simulations, we reasonably assume that if $1995 \leq i \leq 2002$, then $a_m = a_m(i,j)$ and $a_f = a_f(i,j)$ are independent of i and dependent on j. We denote them by $a_m = a_m(j)$ and $a_f = a_f(j)$.

By performing a large number of numerical simulations of the model with the initial data $I_{ma}(1998,j)$ and $I_{fa}(1998,j)$ that are changed variously, with the parameters γ, E, F, b_m, and b_f that are changed variously, and with the infection coefficients $a_m = a_m(j)$ and $a_f = a_f(j)$ that are changed variously in the neighborhood of $\gamma a_m(1983,j)$ and $\gamma a_f(1983,j)$, we determine the initial data, the parameter, and the infection coefficients in such a way that $J_m = J_m(i,j)$ and $J_f = J_f(i,j)$ are as close as possible to $\mathbf{J}_m = \mathbf{J}_m(i,j)$ and $\mathbf{J}_f = \mathbf{J}_f(i,j)$ for $i = 1998,...,2002$ and $j = 15,...,44$. The infection coefficients thus determined are described in Figs. 3-4, the initial data thus determined are described in Figs. 5-6 when $i = 1998$, and the parameters are determined as follows:

$$\gamma = 1.2, E = 5, F = 2, b_m = 0.03, b_f = 0.07. \tag{17}$$

Applying these quantities to (8-12), we obtain Figs. 5-6 when $i = 1999,...,2002$ and Figs. 7-10.

Comparing Figs. 1-2 and Figs. 7-8, we see that $J_m = J_m(i,j)$ and $J_f = J_f(i,j)$ are sufficiently close to $\mathbf{J}_m = \mathbf{J}_m(i,j)$ and $\mathbf{J}_f = \mathbf{J}_f(i,j)$ respectively, i.e., that the model fits the demonstrative data. Furthermore, from the numerical simulations of the model, we see that if the initial data, the infection coefficients, and the parameters leave the functions and the values that are obtained in Figs. 3-4, (17), and Figs. 5-6 with $i = 1998$, then $J_m = J_m(i,j)$ and $J_f = J_f(i,j)$ leave $\mathbf{J}_m = \mathbf{J}_m(i,j)$ and $\mathbf{J}_f = \mathbf{J}_f(i,j)$ respectively. Moreover, from studies by cervical scraping that are performed with pregnant females and nursing students as the control group, we can obtain almost the same result as Fig. 6 when $j = 18,...,23$ (see [7]). Therefore we can conclude that the model can well describe the real-life dynamic phase of chlamydial infection. In particular, by virtue of the numerical simulations of the model, we can grasp the real-life situation of asymptomatic cases, which has been an enigma until now.

Discussion

By inspecting Figs. 3, 4, we see that the infection coefficients have the following properties: (i) The infection coefficient of females is much larger than that of males in the young generation. (ii) The infection coefficient of females of more than 33 years old is almost equal to 0 and is smaller than that of elder males.

The property (i) is explained from the following facts (a) and (b): (a) Chlamydial infection is transmitted more easily from males to females than from females to males. (b) Young persons are more easily infected with chlamydia than elder persons. The fact (a) is attributed to the anatomical difference between males and females. The fact (b) is explained from the following facts (c) and (d): (c) Young persons are more active than elder persons in sexual practice. (d) Young Japanese have few knowledge on chlamydial infection because of lack of sex education (see, e.g., [5]).

The majority of elder persons are married in Japan. Moreover, in extramarital sexual practice married females are much more inactive than married males, which is caused by Japanese sexual customs derived from Confucian culture. The inactivity of married females not only cancels out the effect caused by the fact (a) but also give the property (ii) to the infection coefficients in the elder generation.

From Figs. 5, 6, we see that chlamydial infection is predominant in females. The predominance is derived from the property (i) and the facts (c) and (d). Moreover, inspecting Figs. 5, 6 more fully, we find the following contrast between males and females in chlamydial infection: the numbers of asymptomatically infected males between the ages of 30 and 34 are larger than asymptomatically infected males of other ages, but the number of asymptomatically infected females increases rapidly from 15-year-old to 23-year-old and decrease rapidly from 24-year-old. The contrast is explained from the properties (i-ii) and from the fact that asymptomatically infected females can be detected more easily than asymptomatically infected males for the reason mentioned in obtaining (3).

Inspecting Figs. 5-6, we can conclude that the present situation in chlamydial infection is very critical. For example, about 11% of all 23-year-old females are asymptomatically infected in 2002. We need to regard chlamydial infection not as an exclusive disease of commercial sex workers but as a sexually transmitted infection which has already widely permeated the nation. However, inspecting Figs. 9-10, we see that the future dynamic phase of chlamydial infection is more critical than the present. For example, after ten years about

28% of all 23-year-old females will be asymptomatically infected. As already mentioned in the introduction, the spread of chlamydial infection will make the birthrate decrease. If active countermeasures against the spread of chlamydial infection are not taken in Japan, then the Japanese society will encounter the serious problem in the near future.

References

[1] Y. Kumamoto, T. Tsukamoto, I. Nishiya and et al. *Epidemiological Survey of Sexually Transmitted Disease Prevalence in Japan – Sentinel Surveillance of STD in 1999* -, 11 (1): 72-103, (in Japanese) (2000).

[2] Y. Kumamoto, T. Tsukamoto, I. Nishiya and et al. *Epidemiological Survey of Sexually Transmitted Disease Prevalence in Japan – Sentinel Surveillance of STD in 2000* -, 12 (1): 32-67, (in Japanese) (2001).

[3] Y. Kumamoto, T. Tsukamoto, I. Nishiya and et al. *Epidemiological Survey of Sexually Transmitted Disease Prevalence in Japan – Sentinel Surveillance of STD in 2001* -, 13 (2): 147-167, (in Japanese) (2002).

[4] Y. Kumamoto, T. Tsukamoto, I. Nishiya and et al. *Epidemiological Survey of Sexually Transmitted Disease Prevalence in Japan – Sentinel Surveillance of STD in 2002* -, 15 (1): 17-45, (in Japanese) (2004).

[5] S. Miyadai, "*A Choice of Schoolgirls*", Kodansha, (in Japanese) (1994).

[6] J. D. Murray, "*Mathematical Biology*", I, Third Edition, *Interdisciplinary Applied Mathematics,* Vol. 17, Springer, (2002).

[7] M. Saito, Y. Kimura, Y. Kumamoto, S. Gotoh, K. Miyake, T Ohishi, N Ishibuchi and Y Sotokawa. Relationship between Test Results of Chlamydia Trachomatis Screening by Vaginal Discharges and Sexual Behavior – Study in Nursing Students (Interim Report). *Japanese Journal of Sexually Transmitted Diseases*, **12** (1) : 136-140, (2001).

In: Progress in Nonlinear Analysis Research
Editor: Erik T. Hoffmann, pp. 31-49

ISBN: 978-1-60456-359-7

Chapter 3

ON A NEW CLASS OF NONLINEAR INTEGRAL EQUATIONS WITH LEADS

Natali Hritonenko[1,*] *and Yuri Yatsenko*[2,**]
[1]Department of Mathematics, Prairie View A&M University,
Prairie View, TX 77446, USA
[2]College of Business and Economics, Houston Baptist University,
Houston TX 77074, USA

Abstract

This chapter describes a new class of nonlinear integral equations, which involve endogenous leads presented by the unknown upper limit of integration. Such equations are crucial for a successful investigation of diverse age-dependent mathematical models of significant phenomena in economics, operations research, management sciences, biology, and other scientific areas. Understanding the dynamics of their solutions enhances a progress in solving some important open applied issues. The chapter offers the qualitative analysis and numeric simulation of the integral equations with leads. It answers the question of solvability and describes qualitative properties of the solution. Real-data examples illustrate and confirm presented theoretical outcomes.

1. Introduction

Integral equations have been intensively used in modeling of various applications, namely, in modeling of scientific and engineering processes when a physical force is distributed in a certain area and its impact is not local, but accumulated over the area. For many problems such as "the scattering of waves from an object which is not quite a sphere, the propagation of waves … the use of integral equations offers methods of utility and power" (Morse and Feshbach, 1953). The differential equations are simpler and more common in analytical modeling and numeric simulation, but the integral equations represent a more general tool and

[*] E-mail address: nahritonenko@pvamu.edu
[**] E-mail address: yyatsenko@hbu.edu

describe global situations that can not be modeled by the differential equations. Although the derived equations of motion are often differential, the original general physical conservation laws have an integral form and even a slight modification of the differential equation model may require going back to its original integral equation form. Moreover, if an analysis of differential equation models meets some difficulties, such as complicated boundary geometry or boundary conditions, instability of corresponding numerical algorithms, they are investigated using the integral equations. The integral equations have a better flexibility for accounting new phenomena, better stability of numerical methods, established investigation techniques, and other advantages. The integral equations are also used to prove basic differential equation properties such as solvability and stability.

The *integral dynamic models* take into account the after-effect (hereditary effects or delay) when a continuous sequence of the past states of a dynamic system impacts the future evolution of the system. Such effects cannot be described by the ordinary differential equations. Boltzman applied the integral dynamic models to physics to describe elastic persistence in XIX century. Vito Volterra developed the Boltzman theory and applied the integral models to population ecology. New engineering, economic, biological, and demographical problems have brought a new type of models that involve the optimal control of endogenous delays. The delay is reflected by unknowns in the lower limits of integration and can reflect the age of individuals to be harvested or the lifetime of obsolete equipment to be replaced. A mathematical investigation of corresponding optimal control problems, in particular, their extremum conditions, leads to solving integral equations with the unknown upper limit of integration that represents the future lifetime of system components introduced at current time (Yatsenko, 1995; Yatsenko & Hritonenko, 2005, 2007). Such equations are referred to as the *integral equations with leads*. It is of great importance to develop the qualitative theory and computational tools for solving the integral equations with endogenous delays and leads and demonstrate how they can be used in specific applied areas. In this chapter we concentrate on the following integral equation in x with endogenous leads

$$\int_{t}^{(x(t))^{-1}} \rho(\tau)[q(x(\tau),\tau)-q(t,\tau)]d\tau = \rho(t)p(t), \quad t\in[0,\infty). \tag{1}$$

A similar equation was first obtained in (Malcomson, 1975) but not investigated there due to its complexity. The main novelty of equation (1) is the inverse of an unknown in the upper limit. The equations with unknown "leads" $(x(t))^{-1}$ are always nonlinear. The nonlinearity brings essential mathematical challenges to their investigation. Because of its applied importance, equation (1) is of great theoretical interest, though its investigation has been considered as "simply unbearable" by some scientists.

In this chapter, we develop an investigation technique for the integral equations with endogenous leads. Section 2 considers several applied problems that involve equation (1) in their analysis. Section 3 presents a qualitative analysis of (1) in some special cases relevant to economic and management applications. Numeric solution and computer simulation are given in Section 4. The last section summarizes the obtained results.

2. Applied Importance.

Equation (1) appears in economics, management sciences, operations research, population dynamics, environmental economics, and other applications. This section illustrates its importance to applications and derivation.

2.1. Application to Economics

Modern economic growth is characterized by structural changes based on the introduction of new technologies. Technological renovation in industries undergoing an intensive *technological change* is one of the key aspects of economic development. The technological change is considered "as a driver of global development" (Global Trends 2015) and is among the major trends that will shape the world of 2015 and change a way of thinking (Global Trends 2015; Hritonenko and Yatsenko, 2007; Meadows, et. al., 2004). Technological development has changed human lives and surrounding, for instance, typewriters have been replaced with computers, "cars have created suburbs", and so on. In our high-tech times, it is difficult to believe that 30 years ago, in 1977, the president and founder of Digital Equipment Corporation Ken Olson, said: "there is no reason anyone would want a computer in their home", and earlier, in 1943, the chairman of IBM Thomas Watson stated: "I think there is a world market for may be five computers". Now it is impossible to imagine everyday life, at least in the developed world, without computers. On the other side, according to the Moore law, the efficiency of computers doubles every 18 month. That is, the embodied technological change (Solow et. al., 1966; Cooley et. al., 1997; Boucekkine et. al., 1997; Hritonenko and Yatsenko, 1996b, 1999, 2003) leads to the appearance of new assets[1] that are faster, safer, and more productive due to the latest achievements in science and technology. This determines the necessity of asset replacement. It brings *new questions* such as when and which capital to replace in order to maximize profit and minimize expenses, resources, and pollution. Optimal replacement is in the scope of many disciplines, especially mathematical economics and operations research.

The *vintage capital models*, first proposed by the Nobel Price laureate Robert Solow (Solow et. al., 1966), describe rational replacement of age-structured capital, including optimization of endogenous capital lifetime. The vintage capital models represent a promising direction in theoretical and empirical economic research. They can take into account multiple production factors and resources, energy shortage, environmental impact, endogenous technological change, and other relevant issues. Mathematically, such models involve the optimal control of the non-linear integral equations of Volterra type with unknowns in the integration limits. The models generate highly complex nonlinear dynamics. Despite the applied importance of vintage capital models, just little systematic theoretical study has been offered due to mathematical challenges of their investigation. Many economic issues remain open because "dynamic general equilibrium models with vintage technology often collapse into a mixed delay differential equation system, which cannot be in general solved either mathematically or numerically" (Boucekkine et. al., 1997). Existing economic results about

[1] The terms "capital", "assets", "equipment", and "machines" are used interchangeably in this work as in other relevant literature. The same is true for "economic life, "service life", "life", and "lifetime".

the optimal capital lifetime in vintage models include only the case of constant lifetime. To obtain such results, the constant capital lifetime is postulated or some assumptions under which the constant lifetime is optimal are made. In this chapter a variable lifetime is considered.

An optimal control in the simplest vintage capital model can be presented by the following problem with respect to the unknown variables $x(t)$ and $m(t)$, $t\in[t_0, T)$, $t_0<T\leq\infty$:

$$I = \int_{t_0}^{T} \rho(t)[Q(t) + p(t)m(t)]dt \rightarrow \min, \tag{2}$$

under equality-restrictions

$$P(t) = \int_{x(t)}^{t} m(\tau)d\tau, \tag{3}$$

$$Q(t) = \int_{x(t)}^{t} q(\tau,t)m(\tau)d\tau, \tag{4}$$

inequality-restrictions

$$m_{min}(t) \leq m(t) \leq M(t),\ a'(t) \geq 0,\ a(t) < t,\ t\in[t_0, T), \tag{5}$$

and initial conditions

$$x(t_0) = x_0 < t_0,\ m(\tau) = m_0(\tau),\ \tau\in[a_0, t_0]. \tag{6}$$

The given functions $P(t)$, $p(t)$, $q(\tau,t)$, and $\rho(t)$ are assumed to be Lipschitz continuous at $t\in[t_0,T)$. In the framework of vintage models, the optimal control problem (2)-(6) is applicable to a firm that renovates its capital under technological change. Then, $m(t)$ is the amount of new capital at time t, $x(t)$ is the time of introducing the capital removed at time t, $L(t) = t-x(t)$ is the endogenous lifetime of the capital, $p(t)$ is the unit price of new capital, $P(t)$ is the required recourse, $Q(t)$ is current maintenance expenses, $q(\tau,t)$ is the unit maintenance expense for capital introduced at time τ that decreases in τ (the newer capital is more efficient than the older capital because of technological change), $\rho(t)$ is a discounting factor, $0<\rho(t)\leq 1$, $\rho'(t)\leq 0$. Then the problem (2)-(6) minimizes total operation expenses, which include the maintenance expenses $Q(t)$ and the new investment $p(t)m(t)$.

Various modifications of the basic model (3)-(4) and the optimal control problem (2)-(6) have been investigated by (Malcomson 1975; Hilten 1991; Hritonenko and Yatsenko, 1996, 2003, 2005-2007; Boucekkine et al., 1997, and others). They involve additional balance relations (energy, operating costs, and maintenance), different product outputs, substitution elasticity suggestions, endogenous relations for the rate of technological change, etc. Diverse numerical and analytical investigation methods have been suggested for problems of type (2)-(6). In methods suggested in (Hritonenko and Yatsenko, 1996, 2003, 2005-2007) the problem

(2)-(6) is considered in its original integral setting. The methods are based on obtaining and analyzing extremum conditions.

Theorem 1 (*the necessary and sufficient condition for an extremum*). The functional $I(m)$ in the optimal control problem (2)-(6) is differentiable and its increment is presented in the following form:

$$\delta I = I(m + \delta m) - I(m) = \int_{t_0}^{T} I'(t)\delta m(t)dt + \delta^2 I, \tag{7}$$

where the gradient $I'(t)$ of the functional (2) is

$$I'(t) = \int_{t}^{\overline{x}^{-1}(t)} \rho(\tau)[q(t,\tau) - q(x(\tau),\tau)]d\tau + \rho(t)p(t), \quad t \in [t_0, T), \tag{8}$$

$x^{-1}(t) = \max\{T, x^{-1}(t)\}$, the inverse x^{-1} of x exists because of $x'(t) \geq 0$, and the high-order residual is

$$\delta^2 I = \int_{t_0}^{T} \rho(t) \int_{x(t)}^{x(t)+\delta x(t)} [q(x(t),t) - q(t,\tau)][m(\tau) + \delta m(\tau)]d\tau dt. \tag{9}$$

In order for a function $m^*(t)$, $t \in [t_0, T)$, to be a solution of (2)-(6) it is necessary and sufficient that

$$I'(t) \geq 0 \text{ at } m^*(t) = m_{\min}(t), \tag{10}$$

$$I'(t) \leq 0 \text{ at } m^*(t) = M(t), \tag{11}$$

$$I'(t) \equiv 0 \text{ at } m_{\min}(t) < m^*(t) < M(t),\ t \in [t_0, T). \tag{12}$$

Proof. The proof of the necessary and sufficient extremum conditions (10)-(12) is standard in optimal control theory and can be provided similar to (Ioffe and Tikhomirov 1979). Here we concentrate on the first step of the proof, namely, on the derivation of (8)-(9). The increment in x and m of the objective function (2) after substitution of (4) is presented in the following form:

$$\begin{aligned}\delta I &= I(m + \delta m) - I(m) \\ &= \int_{t_0}^{T} \rho(t)[\int_{x(t)+\delta x(t)}^{t} q(\tau,t)(m(\tau) + \delta m(\tau))d\tau + p(t)(m(t) + \delta m(t))]dt \\ &\quad - \int_{t_0}^{T} \rho(t)[\int_{x(t)}^{t} q(\tau,t)m(\tau)d\tau + p(t)m(t)]dt,\end{aligned}$$

where δx and δm are *admissible* increments of x and m, that is, $x+\delta x$ and $m+\delta m$ satisfy (5)-(6) if x and m satisfy (5)-(6). Adding and subtracting $\rho(t)q(x(t),t)\int_{x(t)}^{x(t)+\delta x(t)}[m(\tau)+\delta m(\tau)]d\tau$ to the last expression and simplifying it, we obtain

$$\begin{aligned}\delta I &= \int_{t_0}^{T}\rho(t)[\int_{x(t)}^{t} q(\tau,t)\delta m(\tau)d\tau - \int_{x(t)}^{x(t)+\delta x(t)} q(\tau,t)(m(\tau)+\delta m(\tau))d\tau + p(t)\delta m(t)]dt = \\ &= \int_{t_0}^{T}\rho(t)\{\int_{x(t)}^{t} q(\tau,t)\delta m(\tau)d\tau - q(x(t),t)\int_{x(t)}^{x(t)+\delta x(t)}(m(\tau)+\delta m(\tau))d\tau + \\ &+ \int_{x(t)}^{x(t)+\delta x(t)}[q(x(t),t)-q(\tau,t)][m(\tau)+\delta m(\tau)]d\tau + \rho(t)p(t)\delta m(t)\}dt.\end{aligned} \tag{13}$$

Let us simplify (13). According to (9), the third integral of (13) is $\delta^2 I$. Constraint (3) of the optimal control problem (2)-(6) can be presented as:

$$P(t) = \int_{x(t)}^{t}\delta m(\tau)d\tau = \int_{x(t)+\delta x(t)}^{t}[m(\tau)+\delta m(\tau)]d\tau.$$

Simplifying the last equality, we obtain

$$\int_{x(t)}^{x(t)+\delta x(t)}[m(\tau)+\delta m(\tau)]d\tau = \int_{x(t)}^{t}\delta m(\tau)d\tau. \tag{14}$$

Applying (14) to the second integral term of (13), interchanging limits in the first and second integrals, and combining terms with δm in (13), we obtain:

$$\delta I = \int_{t_0}^{T}[\int_{t}^{x^{-1}(t)}\rho(\tau)(q(t,\tau)-q(x(\tau),\tau))d\tau + \rho(t)p(t)]\delta m(t)dt - \delta^2 I,$$

that is, the formulas (7)-(9). The rest of the proof of the theorem employs the standard technique of optimal control. An analysis of the second variation (9) is essential to the proof of sufficient conditions. ■

Relation (12) of the extremum conditions (10)-(12) plays an important role in the investigation of the optimal control problem (2)-(6). It represents possible interior regimes and describes internal properties of a process under study. The combination of (8) and (12) leads to equation (1). Hence, the optimality conditions for dynamic systems of type (2)-(6) lead to the integral-functional equations with the unknown function $(x(t))^{-1}$ in the upper integration limits. Since $x(t)$ is the purchase time of the machine being replaced at time t, the inverse function $(x(t))^{-1}$ describes the future service life of the capital bought at t.

An asymptotic analysis of the optimal control problem (2)-(6) reveals interesting patterns of the rational lifetime of capital under technological change. It can be shown that, under certain assumptions, problems of type (2)-(6) possess turnpike properties when their solution (if it exists) is attracted by a special internal trajectory $\widetilde{x}$ of a simpler structure. Such trajectories are called *turnpikes*. The turnpikes do not satisfy all constraints of the optimal control problem but show the best possible dynamics. In many practical cases, the convergence of solutions to turnpikes is fast and can be estimated. In addition, it is easier to find and analyze turnpikes than to solve the optimal control problem itself. The turnpike theorems for (2)-(6) state that under some conditions the optimal capital lifetime $x(t)$ strives to the turnpike $\widetilde{x}(t)$, $t\in[t_0,\infty)$, which is a solution of the integral-functional equation with leads (1) if it exists. Depending on the convergence type, it is possible to prove the turnpike theorems in normal, strong, strongest, and weak forms (Hritonenko and Yatsenko, 1996, 2005). Such theorems are well known for other (non-integral) models in economic growth theory. The existence of turnpike properties is often emphasized as an indicator of the quality of an optimization model. They also produce some important patterns for strategic replacement decisions and are used for a practical analysis of system performance.

2.2. Operations Research, Management Science, and Engineering

One of central problems in this area is the optimal replacement of equipment under improving technology (Regnier et al., 2004; Bylka et al., 1992; Bean et al., 2004; Rogers and Hartman 2005; Jones et al., 1991). Let us consider a production shop that keeps one machine of a particular type for a sufficiently long period of time. The operating and maintenance costs increase as the machine becomes older and deteriorates. New machines appear on the market. They are more productive due to technological change. The shop should determine a policy of selling the machine and buying a new one. In the replacement process in the continuous time t, the *replacement policy* $L=\{L_k, k=1,2,\ldots,\}$ consists of the service lives L_k of the sequentially replaced machines. The policy can be an infinite series $\{L_k\}$, $k=1,\ldots,\infty$, of finite service lives or a finite number of replacements $\{L_k\}$, $k=1,\ldots,N$, $N\geq 0$, with the infinite last service life $L_N=\infty$. Assuming that the process started at $t=0$ and the first machine was purchased at the known time $\tau_0\leq 0$, the replacement times are $\tau_{k+1} = \tau_k + L_{k+1}$, $k=0,1,2,\ldots$. The capital cost $p(t)$ of a machine bought at time t and the operating and maintenance costs $q(t,u)$ at time u for the machine bought at time t, $u \geq t$, are known. The capital cost $p(t)$ takes into account the machine purchase price and installation cost as well as its possible salvage value. Because of deterioration, $q(t,u)$ non-decreases in u at fixed t (when the machine age $x=t-k$ increases). At this point, we make a general assumption of the continuous technological change such that $p(t)$ and $q(t,u)$ decrease in t for any fixed machine age $x=t-u$. Then, the machine replacement problem can be formulated as the *problem* of finding the optimal policy $\pi^*=\{L_k^*, k=1,2,\ldots\}$, that minimizes the discounted value of the total cost of replacement policy over the infinite horizon $[0,\infty)$,

$$J(\pi)\sum_{i=1}^{\infty}[e^{-r\tau_i}p(\tau_i)+\int_{\tau_i}^{\tau_{i+1}}e^{-ru}q(\tau_i,u)du]\to\min_{L_j,j=1,\dots,\infty} \tag{15}$$

where the first term represents the total price of purchased machines (capital cost) and the second term is the total maintenance costs. The parameter r, $r>0$, refers to the discount rate.

The nature of the management problem (15) requires solving nonlinear integer programming problems that possess significant mathematical challenges. Such discrete models have been known since 50's but the related research focuses on their numeric simulation. The analysis is mostly restricted to the existence and properties of the rolling horizons and forecast horizons which are long enough to determine the current replacement decision. Qualitative properties of the optimal machine replacement policies are poorly known. Even the question whether new technologies delay or speed up the optimal replacement is still debated. The equipment replacement naturally involves the age of replaced equipment (machines). Continuous-time versions of the replacement models are similar to the vintage capital models and are promising for the purposes of a qualitative analysis. An investigation of (15) while relaxing one condition, namely, considering time in continuous setting, requires an analysis of (1).

2.3. Population Biology

The optimal maintenance and exploitation of "manageable" age- or size-structured productive populations are important to meat and dairy farms, breeding, forestry, and, in perspective, to bioengineering. The age structure of biological populations can be described by integral equations or partial differential equations. Although harvesting in an age-structured population is one of the most common applied problems, the structure of the optimal harvesting is not completely clear. The integral optimal control problem of maximization of the revenue (net profit) from harvesting in a one-species population can be presented as:

$$I{=}E-Z=h\int_{t_1}^{t_2}p(a(t))X(t-a(t))(1-a'(t))dt-$$

$$-\int_{t_1}^{t_2}[h\int_{t-a(t)}^{t}z(t-u)X(u)du+(1-h)\int_{t-T}^{t}z(t-u)X(u)du]dt\to\max_{a,a',h,X} \tag{16}$$

under the constraints

$$X(t)=h\int_{t-a(t)}^{t}M(t-u)X(u)du+(1-h)\int_{t-T}^{t}M(t-u)X(u)du, \tag{17}$$

$$0\le h\le 1,\ 0\le a(t)\le T,\ X(t)\ge 0,\ t\in[0,T], \tag{18}$$

and the initial conditions

$$X(u) = X_0(u) \geq 0, \qquad t_1\text{-}T \leq u \leq t_1, \tag{19}$$

where X_0 is given. The function $X(t)$ represents the birth intensity in the population with the vital parameters $M(t)$, h and $a(t)$ are the harvesting rate and age of individuals to be harvested, $z(t)$ and $p(t)$ are the expenses to raise the individual up to the age t and its selling price.

From the mathematical point of view, the optimal control problem (16)-(19) is similar to the optimal problem (2)-(6). The difference between applications of (2)-(6) to economics and biology is that populations reproduce themselves while the economy is managed by people. Investigation methods for economic problems have been effectively extended to the harvesting model (16)-(19). An investigation of (16)-(19) also leads to an analysis of integral-functional equations of type (1) in some cases (Hritonenko and Yatsenko, 2006a).

So, we can see that, although models in Sections 2.1-2.4 describe different phenomena, their mathematical investigation involves the integral equation (1).

3. Qualitative Analysis.

As it has been shown in Section 2, the extremum conditions in optimal replacement problems require solving dual integral equations of a special type, which contain the unknown $x^{-1}(t)$ in the upper integration limit. There is no general theory for such equations, and meaningful special cases should be considered. For instance, the assumptions

$$q(\tau, t)= q_0 e^{c_d(t-\tau)} e^{-c_q\tau}, p(t)= p_0 e^{-c_p t}, \rho(t)= e^{-rt}, 0<c_q+c_d<r, c_p+c_d\geq 0. \tag{20}$$

are common in the replacement models of mathematical economics and operations research. The parameter c_d is a deterioration rate, it is larger for the older capital; the parameter c_q represents the effects of technological change on maintenance and operating costs and shows that these costs are lower for the newer capital; the parameter c_p emphasizes the decrease of the price with aging, and q_0, p_0 are initial values of maintenance costs and new equipment price. The exponential technological change (20) means that the maintenance and operating costs (at a fixed age) and the capital cost drop by constant factors after each time period. The exponential deterioration shows that the maintenance and operating costs increase by a constant factor when the machine age increases. The exponential technological change and deterioration correspond to the geometric technological change and deterioration in discrete models (Regnier et al., 2004; Bean et al., 2004; Rogers and Hartman, 2005; Jones et al., 1991). Usually, the maintenance and operating costs and capital costs decrease because of the embodied technological change. In (20), all these costs can increase (c_p or c_q<0) but slower than c_d.

Theorem 2 (*the dynamics of a solution*). Equation (1) in case (20) has a unique solution such that:

(A) If $c_q = c_p = c$, then $t - x(t) = L$, where the constant $L>0$ is uniquely determined from the non-linear equation

$$(r\text{-}c_d) e^{(c+c_d)L} + (c+c_d) e^{-(r-c_d)L} = (r+c)[1+(r\text{-}c_d)p_0/q_0] \tag{21}$$

and, at small $c+c_d$ and r,

$$L \approx \{2p_0/[q_0(c+c_d)]\}^{1/2} \tag{22}$$

(B) If $c_q < c_p$, then the capital lifetime $L(t)=t- x(t)$ monotonically decreases and $x(t)\to t$ at $t\to\infty$.

(C) If $c_q > c_p > -c_d$, then $L(t)$ monotonically increases, $L(t)\to\infty$ and $x(t)\to\infty$ at $t\to\infty$.

(D) If $c_p=-c_d=c$ then the solution exists only if $c>c_q$,

Remark. The solution of the nonlinear integral equation (1) in another special case was considered in (Yatsenko and Hritonenko, 2005). That case was similar to (20) but the case and the proof of cases (A) and (B) did not include the deterioration rate c_d. The deterioration rate was first introduced in (Hritonenko and Yatsenko, 2007b) where its applied importance was emphasized and the proof of Case (A) was briefly sketched. Here we extend the results of (Yatsenko and Hritonenko, 2005; Hritonenko and Yatsenko, 2007b), consider the deterioration rate c_d, expand the proofs of cases (A) and (B), and add two more cases (C) and (D).

Proof. Equation (1) in case (20) can be rewritten in the following form:

$$q_0 \int_t^{x^{-1}(t)} e^{-r\tau}[e^{c_d(\tau-x(\tau))}e^{-c_q x(\tau)} - e^{c_d(\tau-t)}e^{-c_q t}]d\tau = p_0 e^{-rt}e^{-c_p t}, \quad t\in[0,\infty). \tag{23}$$

Case (A). After the direct substitution of t - $x(t) = L$ and $c_q = c_p = c$ to equation (23), it becomes

$$q_0 \int_t^{t+L} e^{-r\tau}[e^{c_d L}e^{-c(\tau-L)} - e^{c_d(\tau-t)}e^{-ct}]d\tau = p_0 e^{-rt}e^{-ct}, \quad t\in[0,\infty).$$

Evaluating the integral in the last equation and simplifying the result, we obtain (21) if the second part of (20) holds. Approximation (22) follows from (21) and the Taylor series for the exponential function at small $c+c_d$ and r. One can show that equation (21) has a unique solution. Indeed, the left side of (21) is $(r+c)$ at $L=0$ and tends to ∞ as $L\to\infty$ and its first derivative $(r-c_d)(c+c_d)\, e^{c_d L}(e^{cL} - e^{-rL})>0$ under the inequalities of (20). Hence, the left side monotonically increases from $(r+c)$ to ∞ and the right side of (21) is a constant greater than is $(r+c)$, which proves the uniqueness of a solution of (21) in Case A.

To analyze other cases, let us differentiate (23) in t:

$$q_0 e^{-rx^{-1}(t)}[e^{c_d(x^{-1}(t)-x(x^{-1}(t)))}e^{-c_q x(x^{-1}(t))} - e^{c_d(x^{-1}(t)-t)}e^{-c_q t}][x^{-1}(t)]'$$
$$-q_0 e^{-rt}[e^{c_d(t-x(t))}e^{-c_q x(t)} - e^{c_d(t-t)}e^{-c_q t}]$$
$$+q_0 \int_t^{x^{-1}(t)} e^{-r\tau}[e^{c_d(\tau-t)}e^{-c_q t}(c_d+c_q)]d\tau = p_0 e^{-rt}e^{-c_p t}(-r-c_d), \qquad t\in[0,\infty).$$

Since $x(x^{-1}(t))=t$, the first term is zero. Evaluating the integral and simplifying the expression, we obtain:

$$(r-c_d)[e^{(c_q+c_d)(t-x(t))}-1]+(c_q+c_d)[e^{-(r-c_d)(x^{-1}(t)-t)}-1]=\frac{p_0}{q_0}(c_p+r)(r-c_d)e^{(c_q-c_p)t}. \quad (24)$$

The solutions of (24) and (23) are equivalent.

Case (*B*). Let $c_q<c_p$. Let us present the solution $L(t)$ of (24) as the sum of an unknown constant part L^0 of the solution and its unknown functional correction $\upsilon(t)$, namely, $L(t)=L^0+\upsilon(t)$. Choosing a certain instant t_a, we determine the constant L^0 from (24) as

$$(r-c_d)[e^{(c_q+c_d)L^0}-1]+(c_q+c_d)[e^{-(r-c_d)L^0}-1]=\frac{p_0}{q_0}(c_p+r)(r-c_d)e^{(c_q-c_p)t_a} \quad (25)$$

The solution L^0 depends on the value t_a and its uniqueness can be proven as in Case (A). Subtracting (25) from equation (24) for $L(t)$, we obtain the equation for the function $\upsilon(t)$:

$$(r-c_d)[e^{(c_q+c_d)\upsilon(t)}-1]e^{(c_q+c_d)L^0}+(c_q+c_d)[e^{-(r-c_d)\upsilon(x^{-1}(t))}-1]e^{-(r-c_d)L^0}$$
$$=\frac{p_0}{q_0}(c_p+r)(r-c_d)[e^{(c_q-c_p)(t-t_a)}-1]e^{(c_q-c_p)t_a} \quad (26)$$

As follows from (26), the values $\upsilon(x^{-1}(t_a))$ and $\upsilon(t_a)$ have different signs, hence there is a point t_b, $t_a<t_b<x^{-1}(t_a)$, such that $\upsilon(t_b)=0$ and $L(t_b)=L^0$. The delay equation (26) connects $\upsilon(x^{-1}(t))$ with $\upsilon(t)$ and can be recurrently solved forward or backward. For the backward solution, equation (26) produces at $t=x(t_b)=t_b-L^0$ the value $\upsilon_d=\upsilon(t_b-L^0)>0$ because $\upsilon(t_b)=0$. Equation (26) has a unique continuous solution $\upsilon(t)$ at $x(t_b-L^0)\le t\le t_b-L^0$ for a suitable initial function $\upsilon(t)$, $t_b-L^0\le t\le t_b$, e.g., for the linear $\upsilon(t)=(t_b-t)\upsilon_d/L^0$. To continue this process to the next interval $[x(x(t_b-L^0)), x(t_b-L^0)]$ and further, it is necessary to prove the convergence of the backward solution process. To do this, let us consider a small variation $\delta\upsilon$ of υ. By (26), we obtain that

$$|\delta\upsilon(t)|\cong\exp[-(c_q+c_d)L(t)-(r-c_3)L(x^{-1}(t))]|\delta\upsilon(x^{-1}(t))|<|\delta\upsilon(x^{-1}(t))| \quad (30)$$

for small values of $|\delta\upsilon(t)|$, hence, the variations $\delta\upsilon(t)$ of $\upsilon(t)$ decrease exponentially when equation (26) is solved backward. Thus, a solution $x(t)$ to (24) can be constructed on any finite interval $[t_0, t_a]$ (starting from a suitable initial condition at the right end $[x(t_a), t_a]$). Then the solution $x(t)$ on the infinite interval $[t_0,\infty)$ is obtained as $t_a \to \infty$. Since $c_q<c_p$, the right side of (24) approaches 0 at $t\to\infty$, hence, $L(t)=t-x(t)\to 0$ as $t\to\infty$.

Case (*C*). If $-c_d< c_p < c_q$, then from (24) at $t\to\infty$

$$e^{(c_q+c_d)L(t)} \approx \frac{p_0}{q_0}(c_p + r)e^{(c_q-c_p)t} . \tag{28}$$

Taking the logarithm of the both sides of (28), we obtain that

$$L(t) \approx \frac{c_q - c_p}{c_q + c_d} t \quad \text{at } t >> 1, \tag{29}$$

that is, under the conditions of Case (C) and inequalities of (20), the capital lifetime $L(t)$ increases as t increases. Substituting $x(t)=t-L(t)$ to (29) we obtain

$$x(t) \approx \frac{c_p + c_d}{c_q + c_d} t \ \text{ and } \ x^{-1}(t) \approx \frac{c_q + c_d}{c_p + c_d} t \ \text{ at } t\to\infty, \tag{30}$$

that justify the rest of the statement of Case (C).

Case (*D*). If c_p=-c_d=c, then (23) leads to

$$\int_t^{x^{-1}(t)} e^{-(r+c)(\tau-t)}[e^{(c-c_q)x(\tau)} - e^{(c-c_q)\tau}]d\tau = \frac{p_0}{q_0}, \quad t \in [0,\infty). \tag{31}$$

Investigating the last equation, we can conclude that the solution of (23) exists only if $c>c_q$ and there is no solution at $c\le c_q$. The behavior of the solution is subject to Case (B) if $c>c_q$. Case (A) shows the solution of (23) or (1) in case (20) if $c_p=c_q$ but with $c_p\neq c_d$ that follows from the inequalities of (20). Case (C) investigates the solution of (23) if $c_p<c_q$, but in this case both c_p and c_q are greater than c_d as follows from (20).

The theorem is proven. ■

Applied interpretation. Theorem 2 studies the interplay of three relevant economic effects: *embodied technological change* c_q, physical *deterioration* of capital ($c_d>0$), and *new capital price* change c_p. The embodied technological change means that newer capital is more efficient and its maintenance is less expensive. The deterioration reflects an increase of maintenance cost with the capital aging. Theorem 2 shows, that the renovation of capital is necessary when deterioration, or technological change, or both occur. So-called *learning* (when c_d <0) plays an important role in recent vintage models literature (Feichtinger et al, 2006). Because of learning, the maintenance cost becomes less expensive for older capital. Case (D) shows that learning diminishes the impact of embodied technological change and the capital renovation is profitable *only if* the increase rate of learning is smaller than the rate of the embodied technological change.

Remark. A similar to (1) nonlinear equation with leads, namely,

$$\int_{t}^{x^{-1}(t)} \rho(\tau)[q(t,\tau)-q(x(\tau),\tau)]d\tau = p(t)\rho(t), \qquad t\in[t_0,\infty) \tag{32}$$

for the increasing in t function $q(t,\tau)$, was investigated in (Hritonenko and Yatsenko, 1996a, 1996b, 2005). Although equations (1) and (32) belong to the same class of integral equations with leads and possess some similar features, their dynamics appears to be quite different.

4. Numeric Simulation

Section 3 investigates the nonlinear integral equation (1) in a special exponential case common in applications. It discusses conditions for the existence of a solution and the dynamics of the solution if it exists. However, a management decision about capital renovation is often necessary to make in a general case. This section offers algorithms for numeric solution of equation (1) in more general cases. The only necessary assumption is the strict monotonicity of $q(t,u)$ in t that reflects the presence of technological change. The functions $p(t)$ and $q(t,u)$ are usually smooth but a case of their discontinuities will be also considered in Section 4.3. The suggested algorithm is implemented in Visual Basic/Excel and can be provided to all interested readers.

4.1. Algorithm

The algorithm for solving the nonlinear integral equation (1) involves two problems:

(i) *The initial problem* consists of finding a solution $x(t)$ of equation (1) on $[t_1, t_0)$ or $[t_0, t_1)$ at a certain given initial monotonic function $x(t)=\xi(t)<t$ at $t\in[t_0, x^{-1}(t_0)]$. Taking $\rho(t)=e^{-rt}$ and differentiating (1) in t, we obtain

$$q(x(t),t)-q(t,t) = -\int_{t}^{x^{-1}(t)} e^{-r(u-t)}\frac{\partial q(t,u)}{\partial t}du + rp(t)-p'(t). \tag{33}$$

The recurrent formula (33) connects $x(t)$ and $x^{-1}(t)$ and can be used for backward or forward solution of (1). More exactly, if the initial function $\xi(t)$, $t\in[t_0, x^{-1}(t_0)]$, satisfies (1) at $t=t_0$, then the recurrent formula (33) produces the solution $x(t)<t$ of (1) for $t<t_0$ and $t>x^{-1}(t_0)$.

The method of the polynomial interpolation can be used to find x from the function $q(x,t)$ for backward solution or from $\int_t^x f(t,u)du$ for forward solution.

It is important to note that the dynamics of the constructed solution $x(t)$ heavily depends on the given functions ξ, p, and q. In the general case, $x(t)$ has discontinuities at points $x^k(t_0)$, $k=0,\pm1, \pm2, \pm3,\ldots$. The function $x(t)$ may also have discontinuities at the points where the derivatives of p and q have jumps. If $x(t)$ becomes non-monotonic at some point ξ, then the inverse $x^{-1}(t_0)$ does not exist at $t=x(\xi)$ and the algorithm stops.

(ii) *The asymptotic problem* aims in solving equation (1) on the infinite interval $[t_0,\infty)$ and includes two steps:

Step 1: finding a proper initial function $x(t)=\xi(t)$, $t\in[x(t_1), t_1]$, and

Step 2: applying the recurrent formula (33) to find the solution $x(t)$ on $[t_0, x(t_1)]$ and extending it to the interval $[t_0,\infty)$.

Although the infinite interval will be considered in the algorithm, this problem needs to be solved in practice on a finite interval $[t_0, t_1]$ in the case when the given functions p and q are known only on $[t_0, t_1]$. Then, the additional assumption is made that the solution dynamics is similar at $t>t_1$.

Step 1 requires a construction of an initial function $x(t)=\xi(t)$ on $[x(t_1), t_1]$, such that (1) is satisfied and $x(t)$ is continuous at $t=x(t_1)$. To implement this, we choose the linear function $\xi(t)=\alpha t-L$ that depends on two parameters α and L. If $\frac{\partial q(t_1,t_1)}{\partial t} > p'(t_1)$, then the initial value $\alpha>1$, otherwise $\alpha<1$. Substituting $\xi(t)$ to (1), we find L and the value $u=x(t_1)=\xi(t_1)$. Then $x(t)$ is found at $t<u$ close to u from (33). If $\lim_{t\to u-0} x(t)<\xi(u)$, then the value of α should be increased, otherwise decreased. The process of finding L,u and calculating $\lim_{t\to u-0} x(t)$ is repeated until $\lim_{t\to u-0} x(t)\approx\xi(u)$.

During Step 2, using the recurrent formula (33) backward we obtain a continuous function $x(t)$ for $[t_0, x(t_1)]$. When t_1- t_0 becomes larger (t_1-$t_0\to\infty$), the obtained $x(t)$ tends to the unique solution of (1) on $[t_0,\infty)$ (if it exists at the given p and q). The proof of this statement is similar to the proof of Theorem 2.

4.2. Numeric Example (Optimal Replacement of Passenger Cars)

To approbate the algorithm, we provide a real-data economic example. (Regnier et al., 2004) used a discrete replacement model with expenses minimization to model the optimal replacement of passenger cars (Honda Accord for years 1985-1998) on the data from various automotive magazines and Internet sources. The corresponding parameters of the continuous model (1) are p_0=\$15,350, q_0=\$91, $c_p=0$, $c_q\approx0.04$, the deterioration rate $c_d\approx0.35$, and the discount rate $r \approx 0.14$. The algorithm of Section 4.1 takes the horizon length T=50 years and the discretization step H=0.1. At the accepted parameter values, the algorithm does not require smaller steps and longer horizons (it is needed up to 300 years and more in (Regnier et al., 2004)). We analyze the optimal replacement dynamics for several scenarios shown in Figure 1. Since $c_p \approx c_q$, the optimal lifetime is constant by Theorem 2 and can be found from the nonlinear equation (21). Its approximate solution is $L^* = 10.57$ years, which is close to the optimal lifetime of L=11 years given in (Regnier et al., 2004) for the discrete model with integer lifetime values.

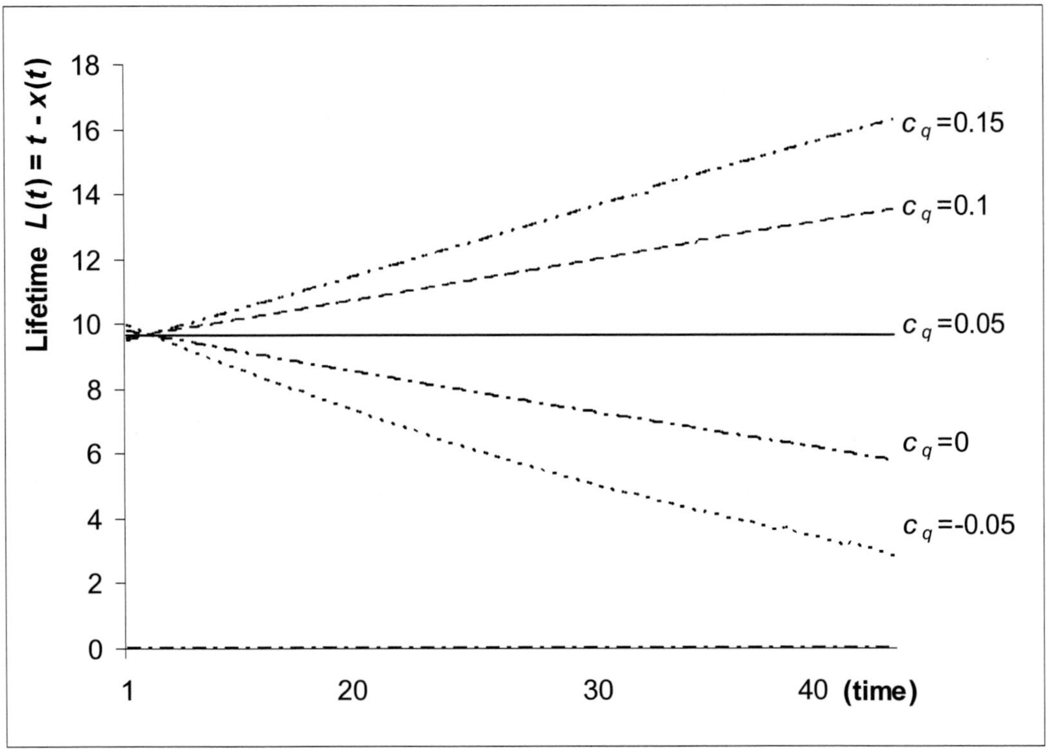

Figure 1. The optimal lifetime $L(t)= t-x(t)$ (where $x(t)$ is the solution of integral equation (23)) at c_d=0.35, c_p=0.05, and different values of c_q= {-0.05, 0, 0.05, 0.1, 0.15}.

If the rates of technological change in the operating costs and capital price are different, $c_p \neq c_q$ in (1), (20), then, by Theorem 2, the optimal machine lifetime decreases or increases depending on the sign of c_p-c_q. Such results for vintage capital models were first obtained in (Hritonenko and Yatsenko 1996a, 1996b). A similar result has been analytically and numerically confirmed for the discrete replacement model (15) in (Regnier et al., 2004) for various scenarios that compute the optimal first lifetime of the machine. We have solved the nonlinear integral equation (1) for the same scenarios. Figure 1 shows the variable optimal lifetime for four different scenarios $c_q > c_p$, $c_q = c_p$, and $c_q < c_p$. The simulation results confirm the analytic conclusions of Theorem 2.

4.3. An Example with Discontinuous Technological Change

An important case occurs when the technological change is discontinuous and implemented in the form of technological breakthroughs (Rogers and Hartman 2005). The dynamics of the optimal replacement in such cases is an open issue. This case is more complicated but it can be also efficiently explored using the technique developed in this chapter. For illustration, we consider a simple case with one technological breakthrough. Namely, let us assume that q and $\hat{p}$ are exponential (20) and the technological change causes the discontinuity in $p(t)$ at instant t_1:

$$p(t)=\begin{cases}\hat{p}(t) & \text{if } t<t_1,\\ B_p\hat{p}(t) & \text{if } t\ge t_1,\end{cases} \qquad B_p<1. \tag{34}$$

Then, the solution $x(t) = t\text{-}L(t)$ of (1) (if it exists) is determined by

$$q_0e^{-(c_q+c_d)x(t)} = q_0e^{-(c_q+c_d)t} + (c_q+c_d)q_0e^{-(c_q+c_d)t}\int_t^{x^{-1}(t)} e^{-(r-c_d)(u-t)}du + e^{-c_d t}(rp(t)-p'(t)) \overset{\text{df}}{=} F(t), \tag{35}$$

$t\in[0,\infty)$. The first derivative of $p(t)$ is

$$p'(t)=\begin{cases}-c_p p_0 e^{-c_p t} & \text{if } t<t_1,\\ p_0e^{-c_p t}[-c_p+(1-B_p)\delta(t-t_1)] & \text{if } t=t_1,\\ -c_pB_pp_0e^{-c_p t} & \text{if } t>t_1,\end{cases} \tag{36}$$

where $\delta(t)$ is the *Dirac delta-function*. Equation (35) is solved sequentially from right to left

$$x(t) = (c_q+c_d)^{-1}[\ \ln q_0 - \ln(F(t))] \tag{37}$$

in some neighborhood of t_1, where $F(t)$ contains $p'(t)$ and includes the delta-function $\delta(t\text{-}t_1)$.

To approbate the algorithms of Section 4.1 on the discontinuous case, we choose parameters B_p=0.35, t_1=25. Other parameters are as in Section 4.2. The simulation results are illustrated in Figure 2. The solid line in Figure 2 demonstrates the optimal $x(t)=t-L(t)$ and the dashed line shows the corresponding inverse $x^{-1}(t)$. The function $x(t)$ and its inverse $x^{-1}(t)$ are symmetric with respect to the (dotted) straight line $x=t$ also shown in Figure 2. The simulated behavior of $x(t)$ is similar to theoretically predicted by (37). The simulation is provided with the finite discretization step h=0.1. Correspondingly, the delta-function in (36) at $t=t_1$ is replaced with the negative $x(t)$ jump of a finite size because of numeric differentiation. The size of the jump essentially depends on the value of h (it will $\to -\infty$ at $h\to 0$). The gray line in Figure 2 shows the left hand side of (35), which is zero at $t>a(t_1)\approx 15$, hence $x(t)$ is optimal and the jump (36) is compensated by the "approximate" delta-function in $x(t)$ while $t>x(t_1)$. The trajectory (37) becomes unacceptable when $t<x(t_1)$, because $x(t)$ is not monotonic at $t=t_1$,

hence, the unique inverse $x^{-1}(t)$ cannot be constructed at $t=x(t_1)$. Indeed, as seen in Figure 2, the left hand side of (35) becomes non-zero and the trajectory $x(t)$ is not optimal at $t<a(t_1)$. From a practical viewpoint, formula (37) is of a little use because it recommends the exploitation of a very old machine at $t=t_1$ for a very short period of time. Therefore, the monotonicity of $x(t)$ needs to be explicitly required as in (Boucekkine 1997, Hritonenko and Yatsenko 1996a, 1996b) to use equation (37) for simulating the optimal life at jumps of technological change.

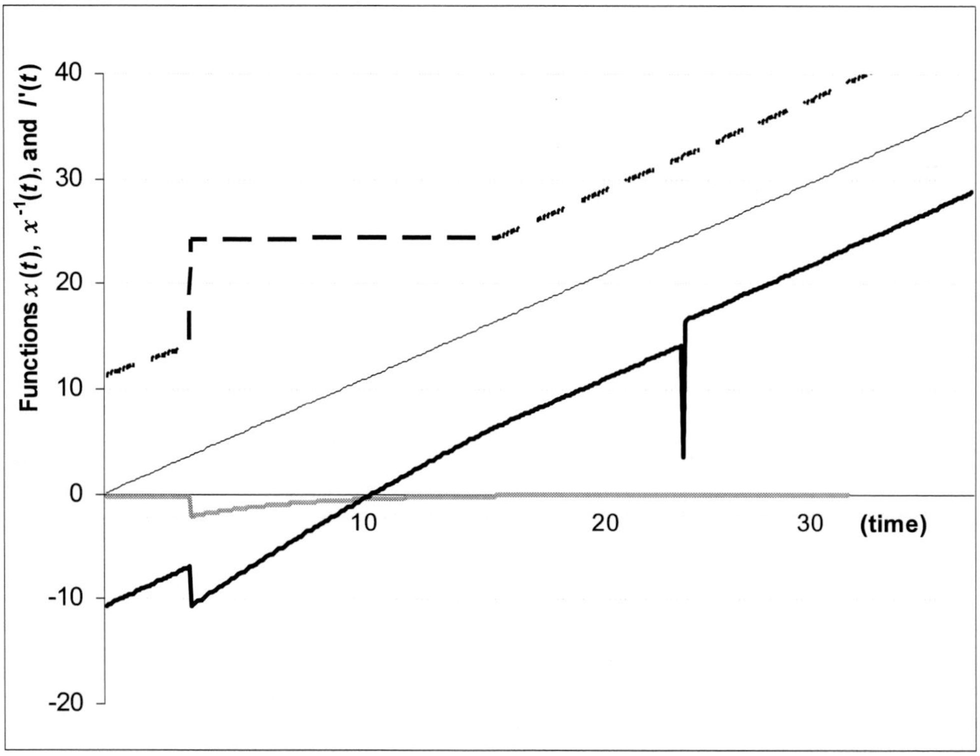

Figure 2. The solution $x(t)$ of integral equation (23) in the case (34) of discontinuous TC at B_p=0.35, t_1=25, c_d=0.35, c_p=c_q=0. The solid line is $x(t)$, the dashed line is the inverse function $x^{-1}(t)$, and the gray line is the gradient (8) in the scale 1000:1.

5. Summary

In this chapter, we have considered a new class of the nonlinear integral equations with leads presented by the unknown upper limit of integration. Its applied importance and qualitative analysis are discussed and a numeric algorithm is offered. Real-data example illustrate and confirm presented theoretical outcomes. Further analysis of the nonlinear equation (1) is relevant and will contribute to important open issues in economics and operations research.

References

Bean, J., Lohmann, J. and Smith, J. (1994) Equipment replacement under technological change, *Naval Research Logistics* **41**, 117-128.

Boucekkine, R., Germain, M. and Licandro, O. (1997) Replacement echoes in the vintage capital growth model, *Journal of Economic Theory* **74**, 333-348.

Bylka, S., Sethi, S.P. and Sorger, G. (1992) Minimal forecast horizons in equipment replacement models with multiple technologies and general switching costs, *Naval Research Logistics* **39**, 487-507.

Cooley, T., Greenwood, J. and Yorukoglu, M. (1997) The replacement problem, *Journal of Monetary Economics* **40**, 457-499.

Feichtinger, G., Hartl, R., Kort, P. and Veliov, V. (2006) Capital accumulation under technological progress and learning: A vintage capital approach, *European Journal of Operations Research* **172**, 293-310.

Global Trends 2015: A Dialogue About the Future With Nongovernment Experts (2000), National Intelligence Council, Washington DC, [http://www.dni.gov/nic/NIC_globaltrend2015.html].

Hilten, O. (1991) The optimal lifetime of capital equipment, *Journal of Economic Theory* **55**, 449-454.

Hritonenko, N. and Yatsenko, Yu. (1996a) Integral-functional equations for optimal renovation problems, *Optimization* **36**, 249-261.

Hritonenko, N. and Yatsenko, Yu. (1996b) *Modeling and Optimization of the Lifetime of Technologies*, Kluwer Academic Publishers, Dordrecht.

Hritonenko, N. and Yatsenko, Yu. (2003) *Applied Mathematical Modeling of Engineering Problems*, Kluwer Academic Publishers, Dordrecht.

Hritonenko, N. and Yatsenko, Yu. (2005) Turnpike properties of optimal delay in integral dynamic models, *Journal of Optimization Theory and Applications* **127**, 109-127.

Hritonenko, N. and Yatsenko, Yu. (2006a) Optimization of harvesting return from age-structured population, *Journal of Bioeconomics*, 8, No 2, 167-179.

Hritonenko, N. and Yatsenko, Yu. (2006b) Optimization of financial and energy structure of productive capital, *IMA Journal of Management Mathematics* **17**, 245-255.

Hritonenko, N. and Yatsenko, Yu. (2007a) Mathematical models of global trends and technological change, *Encyclopedia of Life Support Systems*, Ed. Jerzy A. Filar, Developed under the Auspices of the UNESCO, Eolss Publishers, Oxford, UK [http://www.eolss.net]

Hritonenko, N. and Yatsenko, Yu. (2007b) Optimal equipment replacement without paradoxes: a continuous analysis, *Operations Research Letters*, **35**, 245-250.

Ioffe, A. and Tikhomirov, V. (1979) *Theory of Extremal Problems*, North- Holland, Amsterdam.

Jones, P., Zydiak, J. and Hopp, W. (1991) Parallel machine replacement, *Naval Research Logistics* **38**, 351-365.

Malcomson, J.M. (1975) Replacement and the rental value of capital equipment subject to obsolescence, *Journal of Economic Theory* **10**, 24-41.

Meadows, D., Randers, J., and Meadows, D. (2004) *The Limits To Growth: The 30-Year Update*, White River Junction, VT : Chelsea Green Pub.

Morse, P.M. and Feshbach H. (1953) *Methods of Theoretical Physics*, Volumes I and II, McGraw-Hill Book Company, New York, Toronto, London.

Regnier, E., Sharp, G., and Tovey, C. (2004) Replacement under ongoing technological progress, *IIE Transactions* **36**, 497-508.

Rogers, J. and Hartman, J. (2005) Equipment replacement under continuous and discontinuous technological change, *IMA Journal of Management Mathematics* **16**, 23-36.

R. Solow, Tobin, J., von Weizsacker, C., and Yaari, M. (1966) Neoclassical growth with fixed factor proportions, *Review of Economic Studies* **33**, 79-115.

Yatsenko, Yu. (1995) Volterra integral equations with unknown delay time, *Methods and Applications of Analysis* **2**, 408-419.

Yatsenko, Yu. and Hritonenko, N. (2005) Optimization of the lifetime of capital equipment using integral models, *Journal of Industrial and Management Optimization* **1**, 415-432.

Yatsenko, Yu. and Hritonenko, N. (2007) Network economics and optimal replacement of age-structured IT capital, *Mathematical Methods of Operations Research* **65**, 483-497.

In: Progress in Nonlinear Analysis Research
Editor: Erik T. Hoffmann, pp. 51-60

ISBN 978-1-60456-359-7

Chapter 4

GENERAL CONVERGENCE ANALYSIS FOR A SYSTEM OF NONLINEAR SET-VALUED IMPLICIT VARIATIONAL INCLUSIONS IN REAL BANACH SPACES*

Jian wen Peng†
College of Mathematics and Computer Science,
Chongqing Normal University, Chongqing 400047, PR China
Xin-Bo Yang
Department of Computer and Modem Education Technology,
Chongqing Education College, Chongqing 400067, PR China
Zhang Wei
Department of Computer and Modem Education Technology,
Chongqing Education College, Chongqing 400067, PR China

Abstract

In this paper, we introduce and study a system of nonlinear set-valued implicit variational inclusions with relaxed cocoercive mappings in real Banach spaces. By using the resolvent operator technique for H-accretive operators, we prove the convergence of a new class of perturbed iterative algorithms for solving this system of set-valued implicit variational inclusions in q-uniformly smooth Banach spaces. Our results generalize and improve the corresponding results of recent works.

Key words: system of nonlinear set-valued implicit variational inclusions, resolvent operator, perturbed iterative algorithm, convergence

---

*This research was partially supported by the Natural Science Foundation of China (Grant No. 10171118), the Science and Technology Research Project of Chinese Ministry of Education (Grant No.206123), the Education Committee project Research Foundation of Chongqing (Grant No. 030801), and the Science Committee project Research Foundation of Chongqing (Grant No. 8409).

†E-mail address: jwpeng6@yahoo.com.cn.

1. Introduction

Variational inclusion problems are among the most interesting and intensively studied classes of mathematical problems and have wide applications in the fields of optimization and control, economics and transportation equilibrium, engineering science. For the past years, many existence results and iterative algorithms for various variational inequality and variational inclusion problems have been studied. For details, please see [1-17] and the references therein.

Recently, some new and interesting problems, which are called to be system of variational inclusions were introduced and studied. Kazmi and Bhat [1] introduced a system of nonlinear variational-like inclusions and gave an iterative algorithm for finding its approximate solution. Fang and Huang [2], Verma [3] introduced and studied a system of variational inclusions with H-monotone operators and A-monotone operators, respectively, Yan et al. [4] introduced and studied a system of set-valued variational inclusions which is more general than the model in [2]. Very recently, Chang, Joseph Lee and Chan [10] introduced a system of variational inequalities with relaxed cocoercive mapping in Hilbert spaces.

On the other hand, Huang and Fang [16] introduced H-accretive operators which contains generalized m-accretive operators in [5], H-monotone operators in [2] and the maximal monotone operators as special cases. They also showed some properties of the resolvent operator for H-accretive mappings in Banach spaces.

Inspired and motivated by the above results, we aim in this paper to introduce a new mathematical model, which is called a system of nonlinear set-valued implicit variational inclusions with H-accretive operators and relaxed cocoercive mappings. This new mathematical model contains those in [9, 11, 12, 13, 17] as special cases. By using the resolvent technique for the H-accretive operators, we prove the convergence of a new class of perturbed iterative algorithms for solving this system of set-valued implicit variational inclusions in q-uniformly smooth Banach spaces. Our results generalize and improve the corresponding results of recent works.

2. Preliminaries

Throughout this work we suppose that E is a real Banach space with dual space E^*, $\langle \cdot, \cdot \rangle$ is the dual pair between E and E^*, $CB(E)$ denotes the family of all nonempty closed bounded subsets of E and 2^E denotes the family of all the nonempty subsets of E. The generalized duality mapping $J_q(x) : E \to 2^E$ is defined by

$$J_q(x) = \{f^* \in E^* : \langle x, f^* \rangle = \|x\|^q, \|f^*\| = \|x\|^{q-1}\}$$

where $q > 1$ is a constant. In particular, J_2 is the usual normalized duality mapping. It is known that, in general, $J_q = \|x\|^{q-2} J_2$, for all $x \in E$, and $J_q(x)$ is single-valued if E^* is strictly convex.

The modulus of smoothness of E is the function $\rho_E : [0, +\infty) \to [0, +\infty)$ defined by

$$\rho_E(t) = sup\{\tfrac{1}{2}(\|x+y\| + \|x-y\|) - 1 : \|x\| \leq 1, \|y\| \leq t\}$$

A Banach space E is called uniformly smooth if

$$lim_{t \to 0} \frac{\rho_E(t)}{t} = 0$$

E is called q-uniformly smooth if there exists a constant $c > 0$, such that

$\rho_E(t) \leq ct^q, \quad q > 1.$

Note that J_q is single-valued if E is uniformly smooth. In the study of characteristic inequalities in q-uniformly smooth Banach spaces, Xu [6] proved the following lemma.

Lemma 2.1. Let E be a real uniformly smooth Banach space. Then, E is q-uniformly smooth if and only if there exists a constant $c_q > 0$, such that for all $x, y \in E$

$$\|x + y\|^q \leq \|x\|^q + q\langle y, J_q(x)\rangle + c_q\|y\|^q.$$

Definition 2.1.[14] Let E be a real uniformly smooth Banach space, and $H : E \to E$ be a single valued operators. H is said to be:

(i) accretive, if

$$\langle Hx - Hy, J_q(x - y)\rangle \geq 0, \quad \forall x, y \in E$$

(ii)strictly accretive, if H is accretive and

$$\langle Hx - Hy, J_q(x - y)\rangle = 0, \text{ if and only if } x = y$$

(iii) strongly accretive, if there exists a constant $r > 0$ such that

$$\langle Hx - Hy, J_q(x - y)\rangle \geq r\|x - y\|^q, \qquad \forall x, y \in E$$

Let $\tilde{H}(\cdot, \cdot)$ denote the Hausdorff metric on $CB(E)$ defined by

$$\tilde{H}(A, B) = max\{\sup_{a \in A} d(a, B), \sup_{b \in B} d(A, b)\}, \qquad \forall A, B \in CB(E)$$

where $d(a, B) = inf_{b \in B}\|a - b\|, d(A, b) = inf_{a \in A}\|a - b\|$

Definition 2.2. Let E be a real uniformly smooth Banach space, and $A : E \to CB(E)$ be a set-valued mapping. A is said to be $\tilde{H}$-Lipschitz continuous if there exists a constant $\xi > 0$, such that

$$\tilde{H}(A(x), A(y)) \leq \xi\|x - y\|, \quad \forall x, y \in E$$

Definition 2.3. Let E be a real uniformly smooth Banach space, $T : E \to E$ and $g : E \to E$ be two single-valued mappings, $B : E \to CB(E)$ is a set-valued mapping,

(i) T is said to be μ-Lipschitz continuous if there exists a constant $\mu > 0$ such that

$$\|T(u) - T(v)\| \leq \mu\|u - v\|, \qquad \forall u, v \in E$$

(ii) T is said to be strongly accretive with respect to g and B, if there exists a constant $\gamma > 0$ such that for all $x, y \in E, u \in B(x), v \in B(y)$

$$\langle Tu - Tv, J_q(g(x) - g(y))\rangle \geq \gamma\|x - y\|^q \quad \forall u, v \in E$$

(iii) T is said to be relaxed (ξ, r)-cocoercive with respect to g and B, if there exists a constant $\xi, r > 0$ such that for all $x, y \in E, u \in B(x), v \in B(y)$

$$\langle T(u) - T(v), J_q(g(x) - g(y))\rangle \geq (-\xi)\|T(u) - T(v)\|^q + r\|x - y\|^q, \quad \forall u, v \in E$$

Remark 2.1. If $E = \mathcal{H}$, $g = B = I$ (the identity map on E), then the (ii) and (iii) of definition 2.3 reduce to those of strongly monotonity and relaxed (ξ, r)-cocoercive of T, respectively.

Definition 2.4.[17] Let $H : E \to E$ be a single-valued mapping and $A : E \to 2^E$ be a multi-valued operator. We say that A is H-accretive if A is accretive and $(H+\lambda A)(E) = E$ holds for $\lambda > 0$.

For $i = 1, 2, 3$, let $T_i, H_i, g : E \to E$ be single-valued operators, $B, C, D : E \to CB(E)$ be set-valued mappings and $A_i : E \to 2^E$ be an H-accretive operator. We consider a system of implicit set-valued variational inclusions as follows: to find $(x^*, y^*, z^*, u^*, v^*, w^*)$ with $x^*, y^*, z^* \in E, u^* \in B(y^*), v^* \in C(z^*), w^* \in D(x^*)$ such that

$$\begin{cases} 0 \in \rho T_1(u^*) + H_1(x^*) - g(y^*) + \rho A_1(x^*) \\ 0 \in \eta T_2(v^*) + H_2(y^*) - g(z^*) + \eta A_2(y^*) \\ 0 \in \lambda T_3(w^*) + H_3(z^*) - g(x^*) + \lambda A_3(z^*) \end{cases} \quad (2.1)$$

Below are some special cases of the problem (2.1).

(1) If B, C, D are three single valued mappings, then problem (2.1) reduces to the following problem: finding $(x^*, y^*, z^*) \in E \times E \times E$, such that

$$\begin{cases} 0 \in \rho T_1(B(y^*)) + H_1(x^*) - g(y^*) + \rho A_1(x^*) \\ 0 \in \eta T_2(C(z^*)) + H_2(y^*) - g(z^*) + \eta A_2(y^*) \\ 0 \in \lambda T_3(D(x^*)) + H_3(z^*) - g(x^*) + \lambda A_3(z^*) \end{cases} \quad (2.2)$$

(2) If $x^* = z^*, C = D, \lambda = 0, H_3(z^*) = g(z^*)$, then the problem (2.1) reduces to finding (x^*, y^*, u^*, v^*) with $x^*, y^* \in E, u^* \in B(y^*)$ and $v^* \in C(x^*)$ such that

$$\begin{cases} 0 \in \rho T_1(u^*) + H_1(x^*) - g(y^*) + \rho A_1(x^*) \\ 0 \in \eta T_2(v^*) + H_2(y^*) - g(x^*) + \eta A_2(y^*) \end{cases} \quad (2.3)$$

(3) If B, C are two single valued mappings, then the problem (2.3) reduces to finding $(x^*, y^*) \in E \times E$ such that

$$\begin{cases} 0 \in \rho T_1(B(y^*)) + H_1(x^*) - g(y^*) + \rho A_1(x^*) \\ 0 \in \eta T_2(C(x^*)) + H_2(y^*) - g(x^*) + \eta A_2(y^*) \end{cases} \quad (2.4)$$

(4) If $B = C = I$, then problem (2.4) reduces to finding $(x^*, y^*) \in E \times E$ such that

$$\begin{cases} 0 \in \rho T_1(y^*) + H_1(x^*) - g(y^*) + \rho A_1(x^*) \\ 0 \in \eta T_2(x^*) + H_2(y^*) - g(x^*) + \eta A_2(y^*) \end{cases} \quad (2.5)$$

(5) If $E = \mathcal{H}$ is a Hilbert space, $T_1(y) = A(y) + S(y), T_2(y) = B(x) + T(x)$ and $H_1 = H_2 = g = I, B, D, S, T$ are single valued mappings, then the problem (2.5) reduces to the following system of generalized nonlinear mixed quasi-variational inclusion problem: finding $(x^*, y^*) \in \mathcal{H} \times \mathcal{H}$ such that

$$\begin{cases} 0 \in x - y + \rho(A(y^*) + S(y^*)) + \rho A_1(x^*) \\ 0 \in y - x + \eta(B(x^*) + T(x^*)) + \eta A_2(y^*) \end{cases} \quad (2.6)$$

This was introduced and studied by Agarwal, Huang an Tan [17].

(6) If $E = \mathcal{H}$ is a Hilbert space, $H_1 = H_2 = I, T_1 = T_2$ and $A_1 = A_2 = \partial\phi$, where $\phi : \mathcal{H} \to R \cup \{+\infty\}$ is a proper convex lower semicontinuous function on $\mathcal{H}$ and $\partial\phi$ denotes the subdifferential of function ϕ, then the problem (2.5) is equivalent to find $(x^*, y^*) \in \mathcal{H} \times \mathcal{H}$ such that

$$\begin{cases} \langle \rho T(y^*) + x^* - g(y^*), x - x^* \rangle \geq \phi(x^*) - \phi(y^*) \\ \langle \eta T(x^*) + y^* - g(x^*), x - y^* \rangle \geq \phi(y^*) - \phi(x^*) \end{cases} \quad (2.7)$$

(7) If $E = \mathcal{H}$ is a Hilbert space, $H_1 = H_2 = g = I$ and ϕ is the indicator function of a closed convex subset K in $\mathcal{H}$, that is

$$\phi(u) = I_K(u) = \begin{cases} 0, & u \in K \\ +\infty, & otherwise \end{cases}$$

then the problem (2.7) reduces to finding $(x^*, y^*) \in K \times K$ such that

$$\begin{cases} \langle \rho T(y^*) + x^* - y^*, x - x^* \rangle \geq 0 \\ \langle \eta T(x^*) + y^* - x^*, x - y^* \rangle \geq 0 \end{cases} \quad (2.8)$$

This problem is called a system of nonlinear variational inequality introduced and studied by Vemma [9, 11].

Let $H : E \to E$ be a strictly accretive operator and $A : E \to 2^E$ be a H-accretive operator. Fang and Huang [4] introduced the definition of the resolvent operator $J_H^{A,\lambda} : E \to E$

associated with A, H and λ as follows:

$$J_H^{A,\lambda}(u) = (H + \lambda A)^{-1}(u), \qquad \forall u \in E$$

Lemma 2.1. [14] Let $H : E \to E$ be a strongly accretive operator with constant $\mu > 0$ and $A : E \to 2^E$ be a H-accretive operator, then $J_H^{A,\lambda} : E \to E$ is Lipschitz continuous with constant $\frac{1}{\mu}$, *i.e.*

$$\|J_H^{A,\lambda}(x) - J_H^{A,\lambda}(y)\| \leq \tfrac{1}{\mu}\|x - y\|, \qquad \forall x, y \in E$$

3. Main Result

In this section, we always suppose that E is a q-uniformly smooth Banach space.

Lemma 3.1. Let $A : E \to 2^E$ be H-accretive, then $(x^*, y^*, z^*, u^*, v^*, w^*)$ with $x^*, y^*, z^* \in E, u^* \in B(y^*), v^* \in C(z^*), w^* \in D(x^*)$ is a solution to the SNSIVI (2.1), if and only if $(x^*, y^*, z^*, u^*, v^*, w^*)$ satisfies

$$x^* = J_{H_1}^{A_1,\rho}[g(y^*) - \rho T_1(u^*)] \qquad \text{for } \rho > 0$$
$$y^* = J_{H_2}^{A_2,\eta}[g(z^*) - \eta T_2(v^*)] \qquad \text{for } \eta > 0$$
$$z^* = J_{H_3}^{A_3,\lambda}[g(x^*) - \lambda T_3(w^*)] \qquad \text{for } \lambda > 0.$$

Based on Lemma 3.1, we construct the following general three-step algorithms with error estimates for the problems (2.1).

Algorithm 3.1. For given $x_0 \in E, w_0 \in D(x_0), u_0 \in B(y_0), v_0 \in C(z_0)$, compute the sequence $\{x_n\}, \{y_n\}, \{z_n\}$ such that

$$x_{n+1} = (1 - \alpha_n - d_n)x_n + \alpha_n J_{H_1}^{A_1,\rho}[g(y_n) - \rho T_1(u_n)] + d_n a_n$$
$$y_n = (1 - \beta_n - e_n)x_n + \beta_n J_{H_2}^{A_2,\eta}[g(z_n) - \eta T_2(v_n)] + e_n b_n$$
$$z_n = (1 - \gamma_n - f_n)x_n + \gamma_n J_{H_3}^{A_3,\lambda}[g(x_n) - \lambda T_3(w_n)] + f_n c_n$$
$$u_n \in B(y_n) : \|u_n - u_{n+1}\| \leq \tilde{H}(B(y_n), B(y_{n+1}))$$
$$v_n \in C(z_n) : \|v_n - v_{n+1}\| \leq \tilde{H}(C(z_n), C(z_{n+1}))$$
$$w_n \in D(x_n) : \|w_n - w_{n+1}\| \leq \tilde{H}(D(x_n), D(x_{n+1}))$$

where $0 \leq \alpha_n, \beta_n, \gamma_n, d_n, e_n, f_n \leq 1$, $\{a_n\}, \{b_n\}, \{c_n\}$ are bounded sequence of E.

If $d_n = e_n = f_n = 0$, and B, C, D are three single valued mappings, then Algorithm 3.1 reduces to the following algorithm.

Algorithm 3.2. For given $x_0 \in E, w_0 \in D(x_0), u_0 \in B(y_0), v_0 \in C(z_0)$, compute the sequence $\{x_n\}, \{y_n\}, \{z_n\}$ such that

$$x_{n+1} = (1 - \alpha_n)x_n + \alpha_n J_{H_1}^{A_1,\rho}[g(y_n) - \rho T_1(B(y_n))]$$
$$y_n = (1 - \beta_n)x_n + \beta_n J_{H_2}^{A_2,\eta}[g(z_n) - \eta T_2(C(z_n))]$$
$$z_n = (1 - \gamma_n)x_n + \gamma_n J_{H_3}^{A_3,\lambda}[g(x_n) - \lambda T_3(D(x_n))]$$

where $\{\alpha_n\}, \{\beta_n\}, \{\gamma_n\} \subseteq [0, 1]$, for $n \geq 0$.

If $d_n = e_n = f_n = 0 = \gamma_n$, $C = D$ in Algorithm 3.1, we get

Algorithm 3.3. For given $x_0 \in E, w_0 \in D(x_0), u_0 \in B(y_0)$, compute the sequence $\{x_n\}, \{y_n\}$ such that

$$x_{n+1} = (1 - \alpha_n)x_n + \alpha_n J_{H_1}^{A_1,\rho}[g(y_n) - \rho T_1(u_n)]$$
$$y_n = (1 - \beta_n)x_n + \beta_n J_{H_2}^{A_2,\eta}[g(x_n) - \eta T_2(w_n)]$$
$$u_n \in B(y_n) : \|u_n - u_{n+1}\| \leq \tilde{H}(B(y_n), B(y_{n+1}))$$
$$w_n \in D(x_n) : \|w_n - w_{n+1}\| \leq \tilde{H}(D(x_n), D(x_{n+1}))$$

where $0 \leq \alpha_n, \beta_n \leq 1$.

Lemma 3.2.[13] Let $\{a_n\}$, $\{b_n\}$ and $\{c_n\}$ be three nonnegative real sequences satisfying the following conditions:

$$a_{n+1} \leq (1-\lambda_n)a_n + b_n + c_n, \qquad \forall n \geq n_0$$

where n_0 is some nonnegative integer, $\lambda_n \in (0,1)$ with $\Sigma_{n=0}^{\infty}\lambda_n = \infty, b_n = o(\lambda_n)$ and $\Sigma_{n=0}^{\infty}c_n < \infty$, then $a_n \to 0$ (as $n \to \infty$).

We now present, based on Algorithm 3.1, the approximation-solvability of the problem (2.1) involving the relaxed (ξ_i, r_i)-cocoercive of $T_i : E \to E$.

Theorem 3.1. For $i = 1,2,3$, let E be a real q-uniformly smooth Banach space, $H_i : E \to E$ be a strongly accretive operator with constant $\mu_i > 0$ and $A_i : E \to 2^E$ be an H-accretive operator, T_i be k_i-Lipschitz continuous. Let $g : E \to E$ be Lipschitz continuous with constant $\sigma > 0$. Let $B, C, D : E \to CB(E)$ be $\tilde{H}$-Lipschitz continuous with constants $l_B > 0$, $l_C > 0$ and $l_D > 0$, respectively. Let $T_1 : E \to E$ be relaxed (ξ_1, r_1)-cocoercive with respect to g and B, $T_2 : E \to E$ be relaxed (ξ_2, r_2)-cocoercive with respect to g and C, $T_3 : E \to E$ be relaxed (ξ_3, r_3)-cocoercive with respect to g and D. Supposed that $(x^*, y^*, z^*, w^*, u^*, v^*)$ with $x^*, y^*, z^* \in E, u^* \in B(y^*), v^* \in C(z^*), w^* \in D(x^*)$ is a solution to the problem (2.1) and that $\{x_n\}, \{y_n\}, \{z_n\}, \{w_n\}, \{u_n\}$ and $\{v_n\}$ are the sequences generated by Algorithm 3.1. If the following conditions are satisfied:

$$(3.1)\quad \begin{cases} \sum_{n=0}^{\infty} \alpha_n = \infty, \quad \sum_{n=0}^{\infty}(1-\beta_n) < \infty, \quad e_n \to 0, f_n \to 0, \quad \beta_n, \gamma_n \to 1 \\ 0 < (\sigma^q + \rho q\xi_1 k_1^q l_B^q - \rho q r_1 + \rho^q c_q k_1^q l_B^q)^{\frac{1}{q}} < \mu_1 \\ 0 < (\sigma^q + \eta q\xi_2 k_2^q l_C^q - \eta q r_2 + \eta^q c_q k_2^q l_C^q)^{\frac{1}{q}} < \mu_2 \\ 0 < (\sigma^q + \lambda q\xi_3 k_3^q l_D^q - \lambda q r_3 + \lambda^q c_q k_3^q l_D^q)^{\frac{1}{q}} < \mu_3 \\ r_1 > \xi_1 k_1^q l_B^q, \quad r_2 > \xi_2 k_2^q l_C^q, \quad r_3 > \xi_3 k_3^q l_D^q\}, \qquad \sigma \leq min\{\mu_1, \mu_2, \mu_3\} \end{cases}$$

then the sequences $\{x_n\}, \{y_n\}, \{z_n\}, \{w_n\}, \{u_n\}$ and $\{v_n\}$ converge strongly to x^*, y^*, z^*, w^*, u^*, v^* in E, respectively.

Proof. Since $(x^*, y^*, z^*, w^*, u^*, v^*)$ is a solution to the problem (2.1), by Lemma 3.1, we have

$$\begin{aligned} &\|x_{n+1} - x^*\| \\ &= \|(1-\alpha_n - d_n)(x_n - x^*) + \alpha_n\{J_{H_1}^{A_1,\rho}[g(y_n) - \rho T_1(u_n)] - J_{H_1}^{A_1,\rho}[g(y^*) - \rho T_1(u^*)]\} + d_n(a_n - x^*)\| \\ &\leq (1-\alpha_n - d_n)\|x_n - x^*\| + \tfrac{\alpha_n}{\mu_1}\|g(y_n) - \rho T_1(u_n) - g(y^*) + \rho T_1(u^*)\| + d_n\|a_n - x^*\| \end{aligned} \tag{3.2}$$

Since T_1 is relaxed (ξ_1, r_1)-cocoercive with respect to g and B, and g is σ-Lipschitz continuous, B is $\tilde{H}$-Lipschitz continuous, by Lemma 2.1, we have

$$\begin{aligned} &\|g(y_n) - \rho T_1(u_n) - g(y^*) + \rho T_1(u^*)\|^q \\ &= \|g(y_n) - g(y^*)\|^q - q\rho\langle T_1(u_n) - T_1(u^*), J_q(g(y_n) - g(y^*))\rangle + c_q\rho^q\|T_1(u_n) - T_1(u^*)\|^q \\ &\leq \sigma^q\|y_n - y^*\|^q + \rho q\xi_1\|T_1(u_n) - T_1(u^*)\|^q - \rho q r_1\|y_n - y^*\|^q + c_q\rho^q k_1^q\|u_n - u^*\|^q \\ &\leq [\sigma^q + \rho q\xi_1 k_1^q l_B^q - \rho q r_1 + c_q\rho^q k_1^q l_B^q]\|y_n - y^*\|^q \end{aligned} \tag{3.3}$$

Let

$$L = max\{sup\|a_n - x^*\|, sup\|b_n - y^*\|, sup\|c_n - z^*\|, sup\|x^* - y^*\|, sup\|x^* - z^*\|,\}$$

Substituting (3.3) into (3.2), we obtain

$$\|x_{n+1} - x^*\| \leq (1-\alpha_n - d_n)\|x_n - x^*\| + \theta\alpha_n\|y_n - y^*\| + d_n L \tag{3.4}$$

where $\theta = \frac{(\sigma^q+\rho q\xi_1 k_1^q l_B^q-\rho q r_1+\rho^q c_q k_1^q l_B^q)^{\frac{1}{q}}}{\mu_1} < 1$

Next we make an estimation for $\|y_n - y^*\|$, applying Algorithm 3.1, Lemma 2.1 and 3.1, we have

$$\begin{aligned}
&\|y_n - y^*\| \\
&= \|(1-\beta_n-e_n)(x_n-y^*)+\beta_n\{J_{H_2}^{A_2,\eta}[g(z_n)-\eta T_2(v_n)]-J_{H_2}^{A_2,\eta}[g(z^*)-\eta T_2(v^*)]\}+ \\
&e_n(b_n-y^*)\| \\
&\le (1-\beta_n-e_n)\|x_n-y^*\|+\frac{\beta_n}{\mu_2}\|g(z_n)-\eta T_2(v_n)-g(z^*)+\eta T_2(v^*)\|+e_n\|b_n-y^*\| \\
&\le (1-\beta_n)\|x_n-y^*\|+\beta_n\theta_1\|z_n-z^*\|+e_nL \qquad (3.5)
\end{aligned}$$

where $\theta_1 = \frac{(\sigma^q+\eta q\xi_2 k_2^q l_C^q-\eta q r_2+\eta^q c_q k_2^q l_C^q)^{\frac{1}{q}}}{\mu_2} < 1$

Similarly, we can make an estimation for $\|z_n - z^*\|$, again applying Algorithm 3.1, Lemma 2.1 and 3.1, we have

$$\begin{aligned}
&\|z_n - z^*\| \\
&= \|(1-\gamma_n-f_n)(x_n-z^*)+\gamma_n\{J_{H_3}^{A_3,\lambda}[g(x_n)-\lambda T_3(w_n)]-J_{H_3}^{A_3,\lambda}[g(x^*)-\lambda T_3(w^*)]\}+ \\
&f_n(c_n-z^*)\| \\
&\le (1-\gamma_n-f_n)\|x_n-z^*\|+\frac{\gamma_n}{\mu_3}\|g(x_n)-\lambda T_3(w_n)-g(x^*)+\lambda T_3(w^*)\|+f_n\|c_n-z^*\| \\
&\le (1-\gamma_n)\|x_n-z^*\|+\gamma_n\theta_2\|x_n-x^*\|+f_nL \qquad (3.6)
\end{aligned}$$

where $\theta_2 = \frac{(\sigma^q+\lambda q\xi_3 k_3^q l_D^q-\lambda q r_3+\lambda^q c_q k_3^q l_D^q)^{\frac{1}{q}}}{\mu_3} < 1$

$$\begin{aligned}
\|z_n-z^*\| &\le (1-\gamma_n)\|x_n-x^*\|+(1-\gamma_n)\|x^*-z^*\|+\gamma_n\|x_n-x^*\|+f_nL \\
&\le \|x_n-x^*\|+(1-\gamma_n+f_n)L \qquad (3.7)
\end{aligned}$$

Substituting (3.7) into (3.5), we have

$$\begin{aligned}
\|y_n-y^*\| &\le (1-\beta_n)\|x_n-y^*\|+\beta_n\theta_1\|z_n-z^*\|+e_nL \\
&\le (1-\beta_n)\|x_n-x^*\|+(1-\beta_n)\|x^*-y^*\|+\beta_n[\|x_n-x^*\|+(1-\gamma_n+f_n)L]+e_nL \\
&\le \|x_n-x^*\|+(1-\gamma_n\beta_n+\beta_nf_n+e_n)L \qquad (3.8)
\end{aligned}$$

Substituting (3.8) into (3.4), we get

$$\begin{aligned}
\|x_{n+1}-x^*\| &\le (1-\alpha_n-d_n)\|x_n-x^*\|+\theta\alpha_n\|y_n-y^*\|+d_nL \\
&\le (1-\alpha_n)\|x_n-x^*\|+\theta\alpha_n[\|x_n-x^*\|+(1-\gamma_n\beta_n+\beta_nf_n+e_n)L]+d_nL \\
&\le 1-\alpha_n(1-\theta)\|x_n-x^*\|+\theta\alpha_nL(1-\gamma_n\beta_n+\beta_nf_n+e_n)+d_nL \qquad (3.9)
\end{aligned}$$

Taking $a_n = \|x_n - x^*\|, \lambda_n = \alpha_n(1-\theta), b_n = 0, c_n = L[\theta\alpha_n(1-\gamma_n\beta_n+\beta_nf_n+e_n)+d_n]$

It follows from (3.9) and Lemma 3.2, we obtain
$\|x_n - x^*\| \to 0 \quad i.e. \quad x_n \to x^*.$

Since D is l_D-$\tilde{H}$-Lipschitz continuous, we have

$\|w_n - w^*\| \le \tilde{H}(D(x_n), D(x^*)) \le l_D\|x_n - x^*\| \to 0, \quad (n\to\infty)$

$i.e.\ \ w_n \to w^* \in E, \quad (n \to \infty)$

Similarly , we have that $v_n \to v^*, n\to\infty$ and $w_n \to w^*, n \to \infty$

We now show that $w^* \in D(x^*), u^* \in B(y^*), v^* \in C(z^*)$. In fact,

$$\begin{aligned}
d(w^*, D(x^*)) &\le \|w^*-w_n\|+\tilde{H}(w_n, D(x^*)) \\
&\le \|w^*-w_n\|+\tilde{H}(D(x_n), D(x^*)) \\
&\le \|w^*-w_n\|+l_D\|x_n-x^*\| \to 0 \qquad \text{as } n\to\infty
\end{aligned}$$

where $d(w^*, D(x^*)) = inf\{\|u^* - t\| : t \in D(x^*)$. This implies that $w^* \in D(x^*)$. In a similar way, one can show that $u^* \in B(y^*)$ and $v^* \in C(z^*)$. This completes the proof.

Remark 3.2. If E is 2-uniformly smooth and the following formula is satisfied,

$$\begin{cases} |\rho - \frac{q\xi_1 k_1^2 l_B^2 - qr_1}{2c_2 k_1^2 l_B^2}| < \frac{\sqrt{(q\xi_1 k_1^2 l_B^2 - qr_1)^2 - 4c_2 k_1^2 l_B^2(\sigma^2 - \mu_1^2)}}{2c_2 k_1^2 l_B^2} \\ |\eta - \frac{q\xi_2 k_2^2 l_C^2 - qr_2}{2c_2 k_2^2 l_C^2}| < \frac{\sqrt{(q\xi_2 k_2^2 l_C^2 - qr_2)^2 - 4c_2 k_2^2 l_C^2(\sigma^2 - \mu_2^2)}}{2c_2 k_2^2 l_C^2} \\ |\lambda - \frac{q\xi_3 k_3^2 l_D^2 - qr_3}{2c_2 k_3^2 l_D^2}| < \frac{\sqrt{(q\xi_3 k_3^2 l_D^2 - qr_3)^2 - 4c_2 k_3^2 l_D^2(\sigma^2 - \mu_3^2)}}{2c_2 k_3^2 l_D^2} \end{cases}$$

then the hypothesis (3.1) holds. It is worth noting that Hilbert space and L_p (or l_p) spaces $(2 \leq q \leq \infty)$ are 2-uniformly smooth.

Corollary 3.1. For $i = 1, 2, 3$, let E be a real q-uniformly smooth Banach space, $H_i : E \to E$ be a strongly accretive operator with constant $\mu_i > 0$ and $A_i : E \to 2^E$ be an H_i-accretive operator. Let $g : E \to E$ be σ-Lipschitz continuous, T_i be Lipschitz continuous with constants $k_i > 0$. Let $B, C, D : E \to CB(E)$ be $\tilde{H}$-Lipschitz continuous with constants $l_B > 0$, $l_C > 0$ and $l_D > 0$, respectively. Let $T_1 : E \to E$ be strongly accretive with respect to g and B with constant $\xi_1 > 0$, $T_2 : E \to E$ be strongly accretive with respect to g and C with constant $\xi_2 > 0$, $T_3 : E \to E$ be strongly accretive with respect to g and D with constant $\xi_3 > 0$. Supposed that $(x^*, y^*, z^*, w^*, u^*, v^*)$ such that $x^*, y^*, z^* \in E, u^* \in B(y^*), v^* \in C(z^*), w^* \in D(x^*)$ is a solution to the problem (2.1) and that $\{x_n\}, \{y_n\}, \{z_n\}, \{w_n\}, \{u_n\}$ and $\{v_n\}$ are the sequences by Algorithm 3.1. If the following conditions are satisfied:

$$\begin{cases} \sum_{n=0}^{\infty} \alpha_n = \infty, \quad \sum_{n=0}^{\infty}(1-\beta_n) < \infty, \quad e_n \to 0, f_n \to 0, \quad \beta_n, \gamma_n \to 1 \\ 0 < (\sigma^q - \rho q \xi_1 + \rho^q c_q k_1^q l_B^q)^{\frac{1}{q}} < \mu_1 \\ 0 < (\sigma^q - \eta q \xi_2 + \eta^q c_q k_2^q l_C^q)^{\frac{1}{q}} < \mu_2 \\ (IV) \quad 0 < (\sigma^q - \lambda q \xi_3 + \lambda^q c_q k_3^q l_D^q)^{\frac{1}{q}} < \mu_3 \end{cases}$$

then the sequences $\{x_n\}, \{y_n\}, \{z_n\}, \{w_n\}, \{u_n\}$ and $\{v_n\}$ convergence strongly to $x^*, y^*, z^*, w^*, u^*, v^*$ in E, respectively.

If B, C, D be single valued mapping, then by Theorem 3.1, it is to easy to get following result.

Corollary 3.2. For $i = 1, 2, 3$, let E be a real q-uniformly smooth Banach space, $H_i : E \to E$ be a strongly accretive operator with constant $\mu_i > 0$ and $A_i : E \to 2^E$ be an H-accretive operator. Let $g : E \to E$ be Lipschitz continuous with constant $\sigma > 0$. Assume that $B, C, D : E \to E$ be Lipschitz continuous with constants $l_B > 0$, $l_C > 0$ and $l_D > 0$, respectively, T_i be Lipschitz continuous with constants $k_i > 0$. Let $T_1 : E \to E$ be relaxed (ξ_1, r_1)-cocoercive with respect to g and B, $T_2 : E \to E$ be relaxed (ξ_2, r_2)-cocoercive with respect to g and C, $T_3 : E \to E$ be relaxed (ξ_3, r_3)-cocoercive with respect to g and D. Supposed that $(x^*, y^*, z^*) \in E \times E \times E$ is a solution to the problem (2.2) and that $\{x_n\}, \{y_n\}$ and $\{z_n\}$ are the sequences generated by Algorithm 3.2. If the following conditions are satisfied:

$$\begin{cases} \sum_{n=0}^{\infty} \alpha_n = \infty, \quad \sum_{n=0}^{\infty}(1-\beta_n) < \infty, \quad \beta_n, \gamma_n \to 1 \\ 0 < (\sigma^q + \rho q \xi_1 k_1^q l_B^q - \rho q r_1 + \rho^q c_q k_1^q l_B^q)^{\frac{1}{q}} < \mu_1 \\ 0 < (\sigma^q + \eta q \xi_2 k_2^q l_C^q - \eta q r_2 + \eta^q c_q k_2^q l_C^q)^{\frac{1}{q}} < \mu_2 \\ 0 < (\sigma^q + \lambda q \xi_3 k_3^q l_D^q - \lambda q r_3 + \lambda^q c_q k_3^q l_D^q)^{\frac{1}{q}} < \mu_3 \\ (V) \quad r_1 > \xi_1 k_1^q l_B^q, \quad r_2 > \xi_2 k_2^q l_C^q, \quad r_3 > \xi_3 k_3^q l_D^q\}, \qquad \sigma \leq min\{\mu_1, \mu_2, \mu_3\} \end{cases}$$

then the sequences $\{x_n\}, \{y_n\}, \{z_n\}$ converge strongly to x^*, y^*, z^*, respectively.

Theorem 3.2. For $i = 1, 2$, let E be a real q-uniformly smooth Banach space, $H_i : E \to E$ be a strongly accretive operator with constant $\mu_i > 0$ and $A_i : E \to 2^E$ be an H_i-accretive,

T_i be Lipschitz continuous with constants $k_i > 0$. Let $g : E \to E$ be σ-Lipschitz continuous, $B, D : E \to CB(E)$ be $\tilde{H}$-Lipschitz continuous with constants $l_B > 0$ and $l_D > 0$, respectively, $T_1 : E \to E$ be relaxed (ξ_1, r_1)-cocoercive with respect to g and B, $T_2 : E \to E$ be relaxed (ξ_2, r_2)-cocoercive with respect to g and D. Supposed that (x^*, y^*, u^*, v^*) with $x^*, y^* \in E, u^* \in B(y^*), v^* \in C(x^*)$ is a solution to the problem (2.3), and $\{x_n\}, \{y_n\}, \{u_n\}$ and $\{v_n\}$ are the sequences generated by Algorithm 3.3. If the following conditions are satisfied:

$$\begin{cases} 0 \leq \alpha_n, \beta_n \leq 1; \sum_{n=0}^{\infty} \alpha_n \beta_n = \infty \\ 0 < (\sigma^q + \rho q \xi_1 k_1^q l_B^q - \rho q r_1 + \rho^q c_q k_1^q l_B^q)^{\frac{1}{q}} < \mu_1 \\ 0 < (\sigma^q + \eta q \xi_2 k_2^q l_D^q - \eta q r_2 + \eta^q c_q k_2^q l_D^q)^{\frac{1}{q}} < \mu_2 \\ (IV) \quad r > max\{\xi_1 k_1^q l_B^q, \quad \xi_2 k_2^q l_D^q\}, \quad \sigma \leq min(\mu_1, \mu_2) \end{cases}$$

then the sequences $\{x_n\}, \{y_n\}, \{u_n\}$ and $\{v_n\}$ converge strongly to x^*, y^*, u^*, v^*, respectively.

Remark 3.3. It is easy to see that the above results extend and improve the main results in [9, 11, 12, 13].

References

[1] K.R. Kazmi, M.I. Bhat, Iterative algorithm for a system of nonlinear variational-like inclusions, *Comput. Math. Appl.* **48** (2004) 1929-1935.

[2] Y.P. Fang, N.J. Huang, H-monotone operators and system of variational inclusions, *Commun. Appl. Nonlinear Anal.* **11** (1) (2004) 93-101.

[3] R.U. Verma, A-monotononicity and applications to nonlinear variational inclusion problems, *J. Appl. Math. Stoch. Anal.* **17** (2) (2004) 193-195.

[4] W.Y. Yan, Y.P. Fang, N.J. Huang, A new system of set-valued variational inclusions with H-monotone operators, *Math. Inequal. Appl.* **8** (3) (2005) 537-546.

[5] N.J. Huang, Y.P. Fang, Generalized m-accretive mappings in Banach spaces, *J. Sichuan Univ.* **38** (4) (2001) 591-592.

[6] H.K. Xu, Inequalities in Banach spaces with applications, Nonlinear Anal. 16 (12) (1991) 1127-1138.

[7] K. Deimling. Nonlinear Functional Analysis, Springer-Verlag, Berlin, 1985.

[8] Y.P. Fang, N.J. Huang, H-accretive operators and resolvent operator technique for solving variational inclusions in Banach spaces, *Appl. Math. Lett.* **17** (6) (2004) 647-653.

[9] Ram U. Verma, General convergence analysis for two-step projection methods and applications to variational problems, *Appl. Math. Lett.* **18** (2005) 1286-1292.

[10] S.S. Chang, H.W. Joseph Lee, C.K. Chan, Generalized system for relaxed cocoercive variational inequalities in Hilbert spaces, *Appl. Math. Lett.* **20** (3) (2007) 329-334.

[11] Ram. U. Verma, Projection Methods, Algorithms, and a New System of Nonlinear Variational Inequalities, *Comput. Math. Appl.* **41** (2001) 1025-1031.

[12] H. Nie, Z. Liu, K.H. Kim, S.M. Kang, A system of nonlinear variational inequalities involving strongly monotone and pseudocontractive mappings, *Advances in Nonlinear Variational Inequalities* **6** (2) (2003) 91-99.

[13] R.U. Verma, Nonlinear variational and constrained hemivariational inequalities involving relaxed operators, *Zeitschrift fur Angewandte Mathematik und Mechanik.* **77** (5), (1997) 387-391.

[14] Jian-Wen Peng, Set-valued variational inclusions with T -accretive operators in Banach spaces. *Appl. Math. Lett.* **19** (2006) 273-282.

[15] Heng-you Lan, (A, η)-Accretive mappings and set-valued variational inclusions with relaxed cocoercive mappings in Banach spaces. *Appl. Math. Lett.* **20** (2007) 571-577.

[16] Y.P. Fang, N.J. Huang, Iterative algorithm for a system of variational inclusions involving H-accretive operators in Banach spaces, *Acta Math. Hungar.* **108** (3) (2005) 183-195.

[17] R.P. Agarwal, N.J. Huang, M.Y. Tan, Sensitivity analysis for a new system of generalized nonlinear mixed quasi-variational inclusions, *Appl. Math. Lett.* **17** (2004) 345-352.

In: Progress in Nonlinear Analysis Research
Editor: Erik T. Hoffmann, pp. 61-67
ISBN 978-1-60456-359-7

Chapter 5

On the Homogeneous Monge-Ampère Equation

Yuri Bozhkov*
Departamento de Matemática Aplicada - DMA
Instituto de Matemática, Estatistica e Computação Científica - IMECC
Universidade Estadual de Campinas - UNICAMP
C.P. 6065, 13083-970, Campinas - SP, Brasil

Abstract

The homogeneous Monge-Ampère equation (HMAE) $u_{xx}u_{yy} - u_{xy}^2 = 0$ describes the developable surfaces in $\mathbb{R}^3$. An explicit formula for its general solution was found by Vitaly Ushakov using a series of changes of the involved variables deduced by geometric arguments. In the present work we obtain the general solution of the HMAE by applying a single contact transformation. Further we obtain the general solution of the HMAE in the higher-dimensional case using the same approach. We also discuss the Lie point symmetries of the HMAE.

1. Introduction

It is well known that the homogeneous Monge-Ampère equation (HMAE)

$$u_{xx}u_{yy} - u_{xy}^2 = 0 \tag{1}$$

describes the so-called developable surfaces in $\mathbb{R}^3$, [1], p. 10. (Here the subscripts denote partial derivatives of the function $u = u(x, y)$ with respect to the corresponding independent variable, e.g. $u_x = \frac{\partial u}{\partial x}$.) Clearly, the Gaussian curvature of such surfaces vanishes identically.

In [13] Ushakov found an explicit formula for the general solution of equation (1), supposing that $u_{xx} \neq 0$. His approach is based on the use of a series of changes of the involved variables deduced by geometric arguments. This result can be stated as follows.

*E-mail address: bozhkov@ime.unicamp.br

Theorem 1 (Ushakov, [13]). *The general solution of* (1), *with the assumption* $u_{xx} \neq 0$ *everywhere, is given in the following parametric form:*

$$\begin{cases} x(t,s) &=& g(t) - s\, f'(t), \\ y(t,s) &=& s, \\ u(t,s) &=& t\, g(t) - \int_0^t g(r)dr + s\,\{f(t) - t\, f'(t)\}, \end{cases} \tag{2}$$

where $f \in C^2$, $g \in C^1$, $g' \neq 0$, *are arbitrary functions*.

The main purpose of the present work is to obtain the general solution of the HMAE directly by applying a single *contact* transformation. Namely, in dimension two we make use of the following transformation:

$$\begin{cases} \xi &=& u_x, \\ \eta &=& y, \\ w &=& u - x\, u_x, \\ w_\xi &=& -x, \\ w_\eta &=& u_y. \end{cases} \tag{3}$$

and give another proof of Theorem 1. The advantage of this method is that it can be generalized to higher dimensions.

Let $u = u(x)$, $x = (x_1, ..., x_n) \in \mathbb{R}^n$, $n \geq 2$, be a smooth function (say, $u \in C^2(\mathbb{R}^n)$). We denote by $Hess_n(u)$ the Hessian matrix of u, that is, the matrix whose elements are the second order derivatives of u:

$$Hess_n(u) = \left(\frac{\partial^2 u}{\partial x_i \partial x_j}\right), \quad i,j = 1,...,n.$$

The main result of this paper is the following

Theorem 2. *The general classical solution of the HMAE in* $\mathbb{R}^n$:

$$\det(Hess_n(u)) = 0, \tag{4}$$

with the assumption $Hess_{n-1}(u) \neq 0$ *everywhere, is given in the parametric form:*

$$\begin{cases} x_i(t,s) &=& \dfrac{\partial g}{\partial t_i}(t) - s\,\dfrac{\partial f}{\partial t_i}(t), \quad i = 1,...,n-1, \\ x_n(t,s) &=& s, \\ u(t,s) &=& \displaystyle\sum_{i=1}^{n-1} t_i\,\frac{\partial g}{\partial t_i}(t) - g(t) + s\left\{f(t) - \sum_{i=1}^{n-1} t_i\,\frac{\partial f}{\partial t_i}(t)\right\}, \end{cases} \tag{5}$$

where $t = (t_1, ..., t_{n-1}) \in \mathbb{R}^{n-1}$, $s \in \mathbb{R}$ and $f \in C^2(\mathbb{R}^{n-1})$, $g \in C^2(\mathbb{R}^{n-1})$, $Hess_{n-1}(g) \neq 0$, *are arbitrary functions*.

A transformation of this kind (see (3)) has been used in [5, 6, 11] in the study of the equation:

$$u_{xx}u_{yy} - u_{xy}^2 = 1. \tag{6}$$

Sometimes such a transformation is called partial Legendre transform or Legendre-like transformation. It has been successfully applied in [4,8–10] in the proof of regularity results for some classes of degenerate Monge-Ampère equations.

The main motivation to write this paper is the fact that the finite contact transformation (3) transforms the *linear* equation

$$u_{xx} + u_{yy} = 0$$

into the *fully non-linear* equation (6). See [12]. Although the used here approach is quite similar to that in [5, 6, 10], to the author's knowledge, the usefulness of the contact transformation (3) and its generalization in obtaining the explicit formula for the general solution of the HMAE has not been previously emphasized.

Historical references for the Legendre-like transformation can be found in [10]. For geometric aspects of the HMAE in two dimensions as well as a discussion on removing the assumption $u_{xx} \neq 0$ see [13].

This paper is organized as follows. In section 2 we find the general solution of the HMAE in two dimensions. In section 3 we apply the used approach to the higher-dimensional case and prove Theorem 2. The Lie point symmetries of the HMAE are stated in section 4.

Acknowledgments. We would like to thank Dr. Frank Columbus for having given us the opportunity to participate in the present collection of works on Nonlinear Analysis. We would also like to thank CNPq, Brasil, for partial financial support.

2. The Two-Dimensional Case

In this section we prove Theorem 1. To begin with, we recall the *contact* transformation (3) presented in the Introduction:

$$\begin{cases} \xi &= u_x, \\ \eta &= y, \\ w &= u - x\,u_x, \\ w_\xi &= -x, \\ w_\eta &= u_y. \end{cases}$$

This change of variables is not degenerated since its Jacobian is equal to 1. Its inverse is given by

$$\begin{cases} x &= -w_\xi, \\ y &= \eta, \\ u &= w - \xi\,w_\xi, \\ u_x &= \xi, \\ u_y &= w_\eta. \end{cases} \tag{7}$$

Further, by the chain rule, we obtain:

$$u_{xx} = -\frac{1}{w_{\xi\xi}}, \quad u_{xy} = -\frac{w_{\xi\eta}}{w_{\xi\xi}}, \quad u_{yy} = w_{\eta\eta} - \frac{{w_{\xi\eta}}^2}{w_{\xi\xi}} \tag{8}$$

since $u_{xx} \neq 0$ and hence $w_{\xi\xi} \neq 0$. Then we substitute (8) into (1). In this way the equation (1) for $u = u(x, y)$ is equivalent to the following equation for $w = w(\xi, \eta)$:

$$w_{\eta\eta} = 0. \tag{9}$$

The general solution of (9) can be immediately written as

$$w = f(\xi)\,\eta + h(\xi), \tag{10}$$

where f and h are arbitrary smooth functions of ξ. Now, going back to the original variables x, y and u, the formula (10) assumes the form

$$u - x\,u_x = f(u_x)\,y + h(u_x). \tag{11}$$

We denote $s := y$ and $t := u_x$. Then by (11)

$$u = t\,x + s\,f(t) + h(t). \tag{12}$$

On the other hand

$$x = -w_\xi = -f'(\xi)\,\eta - h'(\xi)$$

from (7) and (10). Hence

$$x = -f'(t)\,s + g(t),$$

where $g = -h'$. Therefore we get the first two relations in (2):

$$\begin{cases} x(t,s) &= g(t) - s\,f'(t), \\ y(t,s) &= s. \end{cases}$$

Thus, we obtain from (12) the third relation in (2):

$$u = [g(t) - s\,f'(t)]\,t + s\,f(t) - \int_0^t g(r)dr$$

$$= t\,g(t) - \int_0^t g(r)dr + s\,\{f(t) - t\,f'(t)\}$$

since $h(t) = -\int_0^t g(r)dr$. This completes the proof.

3. The Higher-Dimensional Case

In this section we prove Theorem 2 following the same arguments of section 2.

We consider the partial Legendre transformation:

$$\begin{cases} \xi_i &= u_{x_i}, \quad ,i = 1, ..., n-1, \\ \xi_n &= x_n, \\ w &= u - \sum_{i=1}^{n-1} x_i\, u_{x_i}, \\ w_{\xi_i} &= -x_i, \\ w_{\xi_n} &= u_{x_n}, \end{cases} \tag{13}$$

which generalizes (3) to higher dimensions (see [8]). We apply (13) to equation (4) and obtain by a tedious calculation that

$$w_{\xi_n \xi_n} = 0. \tag{14}$$

We omit the details merely noting that in this calculation the condition $Hess_{n-1}(u) \neq 0$ has been used.

From (14) we conclude that

$$w = f(\xi_1, ..., \xi_{n-1})\, \xi_n - g(\xi_1, ..., \xi_{n-1}) \tag{15}$$

where f and g are arbitrary smooth functions of $\xi_1, ..., \xi_{n-1}$. From (13) and (15):

$$u - \sum_{i=1}^{n-1} x_i\, u_{x_i} = f(u_{x_1}, ..., u_{x_{n-1}})\, x_n - g(u_{x_1}, ..., u_{x_{n-1}}). \tag{16}$$

We now introduce the parameters: $t_i := u_{x_i}$, $i = 1, ..., n-1$, and $s := x_n$. Then (16) assumes the form:

$$u = \sum_{i=1}^{n-1} t_i x_i + f(t)s - g(t). \tag{17}$$

Further, from (13) and (15) we have

$$x_i = -w_{\xi_i} = -\frac{\partial f}{\partial \xi_i}\, \xi_n + \frac{\partial g}{\partial \xi_i}, \quad i = 1, ..., n-1,$$

that is

$$x_i = \frac{\partial g}{\partial t_i} - s\, \frac{\partial f}{\partial t_i}, \quad i = 1, ..., n-1,$$

which is the first relation in (5). Hence and from (17) we obtain the third relation in (5).

4. The Lie Point Symmetries

In this section we study the Lie point symmetries of the HMAE (1). For this purpose we apply the classical approach of Sophus Lie [12].

Let

$$X = \xi^1(x, y, u)\frac{\partial}{\partial x} + \xi^2(x, y, u)\frac{\partial}{\partial y} + \eta(x, y, u)\frac{\partial}{\partial u}$$

be the infinitesimal generator of a point transformation admitted by equation (1). Then by a straightforward calculation one obtains the following set of determining equations:

$$\xi^1{}_{uu} = \xi^1{}_{yu} = \xi^1{}_{yy} = \xi^2{}_{uu} = \xi^2{}_{xu} = \xi^2{}_{xx} = \eta_{xx} = \eta_{yy} = \eta_{xy} = 0,$$

$$\xi^1{}_{xy} - \eta_{yu} = 0, \quad \xi^1{}_{xx} - 2\eta_{xu} = 0, \quad 2\xi^1{}_{xu} - \eta_{uu} = 0,$$

$$\xi^2{}_{xy} - \eta_{xu} = 0, \quad \xi^2{}_{yy} - 2\eta_{yu} = 0, \quad 2\xi^2{}_{yu} - \eta_{uu} = 0.$$

Hence it is clear that ξ^1 and ξ^2 are linear functions of u and that η is a linear function of x and y. Then another straightforward calculation gives

$$\begin{cases} \xi^1 &= (c_1x + c_2)u + (c_4x + c_5)y + c_4x^2 + c_{12}x + c_{13}, \\ \xi^2 &= (c_1y + c_3)u + (c_7y + c_8)x + c_4y^2 + c_{10}x + c_{11}, \\ \eta &= (c_7u + c_9)x + (c_4u + c_6)y + c_1u^2 + c_{14}u + c_{15}, \end{cases}$$

where c_i, $i = 1, 2, ..., 15$, are arbitrary constants. Thus the HMAE (1) admits a 15-parameter group of Lie point symmetries. This fact is known (see [7], p. 781). We have presented it here for sake of completeness as well as for the fact that it provides a good example to test the wonderful *Mathematica* package SYM created by Stelios Dimas et al. [2, 3]. We would like to observe that the performance of the SYM-package during the computation of the symmetries is excellent.

5. Conclusion

The contact partial Legendre transformation has proved its usefulness in the study of degenerate elliptic non-linear partial differential equations involving the Monge-Amp`ere operator. We belief that further applications of this remarkable transformation are possible.

References

[1] Courant, R.; Hilbert, D.; Methods of Mathematical Physics, Vol. II: Partial Differential Equations; Interscience Publishers (a division of John Wiley & Sons): New York-London, 1962.

[2] Dimas, S; Tsoubelis, D.; SYM:A new symmetry-finding package for Mathematica; *Proceedings of the 10th International Conference in Modern Group Analysis* , 24-30 October 2004, Larnaca, Cyprus, (2004), 64 - 70

[3] Dimas, S; Tsoubelis, D.; A new heuristic algorithm for solving overdetermined systems of PDEs in Mathematica; *6th International Conference on Symmetry in Nonlinear Mathematical Physics*, 20-26 June 2005, Kiev, Ukraine, (2005).

[4] Guan, P; Regularity of a class of quasilinear degenerate elliptic equations; *Adv. Math.* **132** (1997), 24 - 45.

[5] Jörgens, K.; Über die Lösungen der Differentialgleichung $rt - s^2 = 1$; *Math. Ann.* **127** (1954), 130 - 134.

[6] Jörgens, K.; Harmonische Abbidungen und die Differentialgleichung $rt - s^2 = 1$; *Math. Ann.* **129** (1955), 330 - 334.

[7] Manno, G.; Olivieri, F.; Vitolo, R.; On differential equations characterized by their Lie point symmetries; *J. Math. Anal. Appl.* **332** (2007), 767 - 786.

[8] Rios, C.; Sawyer, E. T.; Wheeden, R. L.; A higher-dimensional partial Legendre transform, and regularity of degenerate Monge-Ampère equations; *Adv. Math.* **193** (2005), 373 - 415.

[9] Sawyer, E. T.; Wheeden, R. L.; Regularity of degenerate Monge-Ampère and prescribed Gaussian curvature equations in two dimension; *Potential Anal.* **24** (2006), 267 - 301.

[10] Schulz, F.; Regularity theory for quasilinear elliptic systems and Monge-Ampre equations in two dimensions; *Lecture Notes in Mathematics*, **1445**; Springer-Verlag: Berlin, 1990.

[11] Schulz, F.; Wang, L.; Isolated singularities of Monge-Ampère equations; *Proc. Amer. Math. Soc.* **123** (1995), 3705 - 3708.

[12] Stephani, H.; *Differential equations: their solutions using symmetry* ; Cambridge University Press: Cambridge, 1989.

[13] Ushakov, V.; The explicit general solution of trivial Monge-Ampère equation; *Comment. Math. Helv.* **75** (2000), 125 - 133.

Reviewed by Antonio Carlos Gilli Martins
Departamento de Matemática - DM
Instituto de Matemática, Estatistica e Computação Científica - IMECC
Universidade Estadual de Campinas - UNICAMP
C.P. 6065, 13083-970, Campinas - SP, Brasil
E-mail: gilli@ime.unicamp.br

In: Progress in Nonlinear Analysis Research
Editor: Erik T. Hoffmann, pp. 69-135

ISBN 978-1-60456-359-7

Chapter 6

SOLUTIONS TO SOME OPEN PROBLEMS IN n-DIMENSIONAL FLUID DYNAMICS

Linghai Zhang*
Department of Mathematics, Lehigh University
14 East Packer Avenue, Bethlehem, Pennsylvania 18015 USA

Abstract

The focus of this work is on the solutions to some open problems of the global weak solutions of the Cauchy problems for a general nonlinear dissipative partial differential equation

$$\mathbf{u}_t - \varepsilon\triangle\mathbf{u}_t - \alpha\triangle\mathbf{u} + \mathcal{D}\mathbf{u} + \mathcal{N}(\mathbf{u}, \nabla\mathbf{u}) = \mathbf{0}, \qquad \mathbf{u}(\mathbf{x}, 0) = \mathbf{u}_0(\mathbf{x}),$$

in n-dimensional space, where $n \geq 1$ is an integer, $\alpha > 0$ and $0 \leq \varepsilon \leq 1$ are real constants, and $\triangle = \sum_{k=1}^{n} \dfrac{\partial^2}{\partial x_k^2}$ denotes the classical Laplace operator. More precisely, suppose that the initial function $\mathbf{u}_0 \in L^1(\mathbb{R}^n) \cap H^2(\mathbb{R}^n)$, let $\mathbf{u} = \mathbf{u}(\mathbf{x}, t, \mathbf{u}_0)$ represent the global solutions of the Cauchy problem, we will study the limit

$$\lim_{t\to\infty} \left\{ (1+t)^{2m+\lambda+n/2} \int_{\mathbb{R}^n} \left[|\triangle^m \mathbf{u}(\mathbf{x}, t)|^2 + \varepsilon |\nabla\triangle^m \mathbf{u}(\mathbf{x}, t)|^2 \right] \mathrm{dx} \right\}$$

in terms of the initial function $\mathbf{u}_0$ and the model parameters, such as the dissipation coefficient, where $m \geq 0$ is any integer, $\lambda = 0$ if $\int_{\mathbb{R}^n} \mathbf{u}_0(\mathbf{x})\mathrm{dx} \neq \mathbf{0}$, and $\lambda = 1$ if $\int_{\mathbb{R}^n} \mathbf{u}_0(\mathbf{x})\mathrm{dx} = \mathbf{0}$. The limit problem has been open for a long time. The general model includes the n-dimensional Burgers equation, the n-dimensional Benjamin-Bona-Mahony-Burgers equation, the one-dimensional nonlinear cubic Korteweg-de Vries-Burgers equation, the one-dimensional nonlinear Benjamin-Ono-Burgers equation, the two-dimensional nonlinear nonlocal quasi-geostrophic equation, the n-dimensional incompressible Navier-Stokes equations and the n-dimensional incompressible Magnetohydrodynamics equations as particular examples.

The main ideas in the analysis are Fourier transform, Plancherel's identity, new decomposition of frequency space, lower limit estimate and upper limit estimate.

*E-mail address: liz5@lehigh.edu. Telephone: +1-610-758-4116. Fax: +1-610-758-3767.

Key Words: modified Fourier splitting method, exact limits, decay estimates with sharp rates, dissipative partial differential equations, global weak solutions

Mathematical Subject Classification: 35Q20.

1. Introduction

1.1. The Model Equations

Many dissipative partial differential equations arise from biology, chemistry, physics and fluid dynamics. They have strong nonlinearities, such as the n-dimensional incompressible Navier-Stokes equations and the nonlinear cubic Korteweg-de Vries-Burgers equations. These are evolutionary equations with supercritical nonlinearities. After an extremely long time, the nonlinearity in an equation becomes arbitrarily small compared with the linearity. In another word, the nonlinear partial differential equations can be treated as perturbations of the corresponding linear equations (to get the linear equations, simply drop the nonlinear terms in the nonlinear equations). Therefore, it is very possible to show that the exact limits of the L^2-norms of the global solutions multiplied by $(1+t)^\lambda$, where $\lambda > 0$ is the sharp rate of decay, for the linear and nonlinear differential equations to be the same. The exact limits of these physical quantities, as time approaches infinity, may play significant roles in the evaluations of the Hausdorff dimension and the fractal dimension of the global attractor and the inertial manifold, respectively, of the infinite-dimensional dynamical systems.

Consider the Cauchy problems for a very general dissipative differential equation in n-dimensional space, with the spatial dimension $n \geq 1$:

$$\mathbf{u}_t - \varepsilon\triangle\mathbf{u}_t + \alpha(-\triangle)^\sigma\mathbf{u} + \mathcal{D}\mathbf{u} + \mathcal{N}(\mathbf{u}, \nabla\mathbf{u}) = \mathbf{f}(\mathbf{x}, t) \quad \text{in } \mathbb{R}^n \times \mathbb{R}^+, \tag{1}$$

$$\mathbf{u}(\mathbf{x}, 0) = \mathbf{u}_0(\mathbf{x}) \quad \text{in } \mathbb{R}^n. \tag{2}$$

This model arises from fluid dynamics. In (1), $\alpha > 0$, $0 \leq \varepsilon \leq 1$ and $\sigma > 0$ are constants, $\triangle = \sum_{i=1}^n \dfrac{\partial^2}{\partial x_i^2}$ represents the Laplace operator, $(-\triangle)^\sigma\mathbf{u}$ stands for dissipation, its Fourier transform is given by

$$\widehat{(-\triangle)^\sigma\mathbf{u}}(\xi, t) = |\xi|^{2\sigma}\widehat{\mathbf{u}}(\xi, t).$$

For simplicity, we will only consider the case $\sigma = 1$. In addition, $\mathcal{D}\mathbf{u}$ represents a linear term, its Fourier transform $\widehat{\mathcal{D}\mathbf{u}}(\xi, t) = \mathbf{i}\Gamma(\xi)\widehat{\mathbf{u}}(\xi, t)$, where $\Gamma(\xi)$ stands for a real, odd function of ξ. In this book chapter, $\widehat{\psi}(\xi) = \int_{\mathbb{R}^n} \psi(\mathbf{x})\exp(-\mathbf{i}\mathbf{x}\cdot\xi)d\mathbf{x}$ stands for the Fourier transform of $\psi \in L^1(\mathbb{R}^n)$, $\mathbf{x}\cdot\xi = x_1\xi_1 + x_2\xi_2 + \cdots + x_n\xi_n$ denotes the scalar product of $\mathbf{x} = (x_1, x_2, \cdots, x_n)^T$ and $\xi = (\xi_1, \xi_2, \cdots, \xi_n)^T \in \mathbb{R}^n$. See the Appendix (Subsection 6.3) for the properties of the Fourier transform. By Plancherel's identity, it is straightforward to verify that the integral

$$\int_{\mathbb{R}^n} \phi \cdot (\mathcal{D}\phi)d\mathbf{x} = 0,$$

for any $\phi \in H^2(\mathbb{R}^n)$. In another word, $\mathcal{D}\mathbf{u}$ denotes a dispersion. The nonlinear function $\mathcal{N}$ is sufficiently smooth and the initial data $\mathbf{u}_0 \in L^1(\mathbb{R}^n) \cap H^2(\mathbb{R}^n)$. In this chapter, we are concerned with the long time asymptotic behaviors of the global solutions of (1)-(2).

One of the most challenging problems in applied nonlinear analysis is the limit problem of global solutions of nonlinear evolutionary equations. We are concerned with the exact limits of the L^2-norms divided by the sharp rates of decay of the global solutions, including their derivatives of any order, of problem (1)-(2). It turns out that the Hausdorff dimension and the fractal dimension of the global attractor and the inertial manifold, respectively, depend on the exact limit of the physically important quantity: $\left[(1+t)^{2m+\lambda+n/2}\int_{\mathbb{R}^n}|\Delta^m \mathbf{u}(\mathbf{x},t)|^2 d\mathbf{x}\right]$, as $t\to\infty$, where $m\geq 0$ is any integer and λ depends on the initial function $\mathbf{u}_0$. Let us classify the differential equations to be studied.

Definition of critical nonlinear equations and supercritical nonlinear equations: Consider the following nonlinear and linear equations

$$\begin{aligned}
&\mathbf{u}_t-\varepsilon\Delta\mathbf{u}_t-\alpha\Delta\mathbf{u}+\mathcal{D}\mathbf{u}+\mathcal{N}(\mathbf{u},\nabla\mathbf{u})=0,\\
&\mathbf{v}_t-\varepsilon\Delta\mathbf{v}_t-\alpha\Delta\mathbf{v}+\mathcal{D}\mathbf{v}=0.
\end{aligned}$$

(I) If

$$\lim_{t\to\infty}\frac{\|\mathbf{u}(\cdot,t)-\mathbf{v}(\cdot,t)\|_{L^2(\mathbb{R}^n)}}{\|\mathbf{v}(\cdot,t)\|_{L^2(\mathbb{R}^n)}}=0,$$

then we say the equation (1) is supercritical. After an extremely long time, the nonlinear term is arbitrarily small relative to the linear term, so the nonlinear partial differential equation becomes a perturbation of the linear partial differential equation.

(II) On the other hand, if

$$\lim_{t\to\infty}\frac{\|\mathbf{u}(\cdot,t)-\mathbf{v}(\cdot,t)\|_{L^2(\mathbb{R}^n)}}{\|\mathbf{v}(\cdot,t)\|_{L^2(\mathbb{R}^n)}}>0,$$

then we say the equation (1) is critical. No matter how long time elapses, the decay rates of the nonlinear term and the linear term are the same.

In this work, we will focus on supercritical differential equations.

This definition is probably against the traditional definition of critical or supercritical equations where existence is concerned with. Here our focus is on the optimal rate of decay.

This model is not too abstract because it includes the following fluid dynamics equations when $\varepsilon=0$: **(I)** the one-dimensional cubic Burgers equation

$$u_t+\beta(u^3)_x=\alpha u_{xx},\qquad \text{in } \mathbb{R}\times\mathbb{R}^+,$$

where $\alpha>0$ and $\beta\in\mathbb{R}$ are constants (the one-dimensional Burgers equation involving quadratic nonlinearity $u_t+\beta(u^2)_x=\alpha u_{xx}$ has been solved before), and the n-dimensional quadratic Burgers equation

$$u_t+\beta\cdot\nabla(u^2)=\alpha\Delta u,\qquad \text{in } \mathbb{R}^n\times\mathbb{R}^+,$$

where $\alpha>0$ is a constant and $\beta=(\beta_1,\beta_2,\cdots,\beta_n)^T\in\mathbb{R}^n$ is a constant vector, the spatial dimension $n\geq 2$.

(II) The one-dimensional nonlinear Benjamin-Ono-Burgers equation

$$u_t+u_x+Hu_{xx}+(u^3)_x=\alpha u_{xx},\qquad \text{in } \mathbb{R}\times\mathbb{R}^+,$$

where H stands for the Hilbert transform, defined by $[H\phi](x) = \frac{1}{\pi} P.V. \int_{\mathbb{R}} \frac{\phi(y)}{x-y} dy$, so $\widehat{H\phi}(\xi) = \mathbf{i}\,\text{sign}(\xi)\widehat{\phi}(\xi)$.

(III) The one-dimensional nonlinear cubic Korteweg-de Vries-Burgers equation

$$u_t + u_x + u_{xxx} + (u^3)_x = \alpha u_{xx}, \qquad \text{in } \mathbb{R} \times \mathbb{R}^+.$$

(IV) The two-dimensional nonlinear nonlocal quasi-geostrophic equation

$$u_t + (\psi_y, -\psi_x)^T \cdot \nabla u = \alpha \Delta u, \qquad u = (-\Delta)^{1/2}\psi, \qquad \text{in } \mathbb{R}^2.$$

(V) The n-dimensional incompressible Navier-Stokes equations

$$\begin{aligned} \frac{\partial \mathbf{u}}{\partial t} + (\mathbf{u}\cdot\nabla)\mathbf{u} + \nabla p &= \alpha\Delta\mathbf{u}, \qquad \nabla\cdot\mathbf{u} = 0, \qquad \text{in } \mathbb{R}^n \times \mathbb{R}^+, \\ \mathbf{u}(\mathbf{x},0) &= \mathbf{u}_0(\mathbf{x}), \qquad \nabla\cdot\mathbf{u}_0 = 0, \qquad \text{in } \mathbb{R}^n. \end{aligned}$$

(VI) The n-dimensional Magnetohydrodynamics equations

$$\begin{aligned} \frac{\partial \mathbf{u}}{\partial t} + (\mathbf{u}\cdot\nabla)\mathbf{u} - (\mathbf{A}\cdot\nabla)\mathbf{A} + \nabla\left(p + \frac{\kappa}{2}|\mathbf{A}|^2\right) &= \frac{1}{RE}\Delta\mathbf{u}, \qquad \nabla\cdot\mathbf{u} = 0, \\ \frac{\partial \mathbf{A}}{\partial t} + (\mathbf{u}\cdot\nabla)\mathbf{A} - (\mathbf{A}\cdot\nabla)\mathbf{u} &= \frac{1}{RM}\Delta\mathbf{A}, \qquad \nabla\cdot\mathbf{A} = 0, \end{aligned}$$

where RE represents the Reynolds number and RM denotes the magnetic Reynolds number. Both are positive constants. Here $\kappa = M^2/RE \cdot RM$ is another constant. If we define $P = p + \frac{\kappa}{2}|\mathbf{A}|^2$, then the Magnetohydrodynamics equations become a little simpler. Whenever the general equation represents the Navier-Stokes equations or the Magnetohydrodynamics equations, we require $\nabla\cdot\mathbf{u} = \nabla\cdot\mathbf{A} = 0$ and $\nabla\cdot\mathbf{u}_0 = \nabla\cdot\mathbf{A}_0 = 0$. We will not mention these incompressible conditions explicitly later on for (1). The general equation (1) may contain many other important partial differential equations as well if $\varepsilon > 0$. For example, it includes **(VII)** the general Benjamin-Bona-Mahony-Burgers equations

$$\begin{aligned} u_t - u_{xxt} + u_x + (u^3)_x &= \alpha u_{xx}, \qquad \text{in } \mathbb{R} \times \mathbb{R}^+, \\ u_t - \Delta u_t + \beta\cdot\nabla u + \varphi(u)\cdot\nabla u &= \alpha\Delta u, \qquad \text{in } \mathbb{R}^n \times \mathbb{R}^+, \end{aligned}$$

where, in the second equation, the spatial dimension $n \geq 2$, $\alpha > 0$ is a constant, $\beta = (\beta_1, \beta_2, \cdots, \beta_n)^T$ is a constant vector, and $\varphi(u) = (\varphi_1(u), \varphi_2(u), \cdots, \varphi_n(u))^T$ is a real, nonlinear, vector-valued function of u, satisfying the condition $|\varphi(u)| \leq C|u|$, for all u with $|u| \leq 1$ and for some constant $C > 0$. Almost all of these equations are derived from fluid dynamics. See [3], [19], [22], [33]-[36], [42]-[46], [48], [54], [57], [61]-[62], [72] and [80] for the backgrounds of these model equations.

If the L^2-norm of the global solutions of (1)-(2) is bounded, then we wonder if it converges to zero with an exponential or algebraic rate as $t \to \infty$. We will establish the exact limits with optimal decay rates of the global solutions of the general dissipative partial differential equation. We will show how to use very simple ideas to solve very complicated mathematical problems. We will explore a very general approach so that any other problem will enjoy the same limits as those displayed here in this book chapter as long as some general assumptions are satisfied. The basic hypothesis is that there exists at least a global weak solution, and the nonlinearity and its Fourier transform satisfy certain mild conditions. The main ingredients and technical tools in the analysis are the modified Fourier splitting method, energy estimates, lower and upper limit estimates. We also use various ideas in real analysis, complex analysis and functional analysis. See [38], [39], [49], [73], [74], [79], [84] and [86].

It would not be surprising to expect to see that the L^1-norm remains bounded on $\mathbb{R}^+$ but does not converge to zero at all. For some partial differential equations, there holds the mass conservation law: $\|\mathbf{u}(\cdot,t)\|_{L^1} = \|\mathbf{u}_0\|_{L^1}$, for all $t > 0$. Moreover, due to the energy equation

$$\frac{\mathrm{d}}{\mathrm{d}t}\left[\|\mathbf{u}(\cdot,t)\|_{L^2}^2 + \varepsilon\|\nabla\mathbf{u}(\cdot,t)\|_{L^2}^2\right] + 2\alpha\|\nabla\mathbf{u}(\cdot,t)\|_{L^2}^2 = 0,$$

the L^2-norm of the global solutions converges to zero as time approaches positive infinity, even if we only assume that $\mathbf{u}_0 \in L^2(\mathbb{R}^n)$, but not necessarily in $L^p(\mathbb{R}^n)$, where $1 \le p < 2$. We should point out that there exists no algebraic rate of decay in this L^2 limit. This limit of the L^2-norm is not very interesting because the limit does not carry any valuable information about the initial momentum: i.e. the integral $\int_{\mathbb{R}^n} \mathbf{u}_0(\mathbf{x})\mathrm{d}\mathbf{x}$. When dividing the L^2-norm by the sharp rate of decay, the limit may depend on the initial data.

1.2. The Mathematical Assumptions on Problems (1)-(2)

Hypothesis One: Suppose that the initial data $\mathbf{u}_0 \in L^1(\mathbb{R}^n) \cap H^2(\mathbb{R}^n)$. For certain model equations, such as the n-dimensional incompressible Navier-Stokes equations and the Magnetohydrodynamics equations, due to the incompressible conditions $\nabla\cdot\mathbf{u} = \nabla\cdot\mathbf{u}_0 = 0$ and $\nabla\cdot\mathbf{A} = \nabla\cdot\mathbf{A}_0 = 0$, there must hold

$$\int_{\mathbb{R}^n} \mathbf{u}(\mathbf{x},t)\mathrm{d}\mathbf{x} = 0, \qquad \int_{\mathbb{R}^n} \mathbf{u}_0(\mathbf{x})\mathrm{d}\mathbf{x} = 0,$$

$$\int_{\mathbb{R}^n} \mathbf{A}(\mathbf{x},t)\mathrm{d}\mathbf{x} = 0, \qquad \int_{\mathbb{R}^n} \mathbf{A}_0(\mathbf{x})\mathrm{d}\mathbf{x} = 0.$$

Note that, if the initial data $\mathbf{u}_0$ converges to zero sufficiently fast as $|\mathbf{x}| \to \infty$, then we must have that $|\mathbf{x}|\mathbf{u}_0 \in L^1(\mathbb{R}^n)$. Now, for each of the integers $k = 1, 2, \cdots, n$, we have

$$0 = \int_{\mathbb{R}^n} \nabla\cdot\left(x_k\mathbf{u}_0(\mathbf{x})\right)\mathrm{d}\mathbf{x} = \int_{\mathbb{R}^n} (\mathbf{u}_0)_k(\mathbf{x})\mathrm{d}\mathbf{x} + \int_{\mathbb{R}^n} x_k\nabla\cdot\mathbf{u}_0(x)\mathrm{d}\mathbf{x},$$

where the first equality follows from the divergence theorem, and the last of these integrals vanishes because $\mathbf{u}_0$ is divergence free and hence we find that $\int_{\mathbb{R}^n}(\mathbf{u}_0)_k(\mathbf{x})\mathrm{d}\mathbf{x} = 0$, for each $k = 1, 2, \ldots, n$. Therefore, for any divergence free vector field, there holds $\int_{\mathbb{R}^n} \mathbf{u}_0(\mathbf{x})\mathrm{d}\mathbf{x} = \mathbf{0}$. This result may be true even if $|\mathbf{x}|\mathbf{u}_0 \notin L^1(\mathbb{R}^n)$.

Another quick proof of the same result: Since the initial data is divergence free, $\nabla\cdot\mathbf{u}_0 = 0$, and $\mathbf{u}_0 \in L^1(\mathbb{R}^n)$, taking the Fourier transform yields $\xi\cdot\widehat{\mathbf{u}_0}(\xi) = 0$. From this equation we claim that $\widehat{\mathbf{u}_0}(\mathbf{0}) = \mathbf{0}$, that is, $\int_{\mathbb{R}^n} \mathbf{u}_0(\mathbf{x})\mathrm{d}\mathbf{x} = \mathbf{0}$. In fact, suppose that this is not true, say $\mathbf{a} := \widehat{\mathbf{u}_0}(\mathbf{0}) \neq \mathbf{0}$. Recall that $\mathbf{u}_0 \in L^1(\mathbb{R}^n)$, thus $\widehat{\mathbf{u}_0}(\xi)$ is continuous in ξ. Then by continuity we must have $0 = (\widetilde{\varepsilon}\mathbf{a})\cdot\widehat{\mathbf{u}_0}(\widetilde{\varepsilon}\mathbf{a}) \ge \frac{\widetilde{\varepsilon}}{2}|\mathbf{a}|^2 > 0$, where $0 < \widetilde{\varepsilon} \ll 1$, this is a contradiction.

Let

$$\varphi_{ij}(\mathbf{x}) = \int_{-\infty}^{x_j} u_{0i}(x_1, \cdots, x_{j-1}, y_j, x_{j+1}, \cdots, x_n)\mathrm{d}y_j, \qquad i, j = 1, 2, \cdots, n.$$

Then $\varphi_{ij}(\mathbf{x}) \to 0$ as $|\mathbf{x}| \to \infty$, and

$$\frac{\partial \varphi_{ij}}{\partial x_j}(\mathbf{x}) = u_{0i}(\mathbf{x}).$$

Therefore, for these partial differential equations, motivated by this fact, we will consider those initial data, such that

$$\begin{aligned}
\mathbf{u}_0(\mathbf{x}) &= \left(\sum_{j=1}^{n} \frac{\partial \phi_{1j}}{\partial x_j}(\mathbf{x}), \sum_{j=1}^{n} \frac{\partial \phi_{2j}}{\partial x_j}(\mathbf{x}), \cdots, \sum_{j=1}^{n} \frac{\partial \phi_{nj}}{\partial x_j}(\mathbf{x}) \right)^T, \\
\mathbf{A}_0(\mathbf{x}) &= \left(\sum_{j=1}^{n} \frac{\partial \psi_{1j}}{\partial x_j}(\mathbf{x}), \sum_{j=1}^{n} \frac{\partial \psi_{2j}}{\partial x_j}(\mathbf{x}), \cdots, \sum_{j=1}^{n} \frac{\partial \psi_{nj}}{\partial x_j}(\mathbf{x}) \right)^T,
\end{aligned}$$

for some vector-valued, integrable functions $\Phi_i \equiv (\phi_{i1}, \phi_{i2}, \cdots, \phi_{in})^T$ and $\Psi_i \equiv (\psi_{i1}, \psi_{i2}, \cdots, \psi_{in})^T$, where ϕ_{ij} and $\psi_{ij} \in C^2(\mathbb{R}^n) \cap L^1(\mathbb{R}^n)$, $i, j = 1, 2, \cdots, n$, with

$$\sum_{i=1}^{n}\sum_{j=1}^{n} \frac{\partial^2 \phi_{ij}}{\partial x_i \partial x_j} = 0, \qquad \sum_{i=1}^{n}\sum_{j=1}^{n} \frac{\partial^2 \psi_{ij}}{\partial x_i \partial x_j} = 0,$$

$$\int_{\mathbb{R}^n} |\Phi_i(\mathbf{x})|\mathrm{d}\mathbf{x} < \infty, \qquad \int_{\mathbb{R}^n} |\Psi_i(\mathbf{x})|\mathrm{d}\mathbf{x} < \infty.$$

Hypothesis Two: For some differential equations in higher-dimensional spaces, when the solutions are too weak, we have to make the additional assumption: $\mathbf{u} \in L^q(\mathbb{R}^+, L^p(\mathbb{R}^n))$, where p and q are constants, such that $p > n \geq 3$ and $q > 2$ and $\frac{n}{p} + \frac{2}{q} = 1$. This assumption is satisfied by the global solutions of the equation (1)-(2) with small initial data and it is open for global solutions with large initial data. This assumption can guarantee that the L^2-norm of m-th order derivative is uniformly bounded on $\mathbb{R}^+$, $\forall m \geq 1$.

Hypothesis Three: Suppose that the global solutions and their Fourier transforms satisfy

$$\int_{\mathbb{R}^n} \mathbf{u} \cdot \mathcal{N}(\mathbf{u}, \nabla \mathbf{u})\mathrm{d}\mathbf{x} = 0, \tag{3}$$

$$\left| \int_{\mathbb{R}^n} \mathbf{v} \cdot \mathcal{N}(\mathbf{u}, \nabla \mathbf{u})\mathrm{d}\mathbf{x} \right| \leq C \frac{\|\nabla \mathbf{v}(\cdot, t)\|_{L^\infty(\mathbb{R}^n)}}{(1+t)^{\max\{1, n/2\}}}, \tag{4}$$

$$\left| \widehat{\mathcal{N}(\mathbf{u}, \nabla \mathbf{u})}(\xi, t) \right| \leq \frac{C|\xi|}{(1+t)^{\max\{1, n/2\}}}, \tag{5}$$

$$\left| \widehat{\mathcal{N}(\mathbf{u}, \nabla \mathbf{u})}(\xi, t) \right| \leq \frac{C|\xi|}{(1+t)^{1+n/2}}, \tag{6}$$

where $\mathbf{v}(\cdot, t) \in W^{1,\infty}(\mathbb{R}^n)$, for each fixed $t > 0$. For (6), we need the additional assumption:

$$\mathbf{u}_0(\mathbf{x}) = \left(\sum_{j=1}^{n} \frac{\partial \phi_{1j}}{\partial x_j}(\mathbf{x}), \sum_{j=1}^{n} \frac{\partial \phi_{2j}}{\partial x_j}(\mathbf{x}), \cdots, \sum_{j=1}^{n} \frac{\partial \phi_{nj}}{\partial x_j}(\mathbf{x}) \right)^T,$$

for some real, vector-valued functions $\Phi_i \equiv (\phi_{i1}, \phi_{i2}, \cdots, \phi_{in})^T \in C^2(\mathbb{R}^n) \cap L^1(\mathbb{R}^n)$, where $i = 1, 2, \cdots, n$.

In assumptions (4), (5) and (6), the constant C is independent of (ξ, t) and $\mathbf{u}$, but it may depend on the initial data $\mathbf{u}_0$. Note that all of these assumptions are satisfied by the global solutions of the model equations (I)-(VII) listed above.

1.3. Known Results

A - Existence, uniqueness and regularity: There have been many very important methods and techniques to prove the existence and uniqueness of global strong/weak solutions of nonlinear evolutionary partial differential equations. Perhaps the most comprehensive way is the so-called Leray-Schauder's fixed point principle. See [105]-[108]. To apply this method, the key point is to establish global energy estimates for all possible solutions of the underlying problems. When doing this, maximum principle or conservation laws or nonlinear transforms (e.g. Hopf-Cole transform, see [47]) may play crucial roles. Another method is the well-known Galerkin approximation (consideration of global solutions to initial boundary value problems with periodic boundary conditions, and letting the period approach infinity). Some people use another popular method: semigroup (formal representation of solutions which involves semigroup operators) coupled with integral estimates, so that local solutions can be extended to global solutions step by step. For brevity, we will not provide the tedious and technical proof of the existence, uniqueness, regularity, of the global solutions to the general problem (1)-(2).

On the other hand, the existence of global solutions of (1)-(2) is a completely nontrivial problem, in particular, if the spatial dimension is high. Let us take the Navier-Stokes equations as a concrete example to illustrate the existence and regularity results.

The classical n-dimensional incompressible Navier-Stokes equations

$$\mathbf{u}_t + (\mathbf{u} \cdot \nabla)\mathbf{u} - \Delta \mathbf{u} + \nabla p = \mathbf{f}(\mathbf{x}, t), \qquad \nabla \cdot \mathbf{u} = 0,$$

occupy a central position in modern applied mathematics and in nonlinear partial differential equations. It describes the motion of fluids, such as the flows of the atmosphere and the oceans. The existence, uniqueness and decay estimates of global strong solutions of two-dimensional problems have been solved very well, under the conditions $\mathbf{u}_0 \in H^2(\mathbb{R}^2)$ and $\nabla \cdot \mathbf{u}_0 = 0$, or $\mathbf{u}_0 \in L^1(\mathbb{R}^2) \cap H^2(\mathbb{R}^2)$ and $\nabla \cdot \mathbf{u}_0 = 0$. We call this solution strong or smooth because it satisfies the differential equations in the classical sense and the solution is unique. But for three-dimensional and higher-dimensional problems, the existence and uniqueness of smooth solutions are known only on a finite time interval, unless the initial data is sufficiently small, see Hugo Beirao da Veiga [5]-[8], Kato [50]-[53], Leray [55], etc. Some mathematicians constructed strong solutions $\{\mathbf{u}_m\}_{m=1}^{\infty}$ to certain approximating equations, e.g

$$\mathbf{u}_t - \varepsilon^2 \Delta \mathbf{u}_t - \alpha \Delta \mathbf{u} + \alpha \varepsilon^2 \Delta^2 \mathbf{u} + (\mathbf{u} \cdot \nabla)\mathbf{u} + \nabla p = \mathbf{f}, \qquad \nabla \cdot \mathbf{u} = 0,$$

or

$$\mathbf{u}_t - \varepsilon^2 \Delta \mathbf{u}_t - \alpha \Delta \mathbf{u} + \alpha \varepsilon^2 \Delta^2 \mathbf{u} + (\mathbf{u} \cdot \nabla)(\mathbf{u} - \varepsilon^2 \Delta \mathbf{u}) + \nabla p = \mathbf{f}, \qquad \nabla \cdot \mathbf{u} = 0,$$

which converge weakly to the Navier-Stokes equations as $\varepsilon \to 0$, in the sense of $L^2_{\text{loc}}(\mathbb{R}^+, H^1(\mathbb{R}^n))$. For two-dimensional problems, weak solutions are always strong. However, this is not necessarily true for the solutions of the Navier-Stokes equations in higher-dimensional spaces. Even if we make strong assumptions on the initial data, such as $\mathbf{u}_0 \in L^n(\mathbb{R}^n) \cap H^2(\mathbb{R}^n)$, or $\mathbf{u}_0 \in L^p(\mathbb{R}^n) \cap H^2(\mathbb{R}^n)$ where $p > n$, or $\mathbf{u}_0 \in L^\infty(\mathbb{R}^n) \cap H^n(\mathbb{R}^n)$, the global smooth solutions corresponding to large initial data have not been solved yet. Either the global smooth solution exists for all positive time or the solution blows up at certain finite time. It seems that the traditional method of energy estimates is not sufficient to solve the global existence problem. Nevertheless, from other interesting points of view, many outstanding mathematicians have made significant contributions to the classical Navier-Stokes equations. See Hugo Beirao da Veiga [5], Caffarelli, Kohn and Nirenberg [13], Kato [50], Leray [55], Jindrich Necas, Michael Ruzicka and Vladimir Sverak [60], Maria E. Schonbek and her collaborators [63]-[68], George Sell [69], James Serrin [70]-[71], Roger Temam [75]-[78], Michael Wiegner [81]-[83], etc.

We will list several significant work on the Cauchy problems for the n-dimensional incompressible Navier-Stokes equations. Due to the limit of space, however, many valuable work on the Navier-Stokes equations are not mentioned here.

In 1982, Caffarelli, Kohn and Nirenberg established the partial regularity of suitably weak solutions of the classical Navier-Stokes equations in [13]. Later, Fang-Hua Lin [56] simplified their method of proof and obtained a similar result. The partial regularity of the suitably weak solutions can be summarized as follows: the one-dimensional Hausdorff measure of the singular set (consisting of all irregular points of the suitably weak solutions) is zero. In [70]-[71] James Serrin proved the existence of global smooth solutions under the assumption that $\mathbf{u}$ has some appropriate regularity, i.e. there exist constants p and q: $p > n \geq 3$ and $q > 2$, such that $\frac{n}{p} + \frac{2}{q} < 1$ and

$$\mathbf{u} \in L^q\big(\mathbb{R}^+, L^p(\mathbb{R}^n)\big).$$

For the critical case: $\frac{n}{p} + \frac{2}{q} = 1$, Hugo Beirao da Veiga [5] established the uniqueness and boundedness: $\mathbf{u} \in L^\infty\big(\mathbb{R}^+, L^p(\mathbb{R}^n)\big)$. In [96] by using the same conditions as Hugo Beirao da Veiga did, Zhang proved the global existence by using iteration technique. The iteration technique involves the assumption made by Hugo Beirao da Veiga in [5], the application of the decay estimates established by Michael Wiegner in [81]-[82], and the Hölder's inequality. Actually, Zhang proved that

$$\mathbf{u} \in \left[\bigcap_{p \leq r < \infty} L^{2r/(r-n)}\big(\mathbb{R}^+; L^r(\mathbb{R}^n)\big)\right] \cap \left[\bigcap_{2 \leq s < \infty} L^\infty\big(\mathbb{R}^+; L^s(\mathbb{R}^n)\big)\right].$$

This implies that $\mathbf{u}$ is actually a global smooth solution for the n-dimensional Navier-Stokes equations.

If there exists an initial function $\mathbf{u}_0 \in L^\infty(\mathbb{R}^n) \cap H^2(\mathbb{R}^n)$, such that there hold the above results for the Navier-Stokes equations, then there is a positive number $\delta = \delta(\mathbf{u}_0)$. For any initial data $\widetilde{\mathbf{u}}_0 \in L^p(\mathbb{R}^n)$, if

$$\|\mathbf{u}_0 - \widetilde{\mathbf{u}}_0\|_{L^p(\mathbb{R}^n)} < \delta,$$

then there exists a unique global strong solution $\widetilde{\mathbf{u}} \in L^q\big(\mathbb{R}^+, L^p(\mathbb{R}^n)\big)$ to the Cauchy problem $\widetilde{\mathbf{u}}(\mathbf{x},0) = \widetilde{\mathbf{u}}_0(\mathbf{x})$ for the Navier-Stokes equations. Moreover

$$\sup_{\mathbb{R}^+}\left\{(1+t)^{1+n/2}\int_{\mathbb{R}^n}|\mathbf{u}(\mathbf{x},t)-\widetilde{\mathbf{u}}(\mathbf{x},t)|^2\mathrm{d}\mathbf{x}\right\} \le C\int_{\mathbb{R}^n}|\mathbf{u}_0(\mathbf{x})-\widetilde{\mathbf{u}}_0(\mathbf{x})|^2\mathrm{d}\mathbf{x},$$

where the constant $C = C\big(\int_0^\infty \|\mathbf{u}(\cdot,t)\|^q_{L^p(\mathbb{R}^n)}\mathrm{d}t\big) > 0$ is independent of time. See [95]-[96].

Note: The set consisting of initial functions such that these results hold is open.

In [60], Necas, Ruzicka and Sverak showed a conjecture on self-similar solutions of the form

$$\mathbf{u}(\mathbf{x},t) = \frac{1}{\sqrt{2a(T-t)}}\mathbf{U}\left(\frac{\mathbf{x}}{\sqrt{2a(T-t)}}\right),$$

which was made by Leray in 1934 in [55], is negative, where $T \in \mathbb{R}$ and $a > 0$ are constants. On the other hand, if one of the quantities $\left\|\frac{\partial^{\alpha_1+\alpha_2+\cdots+\alpha_n}\mathbf{u}}{\partial x_1^{\alpha_1}\partial x_2^{\alpha_2}\cdots\partial x_n^{\alpha_n}}(\cdot,t)\right\|^2$ really blows up at some finite time $T > 0$, then the rate is at most equal to $1-\widetilde{\varepsilon}$ with $\widetilde{\varepsilon} \in (0,1)$, i.e. near T, there may hold

$$\left\|\frac{\partial^{\alpha_1+\alpha_2+\cdots+\alpha_n}\mathbf{u}}{\partial x_1^{\alpha_1}\partial x_2^{\alpha_2}\cdots\partial x_n^{\alpha_n}}(\cdot,t)\right\|^2 = \frac{\mathcal{O}(1)}{(T-t)^{1-\widetilde{\varepsilon}}}, \qquad \text{as } t \to T.$$

Let us summarize the existence result of problems (1)-(2).

Theorem 0.1 *Suppose that the initial data* $\mathbf{u}_0 \in H^2(\mathbb{R}^n)$. *Then there exists at least a global weak solution to the Cauchy problem (1)-(2):* $\mathbf{u} \in C_{\text{weak}}\big(\mathbb{R}^+, L^2(\mathbb{R}^n)\big) \cap L^\infty\big(\mathbb{R}^+, L^2(\mathbb{R}^n)\big) \cap L^2_{\text{loc}}\big(\mathbb{R}^+, H^1(\mathbb{R}^n)\big)$, *such that*

$$\lim_{|\mathbf{x}|\to\infty}\mathbf{u}(\mathbf{x},t) = \mathbf{0}, \quad \lim_{|\mathbf{x}|\to\infty}\nabla\mathbf{u}(\mathbf{x},t) = \mathbf{0}, \quad \lim_{|\mathbf{x}|\to\infty}\triangle\mathbf{u}(\mathbf{x},t) = \mathbf{0}, \qquad \forall t \ge 0.$$

Formally, we have the solution representation

$$\begin{aligned}\mathbf{u}(\mathbf{x},t) = &\frac{1}{(2\pi)^n}\int_{\mathbb{R}^n}\exp(+\mathbf{i}\mathbf{x}\cdot\xi)\exp\left[-\frac{\alpha|\xi|^2+\mathbf{i}\Gamma(\xi)}{1+\varepsilon|\xi|^2}t\right]\widehat{\mathbf{u}}_0(\xi)\mathrm{d}\xi\\ &+\frac{1}{(2\pi)^n}\int_0^t\left\{\int_{\mathbb{R}^n}\exp(+\mathbf{i}\mathbf{x}\cdot\xi)\exp\left[-\frac{\alpha|\xi|^2+\mathbf{i}\Gamma(\xi)}{1+\varepsilon|\xi|^2}(t-\tau)\right]\frac{\widehat{\mathbf{f}}(\xi,\tau)}{1+\varepsilon|\xi|^2}\mathrm{d}\xi\right\}\mathrm{d}\tau\\ &-\frac{1}{(2\pi)^n}\int_0^t\left\{\int_{\mathbb{R}^n}\exp(+\mathbf{i}\mathbf{x}\cdot\xi)\exp\left[-\frac{\alpha|\xi|^2+\mathbf{i}\Gamma(\xi)}{1+\varepsilon|\xi|^2}(t-\tau)\right]\frac{\widehat{\mathcal{N}(\mathbf{u},\nabla\mathbf{u})}(\xi,\tau)}{1+\varepsilon|\xi|^2}\mathrm{d}\xi\right\}\mathrm{d}\tau.\end{aligned}$$

If $\Gamma \equiv 0$ *and* $\varepsilon = 0$, *then the global solutions of (1)-(2) can be represented as*

$$\mathbf{u}(\mathbf{x},t) = \frac{1}{(4\pi\alpha t)^{n/2}} \int_{\mathbb{R}^n} \exp\left(-\frac{|\mathbf{x}-\mathbf{y}|^2}{4\alpha t}\right) \mathbf{u}_0(\mathbf{x})\mathrm{d}\mathbf{y}$$
$$+ \int_0^t \left\{ \frac{1}{[4\pi\alpha(t-\tau)]^{n/2}} \int_{\mathbb{R}^n} \exp\left[-\frac{|\mathbf{x}-\mathbf{y}|^2}{4\alpha(t-\tau)}\right] \mathbf{f}(\mathbf{y},\tau)\mathrm{d}\mathbf{y} \right\} \mathrm{d}\tau$$
$$- \int_0^t \left\{ \frac{1}{[4\pi\alpha(t-\tau)]^{n/2}} \int_{\mathbb{R}^n} \exp\left[-\frac{|\mathbf{x}-\mathbf{y}|^2}{4\alpha(t-\tau)}\right] \mathcal{N}(\mathbf{u},\nabla\mathbf{u})(\mathbf{y},\tau)\mathrm{d}\mathbf{y} \right\} \mathrm{d}\tau.$$

B - Decay estimates: There have been many very interesting research results regarding the optimal decay estimates. John Albert [1]-[2] investigated decay of solutions of the following linear and nonlinear Benjamin-Bona-Mahony equation (without dissipation)

$$u_t - u_{xxt} + u_x = 0, \qquad \text{in } \mathbb{R}\times\mathbb{R}^+,$$

$$u_t - u_{xxt} + u_x + u^p u_x = 0, \qquad 4 < p < \infty.$$

Using Fourier transform, inverse Fourier transform and delicate estimates on oscillating integrals, Albert [1]-[2] obtained decay results for the above linear and nonlinear equations, such as

$$(1+t)^{1/3}\|\mathrm{LBBM}_t u_0(\cdot)\|_{L^\infty(\mathbb{R})} \le C\big(\|u_0\|_{L^1(\mathbb{R})} + \|u_0\|_{H^4(\mathbb{R})}\big),$$

where LBBM_t represents the solution operator of the linear equation, defined by

$$[\mathrm{LBBM}_t u_0](x) = \frac{1}{2\pi}\int_{\mathbb{R}} \exp(\mathrm{i}x\xi)\exp\left[-\frac{\mathrm{i}\xi}{1+\xi^2}t\right]\widehat{u_0}(\xi)\mathrm{d}\xi,$$

the initial data is not necessarily small. Moreover, he obtained the decay estimate

$$(1+t)^{1/3}\|u(\cdot,t)\|_{L^\infty(\mathbb{R})} \le C(p,u_0),$$

for the nonlinear equation, provided that the initial data is small.

For the nonlinear critical Benjamin-Bona-Mahony-Burgers equation

$$u_t - u_{xxt} + u_x + uu_x = \alpha u_{xx},$$

by using Cole-Hopf transform, energy estimates, and a maximum principle, Amick, Bona and Schonbek [4] established the exact limit

$$\lim_{t\to\infty}\left[(1+t)^{1/2}\int_{\mathbb{R}}|u(x,t)|^2\mathrm{d}x\right] = \frac{4\alpha^2\mu^2}{2\pi\sqrt{\alpha}}\int_{\mathbb{R}}\exp\big(-2x^2\big)/\left[1+\frac{\mu}{\sqrt{\pi}}\int_x^\infty\exp\big(-\xi^2\big)\mathrm{d}\xi\right]^2\mathrm{d}x,$$

where

$$\mu = \exp\left[-\frac{1}{2\alpha}\int_{\mathbb{R}}u_0(x)\mathrm{d}x\right] - 1.$$

Later, Jerry Bona and Laihan Luo [9]-[12] were concerned with a general Benjamin-Bona-Mahony-Burgers equation

$$u_t - u_{xxt} + u_x + u^p u_x = \alpha u_{xx}, \qquad p \ge 2.$$

They proved that if $u_0 \in L^1(\mathbb{R}) \cap H^2(\mathbb{R})$, then

$$(1+t)^{1/2} \int_{\mathbb{R}} |u(x,t)|^2 \mathrm{d}x \le C_1,$$
$$(1+t)^{3/2} \int_{\mathbb{R}} |u_x(x,t)|^2 \mathrm{d}x \le C_2,$$
$$(1+t)^{1/2} \|u(\cdot,t)\|_{L^\infty(\mathbb{R})} \le C_3,$$
$$\lim_{t\to\infty} \left\{ t^{1/2} \int_{\mathbb{R}} |u(x,t)|^2 \mathrm{d}x \right\} = \frac{1}{(8\pi\alpha)^{1/2}} \left[\int_{\mathbb{R}} u_0(x)\mathrm{d}x \right]^2,$$

where, as before, C_1, C_2 and C_3 are positive constants, independent of time. These results are based on the decay estimates of the solution of the linear Benjamin-Bona-Mahony-Burgers equation: $u_t - u_{xxt} + u_x = \alpha u_{xx}$. The solution of the linear equation is given by

$$[\mathrm{LBBMB}_t u_0](x) = \frac{1}{2\pi} \int_{\mathbb{R}} \exp(\mathbf{i}x\xi) \exp\left[-\frac{\alpha\xi^2 + \mathbf{i}\xi}{1+\xi^2} t \right] \widehat{u_0}(\xi)\mathrm{d}\xi.$$

Daniel Dix [24] first studied self-similar solutions and their role in the temporal asymptotic behaviors of solutions to the complex, quadratic Benjamin-Ono-Burgers equation

$$u_t + Hu_{xx} - \alpha u_{xx} + \overline{u}\overline{u}_x = 0.$$

Later, by delicate analysis, Dix [25] investigated the very general nonlinear evolution equation

$$u_t + \mathcal{P}u_x + \mathcal{Q}u + f(u)_x = 0, \qquad u(x,0) = u_0(x),$$

where $\mathcal{P}$ and $\mathcal{Q}$ are linear differential operators. In particular, $\mathcal{P}\partial_x$ is a dispersive operator and $\mathcal{Q}$ is a dissipative operator. His model incorporates the Korteweg-de Vries-Burgers equation, the Benjamin-Ono-Burgers equation and the Schrödinger-Burgers equations. He proved that the above solutions have the same leading-order, long-time, asymptotic behavior as the solution with the corresponding linearized equation and that the solutions of the nonlinear equations are asymptotically self-similar. He gave sharp rates of decay of various norms of the solutions.

Fokas and Luo [37] investigated the asymptotic behavior of solutions of the Cauchy problem for the Benjamin-Ono-Burgers equation

$$u_t + \beta(u^{p+1})_x + Hu_{xx} = \alpha u_{xx}, \qquad u(x,0) = u_0(x).$$

They established the following result

$$\lim_{t\to\infty} \left\{ t^{3/2} \int_{\mathbb{R}} |u(x,t) - v(x,t)|^2 \mathrm{d}x \right\} = \frac{\beta^2}{8\alpha\sqrt{2\pi\alpha}} \left\{ \int_0^\infty \left[\int_{\mathbb{R}} u^{p+1}(x,t)\mathrm{d}x \right] \mathrm{d}t \right\}^2,$$

where v is the solution of the linear problem: $v_t + Hv_{xx} = \alpha v_{xx}$, $v(x,0) = u_0(x)$.

In 1990, Michael Wiegner obtained the following result in [82]: Suppose that the initial data $\mathbf{u}_0 \in L^\infty(\mathbb{R}^n) \cap H^2(\mathbb{R}^n)$ and that the global weak solutions of the n-dimensional incompressible Navier-Stokes equations $\mathbf{u}_t + (\mathbf{u}\cdot\nabla)\mathbf{u} - \Delta\mathbf{u} + \nabla p = \mathbf{f}(\mathbf{x},t)$, $\nabla\cdot\mathbf{u} = 0$ satisfy

$$\int_0^\infty \left[\int_{\mathbb{R}^n} |\mathbf{u}(\mathbf{x},t)|^p \mathrm{d}\mathbf{x}\right]^{q/p} dt < \infty,$$

where $p > n \geq 3$ and $q > 2$, such that

$$\frac{n}{p} + \frac{2}{q} = 1.$$

Then

$$\int_{\mathbb{R}^n} |\mathbf{u}(\mathbf{x},t) - \mathbf{v}(\mathbf{x},t)|^p \mathrm{d}\mathbf{x} \leq C(1+t)^{-[(n+1)p-n]/2p},$$

where the solution $\mathbf{v}$ of the heat equation $\mathbf{v}_t = \alpha\Delta\mathbf{v}$ is assumed to satisfy the decay estimate

$$\int_{\mathbb{R}^n} |\mathbf{v}(\mathbf{x},t)|^2 \mathrm{d}\mathbf{x} \leq C_0(1+t)^{-\gamma},$$

for some constant $C_0 > 0$, where $\gamma > 1$ is a constant.

Zhang established decay estimates and uniform stability of global solutions of various dissipative differential equations, including the nonlinear cubic Korteweg-de Vries-Burgers equation, the Benjamin-Ono-Burgers equation, the n-dimensional Benjamin-Bona-Mahony-Burgers equation, the n-dimensional incompressible Navier-Stokes equations, the Magnetohydrodynamics equations, and the Schrödinger-Burgers equation. See [87]-[100].

For more decay results, see John Albert [1]-[2], Amick, Bona and Schonbek [4], Beirao da Veiga [5]-[8], Jerry Bona and Laihan Luo [9]-[12], Ana Carpio [14]-[16], Daniel Dix [24]-[26], Gema Duro and Enrique Zuazua [27]-[29], Miguel Escobedo and Enrike Zuazua [30]-[32], Boling Guo and Linghai Zhang [40], Nakao Hayashi, Elena I. Kaikina and Pavel I. Naumkin [41], Ming Mei [58]-[59], Maria Schonbek et al [63]-[68], Michael Wiegner [81]-[83], Zhouping Xin and Ping Zhang [85], Linghai Zhang [87]-[100], Huijiang Zhao [101]-[104], Songmu Zheng [109], and Enrique Zuazua [110]-[111], etc. Roughly speaking, the decay results are due to the property of some linear partial differential operators.

Let us summarize some of the decay results established before.

Theorem 0.2 *Suppose that* $\mathbf{u}_0 \in L^1(\mathbb{R}^n) \cap H^2(\mathbb{R}^n)$. *Then the global weak solutions of these above-mentioned differential equations satisfy*

$$C_1 \leq (1+t)^{n/2} \int_{\mathbb{R}^n} |\mathbf{u}(\mathbf{x},t)|^2 \mathrm{d}\mathbf{x} \leq C_2,$$

if [1]

$$\int_{\mathbb{R}^n} \mathbf{u}_0(\mathbf{x}) \mathrm{d}\mathbf{x} \neq \mathbf{0},$$

[1]The global weak solutions of the Navier-Stokes equations and the Magnetohydrodynamics equations do

where the constants are independent of time, but they do depend on the norms $||\mathbf{u}_0||_{L^1(\mathbb{R}^n)}$ *and* $||\mathbf{u}_0||_{L^2(\mathbb{R}^n)}$*. Moreover*

$$C_3 \leq (1+t)^{1+n/2} \int_{\mathbb{R}^n} |\mathbf{u}(\mathbf{x},t)|^2 \mathrm{d}\mathbf{x} \leq C_4,$$

if there exist smooth, integrable functions $\Phi_i \equiv (\phi_{i1}, \phi_{i2}, \cdots, \phi_{in})^T \in C^1(\mathbb{R}^n) \cap L^1(\mathbb{R}^n)$, *where* $i = 1, 2, \cdots, n$, *such that*

$$\mathbf{u}_0(\mathbf{x}) = \left(\sum_{j=1}^{n} \frac{\partial \phi_{1j}}{\partial x_j}(\mathbf{x}),\ \sum_{j=1}^{n} \frac{\partial \phi_{2j}}{\partial x_j}(\mathbf{x}),\ \cdots,\ \sum_{j=1}^{n} \frac{\partial \phi_{nj}}{\partial x_j}(\mathbf{x}) \right)^T,$$

and at least for one integer $i \in \{1, 2, \cdots, n\}$, *there holds*

$$\int_{\mathbb{R}^n} \Phi_i(\mathbf{x}) \mathrm{d}\mathbf{x} \neq \mathbf{0},$$

where for the Navier-Stokes equations we consider divergence free initial data: $\nabla \cdot \mathbf{u}_0 = 0$, *and for the Magnetohydrodynamics equations we also consider divergence free initial data:* $\nabla \cdot \mathbf{u}_0 = \nabla \cdot \mathbf{A}_0 = 0$.

1.4. Notations

Let $\phi = (\phi_1, \phi_2, \cdots, \phi_n)^T$, define $|\phi|$ by $|\phi|^2 = |\phi_1|^2 + |\phi_2|^2 + \cdots + |\phi_n|^2$. We will apply the regular *Banach* space $L^p(\mathbb{R}^n)$, where $p \in [1, \infty]$. This space is endowed with the L^p-norm, which is defined by

$$\begin{aligned} ||\phi||^p_{L^p(\mathbb{R}^n)} &= \int_{\mathbb{R}^n} |\phi(\mathbf{x})|^p \mathrm{d}\mathbf{x}, && \text{for } 1 \leq p < \infty, \\ ||\phi||_{L^\infty(\mathbb{R}^n)} &= \sup_{\mathbb{R}^n} |\phi(\mathbf{x})|, && \text{for } p = \infty. \end{aligned}$$

As usual, the *Sobolev* space $W^{m,p}(\mathbb{R}^n)$ is endowed with the norm $||\phi||_{W^{m,p}(\mathbb{R})}$, which is defined by

$$\begin{aligned} ||\phi||^p_{W^{m,p}(\mathbb{R}^n)} &= \sum_{\alpha_1+\alpha_2+\cdots+\alpha_n \leq m} \left|\left| \frac{\partial^{\alpha_1+\alpha_2+\cdots+\alpha_n}\phi}{\partial x_1^{\alpha_1} \partial x_2^{\alpha_2} \cdots \partial x_n^{\alpha_n}} \right|\right|^p_{L^p(\mathbb{R}^n)}, && \text{for } 1 \leq p < \infty, \\ ||\phi||_{W^{m,\infty}(\mathbb{R}^n)} &= \sup_{\alpha_1+\alpha_2+\cdots+\alpha_n \leq m} \left|\left| \frac{\partial^{\alpha_1+\alpha_2+\cdots+\alpha_n}\phi}{\partial x_1^{\alpha_1} \partial x_2^{\alpha_2} \cdots \partial x_n^{\alpha_n}} \right|\right|_{L^\infty(\mathbb{R}^n)}, && \text{for } p = \infty. \end{aligned}$$

---

not satisfy these estimates simply because

$$\int_{\mathbb{R}^n} \mathbf{u}_0(\mathbf{x}) \mathrm{d}\mathbf{x} = \mathbf{0}$$

for the Navier-Stokes equations, and

$$\int_{\mathbb{R}^n} \mathbf{u}_0(\mathbf{x}) \mathrm{d}\mathbf{x} = \int_{\mathbb{R}^n} \mathbf{A}_0(\mathbf{x}) \mathrm{d}\mathbf{x} = \mathbf{0}$$

for the Magnetohydrodynamics equations.

In particular, if $\phi = \phi(\cdot, t) \in L^2(\mathbb{R}^n)$, then $\|\phi(\cdot, t)\|$ is a function of t:

$$\|\phi(\cdot, t)\|^2 = \|\phi(\cdot, t)\|^2_{L^2(\mathbb{R}^n)} = \int_{\mathbb{R}^n} |\phi(\mathbf{x}, t)|^2 \mathrm{d}\mathbf{x},$$

and if $\phi \in L^\infty\big(\mathbb{R}^+, W^{m,2}(\mathbb{R}^n)\big)$, then

$$\begin{aligned} \|\phi\|_{L^\infty(\mathbb{R}^+, W^{m,2}(\mathbb{R}^n))} &= \left[\sup_{t\in\mathbb{R}^+} \sum_{\alpha_1+\alpha_2+\cdots+\alpha_n \le m} \left\| \frac{\partial^{\alpha_1+\alpha_2+\cdots+\alpha_n}\phi}{\partial x_1^{\alpha_1}\partial x_2^{\alpha_2}\cdots\partial x_n^{\alpha_n}}(\cdot, t)\right\|^2\right]^{1/2} \\ &= \left[\sup_{t\in\mathbb{R}^+} \sum_{\alpha_1+\alpha_2+\cdots+\alpha_n \le m} \int_{\mathbb{R}^n} \left| \frac{\partial^{\alpha_1+\alpha_2+\cdots+\alpha_n}\phi}{\partial x_1^{\alpha_1}\partial x_2^{\alpha_2}\cdots\partial x_n^{\alpha_n}}(\mathbf{x}, t)\right|^2 \mathrm{d}\mathbf{x}\right]^{1/2}, \end{aligned}$$

where $\alpha_1 \ge 0, \alpha_2 \ge 0, \cdots, \alpha_n \ge 0$ are integers, $\alpha_1 + \alpha_2 + \cdots + \alpha_n \le m$. For convenience, we define two more $functional$ spaces

$$\mathbb{V}(\mathbb{R}^n) = \left\{\phi \in L^1(\mathbb{R}^n) \cap H^2(\mathbb{R}^n) : \int_{\mathbb{R}^n} \phi(\mathbf{x})\mathrm{d}\mathbf{x} \ne \mathbf{0}\right\},$$

and

$$\mathbb{W}(\mathbb{R}^n) = \left\{\phi \in L^1(\mathbb{R}^n) \cap H^2(\mathbb{R}^n) : \int_{\mathbb{R}^n} \phi(\mathbf{x})\mathrm{d}\mathbf{x} = \mathbf{0}\right\}.$$

Note that $\mathbb{W}(\mathbb{R}^n)$ is a vector subspace of $L^1(\mathbb{R}^n) \cap H^2(\mathbb{R}^n)$, but $\mathbb{V}(\mathbb{R}^n)$ is not a vector subspace. Certainly, there holds $\mathbb{V}(\mathbb{R}^n) \cup \mathbb{W}(\mathbb{R}^n) = L^1(\mathbb{R}^n) \cap H^2(\mathbb{R}^n)$. Note that the integrals of the initial data for the Navier-Stokes equations and for the Magnetohydrodynamics equations are equal to zero. For these equations we will use the initial data of the following forms

$$\begin{aligned} \mathbf{u}_0(\mathbf{x}) &= \left(\sum_{j=1}^n \frac{\partial\phi_{1j}}{\partial x_j}(\mathbf{x}), \sum_{j=1}^n \frac{\partial\phi_{2j}}{\partial x_j}(\mathbf{x}), \cdots, \sum_{j=1}^n \frac{\partial\phi_{nj}}{\partial x_j}(\mathbf{x})\right)^T, \\ \mathbf{A}_0(\mathbf{x}) &= \left(\sum_{j=1}^n \frac{\partial\psi_{1j}}{\partial x_j}(\mathbf{x}), \sum_{j=1}^n \frac{\partial\psi_{2j}}{\partial x_j}(\mathbf{x}), \cdots, \sum_{j=1}^n \frac{\partial\psi_{nj}}{\partial x_j}(\mathbf{x})\right)^T, \end{aligned}$$

or

$$\begin{aligned} \mathbf{u}_0(\mathbf{x}) &= \left(\sum_{j=1}^n\sum_{k=1}^n \frac{\partial\phi_{1jk}}{\partial x_j\partial x_k}(\mathbf{x}), \sum_{j=1}^n\sum_{k=1}^n \frac{\partial\phi_{2jk}}{\partial x_j\partial x_k}(\mathbf{x}), \cdots, \sum_{j=1}^n\sum_{k=1}^n \frac{\partial\phi_{njk}}{\partial x_j\partial x_k}(\mathbf{x})\right)^T, \\ \mathbf{A}_0(\mathbf{x}) &= \left(\sum_{j=1}^n\sum_{k=1}^n \frac{\partial\psi_{1jk}}{\partial x_j\partial x_k}(\mathbf{x}), \sum_{j=1}^n\sum_{k=1}^n \frac{\partial\psi_{2jk}}{\partial x_j\partial x_k}(\mathbf{x}), \cdots, \sum_{j=1}^n\sum_{k=1}^n \frac{\partial\psi_{njk}}{\partial x_j\partial x_k}(\mathbf{x})\right)^T. \end{aligned}$$

1.5. The Main Results

Suppose that all of the hypotheses made in Subsection 1.2 are satisfied. Let $\mathbf{f} \equiv \mathbf{0}$ in (1) to simplify the statements of the main results. As usual, we assume that $\nabla \cdot \mathbf{u} = \nabla \cdot \mathbf{u}_0 = 0$ for the Navier-Stokes equations and $\nabla \cdot \mathbf{u} = \nabla \cdot \mathbf{A} = 0$ and $\nabla \cdot \mathbf{u}_0 = \nabla \cdot \mathbf{A}_0 = 0$ for the Magnetohydrodynamics equations. Below $\mathbf{u}_0$, $\mathbf{u}$, $\nabla\mathbf{u}$ and $\triangle\mathbf{u}$ should be replaced with $(\mathbf{u}_0, \mathbf{A}_0)$, $(\mathbf{u}, \mathbf{A})$, $(\nabla\mathbf{u}, \nabla\mathbf{A})$ and $(\triangle\mathbf{u}, \triangle\mathbf{A})$, respectively, in the statement of the results of the Magnetohydrodynamics equations.

Theorem 1. *Suppose that the initial function $\mathbf{u}_0 \in L^1(\mathbb{R}^n) \cap H^2(\mathbb{R}^n)$ and that*

$$\int_{\mathbb{R}^n} \mathbf{u}_0(\mathbf{x})\mathrm{d}\mathbf{x} \neq \mathbf{0}.$$

Then the global solutions of problem (1)-(2) enjoy the limits

$$\lim_{t\to\infty}\left\{(1+t)^{n/2}\int_{\mathbb{R}^n}|\mathbf{u}(\mathbf{x},t)|^2\mathrm{d}\mathbf{x}\right\} = \frac{1}{(2\pi)^n}\left(\frac{\pi}{2\alpha}\right)^{n/2}\left[\int_{\mathbb{R}^n}\mathbf{u}_0(\mathbf{x})\mathrm{d}\mathbf{x}\right]^2,$$

$$\lim_{t\to\infty}\left\{(1+t)^{2m+n/2}\int_{\mathbb{R}^n}|\triangle^m\mathbf{u}(\mathbf{x},t)|^2\mathrm{d}\mathbf{x}\right\} = \frac{1}{(2\pi)^{2m+n}}\left(\frac{\pi}{2\alpha}\right)^{2m+n/2}\prod_{l=1}^{2m}(2l-2+n)\left[\int_{\mathbb{R}^n}\mathbf{u}_0(\mathbf{x})\mathrm{d}\mathbf{x}\right]^2,$$

where the integer $m \geq 1$.

Theorem 2. *Suppose that the initial data $\mathbf{u}_0 \in L^1(\mathbb{R}^n) \cap H^2(\mathbb{R}^n)$ and that*

$$\int_{\mathbb{R}^n} \mathbf{u}_0(\mathbf{x})\mathrm{d}\mathbf{x} = \mathbf{0}.$$

Suppose also that

$$\mathbf{u}_0(\mathbf{x}) = \left(\sum_{j=1}^{n}\frac{\partial\phi_{1j}}{\partial x_j}(\mathbf{x}),\ \sum_{j=1}^{n}\frac{\partial\phi_{2j}}{\partial x_j}(\mathbf{x}),\ \cdots,\ \sum_{j=1}^{n}\frac{\partial\phi_{nj}}{\partial x_j}(\mathbf{x})\right)^T,$$

where $\Phi_i \equiv (\phi_{i1}, \phi_{i2}, \cdots, \phi_{in})^T \in C^1(\mathbb{R}^n) \cap L^1(\mathbb{R}^n)$, $i = 1, 2, \cdots, n$. Then

$$\lim_{t\to\infty}\left\{(1+t)^{1+n/2}\int_{\mathbb{R}^n}|\mathbf{u}(\mathbf{x},t)|^2\mathrm{d}\mathbf{x}\right\} = \frac{1}{(2\pi)^{n+1}}\left(\frac{\pi}{2\alpha}\right)^{1+n/2}\sum_{k=1}^{n}\left[\int_{\mathbb{R}^n}\Phi_k(\mathbf{x})\mathrm{d}\mathbf{x}\right]^2,$$

$$\lim_{t\to\infty}\left\{(1+t)^{2m+1+n/2}\int_{\mathbb{R}^n}|\triangle^m\mathbf{u}(\mathbf{x},t)|^2\mathrm{d}\mathbf{x}\right\} = \frac{1}{(2\pi)^{2m+n+1}}\left(\frac{\pi}{2\alpha}\right)^{2m+1+n/2}\prod_{l=1}^{2m}(2l+n)\sum_{k=1}^{n}\left[\int_{\mathbb{R}^n}\Phi_k(\mathbf{x})\mathrm{d}\mathbf{x}\right]^2,$$

for any integer $m \geq 1$.

The proofs of Theorem 1 and Theorem 2 will be given in Section 3. Some of the analysis of this chapter are very formal since the existence, uniqueness and regularity of global strong solutions to the Cauchy problems for some of the differential equations in higher-dimensional spaces have not been solved yet.

Remark 1. Let

$$\mathbf{f}(\mathbf{x},t)=\frac{\tau(t)}{(4\pi\alpha)^{n/2}(1+t)^{n/2}}\exp\left[-\frac{|\mathbf{x}|^2}{4\alpha(1+t)}\right],$$

where $\tau\in L^1(0,\infty)$ is a vector-valued function of time. Then

$$\widehat{\mathbf{f}}(\xi,t)=\tau(t)\exp\big[-\alpha|\xi|^2(1+t)\big].$$

Define

$$\Lambda=\int_0^\infty\tau(t)\mathrm{d}t.$$

Then we will have very similar results to these given in Theorem 1. The only difference is that the integral $\int_{\mathbb{R}^n}\mathbf{u}_0(\mathbf{x})\mathrm{d}\mathbf{x}$ will be replaced by

$$\int_{\mathbb{R}^n}\mathbf{u}_0(\mathbf{x})\mathrm{d}\mathbf{x}+\Lambda.$$

Outline of the book chapter: In Section 2, we will provide detailed analysis needed for the proofs of the main results of this chapter. In Section 3, we will prove the main results. In Section 4, we will have several applications. In Section 5, we will arrange concluding remarks and open problems, and in Section 6, we provide more known results on the model equations.

2. The Mathematical Analysis of the Model Equations

2.1. Linear Analysis

Consider the following Cauchy problems for the n-dimensional linear differential equations

$$\mathbf{v}_t-\varepsilon\triangle\mathbf{v}_t+\mathcal{D}\mathbf{v}=\alpha\triangle\mathbf{v},\qquad\text{in }\mathbb{R}^n\times\mathbb{R}^+,\tag{7}$$

$$\mathbf{v}(\mathbf{x},0)=\mathbf{v}_0(\mathbf{x}),\qquad\text{in }\mathbb{R}^n,\tag{8}$$

where the spatial dimension $n\geq 1$, $\alpha>0$ and $0\leq\varepsilon\leq 1$ are nonnegative constants. The differential operator $\mathcal{D}$ has been defined earlier. Needless to say, for any initial data $\mathbf{v}_0\in L^1(\mathbb{R}^n)\cap H^2(\mathbb{R}^n)$, there exists a unique global strong solution to the Cauchy problem (7)-(8): $\mathbf{v}\in L^\infty(\mathbb{R}^+,H^2(\mathbb{R}^n))\cap L^2_{\text{loc}}(\mathbb{R}^+,H^3(\mathbb{R}^n))\cap C(\mathbb{R}^+,H^2(\mathbb{R}^n))$, or simply $\mathbf{v}\in C^\infty(\mathbb{R}^+,C^\infty(\mathbb{R}^n))$. This can be proved by means of a priori energy estimates, iteration technique and semigroup method. We will consider very general initial data for problem (7)-(8), in particular, unlike the initial data for the Navier-Stokes equations, they are not necessarily divergence free. Let us now investigate the asymptotic behaviors of the quantity

$$(1+t)^{2m+\lambda+n/2}\int_{\mathbb{R}^n}|\triangle^m\mathbf{v}(\mathbf{x},t)|^2\mathrm{d}\mathbf{x}$$

where λ will be specified as we proceed.

Lemma 2.1. **(I)** *Suppose that the initial data* $\mathbf{v}_0\in L^1(\mathbb{R}^n)\cap H^2(\mathbb{R}^n)$ *and*

$$\int_{\mathbb{R}^n}\mathbf{v}_0(\mathbf{x})\mathrm{d}\mathbf{x}\neq\mathbf{0}.$$

Then there hold the following exact limits for the global solutions of the Cauchy problems (7)-(8)

$$\lim_{t\to\infty}\left[(1+t)^{n/2}\int_{\mathbb{R}^n}|\mathbf{v}(\mathbf{x},t)|^2 d\mathbf{x}\right]=\frac{1}{(2\pi)^n}\left(\frac{\pi}{2\alpha}\right)^{n/2}\left[\int_{\mathbb{R}^n}\mathbf{v}_0(\mathbf{x})d\mathbf{x}\right]^2,$$

$$\lim_{t\to\infty}\left[(1+t)^{2m+n/2}\int_{\mathbb{R}^n}|\triangle^m\mathbf{v}(\mathbf{x},t)|^2 d\mathbf{x}\right]=\frac{1}{(2\pi)^{2m+n}}\left(\frac{\pi}{2\alpha}\right)^{2m+n/2}\prod_{l=1}^{2m}(2l-2+n)\left[\int_{\mathbb{R}^n}\mathbf{v}_0(\mathbf{x})d\mathbf{x}\right]^2,$$

$$\lim_{t\to\infty}\left\{(1+t)\left[(1+t)^{n/2}\int_{\mathbb{R}^n}|\mathbf{v}(\mathbf{x},t)|^2 d\mathbf{x}-\frac{1}{(2\pi)^n}\left(\frac{\pi}{2\alpha}\right)^{n/2}\left(\int_{\mathbb{R}^n}\mathbf{v}_0(\mathbf{x})d\mathbf{x}\right)^2\right]\right\}$$
$$=\frac{1}{(2\pi)^n}\int_{\mathbb{R}^n}\exp\left(-2\alpha|\eta|^2\right)\left\{\sum_{i=1}^n\left|\int_{\mathbb{R}^n}(\mathbf{x}\cdot\eta)\mathbf{v}_{0i}(\mathbf{x})d\mathbf{x}\right|^2-\left[\sum_{i=1}^n\int_{\mathbb{R}^n}(\mathbf{x}\cdot\eta)^2\mathbf{v}_{0i}(\mathbf{x})d\mathbf{x}\right]\left[\sum_{i=1}^n\int_{\mathbb{R}^n}\mathbf{v}_{0i}(\mathbf{x})d\mathbf{x}\right]\right\}d\eta,$$

$$\lim_{t\to\infty}\left\{(1+t)\left[(1+t)^{2m+n/2}\int_{\mathbb{R}^n}|\triangle^m\mathbf{v}(\mathbf{x},t)|^2 d\mathbf{x}\right.\right.$$
$$\left.\left.-\frac{1}{(2\pi)^{2m+n}}\left(\frac{\pi}{2\alpha}\right)^{2m+n/2}\prod_{l=1}^{2m}(2l-2+n)\left(\int_{\mathbb{R}^n}\mathbf{v}_0(\mathbf{x})d\mathbf{x}\right)^2\right]\right\}$$
$$=\frac{1}{(2\pi)^n}\int_{\mathbb{R}^n}|\eta|^{4m}\exp\left(-2\alpha|\eta|^2\right)\left\{\sum_{i=1}^n\left|\int_{\mathbb{R}^n}(\mathbf{x}\cdot\eta)\mathbf{v}_{0i}(\mathbf{x})d\mathbf{x}\right|^2-\left[\sum_{i=1}^n\int_{\mathbb{R}^n}(\mathbf{x}\cdot\eta)^2\mathbf{v}_{0i}(\mathbf{x})d\mathbf{x}\right]\left[\sum_{i=1}^n\int_{\mathbb{R}^n}\mathbf{v}_{0i}(\mathbf{x})d\mathbf{x}\right]\right\}d\eta.$$

There also hold the sharp decay estimates

$$(1+t)^{m+n/2}\|\triangle^m\mathbf{v}(\cdot,t)\|_{L^\infty(\mathbb{R}^n)}\le C_m\big(\|\mathbf{v}_0\|_{L^1(\mathbb{R}^n)},\|\mathbf{v}_0\|_{L^2(\mathbb{R}^n)}\big),\qquad\forall m\ge 0,$$
$$(1+t)^{m+(n+1)/2}\|\nabla\triangle^m\mathbf{v}(\cdot,t)\|_{L^\infty(\mathbb{R}^n)}\le C_m\big(\|\mathbf{v}_0\|_{L^1(\mathbb{R}^n)},\|\mathbf{v}_0\|_{L^2(\mathbb{R}^n)}\big),\qquad\forall m\ge 0.$$

(II) *Suppose that*

$$\int_{\mathbb{R}^n}\mathbf{v}_0(\mathbf{x})d\mathbf{x}=\mathbf{0}$$

and

$$\mathbf{v}_0(\mathbf{x})=\left(\sum_{j=1}^n\frac{\partial\phi_{1j}}{\partial x_j}(\mathbf{x}),\ \sum_{j=1}^n\frac{\partial\phi_{2j}}{\partial x_j}(\mathbf{x}),\ \cdots,\ \sum_{j=1}^n\frac{\partial\phi_{nj}}{\partial x_j}(\mathbf{x})\right)^T,$$

where $\Phi_i\equiv(\phi_{i1},\phi_{i2},\cdots,\phi_{in})^T\in C^1(\mathbb{R}^n)\cap L^1(\mathbb{R}^n)$, $i=1,2,\cdots,n$. *Then*

$$\lim_{t\to\infty}\left[(1+t)^{1+n/2}\int_{\mathbb{R}^n}|\mathbf{v}(\mathbf{x},t)|^2 d\mathbf{x}\right]=\frac{1}{(2\pi)^{n+1}}\left(\frac{\pi}{2\alpha}\right)^{1+n/2}\sum_{k=1}^n\left[\int_{\mathbb{R}^n}\Phi_k(\mathbf{x})d\mathbf{x}\right]^2,$$

$$\lim_{t\to\infty}\left[(1+t)^{2m+1+n/2}\int_{\mathbb{R}^n}|\triangle^m\mathbf{v}(\mathbf{x},t)|^2 d\mathbf{x}\right]=\frac{1}{(2\pi)^{2m+n+1}}\left(\frac{\pi}{2\alpha}\right)^{2m+1+n/2}\prod_{l=1}^{2m}(2l+n)\sum_{k=1}^n\left[\int_{\mathbb{R}^n}\Phi_k(\mathbf{x})d\mathbf{x}\right]^2,$$

$$\lim_{t\to\infty}\left\{(1+t)^2\left[(1+t)^{1+n/2}\int_{\mathbb{R}^n}|\mathbf{v}(\mathbf{x},t)|^2 d\mathbf{x}-\frac{1}{(2\pi)^{n+1}}\left(\frac{\pi}{2\alpha}\right)^{1+n/2}\sum_{k=1}^n\left(\int_{\mathbb{R}^n}\Phi_k(\mathbf{x})d\mathbf{x}\right)^2\right]\right\}$$
$$=\frac{1}{(2\pi)^n}\int_{\mathbb{R}^n}\exp\left(-2\alpha|\eta|^2\right)\left\{\sum_{i=1}^n\left|\sum_{j=1}^n\eta_j\int_{\mathbb{R}^n}(\mathbf{x}\cdot\eta)\phi_{ij}(\mathbf{x})d\mathbf{x}\right|^2\right.$$
$$\left.-\left[\sum_{i=1}^n\sum_{j=1}^n\eta_j\int_{\mathbb{R}^n}(\mathbf{x}\cdot\eta)^2\phi_{ij}(\mathbf{x})d\mathbf{x}\right]\left[\sum_{i=1}^n\sum_{j=1}^n\eta_j\int_{\mathbb{R}^n}\phi_{ij}(\mathbf{x})d\mathbf{x}\right]\right\}d\eta,$$

$$\lim_{t\to\infty}\left\{(1+t)^2\left[(1+t)^{2m+1+n/2}\int_{\mathbb{R}^n}|\triangle^m\mathbf{v}(\mathbf{x},t)|^2 d\mathbf{x}\right.\right.$$
$$\left.\left.-\frac{1}{(2\pi)^{2m+n+1}}\left(\frac{\pi}{2\alpha}\right)^{2m+1+n/2}\prod_{l=1}^{2m}(2l+n)\sum_{k=1}^n\left(\int_{\mathbb{R}^n}\Phi_k(\mathbf{x})d\mathbf{x}\right)^2\right]\right\}$$
$$=\frac{1}{(2\pi)^n}\int_{\mathbb{R}^n}|\eta|^{4m}\exp\left(-2\alpha|\eta|^2\right)\left\{\left[\sum_{i=1}^n\left|\sum_{j=1}^n\eta_j\int_{\mathbb{R}^n}(\mathbf{x}\cdot\eta)\phi_{ij}(\mathbf{x})d\mathbf{x}\right|^2\right]\right.$$
$$\left.-\left[\sum_{i=1}^n\sum_{j=1}^n\eta_j\int_{\mathbb{R}^n}(\mathbf{x}\cdot\eta)^2\phi_{ij}(\mathbf{x})d\mathbf{x}\right]\left[\sum_{i=1}^n\sum_{j=1}^n\eta_j\int_{\mathbb{R}^n}\phi_{ij}(\mathbf{x})d\mathbf{x}\right]\right\}d\eta,$$

and

$$
\begin{aligned}
&(1+t)^{m+(n+1)/2}\|\triangle^m \mathbf{v}(\cdot,t)\|_{L^\infty(\mathbb{R}^n)} \le C_m\big(\|\mathbf{v}_0\|_{L^1(\mathbb{R}^n)}, \|\mathbf{v}_0\|_{L^2(\mathbb{R}^n)}\big), && \forall m \ge 0,\\
&(1+t)^{m+(n+2)/2}\|\nabla\triangle^m \mathbf{v}(\cdot,t)\|_{L^\infty(\mathbb{R}^n)} \le C_m\big(\|\mathbf{v}_0\|_{L^1(\mathbb{R}^n)}, \|\mathbf{v}_0\|_{L^2(\mathbb{R}^n)}\big), && \forall m \ge 0.
\end{aligned}
$$

(III) *Suppose that* $\mathbf{v}_0 \in L^1(\mathbb{R}^n)\cap H^2(\mathbb{R}^n)$ *and*

$$
\mathbf{v}_0(\mathbf{x}) = \left(\sum_{j=1}^n\sum_{k=1}^n \frac{\partial^2\phi_{1jk}}{\partial x_j\partial x_k}(\mathbf{x}),\ \sum_{j=1}^n\sum_{k=1}^n \frac{\partial^2\phi_{2jk}}{\partial x_j\partial x_k}(\mathbf{x}),\ \cdots,\ \sum_{j=1}^n\sum_{k=1}^n \frac{\partial^2\phi_{njk}}{\partial x_j\partial x_k}(\mathbf{x})\right)^T,
$$

where $\phi_{ijk} \in C^2(\mathbb{R}^n)\cap L^1(\mathbb{R}^n)$, *and for some integers* $i,j,k \in \{1,2,\cdots,n\}$, $\int_{\mathbb{R}^n}\phi_{ijk}(\mathbf{x})\mathrm{d}\mathbf{x} \ne 0$, *then*

$$
\lim_{t\to\infty}\left[(1+t)^{2+n/2}\int_{\mathbb{R}^n}|\mathbf{v}(\mathbf{x},t)|^2\mathrm{d}\mathbf{x}\right] = \frac{1}{(2\pi)^n}\int_{\mathbb{R}^n}\exp\left(-2\alpha|\eta|^2\right)\sum_{i=1}^n\left|\sum_{j=1}^n\sum_{k=1}^n \eta_j\eta_k\widehat{\phi_{ijk}}(\mathbf{0})\right|^2\mathrm{d}\eta,
$$

$$
\lim_{t\to\infty}\left[(1+t)^{2m+2+n/2}\int_{\mathbb{R}^n}|\triangle^m\mathbf{v}(\mathbf{x},t)|^2\mathrm{d}\mathbf{x}\right] = \frac{1}{(2\pi)^n}\int_{\mathbb{R}^n}|\eta|^{4m}\exp\left(-2\alpha|\eta|^2\right)\left\{\sum_{i=1}^n\left|\sum_{j=1}^n\sum_{k=1}^n \eta_j\eta_k\widehat{\phi_{ijk}}(\mathbf{0})\right|^2\right\}\mathrm{d}\eta,
$$

$$
\begin{aligned}
&\lim_{t\to\infty}(1+t)^3\left\{(1+t)^{2+n/2}\int_{\mathbb{R}^n}|\mathbf{v}(\mathbf{x},t)|^2\mathrm{d}\mathbf{x} - \frac{1}{(2\pi)^n}\int_{\mathbb{R}^n}\exp\left(-2\alpha|\eta|^2\right)\sum_{i=1}^n\left|\sum_{j=1}^n\sum_{k=1}^n\eta_j\eta_k\widehat{\phi_{ijk}}(\mathbf{0})\right|^2\mathrm{d}\eta\right\}\\
&= \frac{1}{(2\pi)^n}\int_{\mathbb{R}^n}\exp\left(-2\alpha|\eta|^2\right)\left\{\sum_{i=1}^n\left|\sum_{j=1}^n\sum_{k=1}^n\eta_j\eta_k\int_{\mathbb{R}^n}(\mathbf{x}\cdot\eta)\phi_{ijk}(\mathbf{x})\mathrm{d}\mathbf{x}\right|^2\right.\\
&\left.-\left[\sum_{i=1}^n\sum_{j=1}^n\sum_{k=1}^n\eta_j\eta_k\int_{\mathbb{R}^n}(\mathbf{x}\cdot\eta)^2\phi_{ijk}(\mathbf{x})\mathrm{d}\mathbf{x}\right]\left[\sum_{i=1}^n\sum_{j=1}^n\sum_{k=1}^n\eta_j\eta_k\int_{\mathbb{R}^n}\phi_{ijk}(\mathbf{x})\mathrm{d}\mathbf{x}\right]\right\}\mathrm{d}\eta,\\
&\lim_{t\to\infty}(1+t)^3\left\{(1+t)^{2m+2+n/2}\int_{\mathbb{R}^n}|\triangle^m\mathbf{v}(\mathbf{x},t)|^2\mathrm{d}\mathbf{x} - \frac{1}{(2\pi)^n}\int_{\mathbb{R}^n}|\eta|^{4m}\exp\left(-2\alpha|\eta|^2\right)\sum_{i=1}^n\left|\sum_{j=1}^n\sum_{k=1}^n\eta_j\eta_k\widehat{\phi_{ijk}}(\mathbf{0})\right|^2\mathrm{d}\eta\right\}\\
&= \frac{1}{(2\pi)^n}\int_{\mathbb{R}^n}|\eta|^{4m}\exp\left(-2\alpha|\eta|^2\right)\left\{\sum_{i=1}^n\left|\sum_{j=1}^n\sum_{k=1}^n\eta_j\eta_k\int_{\mathbb{R}^n}(\mathbf{x}\cdot\eta)\phi_{ijk}(\mathbf{x})\mathrm{d}\mathbf{x}\right|^2\right.\\
&\left.-\left[\sum_{i=1}^n\sum_{j=1}^n\sum_{k=1}^n\eta_j\eta_k\int_{\mathbb{R}^n}(\mathbf{x}\cdot\eta)^2\phi_{ijk}(\mathbf{x})\mathrm{d}\mathbf{x}\right]\left[\sum_{i=1}^n\sum_{j=1}^n\sum_{k=1}^n\eta_j\eta_k\int_{\mathbb{R}^n}\phi_{ijk}(\mathbf{x})\mathrm{d}\mathbf{x}\right]\right\}\mathrm{d}\eta.
\end{aligned}
$$

There also hold the sharp decay estimates

$$
\begin{aligned}
&(1+t)^{m+(n+2)/2}\|\triangle^m \mathbf{v}(\cdot,t)\|_{L^\infty(\mathbb{R}^n)} \le C_m\big(\|\mathbf{v}_0\|_{L^1(\mathbb{R}^n)}, \|\mathbf{v}_0\|_{L^2(\mathbb{R}^n)}\big), && \forall m \ge 0,\\
&(1+t)^{m+(n+3)/2}\|\nabla\triangle^m \mathbf{v}(\cdot,t)\|_{L^\infty(\mathbb{R}^n)} \le C_m\big(\|\mathbf{v}_0\|_{L^1(\mathbb{R}^n)}, \|\mathbf{v}_0\|_{L^2(\mathbb{R}^n)}\big), && \forall m \ge 0.
\end{aligned}
$$

(IV) *Let* α_1, α_2, $\cdots$, α_n *be nonnegative integers, define* $\mathcal{A} = (\alpha_1,\alpha_2,\cdots,\alpha_n)$ *and* $|\mathcal{A}| = \alpha_1+\alpha_2+\cdots+\alpha_n$. *Set*

$$
\begin{aligned}
\mathbf{v}_0(\mathbf{x}) \;=\; &\left(\sum_{j=1}^n\sum_{k=1}^n\cdots\sum_{l=1}^n \frac{\partial^{|\mathcal{A}|}\phi_{1jk\cdots l}}{\partial x_j\partial x_k\cdots\partial x_l}(\mathbf{x}),\ \sum_{j=1}^n\sum_{k=1}^n\cdots\sum_{l=1}^n \frac{\partial^{|\mathcal{A}|}\phi_{2jk\cdots l}}{\partial x_j\partial x_k\cdots\partial x_l}(\mathbf{x}),\ \cdots,\right.\\
&\left.\sum_{j=1}^n\sum_{k=1}^n\cdots\sum_{l=1}^n \frac{\partial^{|\mathcal{A}|}\phi_{njk\cdots l}}{\partial x_j\partial x_k\cdots\partial x_l}(\mathbf{x})\right)^T,
\end{aligned}
$$

where $\phi_{ijk\cdots l} \in C^{|\mathcal{A}|}(\mathbb{R}^n) \cap L^1(\mathbb{R}^n)$, *and for some integers* $i, j, k, \cdots, l \in \{1, 2, \cdots, n\}$, $\int_{\mathbb{R}^n} \phi_{ijk\cdots l}(\mathbf{x})\mathrm{d}\mathbf{x} \neq 0$, *then*

$$\lim_{t\to\infty} \left[(1+t)^{|\mathcal{A}|+n/2} \int_{\mathbb{R}^n} |\mathbf{v}(\mathbf{x},t)|^2 \mathrm{d}\mathbf{x}\right]$$

$$= \frac{1}{(2\pi)^n} \int_{\mathbb{R}^n} \exp\left(-2\alpha|\eta|^2\right) \sum_{i=1}^{n} \left|\sum_{j=1}^{n}\sum_{k=1}^{n} \cdots \sum_{l=1}^{n} \eta_j\eta_k\cdots\eta_l \widehat{\phi_{ijk\cdots l}}(\mathbf{0})\right|^2 \mathrm{d}\eta,$$

$$\lim_{t\to\infty} \left[(1+t)^{2m+|\mathcal{A}|+n/2} \int_{\mathbb{R}^n} |\triangle^m \mathbf{v}(\mathbf{x},t)|^2 \mathrm{d}\mathbf{x}\right]$$

$$= \frac{1}{(2\pi)^n} \int_{\mathbb{R}^n} |\eta|^{4m} \exp\left(-2\alpha|\eta|^2\right) \sum_{i=1}^{n} \left|\sum_{j=1}^{n}\sum_{k=1}^{n} \cdots \sum_{l=1}^{n} \eta_j\eta_k\cdots\eta_l \widehat{\phi_{ijk\cdots l}}(\mathbf{0})\right|^2 \mathrm{d}\eta,$$

$$\lim_{t\to\infty} (1+t)^{|\mathcal{A}|+1} \left\{(1+t)^{|\mathcal{A}|+n/2} \int_{\mathbb{R}^n} |\mathbf{v}(\mathbf{x},t)|^2 \mathrm{d}\mathbf{x}\right.$$

$$\left. - \frac{1}{(2\pi)^n} \int_{\mathbb{R}^n} \exp\left(-2\alpha|\eta|^2\right) \sum_{i=1}^{n} \left|\sum_{j=1}^{n}\sum_{k=1}^{n} \cdots \sum_{l=1}^{n} \eta_j\eta_k\cdots\eta_l \widehat{\phi_{ijk\cdots l}}(\mathbf{0})\right|^2 \mathrm{d}\eta\right\}$$

$$= \frac{1}{(2\pi)^n} \int_{\mathbb{R}^n} \exp\left(-2\alpha|\eta|^2\right) \left\{\sum_{i=1}^{n} \left|\sum_{j=1}^{n}\sum_{k=1}^{n} \cdots \sum_{l=1}^{n} \eta_j\eta_k\cdots\eta_l \int_{\mathbb{R}^n} (\mathbf{x}\cdot\eta)\phi_{ijk\cdots l}(\mathbf{x})\mathrm{d}\mathbf{x}\right|^2\right.$$

$$- \left[\sum_{i=1}^{n}\sum_{j=1}^{n}\sum_{k=1}^{n} \cdots \sum_{l=1}^{n} \eta_j\eta_k\cdots\eta_l \int_{\mathbb{R}^n} (\mathbf{x}\cdot\eta)^2 \phi_{ijk\cdots l}(\mathbf{x})\mathrm{d}\mathbf{x}\right]$$

$$\left.\times \left[\sum_{i=1}^{n}\sum_{j=1}^{n}\sum_{k=1}^{n} \cdots \sum_{l=1}^{n} \eta_j\eta_k\cdots\eta_l \int_{\mathbb{R}^n} \phi_{ijk\cdots l}(\mathbf{x})\mathrm{d}\mathbf{x}\right]\right\} \mathrm{d}\eta,$$

$$\lim_{t\to\infty} (1+t)^{|\mathcal{A}|+1} \left\{(1+t)^{2m+|\mathcal{A}|+n/2} \int_{\mathbb{R}^n} |\triangle^m \mathbf{v}(\mathbf{x},t)|^2 \mathrm{d}\mathbf{x}\right.$$

$$\left. - \frac{1}{(2\pi)^n} \int_{\mathbb{R}^n} |\eta|^{4m} \exp\left(-2\alpha|\eta|^2\right) \sum_{i=1}^{n} \left|\sum_{j=1}^{n}\sum_{k=1}^{n} \cdots \sum_{l=1}^{n} \eta_j\eta_k\cdots\eta_l \widehat{\phi_{ijk\cdots l}}(\mathbf{0})\right|^2 \mathrm{d}\eta\right\}$$

$$= \frac{1}{(2\pi)^n} \int_{\mathbb{R}^n} |\eta|^{4m} \exp\left(-2\alpha|\eta|^2\right) \left\{\sum_{i=1}^{n} \left|\sum_{j=1}^{n}\sum_{k=1}^{n} \cdots \sum_{l=1}^{n} \eta_j\eta_k\cdots\eta_l \int_{\mathbb{R}^n} (\mathbf{x}\cdot\eta)\phi_{ijk\cdots l}(\mathbf{x})\mathrm{d}\mathbf{x}\right|^2\right.$$

$$- \left[\sum_{i=1}^{n}\sum_{j=1}^{n}\sum_{k=1}^{n} \cdots \sum_{l=1}^{n} \eta_j\eta_k\cdots\eta_l \int_{\mathbb{R}^n} (\mathbf{x}\cdot\eta)^2 \phi_{ijk\cdots l}(\mathbf{x})\mathrm{d}\mathbf{x}\right]$$

$$\left.\times \left[\sum_{i=1}^{n}\sum_{j=1}^{n}\sum_{k=1}^{n} \cdots \sum_{l=1}^{n} \eta_j\eta_k\cdots\eta_l \int_{\mathbb{R}^n} \phi_{ijk\cdots l}(\mathbf{x})\mathrm{d}\mathbf{x}\right]\right\} \mathrm{d}\eta.$$

There also hold the sharp decay estimates

$$(1+t)^{m+(n+|\mathcal{A}|)/2} \|\triangle^m \mathbf{v}(\cdot,t)\|_{L^\infty(\mathbb{R}^n)} \leq C_m\left(\|\mathbf{v}_0\|_{L^1(\mathbb{R}^n)}, \|\mathbf{v}_0\|_{L^2(\mathbb{R}^n)}\right), \quad \forall m \geq 0,$$

$$(1+t)^{m+(n+|\mathcal{A}|+1)/2} \|\nabla\triangle^m \mathbf{v}(\cdot,t)\|_{L^\infty(\mathbb{R}^n)} \leq C_m\left(\|\mathbf{v}_0\|_{L^1(\mathbb{R}^n)}, \|\mathbf{v}_0\|_{L^2(\mathbb{R}^n)}\right), \quad \forall m \geq 0.$$

Proof. Applying the Fourier transform to problem (7)-(8) and using integrating factor idea in ordinary differential equations, we can easily get the Fourier representation

$$\widehat{\mathbf{v}}(\xi,t) = \exp\left[-\frac{\alpha|\xi|^2 + \mathbf{i}\Gamma(\xi)}{1+\varepsilon|\xi|^2} t\right] \widehat{\mathbf{v}}_0(\xi).$$

Define the heat kernel function by

$$G(\mathbf{x},t)=\frac{1}{(4\pi\alpha t)^{n/2}}\exp\left(-\frac{|\mathbf{x}|^2}{4\alpha t}\right),\quad \text{for all } t>0.$$

Then its Fourier transform

$$\widehat{G}(\xi,t)=\exp\left(-\alpha|\xi|^2 t\right).$$

The solution of the Cauchy problem (8) for the linear equation (7) is given by

$$\mathbf{v}(\mathbf{x},t)=\frac{1}{(2\pi)^n}\int_{\mathbb{R}^n}\exp(+\mathbf{i}\mathbf{x}\cdot\xi)\exp\left[-\frac{\alpha|\xi|^2+\mathbf{i}\Gamma(\xi)}{1+\varepsilon|\xi|^2}t\right]\widehat{\mathbf{v}}_0(\xi)d\xi.$$

When $\varepsilon=0$ and $\Gamma(\xi)=\beta\cdot\xi$, where $\beta=(\beta_1,\beta_2,\cdots,\beta_n)^T$ is a constant vector, the solution is given by the convolution product of the heat kernel and the initial data:

$$\mathbf{v}(\mathbf{x},t)\;=\;[G(\cdot,t)*\mathbf{v}_0](\mathbf{x}-\beta t)=\frac{1}{(4\pi\alpha t)^{n/2}}\int_{\mathbb{R}^n}\exp\left(-\frac{|\mathbf{x}-\beta t-\mathbf{y}|^2}{4\alpha t}\right)\mathbf{v}_0(\mathbf{y})d\mathbf{y}.$$

Below the proof is divided into three parts and each part may contain four small parts.

Part One: Exact L^2-limits with sharp rates of decay. Suppose that the initial function $\mathbf{v}_0$ satisfies the condition

$$|\widehat{\mathbf{v}_0}(\xi)|^2=|\xi|^{2k}|\widehat{\Psi}(\xi)|^2,$$

where $k\geq 0$ is an integer, $\Psi\in C^1(\mathbb{R}^n)\cap L^1(\mathbb{R}^n)$ is a complex-valued, continuous function, and

$$\int_{\mathbb{R}^n}\Psi(\mathbf{x})d\mathbf{x}\neq\mathbf{0}.$$

By coupling the Plancherel's identity and the Fourier representation together, we get

$$\begin{aligned}
&\lim_{t\to\infty}\left[t^{k+n/2}\int_{\mathbb{R}^n}|\mathbf{v}(\mathbf{x},t)|^2d\mathbf{x}\right]\\
=\;&\lim_{t\to\infty}\left[\frac{t^{k+n/2}}{(2\pi)^n}\int_{\mathbb{R}^n}|\widehat{\mathbf{v}}(\xi,t)|^2d\xi\right]\\
=\;&\lim_{t\to\infty}\left[\frac{t^{k+n/2}}{(2\pi)^n}\int_{\mathbb{R}^n}\exp\left(-\frac{2\alpha|\xi|^2}{1+\varepsilon|\xi|^2}t\right)|\widehat{\mathbf{v}}_0(\xi)|^2d\xi\right]\\
=\;&\lim_{t\to\infty}\left[\frac{1}{(2\pi)^n}\int_{\mathbb{R}^n}\exp\left(-\frac{2\alpha|\eta|^2}{1+\varepsilon|\eta|^2/t}\right)|\eta|^{2k}\left|\widehat{\Psi}(t^{-1/2}\eta)\right|^2d\eta\right]\\
=\;&\frac{1}{(2\pi)^{n+k}}\left(\frac{\pi}{2\alpha}\right)^{k+n/2}\prod_{l=1}^{k}(2l-2+n)\left|\int_{\mathbb{R}^n}\Psi(\mathbf{x})d\mathbf{x}\right|^2>0,
\end{aligned}$$

where we have made the change of variable $\eta = \sqrt{t}\xi$ for $t > 0$, so that $\mathrm{d}\eta = t^{n/2}\mathrm{d}\xi$. Moreover, let $m \geq 1$ be any integer. Then by using properties of the Fourier transform, we have

$$
\begin{aligned}
& t^{2m+k+n/2} \int_{\mathbb{R}^n} |\triangle^m \mathbf{v}(\mathbf{x},t)|^2 \mathrm{d}\mathbf{x} = \frac{t^{2m+k+n/2}}{(2\pi)^n} \int_{\mathbb{R}^n} |\xi|^{4m} |\widehat{\mathbf{v}}(\xi,t)|^2 \mathrm{d}\xi \\
= \;& \frac{t^{2m+k+n/2}}{(2\pi)^n} \int_{\mathbb{R}^n} |\xi|^{4m} \exp\left(-\frac{2\alpha|\xi|^2}{1+\varepsilon|\xi|^2} t\right) |\widehat{\mathbf{v}_0}(\xi)|^2 \mathrm{d}\xi \\
= \;& \frac{1}{(2\pi)^n} \int_{\mathbb{R}^n} |\eta|^{4m} \exp\left(\frac{-2\alpha|\eta|^2}{1+\varepsilon|\eta|^2/t}\right) |\eta|^{2k} |\widehat{\Psi}(t^{-1/2}\eta)|^2 \mathrm{d}\eta \\
\longrightarrow \;& \frac{1}{(2\pi)^n} \int_{\mathbb{R}^n} |\eta|^{4m+2k} \exp\left(-2\alpha|\eta|^2\right) |\widehat{\Psi}(\mathbf{0})|^2 \mathrm{d}\eta \\
= \;& \frac{1}{(2\pi)^{2m+n+k}} \left(\frac{\pi}{2\alpha}\right)^{2m+k+n/2} \prod_{l=1}^{2m+k} (2l-2+n) \left[\int_{\mathbb{R}^n} \Psi(\mathbf{x}) \mathrm{d}\mathbf{x}\right]^2 > 0,
\end{aligned}
$$

as $t \longrightarrow \infty$.

(I) Suppose that $\mathbf{v}_0 \in L^1(\mathbb{R}^n) \cap H^2(\mathbb{R}^n)$, and

$$\int_{\mathbb{R}^n} \mathbf{v}_0(\mathbf{x}) \mathrm{d}\mathbf{x} \neq \mathbf{0}.$$

Then we have

$$
\begin{aligned}
\lim_{t\to\infty} \left\{(1+t)^{n/2} \int_{\mathbb{R}^n} |\mathbf{v}(\mathbf{x},t)|^2 \mathrm{d}\mathbf{x}\right\} &= \frac{1}{(2\pi)^n} \left(\frac{\pi}{2\alpha}\right)^{n/2} \left[\int_{\mathbb{R}^n} \mathbf{v}_0(\mathbf{x}) \mathrm{d}\mathbf{x}\right]^2, \\
\lim_{t\to\infty} \left\{(1+t)^{2m+n/2} \int_{\mathbb{R}^n} |\triangle^m \mathbf{v}(\mathbf{x},t)|^2 \mathrm{d}\mathbf{x}\right\} &= \frac{1}{(2\pi)^{2m+n}} \left(\frac{\pi}{2\alpha}\right)^{2m+n/2} \prod_{l=1}^{2m} (2l-2+n) \left[\int_{\mathbb{R}^n} \mathbf{v}_0(\mathbf{x}) \mathrm{d}\mathbf{x}\right]^2.
\end{aligned}
$$

(II) Suppose now that

$$\int_{\mathbb{R}^n} \mathbf{v}_0(\mathbf{x}) \mathrm{d}\mathbf{x} = \mathbf{0}.$$

Suppose also that $\Phi_i \equiv (\phi_{i1}, \phi_{i2}, \cdots, \phi_{in})^T$, $\phi_{ij} \in C^2(\mathbb{R}^n) \cap L^1(\mathbb{R}^n)$, such that

$$\mathbf{v}_0(\mathbf{x}) = \left(\sum_{j=1}^n \frac{\partial \phi_{1j}}{\partial x_j}(\mathbf{x}),\ \sum_{j=1}^n \frac{\partial \phi_{2j}}{\partial x_j}(\mathbf{x}),\ \cdots,\ \sum_{j=1}^n \frac{\partial \phi_{nj}}{\partial x_j}(\mathbf{x})\right)^T.$$

Then

$$|\widehat{\mathbf{v}_0}(\xi)|^2 = \sum_{i=1}^n \left|\sum_{j=1}^n \xi_j \widehat{\phi_{ij}}(\xi)\right|^2,$$

and we have the following limits

$$\begin{aligned}
& t^{1+n/2}\int_{\mathbb{R}^n}|\mathbf{v}(\mathbf{x},t)|^2\mathrm{d}\mathbf{x}=\frac{t^{1+n/2}}{(2\pi)^n}\int_{\mathbb{R}^n}|\widehat{\mathbf{v}}(\xi,t)|^2\mathrm{d}\xi \\
= & \frac{t^{1+n/2}}{(2\pi)^n}\int_{\mathbb{R}^n}\exp\left(-\frac{2\alpha|\xi|^2}{1+\varepsilon|\xi|^2}t\right)|\widehat{\mathbf{v}_0}(\xi)|^2\mathrm{d}\xi \\
= & \frac{t^{1+n/2}}{(2\pi)^n}\int_{\mathbb{R}^n}\exp\left(-\frac{2\alpha|\xi|^2}{1+\varepsilon|\xi|^2}t\right)\sum_{i=1}^n\left|\sum_{j=1}^n\xi_j\widehat{\phi_{ij}}(\xi)\right|^2\mathrm{d}\xi \\
= & \frac{1}{(2\pi)^n}\int_{\mathbb{R}^n}\exp\left(-\frac{2\alpha|\eta|^2}{1+\varepsilon|\eta|^2/t}\right)\sum_{i=1}^n\left|\sum_{j=1}^n\eta_j\widehat{\phi_{ij}}(t^{-1/2}\eta)\right|^2\mathrm{d}\eta \\
\to & \frac{1}{(2\pi)^n}\int_{\mathbb{R}^n}\exp\left(-2\alpha|\eta|^2\right)\sum_{i=1}^n\left|\sum_{j=1}^n\eta_j\widehat{\phi_{ij}}(\mathbf{0})\right|^2\mathrm{d}\eta \\
= & \frac{1}{(2\pi)^n}\int_{\mathbb{R}^n}\exp\left(-2\alpha|\eta|^2\right)\sum_{i=1}^n\sum_{j=1}^n\left|\eta_j\widehat{\phi_{ij}}(\mathbf{0})\right|^2\mathrm{d}\eta \\
= & \frac{1}{(2\pi)^n}\left(\frac{\pi}{2\alpha}\right)^{n/2}\frac{1}{4\alpha}\sum_{i=1}^n\sum_{j=1}^n|\widehat{\phi_{ij}}(\mathbf{0})|^2=\frac{1}{(2\pi)^{n+1}}\left(\frac{\pi}{2\alpha}\right)^{1+n/2}\sum_{k=1}^n\left|\int_{\mathbb{R}^n}\Phi_k(\mathbf{x})\mathrm{d}\mathbf{x}\right|^2,
\end{aligned}$$

and

$$\begin{aligned}
& t^{2m+1+n/2}\int_{\mathbb{R}^n}|\triangle^m\mathbf{v}(\mathbf{x},t)|^2\mathrm{d}\mathbf{x} \\
= & \frac{t^{2m+1+n/2}}{(2\pi)^n}\int_{\mathbb{R}^n}|\xi|^{4m}|\widehat{\mathbf{v}}(\xi,t)|^2\mathrm{d}\xi \\
= & \frac{t^{2m+1+n/2}}{(2\pi)^n}\int_{\mathbb{R}^n}|\xi|^{4m}\exp\left(-\frac{2\alpha|\xi|^2}{1+\varepsilon|\xi|^2}t\right)|\widehat{\mathbf{v}_0}(\xi)|^2\mathrm{d}\xi \\
= & \frac{t^{2m+1+n/2}}{(2\pi)^n}\int_{\mathbb{R}^n}|\xi|^{4m}\exp\left(-\frac{2\alpha|\xi|^2}{1+\varepsilon|\xi|^2}t\right)\sum_{i=1}^n\left|\sum_{j=1}^n\xi_j\widehat{\phi_{ij}}(\xi)\right|^2\mathrm{d}\xi \\
= & \frac{1}{(2\pi)^n}\int_{\mathbb{R}^n}|\eta|^{4m}\exp\left(-\frac{2\alpha|\eta|^2}{1+\varepsilon|\eta|^2/t}\right)\sum_{i=1}^n\left|\sum_{j=1}^n\eta_j\widehat{\phi_{ij}}(t^{-1/2}\eta)\right|^2\mathrm{d}\eta \\
\to & \frac{1}{(2\pi)^n}\int_{\mathbb{R}^n}|\eta|^{4m}\exp\left(-2\alpha|\eta|^2\right)\sum_{i=1}^n\left|\sum_{j=1}^n\eta_j\widehat{\phi_{ij}}(\mathbf{0})\right|^2\mathrm{d}\eta \\
= & \frac{1}{(2\pi)^n}\int_{\mathbb{R}^n}|\eta|^{4m}\exp\left(-2\alpha|\eta|^2\right)\sum_{i=1}^n\sum_{j=1}^n\left|\eta_j\widehat{\phi_{ij}}(\mathbf{0})\right|^2\mathrm{d}\eta \\
= & \frac{1}{(2\pi)^n}\left(\frac{\pi}{2\alpha}\right)^{n/2}\left(\frac{1}{4\alpha}\right)^{2m+1}\prod_{l=0}^{2m}(2l+n)\sum_{i=1}^n\sum_{j=1}^n|\widehat{\phi_{ij}}(\mathbf{0})|^2 \\
= & \frac{1}{(2\pi)^{2m+n+1}}\left(\frac{\pi}{2\alpha}\right)^{2m+1+n/2}\prod_{l=1}^{2m}(2l+n)\sum_{k=1}^n\left|\int_{\mathbb{R}^n}\Phi_k(\mathbf{x})\mathrm{d}\mathbf{x}\right|^2,
\end{aligned}$$

as $t \longrightarrow \infty$, where

$$\int_{\mathbb{R}^n} |\eta_j|^2 \exp\left(-2\alpha|\eta|^2\right) \mathrm{d}\eta = \frac{1}{4\alpha}\left(\frac{\pi}{2\alpha}\right)^{n/2},$$

$$\int_{\mathbb{R}^n} |\eta_j|^2 |\eta|^{4m} \exp\left(-2\alpha|\eta|^2\right) \mathrm{d}\eta = \left(\frac{\pi}{2\alpha}\right)^{n/2}\left(\frac{1}{4\alpha}\right)^{2m+1}\prod_{l=1}^{2m}(2l+n).$$

(III) If the initial function

$$\mathbf{v}_0(\mathbf{x}) = \left(\sum_{j=1}^n\sum_{k=1}^n \frac{\partial^2 \phi_{1jk}}{\partial x_j \partial x_k}(\mathbf{x}),\ \sum_{j=1}^n\sum_{k=1}^n \frac{\partial^2 \phi_{2jk}}{\partial x_j \partial x_k}(\mathbf{x}),\ \cdots,\ \sum_{j=1}^n\sum_{k=1}^n \frac{\partial^2 \phi_{njk}}{\partial x_j \partial x_k}(\mathbf{x})\right),$$

then

$$|\widehat{\mathbf{v}_0}(\xi)|^2 = \sum_{i=1}^n \left|\sum_{j=1}^n\sum_{k=1}^n \xi_j\xi_k\widehat{\phi_{ijk}}(\xi)\right|^2.$$

Therefore, we obtain the limits

$$\begin{aligned}(1+t)^{2+n/2}\int_{\mathbb{R}^n}|\mathbf{v}(\mathbf{x},t)|^2\mathrm{d}\mathbf{x} &\rightarrow \frac{1}{(2\pi)^n}\int_{\mathbb{R}^n}\exp\left(-2\alpha|\eta|^2\right)\sum_{i=1}^n\left|\sum_{j=1}^n\sum_{k=1}^n\eta_j\eta_k\widehat{\phi_{ijk}}(\mathbf{0})\right|^2\mathrm{d}\eta,\\ (1+t)^{2m+2+n/2}\int_{\mathbb{R}^n}|\triangle^m\mathbf{v}(\mathbf{x},t)|^2\mathrm{d}\mathbf{x} &\rightarrow \frac{1}{(2\pi)^n}\int_{\mathbb{R}^n}|\eta|^{4m}\exp\left(-2\alpha|\eta|^2\right)\sum_{i=1}^n\left|\sum_{j=1}^n\sum_{k=1}^n\eta_j\eta_k\widehat{\phi_{ijk}}(\mathbf{0})\right|^2\mathrm{d}\eta,\end{aligned}$$

as $t \longrightarrow \infty$.

(IV) If the initial data

$$\mathbf{v}_0(\mathbf{x}) = \left(\sum_{j=1}^n\sum_{k=1}^n\cdots\sum_{l=1}^n \frac{\partial^{|\mathcal{A}|}\phi_{1jk\cdots l}}{\partial x_j\partial x_k\cdots\partial x_l}(\mathbf{x}),\ \cdots,\ \sum_{j=1}^n\sum_{k=1}^n\cdots\sum_{l=1}^n \frac{\partial^{|\mathcal{A}|}\phi_{njk\cdots l}}{\partial x_j\partial x_k\cdots\partial x_l}(\mathbf{x})\right),$$

then

$$|\widehat{\mathbf{v}_0}(\xi)|^2 = \sum_{i=1}^n\left|\sum_{j=1}^n\sum_{k=1}^n\cdots\sum_{l=1}^n\xi_j\xi_k\cdots\xi_l\widehat{\phi_{ijk\cdots l}}(\xi)\right|^2.$$

Now we obtain

$$\begin{aligned}&t^{2m+|\mathcal{A}|+n/2}\int_{\mathbb{R}^n}|\triangle^m\mathbf{v}(\mathbf{x},t)|^2\mathrm{d}\mathbf{x}\\ =\ &\frac{t^{2m+|\mathcal{A}|+n/2}}{(2\pi)^n}\int_{\mathbb{R}^n}|\xi|^{4m}|\widehat{\mathbf{v}}(\xi,t)|^2\mathrm{d}\xi\\ =\ &\frac{t^{2m+|\mathcal{A}|+n/2}}{(2\pi)^n}\int_{\mathbb{R}^n}|\xi|^{4m}\exp\left(-\frac{2\alpha|\xi|^2}{1+\varepsilon|\xi|^2}t\right)|\widehat{\mathbf{v}_0}(\xi)|^2\mathrm{d}\xi\\ =\ &\frac{t^{2m+|\mathcal{A}|+n/2}}{(2\pi)^n}\int_{\mathbb{R}^n}|\xi|^{4m}\exp\left(-\frac{2\alpha|\xi|^2}{1+\varepsilon|\xi|^2}t\right)\sum_{i=1}^n\left|\sum_{j=1}^n\sum_{k=1}^n\cdots\sum_{l=1}^n\xi_j\xi_k\cdots\xi_l\widehat{\phi_{ijk\cdots l}}(\xi)\right|^2\mathrm{d}\xi\\ =\ &\frac{1}{(2\pi)^n}\int_{\mathbb{R}^n}|\eta|^{4m}\exp\left(-\frac{2\alpha|\eta|^2}{1+\varepsilon|\eta|^2/t}\right)\sum_{i=1}^n\left|\sum_{j=1}^n\sum_{k=1}^n\cdots\sum_{l=1}^n\eta_j\eta_k\cdots\eta_l\widehat{\phi_{ijk\cdots l}}(t^{-1/2}\eta)\right|^2\mathrm{d}\eta\\ \rightarrow\ &\frac{1}{(2\pi)^n}\int_{\mathbb{R}^n}|\eta|^{4m}\exp\left(-2\alpha|\eta|^2\right)\sum_{i=1}^n\left|\sum_{j=1}^n\sum_{k=1}^n\cdots\sum_{l=1}^n\eta_j\eta_k\cdots\eta_l\widehat{\phi_{ijk\cdots l}}(\mathbf{0})\right|^2\mathrm{d}\eta,\end{aligned}$$

as $t \to \infty$. The L^2-limits are established.

Part Two: Convergence rates of $(1+t)^{2m+\lambda+n/2}\int_{\mathbb{R}^n}|\triangle^m \mathbf{v}(\mathbf{x},t)|^2 d\mathbf{x}$ **to their limits, as** $t \to \infty$**.** Suppose that the function $\phi \in L^1(\mathbb{R}^n)$ and that

$$\int_{\mathbb{R}^n}(1+|\mathbf{x}|)^3|\phi(\mathbf{x})|d\mathbf{x} < \infty.$$

Note that for any vector $\xi \in \mathbb{R}^n$, we have the formal Taylor's expansion

$$\begin{aligned}
\widehat{\phi}(\xi) &= \int_{\mathbb{R}^n}\exp(-\mathbf{i}\mathbf{x}\cdot\xi)\phi(\mathbf{x})d\mathbf{x} = \sum_{m=0}^{\infty}\frac{(-\mathbf{i})^m}{m!}\int_{\mathbb{R}^n}(\mathbf{x}\cdot\xi)^m\phi(\mathbf{x})d\mathbf{x} \\
&= \int_{\mathbb{R}^n}\phi(\mathbf{x})d\mathbf{x} + \sum_{m=1}^{\infty}\frac{(-1)^m}{(2m)!}\int_{\mathbb{R}^n}(\mathbf{x}\cdot\xi)^{2m}\phi(\mathbf{x})d\mathbf{x} \\
&- \mathbf{i}\int_{\mathbb{R}^n}(\mathbf{x}\cdot\xi)\phi(\mathbf{x})d\mathbf{x} - \mathbf{i}\sum_{m=1}^{\infty}\frac{(-1)^m}{(2m+1)!}\int_{\mathbb{R}^n}(\mathbf{x}\cdot\xi)^{2m+1}\phi(\mathbf{x})d\mathbf{x}.
\end{aligned}$$

Let $t > 0$ and set $\xi = t^{-1/2}\eta$.

(I) If

$$\int_{\mathbb{R}^n}\phi(\mathbf{x})d\mathbf{x} \neq \mathbf{0},$$

then

$$\begin{aligned}
& t\left[|\widehat{\phi}(t^{-1/2}\eta)|^2 - |\widehat{\phi}(\mathbf{0})|^2\right] \\
= & \left[\sum_{m=1}^{\infty}\frac{(-1)^m}{(2m)!}\frac{1}{t^{m-1}}\int_{\mathbb{R}^n}(\mathbf{x}\cdot\eta)^{2m}\phi(\mathbf{x})d\mathbf{x}\right] \\
\times & \left[2\int_{\mathbb{R}^n}\phi(\mathbf{x})d\mathbf{x} + \sum_{m=1}^{\infty}\frac{(-1)^m}{(2m)!}\frac{1}{t^m}\int_{\mathbb{R}^n}(\mathbf{x}\cdot\eta)^{2m}\phi(\mathbf{x})d\mathbf{x}\right] \\
+ & \left|\int_{\mathbb{R}^n}(\mathbf{x}\cdot\eta)\phi(\mathbf{x})d\mathbf{x} + \sum_{m=1}^{\infty}\frac{(-1)^m}{(2m+1)!}\frac{1}{t^m}\int_{\mathbb{R}^n}(\mathbf{x}\cdot\eta)^{2m+1}\phi(\mathbf{x})d\mathbf{x}\right|^2 \\
\to & \left|\int_{\mathbb{R}^n}(\mathbf{x}\cdot\eta)\phi(\mathbf{x})d\mathbf{x}\right|^2 - \left[\int_{\mathbb{R}^n}(\mathbf{x}\cdot\eta)^2\phi(\mathbf{x})d\mathbf{x}\right]\left[\int_{\mathbb{R}^n}\phi(\mathbf{x})d\mathbf{x}\right],
\end{aligned}$$

as $t \to \infty$.

(II) Suppose that the initial function

$$\begin{aligned}
\mathbf{v}_0(\mathbf{x}) = & \left(\sum_{j=1}^{n}\sum_{k=1}^{n}\cdots\sum_{l=1}^{n}\frac{\partial^{|A|}\phi_{1jk\cdots l}}{\partial x_j\partial x_k\cdots\partial x_l}(\mathbf{x}), \sum_{j=1}^{n}\sum_{k=1}^{n}\cdots\sum_{l=1}^{n}\frac{\partial^{|A|}\phi_{2jk\cdots l}}{\partial x_j\partial x_k\cdots\partial x_l}(\mathbf{x}), \cdots,\right. \\
& \left.\sum_{j=1}^{n}\sum_{k=1}^{n}\cdots\sum_{l=1}^{n}\frac{\partial^{|A|}\phi_{njk\cdots l}}{\partial x_j\partial x_k\cdots\partial x_l}(\mathbf{x})\right)^T,
\end{aligned}$$

then

$$\left|\widehat{\mathbf{v}_0}(\xi)\right|^2 = \sum_{i=1}^{n} \left| \sum_{j=1}^{n}\sum_{k=1}^{n}\cdots\sum_{l=1}^{n} \xi_j\xi_k\cdots\xi_l\widehat{\phi_{ijk\cdots l}}(\xi)\right|^2 .$$

Now, if $\xi = t^{-1/2}\eta$, then we obtain the following limit, as $t \to \infty$:

$$\begin{aligned}
& t^{|\mathcal{A}|+1}\left\{\sum_{i=1}^{n}\left|\sum_{j=1}^{n}\sum_{k=1}^{n}\cdots\sum_{l=1}^{n}\xi_j\xi_k\cdots\xi_l\widehat{\phi_{ijk\cdots l}}(\xi)\right|^2-\sum_{i=1}^{n}\left|\sum_{j=1}^{n}\sum_{k=1}^{n}\cdots\sum_{l=1}^{n}\xi_j\xi_k\cdots\xi_l\widehat{\phi_{ijk\cdots l}}(\mathbf{0})\right|^2\right\} \\
= \; & t\left\{\sum_{i=1}^{n}\left|\sum_{j=1}^{n}\sum_{k=1}^{n}\cdots\sum_{l=1}^{n}\eta_j\eta_k\cdots\eta_l\widehat{\phi_{ijk\cdots l}}(t^{-1/2}\eta)\right|^2-\sum_{i=1}^{n}\left|\sum_{j=1}^{n}\sum_{k=1}^{n}\cdots\sum_{l=1}^{n}\eta_j\eta_k\cdots\eta_l\widehat{\phi_{ijk\cdots l}}(\mathbf{0})\right|^2\right\} \\
= \; & t\left\{\sum_{i=1}^{n}\sum_{j=1}^{n}\sum_{k=1}^{n}\cdots\sum_{l=1}^{n}\eta_j\eta_k\cdots\eta_l\left[\int_{\mathbb{R}^n}\phi_{ijk\cdots l}(\mathbf{x})\sum_{m=1}^{\infty}\frac{(-1)^m}{(2m)!}\frac{(\mathbf{x}\cdot\eta)^{2m}}{t^m}\mathrm{d}\mathbf{x}\right]\right\} \\
\times \; & \left\{\sum_{i=1}^{n}\sum_{j=1}^{n}\sum_{k=1}^{n}\cdots\sum_{l=1}^{n}\eta_j\eta_k\cdots\eta_l\left[\int_{\mathbb{R}^n}\phi_{ijk\cdots l}(\mathbf{x})\mathrm{d}\mathbf{x}+\int_{\mathbb{R}^n}\phi_{ijk\cdots l}(\mathbf{x})\sum_{m=0}^{\infty}\frac{(-1)^m}{(2m)!}\frac{(\mathbf{x}\cdot\eta)^{2m}}{t^m}\mathrm{d}\mathbf{x}\right]\right\} \\
+ \; & \left\{\sum_{i=1}^{n}\left|\sum_{j=1}^{n}\sum_{k=1}^{n}\cdots\sum_{l=1}^{n}\eta_j\eta_k\cdots\eta_l\left[\int_{\mathbb{R}^n}\phi_{ijk\cdots l}(\mathbf{x})\sum_{m=0}^{\infty}\frac{(-1)^m}{(2m+1)!}\frac{(\mathbf{x}\cdot\eta)^{2m+1}}{t^m}\mathrm{d}\mathbf{x}\right]\right|^2\right\} \\
= \; & \left\{\sum_{i=1}^{n}\sum_{j=1}^{n}\sum_{k=1}^{n}\cdots\sum_{l=1}^{n}\eta_j\eta_k\cdots\eta_l\left[\int_{\mathbb{R}^n}\phi_{ijk\cdots l}(\mathbf{x})\sum_{m=1}^{\infty}\frac{(-1)^m}{(2m)!}\frac{(\mathbf{x}\cdot\eta)^{2m}}{t^{m-1}}\mathrm{d}\mathbf{x}\right]\right\} \\
\times \; & \left\{\sum_{i=1}^{n}\sum_{j=1}^{n}\sum_{k=1}^{n}\cdots\sum_{l=1}^{n}\eta_j\eta_k\cdots\eta_l\left[\int_{\mathbb{R}^n}\phi_{ijk\cdots l}(\mathbf{x})\mathrm{d}\mathbf{x}+\int_{\mathbb{R}^n}\phi_{ijk\cdots l}(\mathbf{x})\sum_{m=0}^{\infty}\frac{(-1)^m}{(2m)!}\frac{(\mathbf{x}\cdot\eta)^{2m}}{t^m}\mathrm{d}\mathbf{x}\right]\right\} \\
+ \; & \left\{\sum_{i=1}^{n}\left|\sum_{j=1}^{n}\sum_{k=1}^{n}\cdots\sum_{l=1}^{n}\eta_j\eta_k\cdots\eta_l\left[\int_{\mathbb{R}^n}\phi_{ijk\cdots l}(\mathbf{x})\sum_{m=0}^{\infty}\frac{(-1)^m}{(2m+1)!}\frac{(\mathbf{x}\cdot\eta)^{2m+1}}{t^m}\mathrm{d}\mathbf{x}\right]\right|^2\right\} \\
\to \; & \left\{\sum_{i=1}^{n}\left|\sum_{j=1}^{n}\sum_{k=1}^{n}\cdots\sum_{l=1}^{n}\eta_j\eta_k\cdots\eta_l\left[\int_{\mathbb{R}^n}(\mathbf{x}\cdot\eta)\phi_{ijk\cdots l}(\mathbf{x})\mathrm{d}\mathbf{x}\right]\right|^2\right\} \\
- \; & \left\{\sum_{i=1}^{n}\sum_{j=1}^{n}\sum_{k=1}^{n}\cdots\sum_{l=1}^{n}\eta_j\eta_k\cdots\eta_l\left[\int_{\mathbb{R}^n}(\mathbf{x}\cdot\eta)^2\phi_{ijk\cdots l}(\mathbf{x})\mathrm{d}\mathbf{x}\right]\right\} \\
\times \; & \left\{\sum_{i=1}^{n}\sum_{j=1}^{n}\sum_{k=1}^{n}\cdots\sum_{l=1}^{n}\eta_j\eta_k\cdots\eta_l\left[\int_{\mathbb{R}^n}\phi_{ijk\cdots l}(\mathbf{x})\mathrm{d}\mathbf{x}\right]\right\}.
\end{aligned}$$

If we use these limits in the following analysis, we may finish this part immediately.

Without loss of generality, let $\varepsilon = 0$. As $t \to \infty$, we have the limit

$$
\begin{aligned}
& t^{|\mathcal{A}|+1} \Bigg\{ t^{2m+|\mathcal{A}|+n/2} \int_{\mathbb{R}^n} |\Delta^m \mathbf{v}(\mathbf{x},t)|^2 \mathrm{d}\mathbf{x} \\
& - \frac{1}{(2\pi)^n} \int_{\mathbb{R}^n} |\eta|^{4m} \exp\left(-2\alpha|\eta|^2\right) \sum_{i=1}^n \left| \sum_{j=1}^n \sum_{k=1}^n \cdots \sum_{l=1}^n \eta_j \eta_k \cdots \eta_l \widehat{\phi_{ijk\cdots l}}(\mathbf{0}) \right|^2 \mathrm{d}\eta \Bigg\} \\
& = t^{|\mathcal{A}|+1} \Bigg\{ \frac{1}{(2\pi)^n} \int_{\mathbb{R}^n} |\eta|^{4m} \exp\left(-\frac{2\alpha|\eta|^2}{1+\varepsilon|\eta|^2/t} \right) \sum_{i=1}^n \left| \sum_{j=1}^n \sum_{k=1}^n \cdots \sum_{l=1}^n \eta_j \eta_k \cdots \eta_l \widehat{\phi_{ijk\cdots l}}(t^{-1/2}\eta) \right|^2 \mathrm{d}\eta \\
& - \frac{1}{(2\pi)^n} \int_{\mathbb{R}^n} |\eta|^{4m} \exp\left(-2\alpha|\eta|^2\right) \sum_{i=1}^n \left| \sum_{j=1}^n \sum_{k=1}^n \cdots \sum_{l=1}^n \eta_j \eta_k \cdots \eta_l \widehat{\phi_{ijk\cdots l}}(\mathbf{0}) \right|^2 \mathrm{d}\eta \Bigg\} \\
& \to \frac{1}{(2\pi)^n} \int_{\mathbb{R}^n} |\eta|^{4m} \exp\left(-2\alpha|\eta|^2\right) \Bigg\{ \sum_{i=1}^n \left| \sum_{j=1}^n \sum_{k=1}^n \cdots \sum_{l=1}^n \eta_j \eta_k \cdots \eta_l \int_{\mathbb{R}^n} (\mathbf{x}\cdot\eta) \phi_{ijk\cdots l}(\mathbf{x}) \mathrm{d}\mathbf{x} \right|^2 \\
& - \left[\sum_{i=1}^n \sum_{j=1}^n \sum_{k=1}^n \cdots \sum_{l=1}^n \eta_j \eta_k \cdots \eta_l \int_{\mathbb{R}^n} (\mathbf{x}\cdot\eta)^2 \phi_{ijk\cdots l}(\mathbf{x}) \mathrm{d}\mathbf{x} \right] \\
& \times \left[\sum_{i=1}^n \sum_{j=1}^n \sum_{k=1}^n \cdots \sum_{l=1}^n \eta_j \eta_k \cdots \eta_l \int_{\mathbb{R}^n} \phi_{ijk\cdots l}(\mathbf{x}) \mathrm{d}\mathbf{x} \right] \Bigg\} \mathrm{d}\eta.
\end{aligned}
$$

Part Three: Sharp rates of L^∞ decay estimates. First of all, from Part One, we have the following decay estimates

$$
\begin{aligned}
(1+t)^{2m+\lambda+n/2} \int_{\mathbb{R}^n} |\Delta^m \mathbf{v}(\mathbf{x},t)|^2 \mathrm{d}\mathbf{x} &\le C_m, \\
(1+t)^{2m+\lambda+1+n/2} \int_{\mathbb{R}^n} |\nabla\Delta^m \mathbf{v}(\mathbf{x},t)|^2 \mathrm{d}\mathbf{x} &\le C_m,
\end{aligned}
$$

for all $t > 0$, where the constants λ and C_m are independent of time, λ depends on the initial function $\mathbf{u}_0$. By using the Gagliardo-Nirenberg's interpolation inequality (see the Appendix), we get

$$
\|\Delta^m \mathbf{v}(\cdot,t)\|_{L^\infty(\mathbb{R}^n)} \le C \|\Delta^m \mathbf{v}(\cdot,t)\|_{L^2(\mathbb{R}^n)}^{3/4} \|\Delta^{m+n} \mathbf{v}(\cdot,t)\|_{L^2(\mathbb{R}^n)}^{1/4}
$$

and then

$$
\begin{aligned}
& (1+t)^{8m+4n+4\lambda} \|\Delta^m \mathbf{v}(\cdot,t)\|_{L^\infty(\mathbb{R}^n)}^8 \\
\le & C \left[(1+t)^{3(2m+\lambda+n/2)} \|\Delta^m \mathbf{v}(\cdot,t)\|_{L^2(\mathbb{R}^n)}^6 \right] \left[(1+t)^{2m+2n+\lambda+n/2} \|\Delta^{m+n} \mathbf{v}(\cdot,t)\|_{L^2(\mathbb{R}^n)}^2 \right] \le C_m.
\end{aligned}
$$

That is

$$
(1+t)^{m+(n+\lambda)/2} \|\Delta^m \mathbf{v}(\cdot,t)\|_{L^\infty(\mathbb{R}^n)} \le C_m.
$$

Note that $\|\nabla\Delta^m \mathbf{v}(\cdot,t)\|_{L^\infty(\mathbb{R}^n)}$ enjoys a faster decay rate, because

$$
\|\nabla\Delta^m \mathbf{v}(\cdot,t)\|_{L^\infty(\mathbb{R}^n)} \le C \|\nabla\Delta^m \mathbf{v}(\cdot,t)\|_{L^2(\mathbb{R}^n)}^{3/4} \|\nabla\Delta^{m+n} \mathbf{v}(\cdot,t)\|_{L^2(\mathbb{R}^n)}^{1/4},
$$

and then

$$
\begin{aligned}
& (1+t)^{8m+4\lambda+4+4n} \|\nabla\Delta^m \mathbf{v}(\cdot,t)\|_{L^\infty(\mathbb{R}^n)}^8 \\
\le \; & C \left[(1+t)^{3(2m+\lambda+1+n/2)} \|\nabla\Delta^m \mathbf{v}(\cdot,t)\|_{L^2(\mathbb{R}^n)}^6 \right] \left[(1+t)^{2m+2n+\lambda+1+n/2} \|\nabla\Delta^{m+n} \mathbf{v}(\cdot,t)\|_{L^2(\mathbb{R}^n)}^2 \right] \le C_m,
\end{aligned}
$$

i.e.

$$(1+t)^{m+(n+\lambda+1)/2}\|\nabla\triangle^m\mathbf{v}(\cdot,t)\|_{L^\infty(\mathbb{R}^n)}\le C_m,$$

where $\mathbf{v}$ is the global solution of the linear problem (7)-(8).
(I) If

$$\int_{\mathbb{R}^n}\mathbf{v}_0(\mathbf{x})\mathrm{dx}\neq\mathbf{0},$$

then we obtain the following sharp decay estimates

$$(1+t)^{m+n/2}\|\triangle^m\mathbf{v}(\cdot,t)\|_{L^\infty(\mathbb{R}^n)}\le C_m,$$
$$(1+t)^{m+(n+1)/2}\|\nabla\triangle^m\mathbf{v}(\cdot,t)\|_{L^\infty(\mathbb{R}^n)}\le C_m,$$

for some time-independent constant $C_m>0$.
(II) If

$$\mathbf{v}_0(\mathbf{x})=\left(\sum_{j=1}^n\frac{\partial\phi_{1j}}{\partial x_j}(\mathbf{x}),\ \sum_{j=1}^n\frac{\partial\phi_{2j}}{\partial x_j}(\mathbf{x}),\ \cdots,\ \sum_{j=1}^n\frac{\partial\phi_{nj}}{\partial x_j}(\mathbf{x})\right)^T$$

where $\phi_{ij}\in C^2(\mathbb{R}^n)\cap L^1(\mathbb{R}^n)$, $i,j\in\{1,2,\cdots,n\}$, and for some integers $i,j\in\{1,2,\cdots,n\}$, $\int_{\mathbb{R}^n}\phi_{ij}(\mathbf{x})\mathrm{dx}\neq 0$, then

$$(1+t)^{m+(n+1)/2}\|\triangle^m\mathbf{v}(\cdot,t)\|_{L^\infty(\mathbb{R}^n)}\le C_m,$$
$$(1+t)^{m+(n+2)/2}\|\nabla\triangle^m\mathbf{v}(\cdot,t)\|_{L^\infty(\mathbb{R}^n)}\le C_m.$$

(III) If

$$\mathbf{v}_0(\mathbf{x})=\left(\sum_{j=1}^n\sum_{k=1}^n\frac{\partial^2\phi_{1jk}}{\partial x_j\partial x_k}(\mathbf{x}),\ \sum_{j=1}^n\sum_{k=1}^n\frac{\partial^2\phi_{2jk}}{\partial x_j\partial x_k}(\mathbf{x}),\ \cdots,\ \sum_{j=1}^n\sum_{k=1}^n\frac{\partial^2\phi_{njk}}{\partial x_j\partial x_k}(\mathbf{x})\right)^T,$$

where $\phi_{ijk}\in C^2(\mathbb{R}^n)\cap L^1(\mathbb{R}^n)$, and for some integers $i,j,k\in\{1,2,\cdots,n\}$, $\int_{\mathbb{R}^n}\phi_{ijk}(\mathbf{x})\mathrm{dx}\neq 0$, then

$$(1+t)^{m+(n+2)/2}\|\triangle^m\mathbf{v}(\cdot,t)\|_{L^\infty(\mathbb{R}^n)}\le C_m,$$
$$(1+t)^{m+(n+3)/2}\|\nabla\triangle^m\mathbf{v}(\cdot,t)\|_{L^\infty(\mathbb{R}^n)}\le C_m.$$

(IV) If

$$\mathbf{v}_0(\mathbf{x})=\Bigg(\sum_{j=1}^n\sum_{k=1}^n\cdots\sum_{l=1}^n\frac{\partial^{|\mathcal{A}|}\phi_{1jk\cdots l}}{\partial x_j\partial x_k\cdots\partial x_l}(\mathbf{x}),\ \sum_{j=1}^n\sum_{k=1}^n\cdots\sum_{l=1}^n\frac{\partial^{|\mathcal{A}|}\phi_{2jk\cdots l}}{\partial x_j\partial x_k\cdots\partial x_l}(\mathbf{x}),\ \cdots,$$
$$\sum_{j=1}^n\sum_{k=1}^n\cdots\sum_{l=1}^n\frac{\partial^{|\mathcal{A}|}\phi_{njk\cdots l}}{\partial x_j\partial x_k\cdots\partial x_l}(\mathbf{x})\Bigg)^T,$$

where $\phi_{ijk\cdots l}\in C^{|\mathcal{A}|}(\mathbb{R}^n)\cap L^1(\mathbb{R}^n)$, and for some integers $i,j,k,\cdots,l\in\{1,2,\cdots,n\}$, $\int_{\mathbb{R}^n}\phi_{ijk\cdots l}(\mathbf{x})\mathrm{dx}\neq 0$, then

$$(1+t)^{m+(n+|\mathcal{A}|)/2}\|\triangle^m\mathbf{v}(\cdot,t)\|_{L^\infty(\mathbb{R}^n)}\le C_m,\qquad\forall t>0,$$
$$(1+t)^{m+(n+|\mathcal{A}|+1)/2}\|\nabla\triangle^m\mathbf{v}(\cdot,t)\|_{L^\infty(\mathbb{R}^n)}\le C_m,\qquad\forall t>0.$$

The proof of Lemma 2.1 is completed. #

These optimal decay results will be used later when studying the general evolution equation.

2.2. Nonlinear Analysis

Lemma 2.2. *Let the initial function* $\mathbf{u}_0 \in L^1(\mathbb{R}^n) \cap H^2(\mathbb{R}^n)$ *and set* $\mathbf{f} = \mathbf{0}$. **(I)** *Then we have the following elementary estimates for the global solutions of (1)-(2)*

$$\sup_{t\in\mathbb{R}^+} \int_{\mathbb{R}^n} \left[|\mathbf{u}(\mathbf{x},t)|^2 + \varepsilon|\nabla\mathbf{u}(\mathbf{x},t)|^2\right] \mathrm{d}\mathbf{x} \leq \int_{\mathbb{R}^n} \left[|\mathbf{u}_0(\mathbf{x})|^2 + \varepsilon|\nabla\mathbf{u}_0(\mathbf{x})|^2\right] \mathrm{d}\mathbf{x},$$

$$2\alpha \int_0^\infty \left[\int_{\mathbb{R}^n} |\nabla\mathbf{u}(\mathbf{x},t)|^2 \mathrm{d}\mathbf{x}\right] \mathrm{d}t \leq \int_{\mathbb{R}^n} \left[|\mathbf{u}_0(\mathbf{x})|^2 + \varepsilon|\nabla\mathbf{u}_0(\mathbf{x})|^2\right] \mathrm{d}\mathbf{x}.$$

(II) *We also have the Fourier representation for the solutions*

$$\hat{\mathbf{u}}(\xi,t) = \exp\left[-\frac{\alpha|\xi|^2 + \mathbf{i}\Gamma(\xi)}{1+\varepsilon|\xi|^2}t\right]\hat{\mathbf{u}}_0(\xi) - \int_0^t \exp\left[-\frac{\alpha|\xi|^2 + \mathbf{i}\Gamma(\xi)}{1+\varepsilon|\xi|^2}(t-\tau)\right]\frac{\widehat{\mathcal{N}(\mathbf{u},\nabla\mathbf{u})}(\xi,\tau)}{1+\varepsilon|\xi|^2}\mathrm{d}\tau,$$

for all $t > 0$.

Proof. From equation (1), we get the integral equation

$$2\int_0^t \left\{\int_{\mathbb{R}^n} \mathbf{u}\cdot\left[\mathbf{u}_t - \varepsilon\Delta\mathbf{u}_t - \alpha\Delta\mathbf{u} + \mathcal{D}\mathbf{u} + \mathcal{N}(\mathbf{u},\nabla\mathbf{u})\right]\mathrm{d}\mathbf{x}\right\}\mathrm{d}\tau = 0.$$

By using integration by parts and the following boundary conditions

$$\lim_{|\mathbf{x}|\to\infty} \mathbf{u}(\mathbf{x},t) = \mathbf{0}, \qquad \lim_{|\mathbf{x}|\to\infty} \nabla\mathbf{u}(\mathbf{x},t) = \mathbf{0},$$

as well as the assumptions on $\mathcal{D}\mathbf{u}$ and $\mathcal{N}(\mathbf{u},\nabla\mathbf{u})$, we obtain the new equation

$$\int_{\mathbb{R}^n} \left[|\mathbf{u}(\mathbf{x},t)|^2 + \varepsilon|\nabla\mathbf{u}(\mathbf{x},t)|^2\right] \mathrm{d}\mathbf{x} + 2\alpha\int_0^t \left[\int_{\mathbb{R}^n} |\nabla\mathbf{u}(\mathbf{x},\tau)|^2 \mathrm{d}\mathbf{x}\right]\mathrm{d}\tau$$
$$= \int_{\mathbb{R}^n} \left[|\mathbf{u}_0(\mathbf{x})|^2 + \varepsilon|\nabla\mathbf{u}_0(\mathbf{x})|^2\right] \mathrm{d}\mathbf{x},$$

for all time $t > 0$. The first two estimates in Lemma 2.2 follow immediately from this equation. The Fourier representation follows from simple idea in dynamical systems and it is not difficult at all. #

Now, let $\mathbf{u}$ and $\mathbf{v}$ be the solutions of the Cauchy problems (1)-(2) and (7)-(8), respectively, with the initial data $\mathbf{u}(\mathbf{x},0) = \mathbf{u}_0(\mathbf{x}) \in L^1(\mathbb{R}^n) \cap H^2(\mathbb{R}^n)$ and $\mathbf{v}(\mathbf{x},0) = \mathbf{v}_0(\mathbf{x}) \in L^1(\mathbb{R}^n) \cap H^2(\mathbb{R}^n)$. Set $\mathbf{w}(\mathbf{x},t) = \mathbf{u}(\mathbf{x},t) - \mathbf{v}(\mathbf{x},t)$ and $\mathbf{w}_0(\mathbf{x}) = \mathbf{u}_0(\mathbf{x}) - \mathbf{v}_0(\mathbf{x})$. Then

$$\mathbf{w}_t - \varepsilon\Delta\mathbf{w}_t - \alpha\Delta\mathbf{w} + \mathcal{D}\mathbf{w} + \mathcal{N}(\mathbf{u},\nabla\mathbf{u}) = 0 \quad \text{in } \mathbb{R}^n \times \mathbb{R}^+, \tag{9}$$

$$\mathbf{w}(\mathbf{x},0) = \mathbf{w}_0(\mathbf{x}) \quad \text{in } \mathbb{R}^n. \tag{10}$$

Lemma 2.3. (I) *Suppose that* $\mathbf{u}_0 \in L^1(\mathbb{R}^n) \cap H^2(\mathbb{R}^n)$ *and* $\mathbf{v}_0 \in L^1(\mathbb{R}^n) \cap H^2(\mathbb{R}^n)$. *Then*

$$\widehat{\mathbf{w}}(\xi,t) = \exp\left[-\frac{\alpha|\xi|^2 + \mathbf{i}\Gamma(\xi)}{1+\varepsilon|\xi|^2}t\right]\widehat{\mathbf{w}_0}(\xi) - \int_0^t \exp\left[-\frac{\alpha|\xi|^2 + \mathbf{i}\Gamma(\xi)}{1+\varepsilon|\xi|^2}(t-\tau)\right]\frac{\widehat{\mathcal{N}(\mathbf{u},\nabla\mathbf{u})}(\xi,\tau)}{1+\varepsilon|\xi|^2}\mathrm{d}\tau,$$

and

$$\begin{aligned} |\widehat{\mathbf{w}}(\xi,t)| &\leq |\widehat{\mathbf{w}_0}(\xi)| + C_0|\xi|\int_0^t (1+\tau)^{-\max\{1,n/2\}}\mathrm{d}\tau \\ &\leq \|\mathbf{w}_0\|_{L^1(\mathbb{R}^n)} + C_1|\xi|\ln(1+t) && \textit{for } n \leq 2, \\ &\leq \|\mathbf{w}_0\|_{L^1(\mathbb{R}^n)} + C_2|\xi| && \textit{for } n \geq 3, \\ \lim_{|\xi|\to 0}\frac{|\widehat{\mathbf{w}}(\xi,t)|}{|\xi|} &\leq \int_0^t \lim_{|\xi|\to 0}\frac{|\widehat{\mathcal{N}(\mathbf{u},\nabla\mathbf{u})}(\xi,\tau)|}{|\xi|}\mathrm{d}\tau, && \textit{if } \mathbf{w}_0 = \mathbf{0}, \end{aligned}$$

where the constants C_0, C_1 *and* C_2 *are independent of* (ξ,t), *also independent of* $\mathbf{u}$, $\mathbf{v}$ *and* $\mathbf{w}$.

(II) *Let*

$$\int_{\mathbb{R}^n} \mathbf{u}_0(\mathbf{x})\mathrm{d}\mathbf{x} = \mathbf{0}$$

and

$$\mathbf{u}_0(\mathbf{x}) = \left(\sum_{j=1}^n \frac{\partial\phi_{1j}}{\partial x_j}(\mathbf{x}),\ \sum_{j=1}^n \frac{\partial\phi_{2j}}{\partial x_j}(\mathbf{x}),\ \cdots,\ \sum_{j=1}^n \frac{\partial\phi_{nj}}{\partial x_j}(\mathbf{x})\right)^T,$$

where $\Phi_i \equiv (\phi_{i1}, \phi_{i2}, \cdots, \phi_{in})^T \in C^2(\mathbb{R}^n)$, *and*

$$\int_{\mathbb{R}^n} |\Phi_i(\mathbf{x})|\mathrm{d}\mathbf{x} < \infty, \qquad i = 1,2,\cdots,n.$$

Then we have

$$|\widehat{\mathbf{w}}(\xi,t)| \leq \|\mathbf{w}_0\|_{L^1(\mathbb{R}^n)} + C_3|\xi| \qquad \textit{for } n \geq 1,$$

where the constant C_3 *is independent of* (ξ,t), *independent of* $\mathbf{u}$, $\mathbf{v}$ *and* $\mathbf{w}$ *as well.*

Proof. Performing the Fourier transform to equations (9)-(10), we have

$$(1+\varepsilon|\xi|^2)\widehat{\mathbf{w}}_t(\xi,t) + \left[\alpha|\xi|^2 + \mathbf{i}\Gamma(\xi)\right]\widehat{\mathbf{w}}(\xi,t) + \widehat{\mathcal{N}(\mathbf{u},\nabla\mathbf{u})}(\xi,t) = \mathbf{0}.$$

This equation can be regarded as an ordinary differential equation, with ξ being a real vector-valued parameter. Multiplying this equation by the integrating factor $\exp\left[\frac{\alpha|\xi|^2 + \mathbf{i}\Gamma(\xi)}{1+\varepsilon|\xi|^2}t\right]$, we get

$$\frac{\mathrm{d}}{\mathrm{d}t}\left\{\exp\left[\frac{\alpha|\xi|^2 + \mathbf{i}\Gamma(\xi)}{1+\varepsilon|\xi|^2}t\right]\widehat{\mathbf{w}}(\xi,t)\right\} + \exp\left[\frac{\alpha|\xi|^2 + \mathbf{i}\Gamma(\xi)}{1+\varepsilon|\xi|^2}t\right]\frac{\widehat{\mathcal{N}(\mathbf{u},\nabla\mathbf{u})}(\xi,t)}{1+\varepsilon|\xi|^2} = \mathbf{0}.$$

Integrating in time yields

$$\exp\left[\frac{\alpha|\xi|^2 + \mathbf{i}\Gamma(\xi)}{1+\varepsilon|\xi|^2}t\right]\widehat{\mathbf{w}}(\xi,t) + \int_0^t \exp\left[\frac{\alpha|\xi|^2 + \mathbf{i}\Gamma(\xi)}{1+\varepsilon|\xi|^2}\tau\right]\frac{\widehat{\mathcal{N}(\mathbf{u},\nabla\mathbf{u})}(\xi,\tau)}{1+\varepsilon|\xi|^2}\mathrm{d}\tau = \widehat{\mathbf{w}_0}(\xi).$$

Therefore, we obtain

$$\widehat{\mathbf{w}}(\xi,t)=\exp\left[-\frac{\alpha|\xi|^2+\mathbf{i}\Gamma(\xi)}{1+\varepsilon|\xi|^2}t\right]\widehat{\mathbf{w}_0}(\xi)-\int_0^t\exp\left[-\frac{\alpha|\xi|^2+\mathbf{i}\Gamma(\xi)}{1+\varepsilon|\xi|^2}(t-\tau)\right]\frac{\widehat{\mathcal{N}(\mathbf{u},\nabla\mathbf{u})}(\xi,\tau)}{1+\varepsilon|\xi|^2}\mathrm{d}\tau.$$

If we apply the assumption (5), then we find that

$$\begin{aligned}|\widehat{\mathbf{w}}(\xi,t)| &\le \left|\exp\left[-\frac{\alpha|\xi|^2+\mathbf{i}\Gamma(\xi)}{1+\varepsilon|\xi|^2}t\right]\widehat{\mathbf{w}_0}(\xi)\right|\\
&+\left|\int_0^t\exp\left[-\frac{\alpha|\xi|^2+\mathbf{i}\Gamma(\xi)}{1+\varepsilon|\xi|^2}(t-\tau)\right]\frac{\widehat{\mathcal{N}(\mathbf{u},\nabla\mathbf{u})}(\xi,\tau)}{1+\varepsilon|\xi|^2}\mathrm{d}\tau\right|\\
&\le \exp\left(-\frac{\alpha|\xi|^2}{1+\varepsilon|\xi|^2}t\right)|\widehat{\mathbf{w}_0}(\xi)|+\int_0^t\exp\left[-\frac{\alpha|\xi|^2}{1+\varepsilon|\xi|^2}(t-\tau)\right]\left|\frac{\widehat{\mathcal{N}(\mathbf{u},\nabla\mathbf{u})}(\xi,\tau)}{1+\varepsilon|\xi|^2}\right|\mathrm{d}\tau\\
&\le |\widehat{\mathbf{w}_0}(\xi)|+\int_0^t|\widehat{\mathcal{N}(\mathbf{u},\nabla\mathbf{u})}(\xi,\tau)|\mathrm{d}\tau\\
&\le |\widehat{\mathbf{w}_0}(\xi)|+C|\xi|\int_0^t(1+\tau)^{-\max\{1,n/2\}}\mathrm{d}\tau\\
&\le \|\mathbf{w}_0\|_{L^1(\mathbb{R}^n)}+C|\xi|\int_0^t(1+\tau)^{-\max\{1,n/2\}}\mathrm{d}\tau.\end{aligned}$$

Now (I) is proved. If $\mathbf{w}_0\in L^1(\mathbb{R}^n)\cap H^2(\mathbb{R}^n)$ and

$$\int_{\mathbb{R}^n}\mathbf{u}_0(\mathbf{x})\mathrm{d}\mathbf{x}=\mathbf{0},$$

then applying assumption (6), we obtain

$$|\widehat{\mathbf{w}}(\xi,t)|\le\|\mathbf{w}_0\|_{L^1(\mathbb{R}^n)}+C|\xi|\int_0^t(1+\tau)^{-(1+n/2)}\mathrm{d}\tau.$$

Now (II) is proved. Thus, the proof of Lemma 2.3 is completed. #

Lemma 2.4. *Let* $\mathbf{u}_0=\mathbf{v}_0\in L^1(\mathbb{R}^n)\cap H^2(\mathbb{R}^n)$ *and*

$$\int_{\mathbb{R}^n}\mathbf{u}_0(\mathbf{x})\mathrm{d}\mathbf{x}\ne\mathbf{0}.$$

Then

$$(1+t)^{\max\{1,n-1/2\}}\int_{\mathbb{R}^n}\left[|\mathbf{u}(\mathbf{x},t)-\mathbf{v}(\mathbf{x},t)|^2+\varepsilon|\nabla\mathbf{u}(\mathbf{x},t)-\nabla\mathbf{v}(\mathbf{x},t)|^2\right]\mathrm{d}\mathbf{x}\le C,$$

and for any integer $m\ge1$, *we have the decay estimate*

$$(1+t)^{2m+\max\{1,n-1/2\}}\int_{\mathbb{R}^n}\left[|\Delta^m\mathbf{u}(\mathbf{x},t)-\Delta^m\mathbf{v}(\mathbf{x},t)|^2+\varepsilon|\nabla\Delta^m\mathbf{u}(\mathbf{x},t)-\nabla\Delta^m\mathbf{v}(\mathbf{x},t)|^2\right]\mathrm{d}\mathbf{x}\le C_m,$$

where $C>0$ *and* $C_m>0$ *are constants, independent of time.*

Proof. The starting point is the energy equation

$$\frac{\mathrm{d}}{\mathrm{d}t}\int_{\mathbb{R}^n}\left[|\mathbf{w}(\mathbf{x},t)|^2+\varepsilon|\nabla\mathbf{w}(\mathbf{x},t)|^2\right]\mathrm{d}\mathbf{x}$$
$$+2\alpha\int_{\mathbb{R}^n}|\nabla\mathbf{w}(\mathbf{x},t)|^2\mathrm{d}\mathbf{x}+2\int_{\mathbb{R}^n}\mathbf{w}\cdot\left[\mathcal{N}(\mathbf{u},\nabla\mathbf{u})\right]\mathrm{d}\mathbf{x}=0.$$

This equation is obtained by making the scalar product of the vector $2\mathbf{w}$ and the equation (9), integrating with respect to $\mathbf{x}$ over $\mathbb{R}^n$. Note that

$$\int_{\mathbb{R}^n} \mathbf{u} \cdot [\mathcal{N}(\mathbf{u}, \nabla \mathbf{u})] \mathrm{d}\mathbf{x} = 0, \qquad \int_{\mathbb{R}^n} \mathbf{w} \cdot (\mathcal{D}\mathbf{w}) \mathrm{d}\mathbf{x} = 0.$$

For the linear equation: $\mathbf{v}_t - \varepsilon \triangle \mathbf{v}_t + \mathcal{D}\mathbf{v} = \alpha \triangle \mathbf{v}$, if the initial data is divergence free, i.e. $\nabla \cdot \mathbf{v}_0 = 0$, then the solution of the Cauchy problem $\mathbf{v}(\mathbf{x}, 0) = \mathbf{v}_0(\mathbf{x})$ is also divergence free, that is, $\nabla \cdot \mathbf{v} = 0$, because

$$\xi \cdot \widehat{\mathbf{v}}(\xi, t) = \exp\left[-\frac{\alpha|\xi|^2 + \mathbf{i}\Gamma(\xi)}{1 + \varepsilon|\xi|^2} t\right] [\xi \cdot \widehat{\mathbf{v}}_0(\xi)] = 0,$$

for all $(\xi, t) \in \mathbb{R}^n \times \mathbb{R}^+$. This fact plays a role when studying the general model equation, which includes the Navier-Stokes equations and the Magnetohydrodynamics equations. Now the above energy equation becomes

$$\frac{\mathrm{d}}{\mathrm{d}t} \int_{\mathbb{R}^n} \left[|\mathbf{w}(\mathbf{x}, t)|^2 + \varepsilon |\nabla \mathbf{w}(\mathbf{x}, t)|^2\right] \mathrm{d}\mathbf{x} + 2\alpha \int_{\mathbb{R}^n} |\nabla \mathbf{w}(\mathbf{x}, t)|^2 \mathrm{d}\mathbf{x} = 2 \int_{\mathbb{R}^n} \mathbf{v} \cdot [\mathcal{N}(\mathbf{u}, \nabla \mathbf{u})] \mathrm{d}\mathbf{x}.$$

Applying the Plancherel's identity and then multiplying it by $(1 + t)^{2n+2}$ yield

$$\begin{aligned} &\frac{\mathrm{d}}{\mathrm{d}t}\left[(1+t)^{2n+2} \int_{\mathbb{R}^n} (1 + \varepsilon|\xi|^2)|\widehat{\mathbf{w}}(\xi, t)|^2 \mathrm{d}\xi\right] + 2\alpha(1+t)^{2n+2} \int_{\mathbb{R}^n} |\xi|^2 |\widehat{\mathbf{w}}(\xi, t)|^2 \mathrm{d}\xi \\ &= (2n+2)(1+t)^{2n+1} \int_{\mathbb{R}^n} (1 + \varepsilon|\xi|^2)|\widehat{\mathbf{w}}(\xi, t)|^2 \mathrm{d}\xi + 2(1+t)^{2n+2} \int_{\mathbb{R}^n} \overline{\widehat{\mathbf{v}}}(\xi, t) \widehat{\mathcal{N}(\mathbf{u}, \nabla \mathbf{u})}(\xi, t) \mathrm{d}\xi. \end{aligned}$$

Here we develop the modified Fourier splitting technique. Splitting the whole space $\mathbb{R}^n$ into two time-dependent subspaces: $\mathbb{R}^n = B(t) \cup B(t)^c$ and $\mathbb{R}^n = \Omega(t) \cup \Omega(t)^c$, where $\alpha(1+t) > (n+1)\varepsilon$, the small balls in $\mathbb{R}^n$ are defined by

$$\begin{aligned} B(t) &= \left\{\xi \in \mathbb{R}^n : \alpha(1+t)|\xi|^2 \le (n+1)(1 + \varepsilon|\xi|^2)\right\}, \\ \Omega(t) &= \left\{\xi \in \mathbb{R}^n : \alpha(1+t)^2|\xi|^2 \le (n+1)(1 + \varepsilon|\xi|^2)\right\}, \end{aligned}$$

the superscript c in $B(t)^c$ and $\Omega(t)^c$ denotes complementary. Before we finish the proof of Lemma 2.4, let us prove some relevant results.

An Observation: (I) *Let $r > 0$ be a given constant. Then*

$$\int_{\{\xi \in \mathbb{R}^n : |\xi| \le r\}} |\xi|^m \mathrm{d}\xi = r^{m+n} \int_{\{\eta \in \mathbb{R}^n : |\eta| \le 1\}} |\eta|^m \mathrm{d}\eta = r^{m+n} C(m, n),$$

where $\xi = r\eta$, and $C(m, n)$ is a constant, depending only on the integers m and n.

(II) *Let the constants r_1 and r_2 satisfy $0 < r_1 < r_2$ and $\left(\frac{r_1}{r_2}\right)^{m+n} \le \frac{1}{2}$. Then*

$$\int_{\{\xi \in \mathbb{R}^n : r_1 \le |\xi| \le r_2\}} |\xi|^m \mathrm{d}\xi \ge \frac{1}{2} {r_2}^{m+n} C(m, n).$$

Proof. (I) is easy to prove. Let us consider (II). Note that

$$\int_{\{\xi \in \mathbb{R}^n : r_1 \le |\xi| \le r_2\}} |\xi|^m \mathrm{d}\xi = {r_2}^{m+n} \int_{\{\eta \in \mathbb{R}^n : \frac{r_1}{r_2} \le |\eta| \le 1\}} |\eta|^m \mathrm{d}\eta.$$

We have the following estimates

$$
\begin{aligned}
\int_{\{\eta\in\mathbb{R}^n:\frac{r_1}{r_2}\le|\eta|\le1\}}|\eta|^m\mathrm{d}\eta &= \int_{\{\eta\in\mathbb{R}^n:|\eta|\le1\}}|\eta|^m\mathrm{d}\eta-\int_{\{\eta\in\mathbb{R}^n:|\eta|\le\frac{r_1}{r_2}\}}|\eta|^m\mathrm{d}\eta\\
&\ge \int_{\{\eta\in\mathbb{R}^n:|\eta|\le1\}}|\eta|^m\mathrm{d}\eta-\left(\frac{r_1}{r_2}\right)^{m+n}\int_{\{\eta\in\mathbb{R}^n:|\eta|\le1\}}|\eta|^m\mathrm{d}\eta\\
&= \left[1-\left(\frac{r_1}{r_2}\right)^{m+n}\right]\int_{\{\eta\in\mathbb{R}^n:|\eta|\le1\}}|\eta|^m\mathrm{d}\eta\\
&\ge \frac{1}{2}\int_{\{\eta\in\mathbb{R}^n:|\eta|\le1\}}|\eta|^m\mathrm{d}\eta.
\end{aligned}
$$

Therefore

$$
\int_{\{\xi\in\mathbb{R}^n:r_1\le|\xi|\le r_2\}}|\xi|^m\mathrm{d}\xi\ge\frac{1}{2}{r_2}^{m+n}\int_{\{\eta\in\mathbb{R}^n:|\eta|\le1\}}|\eta|^m\mathrm{d}\eta=\frac{1}{2}{r_2}^{m+n}C(m,n).
$$

The proof of the observation is finished. #

A fine technical lemma: *Suppose that*

$$
\lim_{t\to\infty}\left[\sup_{|\xi|\le t^{-1/4}}\frac{|\widehat{\mathbf{w}}(\xi,t)|^2}{|\xi|^2}\right]=\left\{\int_0^\infty\left[\int_{\mathbb{R}^n}|\mathbf{u}(\mathbf{x},t)|^2\mathrm{d}\mathbf{x}\right]\mathrm{d}t\right\}\mathcal{O}(1).
$$

Then there exists a sufficiently large number $\widetilde{T}$, such that for all $t>\widetilde{T}$, we have

$$
\begin{aligned}
&(2n+2)(1+t)^{2n+1}\int_{\mathbb{R}^n}(1+\varepsilon|\xi|^2)|\widehat{\mathbf{w}}(\xi,t)|^2\mathrm{d}\xi\\
&\le 2\alpha(1+t)^{2n+2}\int_{\mathbb{R}^n}|\xi|^2|\widehat{\mathbf{w}}(\xi,t)|^2\mathrm{d}\xi+C(1+t)^{n-1}[\ln(1+t)]^{2-n_*},
\end{aligned}
$$

for some positive, time-independent constant C, where $n_*=0$ if $1\le n\le 2$, and $n_*=2$ if $n\ge 3$.

Proof. Let $t\gg1$ be sufficiently large. Then by using the Observation, we have the following estimates

$$
\begin{aligned}
& \mu(t) \\
\equiv\ & \int_{\Omega(t)^c} \left[2\alpha(1+t)^{2n+2}|\xi|^2 - (2n+2)(1+t)^{2n+1}(1+\varepsilon|\xi|^2)\right] |\widehat{\mathbf{w}}(\xi,t)|^2 \mathrm{d}\xi \\
=\ & \int_{B(t)^c} \left[2\alpha(1+t)^{2n+2}|\xi|^2 - (2n+2)(1+t)^{2n+1}(1+\varepsilon|\xi|^2)\right] |\widehat{\mathbf{w}}(\xi,t)|^2 \mathrm{d}\xi \\
+\ & \int_{B(t)\cap\Omega(t)^c} \left[2\alpha(1+t)^{2n+2}|\xi|^2 - (2n+2)(1+t)^{2n+1}(1+\varepsilon|\xi|^2)\right] |\widehat{\mathbf{w}}(\xi,t)|^2 \mathrm{d}\xi \\
\geq\ & \int_{B(t)^c\cap\{\xi\in\mathbb{R}^n:|\xi|\leq t^{-1/4}\}} \left[2\alpha(1+t)^{2n+2}|\xi|^2 - (2n+2)(1+t)^{2n+1}(1+\varepsilon|\xi|^2)\right] |\widehat{\mathbf{w}}(\xi,t)|^2 \mathrm{d}\xi \\
+\ & \int_{B(t)\cap\Omega(t)^c} \left[2\alpha(1+t)^{2n+2}|\xi|^2 - (2n+2)(1+t)^{2n+1}(1+\varepsilon|\xi|^2)\right] |\widehat{\mathbf{w}}(\xi,t)|^2 \mathrm{d}\xi \\
\geq\ & \left\{\int_{B(t)^c\cap\{\xi\in\mathbb{R}^n:|\xi|\leq t^{-1/4}\}} \left[2\alpha(1+t)^{2n+2}|\xi|^2 - (2n+2)(1+t)^{2n+1}(1+\varepsilon|\xi|^2)\right] |\xi|^2 \mathrm{d}\xi\right\} \\
& \times \left\{\min_{\xi\in B(t)^c\cap\{\xi\in\mathbb{R}^n:|\xi|\leq t^{-1/4}\}} \frac{|\widehat{\mathbf{w}}(\xi,t)|^2}{|\xi|^2}\right\} \\
+\ & \left\{\int_{B(t)\cap\Omega(t)^c} \left[2\alpha(1+t)^{2n+2}|\xi|^2 - (2n+2)(1+t)^{2n+1}(1+\varepsilon|\xi|^2)\right] |\xi|^2 \mathrm{d}\xi\right\} \\
& \times \left\{\max_{\xi\in B(t)\cap\Omega(t)^c} \frac{|\widehat{\mathbf{w}}(\xi,t)|^2}{|\xi|^2}\right\} \\
\geq\ & (1+t)^{7n/4}\left\{[\alpha(1+t)-(n+1)\varepsilon]C(4,n)-(n+1)C(2,n)]\right\} \\
\times\ & \left\{\min_{\xi\in B(t)^c\cap\{\xi\in\mathbb{R}^n:|\xi|\leq t^{-1/4}\}} \frac{|\widehat{\mathbf{w}}(\xi,t)|^2}{|\xi|^2}\right\} \\
+\ & 2(1+t)^{3n/2}\left\{[\alpha(1+t)-(n+1)\varepsilon]C(4,n)-(n+1)C(2,n)]\right\} \\
\times\ & \left[\max_{\xi\in B(t)\cap\Omega(t)^c} \frac{|\widehat{\mathbf{w}}(\xi,t)|^2}{|\xi|^2}\right].
\end{aligned}
$$

Note that

$$
\begin{aligned}
& \lim_{\xi\to 0} \frac{|\widehat{\mathbf{w}}(\xi,t)|^2}{|\xi|^2} \leq C\left|\int_0^t \left[\int_{\mathbb{R}} |\mathbf{u}(x,\tau)|^{2+\kappa}\mathbf{dx}\right]\mathrm{d}\tau\right|^2, \qquad \forall t>0, \\
& \lim_{t\to\infty}\left[\max_{\xi\in B(t)\cap\Omega(t)^c} \frac{|\widehat{\mathbf{w}}(\xi,t)|^2}{|\xi|^2}\right] \leq C\left|\int_0^\infty \left[\int_{\mathbb{R}} |\mathbf{u}(x,t)|^{2+\kappa}\mathbf{dx}\right]\mathrm{d}t\right|^2, \\
& \lim_{t\to\infty}\left[\min_{\xi\in B(t)^c\cap\{\xi\in\mathbb{R}^n:|\xi|\leq t^{-1/4}\}} \frac{|\widehat{\mathbf{w}}(\xi,t)|^2}{|\xi|^2}\right] \leq C\left|\int_0^\infty \left[\int_{\mathbb{R}} |\mathbf{u}(x,t)|^{2+\kappa}\mathbf{dx}\right]\mathrm{d}t\right|^2,
\end{aligned}
$$

where $\kappa = 1$ if $n = 1$, and $\kappa = 0$ if $n \geq 2$; $C > 0$ is an absolute constant.

For an appropriate positive number $\widetilde{\varepsilon} > 0$, there exists a sufficiently large number $\widetilde{T}(\widetilde{\varepsilon}) \gg 1$, such that

$$
-\widetilde{\varepsilon} < \max_{\xi\in B(t)\cap\Omega(t)^c} \frac{|\widehat{\mathbf{w}}(\xi,t)|^2}{|\xi|^2} - \min_{\xi\in B(t)^c\cap\{\xi\in\mathbb{R}^n:|\xi|\leq t^{-1/4}\}} \frac{|\widehat{\mathbf{w}}(\xi,t)|^2}{|\xi|^2} < \widetilde{\varepsilon},
$$

whenever $t > \widetilde{T}(\widetilde{\varepsilon})$. Overall, we find that $\mu(t) \geq 0$ if $t > \widetilde{T}(\widetilde{\varepsilon})$. Now we get the following estimates

$$\begin{aligned}
&2\alpha(1+t)^{2n+2}\int_{\mathbb{R}^n}|\xi|^2|\widehat{\mathbf{w}}(\xi,t)|^2\mathrm{d}\xi\\
&=2\alpha(1+t)^{2n+2}\int_{\Omega(t)}|\xi|^2|\widehat{\mathbf{w}}(\xi,t)|^2\mathrm{d}\xi+2\alpha(1+t)^{2n+2}\int_{\Omega(t)^c}|\xi|^2|\widehat{\mathbf{w}}(\xi,t)|^2\mathrm{d}\xi\\
&\geq (2n+2)(1+t)^{2n+1}\int_{\Omega(t)^c}(1+\varepsilon|\xi|^2)|\widehat{\mathbf{w}}(\xi,t)|^2\mathrm{d}\xi\\
&=(2n+2)(1+t)^{2n+1}\int_{\mathbb{R}^n}(1+\varepsilon|\xi|^2)|\widehat{\mathbf{w}}(\xi,t)|^2\mathrm{d}\xi-(2n+2)(1+t)^{2n+1}\int_{\Omega(t)}(1+\varepsilon|\xi|^2)|\widehat{\mathbf{w}}(\xi,t)|^2\mathrm{d}\xi.
\end{aligned}$$

By using Lemma 2.3, it is easy to get the following estimates

$$\begin{aligned}
&(2n+2)(1+t)^{2n+1}\int_{\Omega(t)}(1+\varepsilon|\xi|^2)|\widehat{\mathbf{w}}(\xi,t)|^2\mathrm{d}\xi\\
\leq\ & C(1+t)^{n-1}[\ln(1+t)]^2, \qquad \text{if } n\leq 2,\\
\leq\ & C(1+t)^{n-1}, \qquad \text{if } n\geq 3.
\end{aligned}$$

The proof of the fine technical lemma is finished. #

Now we come back to the proof of Lemma 2.4. Assumption (4) and the decay estimate of Lemma 2.1 imply that

$$\begin{aligned}
&2(1+t)^{2n+2}\left|\int_{\mathbb{R}^n}\overline{\widehat{\mathbf{v}}}(\xi,t)\widehat{\mathcal{N}(\mathbf{u},\nabla\mathbf{u})}(\xi,t)\mathrm{d}\xi\right|\\
\leq\ & C(1+t)^{2n+2}\|\nabla\mathbf{v}(\cdot,t)\|_{L^\infty(\mathbb{R}^n)}(1+t)^{-\max\{1,n/2\}}\leq C(1+t)^{(3n+1)/2}, \text{ if } n\leq 2,\\
\leq\ & C(1+t)^{n+3/2}, \text{ if } n\geq 3.
\end{aligned}$$

From the energy inequality

$$\begin{aligned}
&\frac{\mathrm{d}}{\mathrm{d}t}\left[(1+t)^{2n+2}\int_{\mathbb{R}^n}(1+\varepsilon|\xi|^2)|\widehat{\mathbf{w}}(\xi,t)|^2\mathrm{d}\xi\right]+2\alpha(1+t)^{2n+2}\int_{\mathbb{R}^n}|\xi|^2|\widehat{\mathbf{w}}(\xi,t)|^2\mathrm{d}\xi\\
\leq\ & (2n+2)(1+t)^{2n+1}\int_{\mathbb{R}^n}(1+\varepsilon|\xi|^2)|\widehat{\mathbf{w}}(\xi,t)|^2\mathrm{d}\xi+2(1+t)^{2n+2}\left|\int_{\mathbb{R}^n}\overline{\widehat{\mathbf{v}}}(\xi,t)\widehat{\mathcal{N}(\mathbf{u},\nabla\mathbf{u})}(\xi,t)\mathrm{d}\xi\right|,
\end{aligned}$$

and the fine technical lemma, we have

$$\begin{aligned}
&\frac{\mathrm{d}}{\mathrm{d}t}\left[(1+t)^{2n+2}\int_{\mathbb{R}^n}(1+\varepsilon|\xi|^2)|\widehat{\mathbf{w}}(\xi,t)|^2\mathrm{d}\xi\right]\\
\leq\ & C(1+t)^{(3n+1)/2}+C(1+t)^{n-1}[\ln(1+t)]^2 \text{ for } n\leq 2,\\
\leq\ & C(1+t)^{n+3/2}+C(1+t)^{n-1} \qquad \text{for } n\geq 3.
\end{aligned}$$

Integrating this above differential inequality in time yields the decay estimate

$$\begin{aligned}
&(1+t)^{2n+2}\int_{\mathbb{R}^n}(1+\varepsilon|\xi|^2)|\widehat{\mathbf{w}}(\xi,t)|^2\mathrm{d}\xi\\
\leq\ & \int_{\mathbb{R}^n}(1+\varepsilon|\xi|^2)|\widehat{\mathbf{w}_0}(\xi)|^2\mathrm{d}\xi+C(1+t)^{(3n+3)/2}+C(1+t)^n[\ln(1+t)]^2 \text{ for } n\leq 2,\\
\leq\ & \int_{\mathbb{R}^n}(1+\varepsilon|\xi|^2)|\widehat{\mathbf{w}_0}(\xi)|^2\mathrm{d}\xi+C(1+t)^{n+5/2}+C(1+t)^n, \text{ for } n\geq 3.
\end{aligned}$$

Finally, noting that $\mathbf{u}_0 = \mathbf{v}_0$, we get the estimate

$$(1+t)^{2n+2} \int_{\mathbb{R}^n} \left[|\mathbf{u}(\mathbf{x},t) - \mathbf{v}(\mathbf{x},t)|^2 + \varepsilon|\nabla\mathbf{u}(\mathbf{x},t) - \nabla\mathbf{v}(\mathbf{x},t)|^2\right] \mathrm{d}\mathbf{x} \le C(1+t)^{(3n+3)/2},$$

if $n \le 2$, where C is a positive constant independent of t, and

$$(1+t)^{2n+2} \int_{\mathbb{R}^n} \left[|\mathbf{u}(\mathbf{x},t) - \mathbf{v}(\mathbf{x},t)|^2 + \varepsilon|\nabla\mathbf{u}(\mathbf{x},t) - \nabla\mathbf{v}(\mathbf{x},t)|^2\right] \mathrm{d}\mathbf{x} \le C(1+t)^{n+5/2},$$

if $n \ge 3$. The second estimate can be proved very similarly. Let $m \ge 1$ be a fixed integer. From equation (9), we may get the integral equation

$$2\int_{\mathbb{R}^n} \triangle^{2m}\mathbf{w} \cdot [\mathbf{w}_t - \varepsilon\triangle\mathbf{w}_t - \alpha\triangle\mathbf{w} + \mathcal{D}\mathbf{w} + \mathcal{N}(\mathbf{u}, \nabla\mathbf{u})]\, \mathrm{d}\mathbf{x} = 0.$$

Using integration by parts, we find that

$$\begin{aligned}&\frac{\mathrm{d}}{\mathrm{d}t}\int_{\mathbb{R}^n} \left[|\triangle^m\mathbf{w}(\mathbf{x},t)|^2 + \varepsilon|\nabla\triangle^m\mathbf{w}(\mathbf{x},t)|^2\right] \mathrm{d}\mathbf{x} + 2\alpha\int_{\mathbb{R}^n} |\nabla\triangle^m\mathbf{w}(\mathbf{x},t)|^2\mathrm{d}\mathbf{x}\\ &= 2\int_{\mathbb{R}^n} \nabla\triangle^m\mathbf{w} \cdot \nabla\triangle^{m-1}\mathcal{N}(\mathbf{u}, \nabla\mathbf{u})\mathrm{d}\mathbf{x},\end{aligned}$$

where

$$\int_{\mathbb{R}^n} \triangle^{2m}\mathbf{w} \cdot \mathcal{D}\mathbf{w}\mathrm{d}\mathbf{x} = 0.$$

Let $\mathbf{f} = (f_1, f_2, \cdots, f_n)^T$ and $\mathbf{g} = (g_1, g_2, \cdots, g_n)^T \in C^1(\mathbb{R}^n, \mathbb{R}^n)$. We define the scalar product $\nabla\mathbf{f} \cdot \nabla\mathbf{g}$ in this way:

$$\nabla\mathbf{f} \cdot \nabla\mathbf{g} = \sum_{i=1}^{n}\sum_{j=1}^{n} \frac{\partial f_i}{\partial x_j}\frac{\partial g_i}{\partial x_j}.$$

Obviously

$$2\left|\int_{\mathbb{R}^n} \nabla\triangle^m\mathbf{w} \cdot \nabla\triangle^{m-1}\mathcal{N}(\mathbf{u}, \nabla\mathbf{u})\mathrm{d}\mathbf{x}\right| \le \alpha\int_{\mathbb{R}^n} |\nabla\triangle^m\mathbf{w}|^2\mathrm{d}\mathbf{x} + \frac{1}{\alpha}\int_{\mathbb{R}^n} |\nabla\triangle^{m-1}\mathcal{N}(\mathbf{u}, \nabla\mathbf{u})|^2\mathrm{d}\mathbf{x}.$$

Therefore

$$\begin{aligned}\frac{\mathrm{d}}{\mathrm{d}t}\int_{\mathbb{R}^n} \left[|\triangle^m\mathbf{w}(\mathbf{x},t)|^2 + \varepsilon|\nabla\triangle^m\mathbf{w}(\mathbf{x},t)|^2\right] \mathrm{d}\mathbf{x} + \alpha\int_{\mathbb{R}^n} |\nabla\triangle^m\mathbf{w}(\mathbf{x},t)|^2\mathrm{d}\mathbf{x}&\\ \le \frac{1}{\alpha}\int_{\mathbb{R}^n} |\nabla\triangle^{m-1}\mathcal{N}(\mathbf{u}, \nabla\mathbf{u})|^2\mathrm{d}\mathbf{x}.&\end{aligned}$$

We may have to apply Newton-Leibniz's differentiation rule, Hölder's inequality, Gagliardo-Nirenberg's interpolation inequality, and decay results of Lemma 2.1 to control the integral

$$\frac{1}{\alpha}\int_{\mathbb{R}^n} |\nabla\triangle^{m-1}\mathcal{N}(\mathbf{u}, \nabla\mathbf{u})|^2\mathrm{d}\mathbf{x}$$

in terms of

$$\int_{\mathbb{R}^n} |\mathbf{u}(\mathbf{x},t)|^{2\rho} d\mathbf{x} \qquad \text{and} \qquad \int_{\mathbb{R}^n} |\nabla \triangle^m \mathbf{u}(\mathbf{x},t)|^2 d\mathbf{x}.$$

Let us just pick up a typical term and give the details. Suppose that M and N are integers, satisfying the conditions $0 \le M, N \le 2m$ and $M + N = 2m$. Without loss of generality, we may assume that $0 \le M \le m$ and $m \le N \le 2m$. Then we have the following estimates

$$\begin{aligned}
&\int_{\mathbb{R}^n} \left[\sum_{\alpha_1+\alpha_2+\cdots+\alpha_n=M} \left| \frac{\partial^{\alpha_1+\alpha_2+\cdots+\alpha_n} \mathbf{u}}{\partial x_1^{\alpha_1} \partial x_2^{\alpha_2} \cdots \partial x_n^{\alpha_n}}(\mathbf{x},t) \right|^2 \right] \left[\sum_{\beta_1+\beta_2+\cdots+\beta_n=N} \left| \frac{\partial^{\beta_1+\beta_2+\cdots+\beta_n} \mathbf{u}}{\partial x_1^{\beta_1} \partial x_2^{\beta_2} \cdots \partial x_n^{\beta_n}}(\mathbf{x},t) \right|^2 \right] d\mathbf{x} \\
&\le \left\{ \int_{\mathbb{R}^n} \left[\sum_{\alpha_1+\alpha_2+\cdots+\alpha_n=M} \left| \frac{\partial^{\alpha_1+\alpha_2+\cdots+\alpha_n} \mathbf{u}}{\partial x_1^{\alpha_1} \partial x_2^{\alpha_2} \cdots \partial x_n^{\alpha_n}}(\mathbf{x},t) \right|^2 \right]^\rho d\mathbf{x} \right\}^{1/\rho} \\
&\times \left\{ \int_{\mathbb{R}^n} \left[\sum_{\beta_1+\beta_2+\cdots+\beta_n=N} \left| \frac{\partial^{\beta_1+\beta_2+\cdots+\beta_n} \mathbf{u}}{\partial x_1^{\beta_1} \partial x_2^{\beta_2} \cdots \partial x_n^{\beta_n}}(\mathbf{x},t) \right|^2 \right]^\sigma d\mathbf{x} \right\}^{1/\sigma} \\
&\le C \left\{ \left[\int_{\mathbb{R}^n} |\mathbf{u}(\mathbf{x},t)|^{2\rho'} d\mathbf{x} \right]^{(1-\theta)/\rho'} \left[\int_{\mathbb{R}^n} |\nabla \triangle^m \mathbf{u}(\mathbf{x},t)|^2 d\mathbf{x} \right]^\theta \right\}^{1/\rho} \\
&\times \left\{ \left[\int_{\mathbb{R}^n} |\mathbf{u}(\mathbf{x},t)|^{2\sigma'} d\mathbf{x} \right]^{(1-\Theta)/\sigma'} \left[\int_{\mathbb{R}^n} |\nabla \triangle^m \mathbf{u}(\mathbf{x},t)|^2 d\mathbf{x} \right]^\Theta \right\}^{1/\sigma},
\end{aligned}$$

where

$$\begin{aligned}
\frac{1}{\rho} + \frac{1}{\sigma} &= 1, \\
M &> \frac{n}{2}\left(\frac{1}{\rho} - \frac{1}{\rho'}\right), \\
2m + 1 - M &> \frac{n}{2}\left(1 - \frac{1}{\rho}\right), \\
\theta &= \frac{M - \frac{n}{2}\left(\frac{1}{\rho} - \frac{1}{\rho'}\right)}{2m + 1 - \frac{n}{2}\left(1 - \frac{1}{\rho'}\right)}, \\
N &> \frac{n}{2}\left(\frac{1}{\sigma} - \frac{1}{\sigma'}\right), \\
2m + 1 - N &> \frac{n}{2}\left(1 - \frac{1}{\sigma}\right), \\
\Theta &= \frac{N - \frac{n}{2}\left(\frac{1}{\sigma} - \frac{1}{\sigma'}\right)}{2m + 1 - \frac{n}{2}\left(1 - \frac{1}{\sigma'}\right)}.
\end{aligned}$$

Note that

$$\frac{\theta}{\rho} + \frac{\Theta}{\sigma} < 1.$$

Following Michael Wiegner [82], we have the following decay estimates

$$\begin{aligned}
(1+t)^{(2\rho'-1)n/2} \int_{\mathbb{R}^n} |\mathbf{u}(\mathbf{x},t)|^{2\rho'} d\mathbf{x} &\le C(\rho), \\
(1+t)^{(2\sigma'-1)n/2} \int_{\mathbb{R}^n} |\mathbf{u}(\mathbf{x},t)|^{2\sigma'} d\mathbf{x} &\le C(\sigma).
\end{aligned}$$

Applying the Plancherel's identity and multiplying the resulting equation by $(1+t)^{2m+2n+2}$, we have

$$\frac{\mathrm{d}}{\mathrm{d}t}\left\{(1+t)^{2m+2n+2}\int_{\mathbb{R}^n}\left[|\xi|^{4m}|\widehat{\mathbf{w}}(\xi,t)|^2+\varepsilon|\xi|^{4m+2}|\widehat{\mathbf{w}}(\xi,t)|^2\right]\mathrm{d}\xi\right\}$$
$$+\alpha(1+t)^{2m+2n+2}\int_{\mathbb{R}^n}|\xi|^{4m+2}|\widehat{\mathbf{w}}(\xi,t)|^2\mathrm{d}\xi$$
$$\leq(2m+2n+2)(1+t)^{2m+2n+1}\int_{\mathbb{R}^n}\left[|\xi|^{4m}|\widehat{\mathbf{w}}(\xi,t)|^2+\varepsilon|\xi|^{4m+2}|\widehat{\mathbf{w}}(\xi,t)|^2\right]\mathrm{d}\xi$$
$$+\frac{(1+t)^{2m+2n+2}}{\alpha}\int_{\mathbb{R}^n}|\xi|^{4m-2}|\widehat{\mathcal{N}(\mathbf{u},\nabla\mathbf{u})}(\xi,t)|^2\mathrm{d}\xi.$$

Other details are omitted. #

To improve the decay results for all dimension $n\geq 1$, we may repeat the same procedure as in Lemma 2.4 and using the additional condition

$$\mathbf{u}_0(\mathbf{x})=\left(\sum_{j=1}^n\frac{\partial\phi_{1j}}{\partial x_j}(\mathbf{x}),\ \sum_{j=1}^n\frac{\partial\phi_{2j}}{\partial x_j}(\mathbf{x}),\ \cdots,\ \sum_{j=1}^n\frac{\partial\phi_{nj}}{\partial x_j}(\mathbf{x})\right)^T.$$

We decompose the entire space as before: $\mathbb{R}^n=B(t)\cup B(t)^c$ and $\mathbb{R}^n=\Omega(t)\cup\Omega(t)^c$, where $B(t)$ and $\Omega(t)$ are small balls, given by

$$B(t)\equiv\left\{\xi\in\mathbb{R}^n:\alpha(1+t)|\xi|^2\leq(n+1)(1+\varepsilon|\xi|^2)\right\},$$
$$\Omega(t)\equiv\left\{\xi\in\mathbb{R}^n:\alpha(1+t)^2|\xi|^2\leq(n+1)(1+\varepsilon|\xi|^2)\right\}.$$

Lemma 2.5. *Let* $\mathbf{u}_0=\mathbf{v}_0\in L^1(\mathbb{R}^n)\cap H^2(\mathbb{R}^n)$ *and*

$$\int_{\mathbb{R}^n}\mathbf{u}_0(\mathbf{x})\mathrm{d}\mathbf{x}=\mathbf{0},$$

$$\mathbf{u}_0(\mathbf{x})=\left(\sum_{j=1}^n\frac{\partial\phi_{1j}}{\partial x_j}(\mathbf{x}),\ \sum_{j=1}^n\frac{\partial\phi_{2j}}{\partial x_j}(\mathbf{x}),\ \cdots,\ \sum_{j=1}^n\frac{\partial\phi_{nj}}{\partial x_j}(\mathbf{x})\right)^T.$$

Suppose that $\Phi_i\equiv(\phi_{i1},\phi_{i2},\cdots,\phi_{in})^T\in C^1(\mathbb{R}^n)\cap L^1(\mathbb{R}^n)$, $i=1,2,\cdots,n$. *Then*

$$(1+t)^{n+1}\int_{\mathbb{R}^n}\left[|\mathbf{u}(\mathbf{x},t)-\mathbf{v}(\mathbf{x},t)|^2+\varepsilon|\nabla\mathbf{u}(\mathbf{x},t)-\nabla\mathbf{v}(\mathbf{x},t)|^2\right]\mathrm{d}\mathbf{x}\leq C,$$

and

$$(1+t)^{2m+n+1}\int_{\mathbb{R}^n}\left[|\Delta^m\mathbf{u}(\mathbf{x},t)-\Delta^m\mathbf{v}(\mathbf{x},t)|^2+\varepsilon|\nabla\Delta^m\mathbf{u}(\mathbf{x},t)-\nabla\Delta^m\mathbf{v}(\mathbf{x},t)|^2\right]\mathrm{d}\mathbf{x}\leq C_m,$$

where $C>0$ *and* $C_m>0$ *are constants, independent of time.*

Proof. It is easy to see that

$$2\alpha(1+t)^{2n+2}\int_{B(t)}|\xi|^2|\widehat{\mathbf{w}}(\xi,t)|^2\mathrm{d}\xi\ \leq\ (2n+2)(1+t)^{2n+1}\int_{B(t)}(1+\varepsilon|\xi|^2)|\widehat{\mathbf{w}}(\xi,t)|^2\mathrm{d}\xi,$$
$$2\alpha(1+t)^{2n+2}\int_{B(t)^c}|\xi|^2|\widehat{\mathbf{w}}(\xi,t)|^2\mathrm{d}\xi\ \geq\ (2n+2)(1+t)^{2n+1}\int_{B(t)^c}(1+\varepsilon|\xi|^2)|\widehat{\mathbf{w}}(\xi,t)|^2\mathrm{d}\xi.$$

Define the function $\mu = \mu(t)$ as before and use the results in the observation and the fine technical lemma. Overall, we find that $\mu(t) \geq 0$ if $t > \widetilde{T}(\widetilde{\varepsilon})$. Therefore

$$\begin{aligned}
& 2\alpha(1+t)^{2n+2}\int_{\mathbb{R}^n}|\xi|^2|\widehat{\mathbf{w}}(\xi,t)|^2\mathrm{d}\xi \\
= \;& 2\alpha(1+t)^{2n+2}\int_{\Omega(t)}|\xi|^2|\widehat{\mathbf{w}}(\xi,t)|^2\mathrm{d}\xi + 2\alpha(1+t)^{2n+2}\int_{\Omega(t)^c}|\xi|^2|\widehat{\mathbf{w}}(\xi,t)|^2\mathrm{d}\xi \\
\geq \;& (2n+2)(1+t)^{2n+1}\int_{\Omega(t)^c}(1+\varepsilon|\xi|^2)|\widehat{\mathbf{w}}(\xi,t)|^2\mathrm{d}\xi \\
= \;& (2n+2)(1+t)^{2n+1}\int_{\mathbb{R}^n}(1+\varepsilon|\xi|^2)|\widehat{\mathbf{w}}(\xi,t)|^2\mathrm{d}\xi \\
- \;& (2n+2)(1+t)^{2n+1}\int_{\Omega(t)}(1+\varepsilon|\xi|^2)|\widehat{\mathbf{w}}(\xi,t)|^2\mathrm{d}\xi.
\end{aligned}$$

Then we have

$$\begin{aligned}
& \frac{\mathrm{d}}{\mathrm{d}t}\left[(1+t)^{2n+2}\int_{\mathbb{R}^n}(1+\varepsilon|\xi|^2)|\widehat{\mathbf{w}}(\xi,t)|^2\mathrm{d}\xi\right] \\
\leq \;& (2n+2)(1+t)^{2n+1}\int_{\Omega(t)}(1+\varepsilon|\xi|^2)|\widehat{\mathbf{w}}(\xi,t)|^2\mathrm{d}\xi + 2(1+t)^{2n+2}\left|\int_{\mathbb{R}^n}\overline{\widehat{\mathbf{v}}}(\xi,t)\widehat{\mathcal{N}(\mathbf{u},\nabla\mathbf{u})}(\xi,t)\mathrm{d}\xi\right|.
\end{aligned}$$

By Lemma 2.3, it is easy to get the following estimate

$$\begin{aligned}
& (2n+2)(1+t)^{2n+1}\int_{\Omega(t)}(1+\varepsilon|\xi|^2)|\widehat{\mathbf{w}}(\xi,t)|^2\mathrm{d}\xi \\
\leq \;& C(1+t)^{n-1}, \qquad \text{for all dimension } n \geq 1.
\end{aligned}$$

Moreover

$$\begin{aligned}
& 2(1+t)^{2n+2}\left|\int_{\mathbb{R}^n}\overline{\widehat{\mathbf{v}}}(\xi,t)\widehat{\mathcal{N}(\mathbf{u},\nabla\mathbf{u})}(\xi,t)\mathrm{d}\xi\right| \\
& \leq C(1+t)^{2n+2}\|\nabla\mathbf{v}(\cdot,t)\|_{L^\infty(\mathbb{R}^n)}(1+t)^{-(1+n/2)} \leq C(1+t)^n.
\end{aligned}$$

Therefore, we have

$$\frac{\mathrm{d}}{\mathrm{d}t}\left\{(1+t)^{2n+2}\int_{\mathbb{R}^n}(1+\varepsilon|\xi|^2)|\widehat{\mathbf{w}}(\xi,t)|^2\mathrm{d}\xi\right\} \leq C(1+t)^{n-1} + C(1+t)^n.$$

Integrating in time, we find

$$(1+t)^{2n+2}\int_{\mathbb{R}^n}(1+\varepsilon|\xi|^2)|\widehat{\mathbf{w}}(\xi,t)|^2\mathrm{d}\xi \leq \int_{\mathbb{R}^n}(1+\varepsilon|\xi|^2)|\widehat{\mathbf{w}_0}(\xi,t)|^2\mathrm{d}\xi + C(1+t)^{n+1}.$$

Finally, using $\mathbf{w}_0 = \mathbf{0}$, we obtain

$$(1+t)^{1+n/2}\int_{\mathbb{R}^n}\left[|\mathbf{u}(\mathbf{x},t)-\mathbf{v}(\mathbf{x},t)|^2 + \varepsilon|\nabla\mathbf{u}(\mathbf{x},t)-\nabla\mathbf{v}(\mathbf{x},t)|^2\right]\mathrm{d}\mathbf{x} \leq \frac{C}{(1+t)^{n/2}}.$$

The decay estimates for the case $m \geq 1$ can be proved very similarly. The proof of Lemma 2.5 is finished. #

Lemma 2.6. **(I)** *Suppose that the initial functions satisfy* $\mathbf{u}_0 = \mathbf{v}_0 \in L^1(\mathbb{R}^n) \cap H^2(\mathbb{R}^n)$ *and*

$$\int_{\mathbb{R}^n} \mathbf{u}_0(\mathbf{x})\mathrm{d}\mathbf{x} \neq \mathbf{0}.$$

Then

$$\lim_{t\to\infty} \left\{(1+t)^{n/2} \int_{\mathbb{R}^n} \left[|\mathbf{u}(\mathbf{x},t)|^2 + \varepsilon|\nabla \mathbf{u}(\mathbf{x},t)|^2\right] \mathrm{d}\mathbf{x}\right\}$$
$$= \lim_{t\to\infty} \left\{(1+t)^{n/2} \int_{\mathbb{R}^n} \left[|\mathbf{v}(\mathbf{x},t)|^2 + \varepsilon|\nabla \mathbf{v}(\mathbf{x},t)|^2\right] \mathrm{d}\mathbf{x}\right\}$$

exists, and

$$\lim_{t\to\infty} \left\{(1+t)^{2m+n/2} \int_{\mathbb{R}^n} \left[|\triangle^m \mathbf{u}(\mathbf{x},t)|^2 + \varepsilon|\nabla\triangle^m \mathbf{u}(\mathbf{x},t)|^2\right] \mathrm{d}\mathbf{x}\right\}$$
$$= \lim_{t\to\infty} \left\{(1+t)^{2m+n/2} \int_{\mathbb{R}^n} \left[|\triangle^m \mathbf{v}(\mathbf{x},t)|^2 + \varepsilon|\nabla\triangle^m \mathbf{v}(\mathbf{x},t)|^2\right] \mathrm{d}\mathbf{x}\right\}$$

exists, where $m \geq 1$ *is any integer.*
(II) *Suppose that the initial functions satisfy* $\mathbf{u}_0 = \mathbf{v}_0 \in L^1(\mathbb{R}^n) \cap H^2(\mathbb{R}^n)$ *and*

$$\int_{\mathbb{R}^n} \mathbf{u}_0(\mathbf{x})\mathrm{d}\mathbf{x} = \mathbf{0}.$$

Let the scalar functions $\phi_{ij} \in C^2(\mathbb{R}^n) \cap L^1(\mathbb{R}^n)$, $i, j \in \{1, 2, \cdots, n\}$. *Define the real vector-valued functions* $\Phi_i = (\phi_{i1}, \phi_{i2}, \cdots, \phi_{in})^T$. *Let the initial data*

$$\mathbf{u}_0(\mathbf{x}) = \left(\sum_{j=1}^{n} \frac{\partial \phi_{1j}}{\partial x_j}(\mathbf{x}),\ \sum_{j=1}^{n} \frac{\partial \phi_{2j}}{\partial x_j}(\mathbf{x}),\ \cdots,\ \sum_{j=1}^{n} \frac{\partial \phi_{nj}}{\partial x_j}(\mathbf{x})\right)^T.$$

Then

$$\lim_{t\to\infty} \left\{(1+t)^{1+n/2} \int_{\mathbb{R}^n} \left[|\mathbf{u}(\mathbf{x},t)|^2 + \varepsilon|\nabla \mathbf{u}(\mathbf{x},t)|^2\right] \mathrm{d}\mathbf{x}\right\}$$
$$= \lim_{t\to\infty} \left\{(1+t)^{1+n/2} \int_{\mathbb{R}^n} \left[|\mathbf{v}(\mathbf{x},t)|^2 + \varepsilon|\nabla \mathbf{v}(\mathbf{x},t)|^2\right] \mathrm{d}\mathbf{x}\right\}$$

exists, and

$$\lim_{t\to\infty} \left\{(1+t)^{2m+1+n/2} \int_{\mathbb{R}^n} \left[|\triangle^m \mathbf{u}(\mathbf{x},t)|^2 + \varepsilon|\nabla\triangle^m \mathbf{u}(\mathbf{x},t)|^2\right] \mathrm{d}\mathbf{x}\right\}$$
$$= \lim_{t\to\infty} \left\{(1+t)^{2m+1+n/2} \int_{\mathbb{R}^n} \left[|\triangle^m \mathbf{v}(\mathbf{x},t)|^2 + \varepsilon|\nabla\triangle^m \mathbf{v}(\mathbf{x},t)|^2\right] \mathrm{d}\mathbf{x}\right\}$$

exists, where $m \geq 1$ *is any integer.*

Proof. Let $\mathbf{u} = \mathbf{u}(\mathbf{x}, t, \mathbf{u}_0)$ and $\mathbf{v} = \mathbf{v}(\mathbf{x}, t, \mathbf{u}_0)$ be the solutions of the nonlinear differential equation (1) and the linear differential equation (7), respectively, with the same initial data

$\mathbf{u}(\mathbf{x}, 0, \mathbf{u}_0) = \mathbf{v}(\mathbf{x}, 0, \mathbf{u}_0) = \mathbf{u}_0(\mathbf{x})$. First of all, let us consider the case $\varepsilon = 0$. Suppose that the integer $m \geq 0$. By using triangle inequality, we have the upper bound estimate

$$\begin{aligned} & (1+t)^{m+n/4}\|\triangle^m \mathbf{u}(\cdot,t)\|_{L^2(\mathbb{R}^n)} \\ = & (1+t)^{m+n/4}\|\triangle^m \mathbf{v}(\cdot,t) + \triangle^m \mathbf{u}(\cdot,t) - \triangle^m \mathbf{v}(\cdot,t)\|_{L^2(\mathbb{R}^n)} \\ \leq & (1+t)^{m+n/4}\|\triangle^m \mathbf{v}(\cdot,t)\|_{L^2(\mathbb{R}^n)} + (1+t)^{m+n/4}\|\triangle^m \mathbf{u}(\cdot,t) - \triangle^m \mathbf{v}(\cdot,t)\|_{L^2(\mathbb{R}^n)} \\ \leq & (1+t)^{m+n/4}\|\triangle^m \mathbf{v}(\cdot,t)\|_{L^2(\mathbb{R}^n)} + C_m(1+t)^{-\max\{1,n-1\}/4}, \end{aligned}$$

where we have taken the results of Lemma 2.4 into account. Now we get a bound for the upper limit:

$$\limsup_{t\to\infty}\left[(1+t)^{m+n/4}\|\triangle^m \mathbf{u}(\cdot,t)\|_{L^2(\mathbb{R}^n)}\right] \leq \lim_{t\to\infty}\left[(1+t)^{m+n/4}\|\triangle^m \mathbf{v}(\cdot,t)\|_{L^2(\mathbb{R}^n)}\right].$$

On the other hand, we have the lower bound estimate

$$\begin{aligned} & (1+t)^{m+n/4}\|\triangle^m \mathbf{u}(\cdot,t)\|_{L^2(\mathbb{R}^n)} \\ = & (1+t)^{m+n/4}\|\triangle^m \mathbf{v}(\cdot,t) + \triangle^m \mathbf{u}(\cdot,t) - \triangle^m \mathbf{v}(\cdot,t)\|_{L^2(\mathbb{R}^n)} \\ \geq & (1+t)^{m+n/4}\|\triangle^m \mathbf{v}(\cdot,t)\|_{L^2(\mathbb{R}^n)} - (1+t)^{m+n/4}\|\triangle^m \mathbf{u}(\cdot,t) - \triangle^m \mathbf{v}(\cdot,t)\|_{L^2(\mathbb{R}^n)} \\ \geq & (1+t)^{m+n/4}\|\triangle^m \mathbf{v}(\cdot,t)\|_{L^2(\mathbb{R}^n)} - C_m(1+t)^{-max\{1,n-1\}/4}. \end{aligned}$$

Therefore, we also get a bound for the lower limit:

$$\liminf_{t\to\infty}\left[(1+t)^{m+n/4}\|\triangle^m \mathbf{u}(\cdot,t)\|_{L^2(\mathbb{R}^n)}\right] \geq \lim_{t\to\infty}\left[(1+t)^{m+n/4}\|\triangle^m \mathbf{v}(\cdot,t)\|_{L^2(\mathbb{R}^n)}\right].$$

By coupling these two estimates together, we obtain

$$\limsup_{t\to\infty}\left[(1+t)^{m+n/4}\|\triangle^m \mathbf{u}(\cdot,t)\|_{L^2(\mathbb{R}^n)}\right] \leq \liminf_{t\to\infty}\left[(1+t)^{m+n/4}\|\triangle^m \mathbf{u}(\cdot,t)\|_{L^2(\mathbb{R}^n)}\right].$$

Automatically, there holds

$$\liminf_{t\to\infty}\left[(1+t)^{m+n/4}\|\triangle^m \mathbf{u}(\cdot,t)\|_{L^2(\mathbb{R}^n)}\right] \leq \limsup_{t\to\infty}\left[(1+t)^{m+n/4}\|\triangle^m \mathbf{u}(\cdot,t)\|_{L^2(\mathbb{R}^n)}\right].$$

Therefore, there exists the following limit

$$\lim_{t\to\infty}\left\{(1+t)^{m+n/4}\|\triangle^m \mathbf{u}(\cdot,t)\|_{L^2(\mathbb{R}^n)}\right\} = \lim_{t\to\infty}\left\{(1+t)^{m+n/4}\|\triangle^m \mathbf{v}(\cdot,t)\|_{L^2(\mathbb{R}^n)}\right\}.$$

The idea in the analysis for the case $\varepsilon > 0$ is the same as the idea in the case $\varepsilon = 0$. Therefore

$$\begin{aligned} & \lim_{t\to\infty}\left\{(1+t)^{2m+n/2}\left[\|\triangle^m \mathbf{u}(\cdot,t)\|^2_{L^2(\mathbb{R}^n)} + \varepsilon\|\nabla\triangle^m \mathbf{u}(\cdot,t)\|^2_{L^2(\mathbb{R}^n)}\right]\right\} \\ = & \lim_{t\to\infty}\left\{(1+t)^{2m+n/2}\left[\|\triangle^m \mathbf{v}(\cdot,t)\|^2_{L^2(\mathbb{R}^n)} + \varepsilon\|\nabla\triangle^m \mathbf{v}(\cdot,t)\|^2_{L^2(\mathbb{R}^n)}\right]\right\} \\ = & \lim_{t\to\infty}\left[(1+t)^{2m+n/2}\|\triangle^m \mathbf{v}(\cdot,t)\|^2_{L^2(\mathbb{R}^n)}\right] \\ = & \frac{1}{(2\pi)^{2m+n}}\left(\frac{\pi}{2\alpha}\right)^{2m+n/2}\prod_{l=1}^{2m}(2l-2+n)\left|\int_{\mathbb{R}^n}\mathbf{u}_0(\mathbf{x})d\mathbf{x}\right|^2. \end{aligned}$$

The other case can be accomplished very similarly, where the key estimate is

$$(1+t)^{2m+1+n/2}\int_{\mathbb{R}^n}\left[|\triangle^m\mathbf{u}(\mathbf{x},t)-\triangle^m\mathbf{v}(\mathbf{x},t)|^2+\varepsilon|\nabla\triangle^m\mathbf{u}(\mathbf{x},t)-\nabla\triangle^m\mathbf{v}(\mathbf{x},t)|^2\right]\mathrm{d}\mathbf{x}\le\frac{C_m}{(1+t)^{n/2}},$$

if

$$\int_{\mathbb{R}^n}\mathbf{u}_0(\mathbf{x})\mathrm{d}\mathbf{x}=\mathbf{0}.$$

The proof of Lemma 2.6 is finished. #

3. The Proofs of the Main Results

Theorem 3. *Let the initial data* $\mathbf{u}_0\in L^1(\mathbb{R}^n)\cap H^2(\mathbb{R}^n)$ *and*

$$\int_{\mathbb{R}^n}\mathbf{u}_0(\mathbf{x})\mathrm{d}\mathbf{x}\ne\mathbf{0}.$$

Then

$$\lim_{t\to\infty}\left\{(1+t)^{n/2}\int_{\mathbb{R}^n}\left[|\mathbf{u}(\mathbf{x},t)|^2+\varepsilon|\nabla\mathbf{u}(\mathbf{x},t)|^2\right]\mathrm{d}\mathbf{x}\right\}=\frac{1}{(2\pi)^n}\left(\frac{\pi}{2\alpha}\right)^{n/2}\left[\int_{\mathbb{R}^n}\mathbf{u}_0(\mathbf{x})\mathrm{d}\mathbf{x}\right]^2,$$

$$\lim_{t\to\infty}\left\{(1+t)^{2m+n/2}\int_{\mathbb{R}^n}\left[|\triangle^m\mathbf{u}(\mathbf{x},t)|^2+\varepsilon|\nabla\triangle^m\mathbf{u}(\mathbf{x},t)|^2\right]\mathrm{d}\mathbf{x}\right\}$$
$$=\frac{1}{(2\pi)^{2m+n}}\left(\frac{\pi}{2\alpha}\right)^{2m+n/2}\prod_{l=1}^{2m}(2l-2+n)\left[\int_{\mathbb{R}^n}\mathbf{u}_0(\mathbf{x})\mathrm{d}\mathbf{x}\right]^2.$$

Proof. Note that for any integer $m\ge 0$, we have

$$\lim_{t\to\infty}\left[(1+t)^{2m+n/2}\int_{\mathbb{R}^n}|\triangle^{m+1}\mathbf{u}(\mathbf{x},t)|^2\mathrm{d}\mathbf{x}\right]=\lim_{t\to\infty}\left[(1+t)^{2m+n/2}\int_{\mathbb{R}^n}|\triangle^{m+1}\mathbf{v}(\mathbf{x},t)|^2\mathrm{d}\mathbf{x}\right]=0.$$

Therefore, there exists the limit

$$\lim_{t\to\infty}\left[(1+t)^{2m+n/2}\|\triangle^m\mathbf{u}(\cdot,t)\|_\varepsilon^2\right]=\lim_{t\to\infty}\left[(1+t)^{2m+n/2}\|\triangle^m\mathbf{v}(\cdot,t)\|_\varepsilon^2\right]$$
$$=\lim_{t\to\infty}\left[(1+t)^{2m+n/2}\|\triangle^m\mathbf{v}(\cdot,t)\|_{L^2(\mathbb{R}^n)}^2\right]$$
$$=\frac{1}{(2\pi)^{2m+n}}\left(\frac{\pi}{2\alpha}\right)^{2m+n/2}\prod_{l=1}^{2m}(2l-2+n)\left|\int_{\mathbb{R}^n}\mathbf{u}_0(\mathbf{x})\mathrm{d}\mathbf{x}\right|^2.$$

The proof of Theorem 3 is finished. #

Theorem 4. *Let the initial data* $\mathbf{u}_0\in L^1(\mathbb{R}^n)\cap H^2(\mathbb{R}^n)$, *satisfying the following conditions*

$$\int_{\mathbb{R}^n}\mathbf{u}_0(\mathbf{x})\mathrm{d}\mathbf{x}=\mathbf{0},$$

and

$$\mathbf{u}_0(\mathbf{x})=\left(\sum_{j=1}^n\frac{\partial\phi_{1j}}{\partial x_j}(\mathbf{x}),\ \sum_{j=1}^n\frac{\partial\phi_{2j}}{\partial x_j}(\mathbf{x}),\ \cdots,\ \sum_{j=1}^n\frac{\partial\phi_{nj}}{\partial x_j}(\mathbf{x})\right),$$

where $\Phi_i \equiv (\phi_{i1}, \phi_{i2}, \cdots, \phi_{in})^T \in C^2(\mathbb{R}^n) \cap L^1(\mathbb{R}^n)$, *for all* $i = 1, 2, \cdots, n$. *Then*

$$\lim_{t\to\infty} \left\{ (1+t)^{1+n/2} \int_{\mathbb{R}^n} \left[|\mathbf{u}(\mathbf{x},t)|^2 + \varepsilon |\nabla \mathbf{u}(\mathbf{x},t)|^2 \right] \mathrm{d}\mathbf{x} \right\}$$
$$= \frac{1}{(2\pi)^{n+1}} \left(\frac{\pi}{2\alpha}\right)^{1+n/2} \sum_{k=1}^{n} \left[\int_{\mathbb{R}^n} \Phi_k(\mathbf{x}) \mathrm{d}\mathbf{x} \right]^2 .$$

For any integer $m \geq 1$, *we have*

$$\lim_{t\to\infty} \left\{ (1+t)^{2m+1+n/2} \int_{\mathbb{R}^n} \left[|\triangle^m \mathbf{u}(\mathbf{x},t)|^2 + \varepsilon |\nabla \triangle^m \mathbf{u}(\mathbf{x},t)|^2 \right] \mathrm{d}\mathbf{x} \right\}$$
$$= \frac{1}{(2\pi)^{2m+n+1}} \left(\frac{\pi}{2\alpha}\right)^{2m+1+n/2} \prod_{l=1}^{2m} (2l+n) \sum_{k=1}^{n} \left[\int_{\mathbb{R}^n} \Phi_k(\mathbf{x}) \mathrm{d}\mathbf{x} \right]^2 .$$

Proof. The proof of Theorem 4 is very similar to the proof of Theorem 3 and is omitted. #

The proofs of Theorem 1 and Theorem 2: They may be accomplished by coupling the results of Theorem 3 and Theorem 4, respectively, and the following limits

$$\lim_{t\to\infty} \left\{ (1+t)^{2m+n/2} \int_{\mathbb{R}} |\nabla \triangle^m \mathbf{u}(\mathbf{x},t)|^2 \mathrm{d}\mathbf{x} \right\} = 0 \qquad \text{if} \qquad \int_{\mathbb{R}} \mathbf{u}_0(\mathbf{x}) \mathrm{d}\mathbf{x} \neq \mathbf{0},$$

and

$$\lim_{t\to\infty} \left\{ (1+t)^{2m+1+n/2} \int_{\mathbb{R}} |\nabla \triangle^m \mathbf{u}(\mathbf{x},t)|^2 \mathrm{d}\mathbf{x} \right\} = 0 \qquad \text{if} \qquad \int_{\mathbb{R}} \mathbf{u}_0(\mathbf{x}) \mathrm{d}\mathbf{x} = \mathbf{0},$$

respectively. #

Remark 2. By using the energy equation

$$\int_{\mathbb{R}^n} \left[|\mathbf{u}(\mathbf{x},t)|^2 + \varepsilon |\nabla \mathbf{u}(\mathbf{x},t)|^2 \right] \mathrm{d}\mathbf{x} + 2\alpha \int_0^t \left[\int_{\mathbb{R}^n} |\nabla \mathbf{u}(\mathbf{x},\tau)|^2 \mathrm{d}\mathbf{x} \right] \mathrm{d}\tau$$
$$= \int_{\mathbb{R}^n} \left[|\mathbf{u}_0(\mathbf{x})|^2 + \varepsilon |\nabla \mathbf{u}_0(\mathbf{x})|^2 \right] \mathrm{d}\mathbf{x},$$

we may find the following limit

$$\lim_{t\to\infty} (1+t)^{n/2} \left\{ \int_{\mathbb{R}^n} \left[|\mathbf{u}_0(\mathbf{x})|^2 + \varepsilon |\nabla \mathbf{u}_0(\mathbf{x})|^2 \right] \mathrm{d}\mathbf{x} - 2\alpha \int_0^t \left[\int_{\mathbb{R}^n} |\nabla \mathbf{u}(\mathbf{x},\tau)|^2 \mathrm{d}\mathbf{x} \right] \mathrm{d}\tau \right\}$$
$$= \frac{1}{(2\pi)^n} \left(\frac{\pi}{2\alpha}\right)^{n/2} \left[\int_{\mathbb{R}^n} \mathbf{u}_0(\mathbf{x}) \mathrm{d}\mathbf{x} \right]^2 .$$

This limit is positive if

$$\int_{\mathbb{R}^n} \mathbf{u}_0(\mathbf{x}) \mathrm{d}\mathbf{x} \neq 0.$$

Remark 3. Suppose that $\alpha_0 \geq 0, \alpha_1 \geq 0, \alpha_2 \geq 0, \cdots, \alpha_n \geq 0$ are nonnegative integers. Then, by Plancherel's identity, we find that for problem (7)-(8)

$$\begin{aligned}
& t^{2\alpha_0+\alpha_1+\alpha_2+\cdots+\alpha_n+n/2} \int_{\mathbb{R}^n} \left| \frac{\partial^{\alpha_0+\alpha_1+\alpha_2+\cdots+\alpha_n} \mathbf{v}}{\partial t^{\alpha_0} \partial x_1^{\alpha_1} \partial x_2^{\alpha_2} \cdots \partial x_n^{\alpha_n}} \right|^2 \mathrm{d}\mathbf{x} \\
= & \frac{t^{2\alpha_0+\alpha_1+\alpha_2+\cdots+\alpha_n+n/2}}{(2\pi)^n} \int_{\mathbb{R}^n} |\xi_1|^{2\alpha_1} |\xi_2|^{2\alpha_2} \cdots |\xi_n|^{2\alpha_n} \left| \frac{\alpha|\xi|^2 + \mathbf{i}\Gamma(\xi)}{1+\varepsilon|\xi|^2} \right|^{2\alpha_0} |\hat{\mathbf{v}}(\xi, t)|^2 \mathrm{d}\xi \\
= & \frac{t^{2\alpha_0+\alpha_1+\alpha_2+\cdots+\alpha_n+n/2}}{(2\pi)^n} \int_{\mathbb{R}^n} |\xi_1|^{2\alpha_1} |\xi_2|^{2\alpha_2} \cdots |\xi_n|^{2\alpha_n} \left| \frac{\alpha|\xi|^2 + \mathbf{i}\Gamma(\xi)}{1+\varepsilon|\xi|^2} \right|^{2\alpha_0} \exp\left(-\frac{2\alpha|\xi|^2}{1+\varepsilon|\xi|^2} t \right) |\hat{\mathbf{v}}_0(\xi)|^2 \mathrm{d}\xi \\
= & \frac{1}{(2\pi)^n} \int_{\mathbb{R}^n} |\eta_1|^{2\alpha_1} |\eta_2|^{2\alpha_2} \cdots |\eta_n|^{2\alpha_n} \left| \frac{\alpha|\eta|^2 + t\Gamma(\eta/\sqrt{t})\mathbf{i}}{1+\varepsilon|\eta|^2/t} \right|^{2\alpha_0} \exp\left(-\frac{2\alpha|\eta|^2}{1+\varepsilon|\eta|^2/t} \right) |\hat{\mathbf{v}}_0(t^{-1/2}\eta)|^2 \mathrm{d}\eta \\
\to & \frac{(2\alpha)^{2\alpha_0}}{(2\pi)^n} \left(\frac{\pi}{2\alpha} \right)^{n/2} \frac{1}{(4\alpha)^{2\alpha_0+\alpha_1+\alpha_2+\cdots+\alpha_n}} \left[\prod_{i=1}^n (\alpha_i)!! \right] \\
\times & \left[\prod_{j=1}^{2\alpha_0} \left(j - 1 + \frac{n}{2} + \alpha_1 + \alpha_2 + \cdots + \alpha_n \right) \right] \left[\int_{\mathbb{R}^n} \mathbf{v}_0(\mathbf{x}) \mathrm{d}\mathbf{x} \right]^2,
\end{aligned}$$

as $t \to \infty$, where we need the following additional assumption, only when $\alpha_0 \geq 1$:

$$\lim_{t\to\infty} \left[t\Gamma(\eta/\sqrt{t}) \right] = 0 \qquad \text{for all } \eta \in \mathbb{R}^n.$$

We also used the definitions: $(\alpha_i)!! = 1$ if $\alpha_i = 0$ and

$$\prod_{j=1}^{2\alpha_0} \left(j - 1 + \frac{n}{2} + \alpha_1 + \alpha_2 + \cdots + \alpha_n \right) = 1$$

if $\alpha_0 = 0$. Now, suppose that

$$\int_{\mathbb{R}^n} \mathbf{v}_0(\mathbf{x}) \mathrm{d}\mathbf{x} = \mathbf{0},$$

and that

$$\mathbf{v}_0(\mathbf{x}) = \left(\sum_{j=1}^n \frac{\partial \phi_{1j}}{\partial x_j}(\mathbf{x}), \ \sum_{j=1}^n \frac{\partial \phi_{2j}}{\partial x_j}(\mathbf{x}), \ \cdots, \ \sum_{j=1}^n \frac{\partial \phi_{nj}}{\partial x_j}(\mathbf{x}) \right)^T,$$

where $\Phi_i \equiv (\phi_{i1}, \phi_{i2}, \cdots, \phi_{in})^T \in C^2(\mathbb{R}^n) \cap L^1(\mathbb{R}^n), i = 1, 2, \cdots, n$. Then

$$\begin{aligned}
& (1+t)^{2\alpha_0+\alpha_1+\alpha_2+\cdots+\alpha_n+1+n/2} \int_{\mathbb{R}^n} \left| \frac{\partial^{\alpha_0+\alpha_1+\alpha_2+\cdots+\alpha_n} \mathbf{v}}{\partial t^{\alpha_0} \partial x_1^{\alpha_1} \partial x_2^{\alpha_2} \cdots \partial x_n^{\alpha_n}}(\mathbf{x}, t) \right|^2 \mathrm{d}\mathbf{x} \\
\to & \frac{(2\alpha)^{2\alpha_0}}{(2\pi)^{n+1}} \left(\frac{\pi}{2\alpha} \right)^{1+n/2} \frac{1}{(4\alpha)^{2\alpha_0+\alpha_1+\alpha_2+\cdots+\alpha_n}} \left[(1+\alpha_k)!! \prod_{i=1, i\neq k}^n (\alpha_i)!! \right] \\
\times & \left[\prod_{j=1}^{2\alpha_0} \left(j + \frac{n}{2} + \alpha_1 + \alpha_2 + \cdots + \alpha_n \right) \right] \sum_{k=1}^n \left| \int_{\mathbb{R}^n} \Phi_k(\mathbf{x}) \mathrm{d}\mathbf{x} \right|^2,
\end{aligned}$$

as $t \to \infty$, where we have applied the following formulas

$$\int_{\mathbb{R}} |\xi_i|^{2\alpha_i} \exp\big(-2\alpha|\xi_i|^2 t\big) d\xi_i = \frac{\alpha_i!!}{(4\alpha t)^{\alpha_i}} \sqrt{\frac{\pi}{2\alpha t}},$$

$$\int_{\mathbb{R}^n} |\xi_1|^{2\alpha_1} |\xi_2|^{2\alpha_2} \cdots |\xi_n|^{2\alpha_n} |\xi|^{4\alpha_0} \exp\left(-2\alpha|\xi|^2 t\right) d\xi$$

$$= 2^{2\alpha_0} \left(\frac{\pi}{2\alpha t}\right)^{n/2} \frac{1}{(4\alpha t)^{2\alpha_0+\alpha_1+\alpha_2+\cdots+\alpha_n}} \left[\prod_{i=1}^{n} (\alpha_i!!)\right]$$

$$\times \left[\prod_{j=1}^{2\alpha_0} \left(j - 1 + \frac{n}{2} + \alpha_1 + \alpha_2 + \cdots + \alpha_n\right)\right].$$

The last factor is equal to one if $\alpha_0 = 0$. Recall that $0!! = 1$ and $m!! = 1 \cdot 3 \cdot 5 \cdots (2m - 3) \cdot (2m - 1)$.

Remark 4. For any nonnegative integers $\alpha_0 \geq 0$, $\alpha_1 \geq 0$, $\alpha_2 \geq 0$, $\cdots$, $\alpha_n \geq 0$, with $\alpha_1 + \alpha_2 + \cdots + \alpha_n \geq 1$, based on Remark 3 there hold the exact limits

$$\lim_{t \to \infty} \Bigg\{ (1+t)^{2\alpha_0+\alpha_1+\alpha_2+\cdots+\alpha_n+1+n/2}$$

$$\times \int_{\mathbb{R}^n} \left[\left| \frac{\partial^{\alpha_0+\alpha_1+\alpha_2+\cdots+\alpha_n} \mathbf{u}}{\partial t^{\alpha_0} \partial x_1^{\alpha_1} \partial x_2^{\alpha_2} \cdots \partial x_n^{\alpha_n}} (\mathbf{x}, t) \right|^2 + \varepsilon \left| \frac{\partial^{\alpha_0+\alpha_1+\alpha_2+\cdots+\alpha_n} \nabla \mathbf{u}}{\partial t^{\alpha_0} \partial x_1^{\alpha_1} \partial x_2^{\alpha_2} \cdots \partial x_n^{\alpha_n}} (\mathbf{x}, t) \right|^2 \right] d\mathbf{x} \Bigg\}$$

$$= \frac{(2\alpha)^{2\alpha_0}}{(2\pi)^n} \left(\frac{\pi}{2\alpha}\right)^{n/2} \frac{1}{(4\alpha)^{2\alpha_0+\alpha_1+\alpha_2+\cdots+\alpha_n}} \left[\prod_{i=1}^{n} (\alpha_i)!!\right]$$

$$\times \left[\prod_{j=1}^{2\alpha_0} \left(j - 1 + \frac{n}{2} + \alpha_1 + \alpha_2 + \cdots + \alpha_n\right)\right] \left[\int_{\mathbb{R}^n} \mathbf{u}_0(\mathbf{x}) d\mathbf{x}\right]^2.$$

Given two initial functions $\mathbf{u}_0$ and $\widetilde{\mathbf{u}_0}$ in $L^1(\mathbb{R}^n) \cap H^2(\mathbb{R}^n)$, there exist two solutions $\mathbf{u} = \mathbf{u}(\mathbf{x}, t, \mathbf{u}_0)$ and $\widetilde{\mathbf{u}} = \widetilde{\mathbf{u}}(\mathbf{x}, t, \widetilde{\mathbf{u}}_0)$ of problem (1)-(2) corresponding to $\mathbf{u}_0$ and $\widetilde{\mathbf{u}_0}$.

Theorem 5. Suppose that the initial functions $\mathbf{u}_0$ and $\widetilde{\mathbf{u}_0} \in L^1(\mathbb{R}^n) \cap H^2(\mathbb{R}^n)$ and that

$$\int_{\mathbb{R}^n} \left[\mathbf{u}_0(\mathbf{x}) - \widetilde{\mathbf{u}_0}(\mathbf{x})\right] d\mathbf{x} \neq \mathbf{0}.$$

Then the global solutions $\mathbf{u}$ and $\widetilde{\mathbf{u}}$ of (1)-(2) satisfy

$$\lim_{t \to \infty} \left\{ (1+t)^{2m+n/2} \int_{\mathbb{R}^n} |\Delta^m \mathbf{u}(\mathbf{x}, t) - \Delta^m \widetilde{\mathbf{u}}(\mathbf{x}, t)|^2 d\mathbf{x} \right\}$$

$$= \frac{1}{(2\pi)^{2m+n}} \left(\frac{\pi}{2\alpha}\right)^{2m+n/2} \prod_{l=1}^{2m} (2l - 2 + n) \left\{ \int_{\mathbb{R}^n} [\mathbf{u}_0(\mathbf{x}) - \widetilde{\mathbf{u}}_0(\mathbf{x})] d\mathbf{x} \right\}^2.$$

Proof. The idea of the proof of Theorem 5 is very similar to those presented in the proofs of Lemma 2.3, Lemma 2.4, Lemma 2.5, Theorem 3 and Theorem 4. Here we only sketch the proof. Define $\Psi(\mathbf{x},t) = \mathbf{u}(\mathbf{x},t) - \widetilde{\mathbf{u}}(\mathbf{x},t)$, $\Phi(\mathbf{x},t) = \mathbf{v}(\mathbf{x},t) - \widetilde{\mathbf{v}}(\mathbf{x},t)$, where $\mathbf{v}$ and $\widetilde{\mathbf{v}}$ are the global solutions corresponding to the initial functions $\mathbf{u}_0$ and $\widetilde{\mathbf{u}}_0$, respectively, of the linear problems (7)-(8), and $\Psi_0(\mathbf{x}) = \Phi_0(\mathbf{x}) = \mathbf{u}_0(\mathbf{x}) - \widetilde{\mathbf{u}_0}(\mathbf{x})$. Then Ψ and Φ satisfy

$$\begin{aligned} \Psi_t - \varepsilon\triangle\Psi_t + \mathcal{D}\Psi + \left[\mathcal{N}(\mathbf{u},\nabla\mathbf{u}) - \mathcal{N}(\widetilde{\mathbf{u}},\nabla\widetilde{\mathbf{u}})\right] &= \alpha\triangle\Psi, \\ \Psi(\mathbf{x},0) &= \Psi_0(\mathbf{x}), \end{aligned}$$

and

$$\begin{aligned} \Phi_t - \varepsilon\triangle\Phi_t + \mathcal{D}\Phi &= \alpha\triangle\Phi, \\ \Phi(\mathbf{x},0) &= \Phi_0(\mathbf{x}), \end{aligned}$$

respectively. Then, very similar to before, we obtain the estimates

$$(1+t)^{2m+\max\{1,n-1/2\}}\int_{\mathbb{R}^n}\left[|\triangle^m\Psi(\mathbf{x},t) - \triangle^m\Phi(\mathbf{x},t)|^2 + \varepsilon|\nabla\triangle^m\Psi(\mathbf{x},t) - \nabla\triangle^m\Phi(\mathbf{x},t)|^2\right]d\mathbf{x} \le C_m,$$

if

$$\int_{\mathbb{R}^n}\left[\mathbf{u}_0(\mathbf{x}) - \widetilde{\mathbf{u}_0}(\mathbf{x})\right]d\mathbf{x} \neq \mathbf{0}.$$

The other details are very similar to before and they are omitted. #

The results of Theorem 5 imply that for each fixed integer $m \ge 0$, the nonlinear operator

$$\mathcal{L}_m : \mathbf{u}_0| \to \lim_{t\to\infty}\left\{(1+t)^{2m+n/2}\int_{\mathbb{R}^n}\left[|\triangle^m\mathbf{u}(\mathbf{x},t)|^2 + \varepsilon|\nabla\triangle^m\mathbf{u}(\mathbf{x},t)|^2\right]d\mathbf{x}\right\}^{1/2}$$

is Lipschitz continuous. Note that the global solution $\mathbf{u}$ depends on the initial data $\mathbf{u}_0$. Indeed, for any two initial functions $\mathbf{u}_0$ and $\widetilde{\mathbf{u}_0}$ in $L^1(\mathbb{R}^n)\cap H^2(\mathbb{R}^n)$, there holds the estimate

$$|\mathcal{L}_m\mathbf{u}_0 - \mathcal{L}_m\widetilde{\mathbf{u}_0}| \le \left[\frac{1}{(2\pi)^{2m+n}}\left(\frac{\pi}{2\alpha}\right)^{2m+n/2}\prod_{l=1}^{2m}(2l-2+n)\right]^{1/2}\left[\int_{\mathbb{R}^n}|\mathbf{u}_0(\mathbf{x}) - \widetilde{\mathbf{u}_0}(\mathbf{x})|d\mathbf{x}\right].$$

#

Theorem 6. Suppose that the initial data $\mathbf{u}_0$ and $\widetilde{\mathbf{u}_0} \in L^1(\mathbb{R}^n)\cap H^2(\mathbb{R}^n)$ and that

$$\int_{\mathbb{R}^n}\left[\mathbf{u}_0(\mathbf{x}) - \widetilde{\mathbf{u}_0}(\mathbf{x})\right]d\mathbf{x} = \mathbf{0},$$

$$\mathbf{u}_0(\mathbf{x}) - \widetilde{\mathbf{u}}_0(\mathbf{x}) = \left(\sum_{j=1}^n\frac{\partial\phi_{1j}}{\partial x_j}(\mathbf{x}),\ \sum_{j=1}^n\frac{\partial\phi_{2j}}{\partial x_j}(\mathbf{x})\ \cdots,\ \sum_{j=1}^n\frac{\partial\phi_{nj}}{\partial x_j}(\mathbf{x})\right)^T.$$

Suppose that $\Phi_i \equiv (\phi_{i1},\phi_{i2},\cdots,\phi_{in})^T \in C^2(\mathbb{R}^n)\cap L^1(\mathbb{R}^n)$, where $i = 1,2,\cdots,n$. Then the global solutions $\mathbf{u}$ and $\widetilde{\mathbf{u}}$ satisfy

$$\begin{aligned} &\lim_{t\to\infty}\left\{(1+t)^{2m+1+n/2}\int_{\mathbb{R}^n}|\triangle^m\mathbf{u}(\mathbf{x},t) - \triangle^m\widetilde{\mathbf{u}}(\mathbf{x},t)|^2 d\mathbf{x}\right\} \\ &= \frac{1}{(2\pi)^{2m+n+1}}\left(\frac{\pi}{2\alpha}\right)^{2m+1+n/2}\prod_{l=1}^{2m}(2l+n)\sum_{k=1}^n\left\{\int_{\mathbb{R}^n}\Phi_k(\mathbf{x})d\mathbf{x}\right\}^2. \end{aligned}$$

Proof. The proof of Theorem 6 is very similar to the proof of Theorem 5. The only difference is the rate of decay

$$(1+t)^{2m+n+1}\int_{\mathbb{R}^n}\left[|\triangle^m\Psi(\mathbf{x},t)-\triangle^m\Phi(\mathbf{x},t)|^2+\varepsilon|\nabla\triangle^m\Psi(\mathbf{x},t)-\nabla\triangle^m\Phi(\mathbf{x},t)|^2\right]\mathrm{d}\mathbf{x} \leq C_m,$$

if

$$\int_{\mathbb{R}^n}\left[\mathbf{u}_0(\mathbf{x})-\widetilde{\mathbf{u}_0}(\mathbf{x})\right]\mathrm{d}\mathbf{x}=\mathbf{0}.$$

All other details are omitted. #

4. Applications

In this section, we are going to provide several applications of Theorem 1 and Theorem 2, without offering the proofs for obvious reasons. The equations we are going to study in this section are good examples of (1). As we can easily see, all of the general assumptions (3)-(4)-(5)-(6) made in Section 1 are satisfied. These limits have not been obtained before. Therefore, they are new.

4.1. The n-dimensional Burgers Equations

Consider the one-dimensional cubic Burgers equation

$$\begin{aligned} u_t+u_x+\beta(u^3)_x &= \alpha u_{xx} && \text{in } \mathbb{R}\times\mathbb{R}^+,\\ u(x,0) &= u_0(x) && \text{in } \mathbb{R}, \end{aligned}$$

and the n-dimensional quadratic Burgers equation

$$\begin{aligned} u_t+\gamma\cdot\nabla u+\beta\cdot\nabla(u^2) &= \alpha\triangle u && \text{in } \mathbb{R}^n\times\mathbb{R}^+,\\ u(\mathbf{x},0) &= u_0(\mathbf{x}) && \text{in } \mathbb{R}^n. \end{aligned}$$

Theorem 4.1. (I) Suppose that $u_0\in L^1(\mathbb{R})\cap L^2(\mathbb{R})$. Then the global solution of the one-dimensional cubic Burgers equation enjoys the limit

$$\lim_{t\to\infty}\left[(1+t)^{1/2}\int_{\mathbb{R}}|u(x,t)|^2\mathrm{d}x\right]=\frac{1}{2\pi}\left(\frac{\pi}{2\alpha}\right)^{1/2}\left[\int_{\mathbb{R}}u_0(x)\mathrm{d}x\right]^2.$$

Additionally, if $u_0(x)=\phi'(x)$ and $\phi\in C^1(\mathbb{R})\cap L^1(\mathbb{R})$, then

$$\lim_{t\to\infty}\left[(1+t)^{3/2}\int_{\mathbb{R}}|u(x,t)|^2\mathrm{d}x\right]=\frac{1}{8\pi\alpha}\left(\frac{\pi}{2\alpha}\right)^{1/2}\left[\int_{\mathbb{R}}\phi(x)\mathrm{d}x\right]^2.$$

(II) Suppose that $u_0\in L^1(\mathbb{R}^n)\cap L^2(\mathbb{R}^n)$ and the dimension $n\geq 2$. Then the global solution of the n-dimensional Burgers equation enjoys the limit

$$\lim_{t\to\infty}\left[(1+t)^{n/2}\int_{\mathbb{R}^n}|u(\mathbf{x},t)|^2\mathrm{d}\mathbf{x}\right]=\frac{1}{(2\pi)^n}\left(\frac{\pi}{2\alpha}\right)^{n/2}\left[\int_{\mathbb{R}^n}u_0(\mathbf{x})\mathrm{d}\mathbf{x}\right]^2.$$

Moreover, if $u_0(\mathbf{x}) = \sum_{k=1}^{n} \dfrac{\partial \phi_k}{\partial x_k}(\mathbf{x})$, where $\phi_k \in C^1(\mathbb{R}^n) \cap L^1(\mathbb{R}^n)$, for all integers $1 \le k \le n$, then

$$\lim_{t\to\infty}\left[(1+t)^{1+n/2}\int_{\mathbb{R}^n}|u(\mathbf{x},t)|^2\mathrm{d}\mathbf{x}\right] = \frac{1}{(2\pi)^{n+1}}\left(\frac{\pi}{2\alpha}\right)^{1+n/2}\sum_{k=1}^{n}\left[\int_{\mathbb{R}^n}\phi_k(\mathbf{x})\mathrm{d}\mathbf{x}\right]^2.$$

4.2. The One-Dimensional Nonlinear Benjamin-Ono-Burgers Equation

Consider

$$u_t + u_x + Hu_{xx} + (u^3)_x = \alpha u_{xx} \qquad \text{in } \mathbb{R}\times\mathbb{R}^+.$$

Theorem 4.2. Suppose that $u_0 \in L^1(\mathbb{R}) \cap L^2(\mathbb{R})$. Then the global solution of the Benjamin-Ono-Burgers equation enjoys the limit

$$\lim_{t\to\infty}\left[(1+t)^{1/2}\int_{\mathbb{R}}|u(x,t)|^2\mathrm{d}x\right] = \frac{1}{2\pi}\left(\frac{\pi}{2\alpha}\right)^{1/2}\left[\int_{\mathbb{R}}u_0(x)\mathrm{d}x\right]^2.$$

Additionally, if $u_0(x) = \phi'(x)$ and $\phi \in C^1(\mathbb{R}) \cap L^1(\mathbb{R})$, then

$$\lim_{t\to\infty}\left[(1+t)^{3/2}\int_{\mathbb{R}}|u(x,t)|^2\mathrm{d}x\right] = \frac{1}{8\pi\alpha}\left(\frac{\pi}{2\alpha}\right)^{1/2}\left[\int_{\mathbb{R}}\phi(x)\mathrm{d}x\right]^2.$$

4.3. The One-Dimensional Nonlinear Cubic Korteweg-de Vries-Burgers Equation

Consider

$$u_t + u_x + u_{xxx} + (u^3)_x - \alpha u_{xx} = 0, \qquad \text{in } \mathbb{R}^n\times\mathbb{R}^+.$$

Theorem 4.3. Suppose that $u_0 \in L^1(\mathbb{R}) \cap L^2(\mathbb{R})$. Then the global solution of the cubic Korteweg-de Vries-Burgers equation enjoys the limit

$$\lim_{t\to\infty}\left[(1+t)^{1/2}\int_{\mathbb{R}}|u(x,t)|^2\mathrm{d}x\right] = \frac{1}{2\pi}\left(\frac{\pi}{2\alpha}\right)^{1/2}\left[\int_{\mathbb{R}}u_0(x)\mathrm{d}x\right]^2.$$

Additionally, if $u_0(x) = \phi'(x)$ and $\phi \in C^1(\mathbb{R}) \cap L^1(\mathbb{R})$, then

$$\lim_{t\to\infty}\left[(1+t)^{3/2}\int_{\mathbb{R}}|u(x,t)|^2\mathrm{d}x\right] = \frac{1}{8\pi\alpha}\left(\frac{\pi}{2\alpha}\right)^{1/2}\left[\int_{\mathbb{R}}\phi(x)\mathrm{d}x\right]^2.$$

4.4. The Two-Dimensional Nonlinear Nonlocal Quasi-geostrophic Equation

Consider

$$u_t + \omega\cdot\nabla u = \alpha\triangle u, \qquad \omega = (\psi_y, -\psi_x), \qquad u = (-\triangle)^{1/2}\psi, \qquad \text{in } \mathbb{R}^2.$$

The Fourier transform of ω is

$$\widehat{\omega}(\xi,\eta,t) = \mathbf{i}\frac{(\eta, \ -\xi)}{\sqrt{\xi^2+\eta^2}}\widehat{u}(\xi,\eta,t).$$

Theorem 4.4. Suppose that $u_0 \in L^1(\mathbb{R}^2) \cap H^2(\mathbb{R}^2)$. Then the global solution of the two-dimensional quasi-geostrophic equation enjoys the limit

$$\lim_{t\to\infty}\left[(1+t)\int_{\mathbb{R}^2}|u(\mathbf{x},t)|^2 d\mathbf{x}\right] = \frac{1}{8\pi\alpha}\left[\int_{\mathbb{R}^2} u_0(\mathbf{x})d\mathbf{x}\right]^2.$$

Moreover, if $u_0(\mathbf{x}) = \sum_{k=1}^2 \frac{\partial\phi_k}{\partial x_k}(\mathbf{x})$, where $\phi_k \in C^1(\mathbb{R}^2)\cap L^1(\mathbb{R}^2)$, $k = 1, 2$, then

$$\lim_{t\to\infty}\left[(1+t)^2\int_{\mathbb{R}^2}|u(\mathbf{x},t)|^2 d\mathbf{x}\right] = \frac{1}{32\pi\alpha^2}\sum_{k=1}^{2}\left[\int_{\mathbb{R}^2} \phi_k(\mathbf{x})d\mathbf{x}\right]^2.$$

4.5. The n-dimensional Incompressible Navier-Stokes Equations, with Arbitrarily Large Initial Data in $L^1(\mathbb{R}^n)\cap H^2(\mathbb{R}^n)$.

Consider

$$\begin{aligned}
\mathbf{u}_t - \alpha\Delta\mathbf{u} + (\mathbf{u}\cdot\nabla)\mathbf{u} + \nabla p &= 0, \qquad \nabla\cdot\mathbf{u} = 0, \qquad \text{in } \mathbb{R}^n\times\mathbb{R}^+,\\
\mathbf{u}(\mathbf{x},0) &= \mathbf{u}_0(\mathbf{x}), \qquad \nabla\cdot\mathbf{u}_0 = 0, \qquad \text{in } \mathbb{R}^n.
\end{aligned}$$

Theorem 4.5. Suppose that $\mathbf{u}_0 \in L^1(\mathbb{R}^n)\cap H^2(\mathbb{R}^n)$ and $\nabla\cdot\mathbf{u}_0 = 0$. Suppose also that there exist functions $\Phi_i \equiv (\phi_{i1},\phi_{i2},\cdots,\phi_{in})^T \in C^1(\mathbb{R}^n)\cap L^1(\mathbb{R}^n)$, for $i = 1,2,\cdots,n$, such that

$$\mathbf{u}_0(\mathbf{x}) = \left(\sum_{j=1}^n \frac{\partial\phi_{1j}}{\partial x_j}(\mathbf{x}),\ \sum_{j=1}^n \frac{\partial\phi_{2j}}{\partial x_j}(\mathbf{x}),\ \cdots,\ \sum_{j=1}^n \frac{\partial\phi_{nj}}{\partial x_j}(\mathbf{x})\right)^T,$$

and

$$\sum_{i=1}^n\sum_{j=1}^n \frac{\partial^2\phi_{ij}}{\partial x_i\partial x_j} = 0.$$

Then the global solutions of the Navier-Stokes equations enjoy the limit

$$\lim_{t\to\infty}\left[(1+t)^{1+n/2}\int_{\mathbb{R}^n}|\mathbf{u}(\mathbf{x},t)|^2 d\mathbf{x}\right] = \frac{1}{(2\pi)^{n+1}}\left(\frac{\pi}{2\alpha}\right)^{1+n/2}\sum_{k=1}^{n}\left[\int_{\mathbb{R}^n}\Phi_k(\mathbf{x})d\mathbf{x}\right]^2.$$

4.6. The n-dimensional Magnetohydrodynamics Equations

Without loss of generality, suppose that the Reynolds number $RE = 1$ and the magnetic Reynolds number $RM = 1$. Consider

$$\begin{aligned}
\frac{\partial \mathbf{u}}{\partial t} + (\mathbf{u}\cdot\nabla)\mathbf{u} - (\mathbf{A}\cdot\nabla)\mathbf{A} + \nabla\left(p + \frac{\kappa}{2}|\mathbf{A}|^2\right) &= \triangle\mathbf{u}, \qquad \nabla\cdot\mathbf{u} = 0 \qquad \text{in } \mathbb{R}^n\times\mathbb{R}^+,\\
\frac{\partial \mathbf{A}}{\partial t} + (\mathbf{u}\cdot\nabla)\mathbf{A} - (\mathbf{A}\cdot\nabla)\mathbf{u} &= \triangle\mathbf{A}, \qquad \nabla\cdot\mathbf{A} = 0 \qquad \text{in } \mathbb{R}^n\times\mathbb{R}^+,\\
\mathbf{u}(\mathbf{x},0) &= \mathbf{u}_0(\mathbf{x}) \qquad \nabla\cdot\mathbf{u}_0 = 0 \qquad \text{in } \mathbb{R}^n,\\
\mathbf{A}(\mathbf{x},0) &= \mathbf{A}_0(\mathbf{x}) \qquad \nabla\cdot\mathbf{A}_0 = 0 \qquad \text{in } \mathbb{R}^n.
\end{aligned}$$

Theorem 4.6. Suppose that $(\mathbf{u}_0, \mathbf{A}_0) \in L^1(\mathbb{R}^n)\cap H^2(\mathbb{R}^n)$ and $\nabla\cdot\mathbf{u}_0 = \nabla\cdot\mathbf{A}_0 = 0$. Suppose also that

$$\mathbf{u}_0(\mathbf{x}) = \left(\sum_{j=1}^n \frac{\partial\phi_{1j}}{\partial x_j}(\mathbf{x}),\ \sum_{j=1}^n \frac{\partial\phi_{2j}}{\partial x_j}(\mathbf{x}),\ \cdots,\ \sum_{j=1}^n \frac{\partial\phi_{nj}}{\partial x_j}(\mathbf{x})\right)^T, \qquad \sum_{i=1}^n \frac{\partial^2\phi_{ij}}{\partial x_i\partial x_j} = 0,$$

$$\mathbf{A}_0(\mathbf{x}) = \left(\sum_{j=1}^n \frac{\partial\psi_{1j}}{\partial x_j}(\mathbf{x}),\ \sum_{j=1}^n \frac{\partial\psi_{2j}}{\partial x_j}(\mathbf{x}),\ \cdots,\ \sum_{j=1}^n \frac{\partial\psi_{nj}}{\partial x_j}(\mathbf{x})\right)^T, \qquad \sum_{j=1}^n \frac{\partial^2\psi_{ij}}{\partial x_i\partial x_j} = 0,$$

where $\Phi_i \equiv (\phi_{i1}, \phi_{i2}, \cdots, \phi_{in})^T$ and $\Psi_i \equiv (\psi_{i1}, \psi_{i2}, \cdots, \psi_{in})^T \in C^2(\mathbb{R}^n)\cap L^1(\mathbb{R}^n)$, and $i = 1, 2, \cdots, n$, then the global solutions of the Magnetohydrodynamics equations satisfy

$$\begin{aligned}
&\lim_{t\to\infty}\left\{(1+t)^{1+n/2}\int_{\mathbb{R}^n}\left[|\mathbf{u}(\mathbf{x},t)|^2 + |\mathbf{A}(\mathbf{x},t)|^2\right]\mathrm{d}\mathbf{x}\right\}\\
&= \frac{1}{(2\pi)^{n+1}}\left(\frac{\pi}{2\alpha}\right)^{1+n/2}\left\{\sum_{k=1}^n\left[\int_{\mathbb{R}}\Phi_k(\mathbf{x})\mathrm{d}\mathbf{x}\right]^2 + \sum_{k=1}^n\left[\int_{\mathbb{R}}\Psi_k(\mathbf{x})\mathrm{d}\mathbf{x}\right]^2\right\}.
\end{aligned}$$

4.7. The n-dimensional Nonlinear Benjamin-Bona-Mahony-Burgers Equations

$$\begin{aligned}
u_t - u_{xxt} + u_x + (u^3)_x &= \alpha u_{xx} && \text{in } \mathbb{R}\times\mathbb{R}^+,\\
u_t - \triangle u_t + \beta\cdot\nabla u + \varphi(u)\cdot\nabla u &= \alpha\triangle u && \text{in } \mathbb{R}^n\times\mathbb{R}^+.
\end{aligned}$$

Theorem 4.7. **(I)** Suppose that $u_0 \in L^1(\mathbb{R})\cap H^2(\mathbb{R})$. Then the global solution of the one-dimensional nonlinear Benjamin-Bona-Mahony-Burgers equation enjoys the limit

$$\lim_{t\to\infty}\left[(1+t)^{1/2}\int_{\mathbb{R}}|u(x,t)|^2\mathrm{d}x\right] = \frac{1}{2\pi}\left(\frac{\pi}{2\alpha}\right)^{1/2}\left[\int_{\mathbb{R}} u_0(x)\mathrm{d}x\right]^2.$$

Moreover, if $u_0(x) = \phi'(x)$, for some function $\phi \in C^1(\mathbb{R})\cap L^1(\mathbb{R})$, then

$$\lim_{t\to\infty}\left[(1+t)^{3/2}\int_{\mathbb{R}}|u(x,t)|^2\mathrm{d}x\right] = \frac{1}{8\pi\alpha}\left(\frac{\pi}{2\alpha}\right)^{1/2}\left[\int_{\mathbb{R}}\phi(x)\mathrm{d}x\right]^2.$$

(II) Suppose that $u_0 \in L^1(\mathbb{R}^n) \cap H^2(\mathbb{R}^n)$. Suppose also that $|\varphi(u) \le C|u|$ for all u with $|u| \le 1$, and for some constant $C > 0$. Then the global solution of the n-dimensional nonlinear Benjamin-Bona-Mahony-Burgers equation enjoys the limit

$$\lim_{t\to\infty}\left[(1+t)^{n/2}\int_{\mathbb{R}^n}|u(\mathbf{x},t)|^2\mathrm{d}\mathbf{x}\right] = \frac{1}{(2\pi)^n}\left(\frac{\pi}{2\alpha}\right)^{n/2}\left[\int_{\mathbb{R}^n}u_0(\mathbf{x})\mathrm{d}\mathbf{x}\right]^2.$$

Moreover, if $u_0(\mathbf{x}) = \sum_{k=1}^{n}\dfrac{\partial\phi_k}{\partial x_k}(\mathbf{x})$, where $\phi_k \in C^1(\mathbb{R}^n)\cap L^1(\mathbb{R}^n)$, for all integers $k \in \{1,2,\cdots,n\}$, then

$$\lim_{t\to\infty}\left[(1+t)^{1+n/2}\int_{\mathbb{R}^n}|u(\mathbf{x},t)|^2\mathrm{d}\mathbf{x}\right] = \frac{1}{(2\pi)^{n+1}}\left(\frac{\pi}{2\alpha}\right)^{1+n/2}\sum_{k=1}^{n}\left[\int_{\mathbb{R}^n}\phi_k(\mathbf{x})\mathrm{d}\mathbf{x}\right]^2.$$

5. Concluding Remarks and Open Problems

There are certain equations which do not take the form of (1). Consider

$$\mathbf{u}_t + \alpha\mathbf{u} + \mathcal{D}\mathbf{u} + \mathcal{N}(\mathbf{u},\nabla\mathbf{u}) = 0, \qquad \mathbf{u}(\mathbf{x},0) = \mathbf{u}_0(\mathbf{x}).$$

As before, we may get

$$\frac{\mathrm{d}}{\mathrm{d}t}\int_{\mathbb{R}^n}|\mathbf{u}(\mathbf{x},t)|^2\mathrm{d}\mathbf{x} + 2\alpha\int_{\mathbb{R}^n}|\mathbf{u}(\mathbf{x},t)|^2\mathrm{d}\mathbf{x} = 0.$$

Integrating in time yields the exponential decay

$$\exp(2\alpha t)\int_{\mathbb{R}^n}|\mathbf{u}(\mathbf{x},t)|^2\mathrm{d}\mathbf{x} = \int_{\mathbb{R}^n}|\mathbf{u}_0(\mathbf{x})|^2\mathrm{d}\mathbf{x}.$$

The interesting point of this problem is that the rate of decay is independent of the spatial dimension n.

Let $\rho > 0$ and $\sigma > 0$ be positive constants. Using similar ideas, we may establish the exact L^2 limits of the global weak solutions of the following model equation

$$\mathbf{u}_t + \varepsilon(-\triangle)^\rho\mathbf{u}_t + \alpha(-\triangle)^\sigma\mathbf{u} + \mathcal{D}\mathbf{u} + \mathcal{N}(\mathbf{u},\nabla\mathbf{u}) = 0, \qquad \mathbf{u}(\mathbf{x},0) = \mathbf{u}_0(\mathbf{x}).$$

5.1. A General Result and an Open Problem

Consider the Cauchy problems for the differential equations (1) and (7).

Theorem 5.0 *Suppose that the initial data $\mathbf{u}_0 \in L^1(\mathbb{R}^n)\cap H^2(\mathbb{R}^n)$. Let $\{f_m\}$ and $\{g_m\}$ be two sequences of functions, where $\{f_m\}$ are defined on $[0,\infty)$, such that $f_m(0) = 1$, $f_m' > 0$ and that $f_m(t) \to \infty$, as $t \to \infty$, and $\{g_m\}$ are defined on $\mathbb{R}^n$. Suppose that the solutions of the nonlinear equation (1) and the linear equation (7) satisfy:* **(I)**

$$\lim_{t\to\infty}\left[f_m(t)\|\Delta^m\mathbf{v}(\cdot,t)\|_{L^p(\mathbb{R}^n)}\right] = g_m(\mathbf{u}_0),$$

and **(II)**

$$\lim_{t\to\infty}\left[f_m(t)\|\Delta^m\mathbf{u}(\cdot,t) - \Delta^m\mathbf{v}(\cdot,t)\|_{L^p(\mathbb{R}^n)}\right] = 0,$$

where $p \in (1, \infty)$. *Then*

$$\lim_{t\to\infty} \left[f_m(t) \|\Delta^m \mathbf{u}(\cdot, t)\|_{L^p(\mathbb{R}^n)} \right] = g_m(\mathbf{u}_0).$$

The proof of Theorem 5.0 is almost identical to the proof of Theorem 1 and is omitted.
#

An open problem: The following limit may exist. But can we compute the limit in terms of the initial data and the coefficients of the differential equations (1)-(2)? The limit problem is

$$\lim_{t\to\infty} \left\{ (1+t)^{[m+(n+\lambda)/2]p-n/2} \int_{\mathbb{R}^n} |\Delta^m \mathbf{u}(\mathbf{x}, t)|^p d\mathbf{x} \right\}^{1/p},$$

where $1 < p < \infty$ and $m \geq 0$ is any integer.

We believe that the limit is equal to zero as $p \to \infty$.

The exact limits may be used to study the Hausdorff dimension of the global attractors of the equations.

5.2. More Equations

Consider the following model equations.

1. The two-dimensional fluid dynamics equation arising from geophysics

$$u_t - \Delta u_t + \beta \cdot \nabla u + \gamma \cdot \nabla \Delta u + \Delta^2 u + u_x \Delta u_y - u_y \Delta u_x + \delta \cdot \nabla(u^3) = 0, \quad \text{in } \mathbb{R}^2 \times \mathbb{R}^+,$$

where $\beta = (\beta_1, \beta_2)^T$, $\gamma = (\gamma_1, \gamma_2)^T$ and $\delta = (\delta_1, \delta_2)^T$ are real constant vectors.

2. The two-dimensional nonlinear nonlocal quasi-geostrophic equation

$$u_t + (-\Delta)^\sigma u + (\psi_y, -\psi_x) \cdot \nabla u = 0, \qquad u = (-\Delta)^{1/2}\psi \qquad \text{in } \mathbb{R}^2 \times \mathbb{R}^+,$$

where $0 < \sigma < 1$ is a constant. The Fourier transform of the dissipation is equal to

$$\widehat{(-\Delta)^\sigma u}(\xi, \eta, t) = \left(|\xi|^2 + |\eta|^2\right)^\sigma \widehat{u}(\xi, \eta, t).$$

3. The n-dimensional Cahn-Hilliard equation

$$u_t + \alpha \Delta^2 u = \beta \Delta(u^2).$$

4. The n-dimensional non-degenerate system of filtration type

$$\mathbf{u}_t - \alpha \Delta_{\mathbf{x}} \mathbf{u} = \Delta_{\mathbf{x}} [\nabla_{\mathbf{u}} \varphi(\mathbf{u})],$$

where

$$\nabla_{\mathbf{u}} \varphi(\mathbf{u}) = \left(\frac{\partial \varphi}{\partial u_1}(\mathbf{u}), \frac{\partial \varphi}{\partial u_2}(\mathbf{u}), \cdots, \frac{\partial \varphi}{\partial u_n}(\mathbf{u}) \right)^T.$$

5. The n-dimensional Landau-Lifschitz system

$$\begin{aligned}\frac{\partial \mathbf{Z}}{\partial t} &= \mathbf{Z}\times\triangle\mathbf{Z}+\mathbf{Z}\times(\mathbf{Z}\times\triangle\mathbf{Z}),\\ \mathbf{Z}(\mathbf{x},0) &= \mathbf{Z}_0(\mathbf{x}), \qquad |\mathbf{Z}_0(\mathbf{x})|=1, \text{ for all } \mathbf{x}\in\mathbb{R}^n,\end{aligned}$$

where $\mathbf{Z} = \mathbf{Z}(\mathbf{x},t) = (Z_1,Z_2,Z_3) \in \mathbb{R}^3$ is a real vector-valued function of $\mathbf{x} = (x_1,x_2,\cdots,x_n)^T \in \mathbb{R}^n$ and $t>0$.

6. The n-dimensional Kuramoto-Sivashinsky equation

$$u_t+\gamma u+\triangle u+\triangle^2 u+\Phi(u)\cdot\nabla u=0.$$

7. The nonlinear Schrödinger-Burgers equation

$$\mathbf{u}_t+\mathbf{i}\triangle\mathbf{u}+\beta\cdot\nabla(|\mathbf{u}|^p\mathbf{u})=\alpha\triangle\mathbf{u},$$

where $\alpha>0$ is a constant and $\beta=(\beta_1,\beta_2,\cdots,\beta_n)^T\in\mathbb{C}^n$ is a complex-valued constant vector.

8. The Zakharov-Burgers system

$$\triangle u_t-\alpha\triangle^2 u=\mathbf{i}\triangle^2 u-\mathbf{i}\nabla\cdot(v\nabla u), \qquad v_t=\triangle(v+|\nabla u|^2),$$

where u is a complex-valued function of $(\mathbf{x},t)$ and v is a real-valued function of $(\mathbf{x},t)$.

All of these model equations may be included in the more general system

$$\mathcal{P}\mathbf{u}_t+\mathcal{Q}\mathbf{u}+\mathcal{R}\mathbf{u}+\mathcal{F}(\mathbf{u},\nabla\mathbf{u},\triangle\mathbf{u})=\mathbf{0}, \qquad \text{in } \mathbb{R}^n\times\mathbb{R}^+,$$

where $\mathcal{P}$, $\mathcal{Q}$, and $\mathcal{R}$ are linear differential operators. In particular, $\mathcal{P}$ and $\mathcal{Q}$ are dissipative operators and $\mathcal{R}$ is a dispersion operator. $\mathcal{F}$ is a nonlinear smooth function of $\mathbf{u}$, $\nabla\mathbf{u}$, and $\triangle\mathbf{u}$.

If we use the same ideas as in this chapter, we can obtain similar results. Here we state some of them.

Theorem 5.1. Suppose that $u_0\in L^1(\mathbb{R}^n)\cap H^2(\mathbb{R}^n)$. Then the global solution of the n-dimensional Cahn-Hilliard equation $u_t+\alpha\triangle^2 u=\beta\triangle(u^2)$ enjoys the following limit

$$\lim_{t\to\infty}\left[(1+t)^{n/4}\int_{\mathbb{R}^2}|u(\mathbf{x},t)|^2\mathrm{d}\mathbf{x}\right]=\frac{1}{(2\pi)^n}\left[\int_{\mathbb{R}^n}\exp\big(-\alpha|\xi|^4\big)\mathrm{d}\xi\right]\left[\int_{\mathbb{R}^n}u_0(\mathbf{x})\mathrm{d}\mathbf{x}\right]^2.$$

Theorem 5.2. Suppose that $\mathbf{u}_0\in L^1(\mathbb{R}^n)\cap H^2(\mathbb{R}^n)$. Suppose also that $|\nabla_{\mathbf{u}}\varphi(\mathbf{u})|\le C|\mathbf{u}|^2$, for all $\mathbf{u}$ with $|\mathbf{u}|\le 1$ and some constant $C>0$. Then the global solution of the n-dimensional non-degenerate system of filtration type equation $\mathbf{u}_t-\alpha\triangle_{\mathbf{x}}\mathbf{u}=\triangle_{\mathbf{x}}[\nabla_{\mathbf{u}}\varphi(\mathbf{u})]$ enjoy the limit

$$\lim_{t\to\infty}\left[(1+t)^{n/2}\int_{\mathbb{R}^n}|\mathbf{u}(\mathbf{x},t)|^2\mathrm{d}\mathbf{x}\right]=\frac{1}{(2\pi)^n}\left(\frac{\pi}{2\alpha}\right)^{n/2}\left[\int_{\mathbb{R}^n}\mathbf{u}_0(\mathbf{x})\mathrm{d}\mathbf{x}\right]^2.$$

5.3. An Open Problem for a Non-degenerate Parabolic System

Certainly, the most important open problem is the global existence of strong or smooth solutions of (1)-(2).

Suppose that the initial data satisfy the following conditions

$$\widehat{\mathbf{u}}_0(\mathbf{0}) = \mathbf{0},$$
$$\frac{\partial \widehat{\mathbf{u}}_0}{\partial \xi_j}(\mathbf{0}) = \mathbf{0},$$
$$\frac{\partial^2 \widehat{\mathbf{u}}_0}{\partial \xi_j \partial \xi_k}(\mathbf{0}) = \mathbf{0},$$

for all integers $j, k = 1, 2, \cdots, n$ and

$$\frac{\partial^3 \widehat{\mathbf{u}}_0}{\partial \xi_j \partial \xi_k \partial \xi_l}(\mathbf{0}) \neq \mathbf{0},$$

for some $j, k, l = 1, 2, \cdots, n$. Can we improve the decay results of the global weak solutions of the non-degenerate parabolic system of equations: $\mathbf{u}_t = \alpha \triangle \mathbf{u} + \triangle[\nabla_{\mathbf{u}} \varphi(\mathbf{u})]$?

5.4. An Open Problem for a Critical Equation

What we established in this chapter are decay results with sharp rates for global solutions of nonlinear evolution equations with supercritical nonlinearity. The following model is a dissipative partial differential equation with critical nonlinearity:

$$u_t - \varepsilon u_{xxt} + \delta u_x + \gamma H u_{xx} + \beta u_{xxx} + (u^2)_x = \alpha u_{xx}, \qquad \text{in } \mathbb{R} \times \mathbb{R}^+,$$

where $\alpha > 0$, and $\beta, \gamma, \delta, \varepsilon \in \mathbb{R}$ are nonnegative constants. Consider the initial value problems $u(x,0) = u_0(x)$. Suppose that the initial data $u_0 \in L^1(\mathbb{R}) \cap H^2(\mathbb{R})$, such that the Cauchy problems for the nonlinear equation supports a global strong solution $u \in L^\infty(\mathbb{R}^+, H^2(\mathbb{R})) \cap L^2_{\text{loc}}(\mathbb{R}^+, H^3(\mathbb{R})) \cap C(\mathbb{R}^+, H^2(\mathbb{R}))$. Then one can prove that

$$C_1 \leq \sup_{0<t<\infty} \left[(1+t)^{1/2} \int_{\mathbb{R}} |u(x,t)|^2 \mathrm{d}x \right] \leq C_2, \text{ if } \int_{\mathbb{R}} u_0(x) \mathrm{d}x \neq 0,$$

where $C_1 > 0$ and $C_2 > 0$ are constants independent of time and u, but they depend on the initial data u_0.

Suppose that $u_0(x) = \phi'(x)$ and $\phi \in C^1(\mathbb{R}) \cap L^1(\mathbb{R})$. The following limits

$$\lim_{t \to \infty} \left[(1+t)^{m+3/2} \int_{\mathbb{R}} |u_{x^m}(x,t)|^2 \mathrm{d}x \right],$$

in terms of the constant α and the integral of the initial data, $\int_{\mathbb{R}} u_0(x) \mathrm{d}x$, is open. To solve these limit problems, we believe that the well-known Cole-Hopf transform and the linear equation

$$v_t - \varepsilon v_{xxt} + \delta v_x + \gamma H v_{xx} + \beta v_{xxx} = \alpha v_{xx}$$

may play very important roles.

6. Appendices - More Known Results

6.1. A General Nonlinear Dissipative Dispersive Wave Equation

Consider the Cauchy problems

$$u_t - \varepsilon u_{xxt} + \delta u_x + \gamma H u_{xx} + \beta u_{xxx} + (u^3)_x = \alpha u_{xx}, \qquad u(x,0) = u_0(x).$$

Theorem 6.1 *Suppose that the initial data $u_0 \in L^1(\mathbb{R}) \cap H^2(\mathbb{R})$. Then there exists a unique global strong solution $u \in L^\infty(\mathbb{R}^+; H^2(\mathbb{R})) \cap L^2_{\mathrm{loc}}(\mathbb{R}^+; H^3(\mathbb{R}))$ to the Cauchy problem $u(x,0) = u_0(x)$, and that*

$$\lim_{x\to\pm\infty} u(x,t) = \lim_{x\to\pm\infty} u_x(x,t) = \lim_{x\to\pm\infty} u_{xx}(x,t) = 0, \qquad \forall t \geq 0.$$

Moreover, there hold the following estimates:

(I)

$$\sup_{t\in\mathbb{R}^+} \int_{\mathbb{R}} \left[|u(x,t)|^2 + \varepsilon|u_x(x,t)|^2\right] \mathrm{d}x \leq \int_{\mathbb{R}} \left[|u_0(x)|^2 + \varepsilon|u_{0x}(x)|^2\right] \mathrm{d}x,$$

$$2\alpha \int_0^\infty \left[\int_{\mathbb{R}} |u_x(x,t)|^2 \mathrm{d}x\right] \mathrm{d}t \leq \int_{\mathbb{R}} \left[|u_0(x)|^2 + \varepsilon|u_{0x}(x)|^2\right] \mathrm{d}x.$$

(II)

$$\sup_{t\in\mathbb{R}^+} \int_{\mathbb{R}} \left[|u_x(x,t)|^2 + \varepsilon|u_{xx}(x,t)|^2\right] \mathrm{d}x$$

$$\leq \int_{\mathbb{R}} \left[|u_{0x}(x)|^2 + \varepsilon|u_{0x}(x)|^2\right] \mathrm{d}x \exp\left\{\frac{9}{2\alpha^2}\left[\int_{\mathbb{R}} \left(|u_0(x)|^2 + \varepsilon|u_{0x}(x)|^2\right) \mathrm{d}x\right]^2\right\},$$

$$\alpha \int_0^\infty \left[\int_{\mathbb{R}} |u_{xx}(x,t)|^2 \mathrm{d}x\right] \mathrm{d}t$$

$$\leq \int_{\mathbb{R}} \left[|u_{0x}(x)|^2 + \varepsilon|u_{0x}(x)|^2\right] \mathrm{d}x \exp\left\{\frac{9}{2\alpha^2}\left[\int_{\mathbb{R}} \left(|u_0(x)|^2 + \varepsilon|u_{0x}(x)|^2\right) \mathrm{d}x\right]^2\right\}.$$

(III)

$$(1+t)^{1/2} \int_{\mathbb{R}} |u(x,t)|^2 \mathrm{d}x + (1+t)^{3/2} \int_{\mathbb{R}} |u_x(x,t)|^2 \mathrm{d}x \leq C(u_0),$$

$$\lim_{t\to\infty} \left[t^{1/2}\|u(\cdot,t)\|_{L^\infty(\mathbb{R})}\right] + \lim_{t\to\infty} \left[t^{3/2}\|u_x(\cdot,t)\|_{L^\infty(\mathbb{R})}\right] = 0.$$

(IV)

$$\left|\widehat{(u^3)_x}(\xi,t)\right| = |\xi| \left|\widehat{u^3}(\xi,t)\right|, \qquad \int_{\mathbb{R}} u(u^3)_x \mathrm{d}x = 0,$$

$$\left|\int_{\mathbb{R}} v(u^3)_x \mathrm{d}x\right| = \left|\int_{\mathbb{R}} v_x u^3 \mathrm{d}x\right| \leq \|v_x(\cdot,t)\|_{L^\infty(\mathbb{R})} \int_{\mathbb{R}} |u(x,t)|^3 \mathrm{d}x$$

$$\leq \|v_x(\cdot,t)\|_{L^\infty(\mathbb{R})} \left[\int_{\mathbb{R}} |u(x,t)|^2 \mathrm{d}x\right]^{5/4} \left[\int_{\mathbb{R}} |u_x(x,t)|^2 \mathrm{d}x\right]^{1/4}.$$

(V)

$$
\begin{aligned}
\widehat{u}(\xi,t) &= \exp\left[-\frac{\alpha|\xi|^2 - \mathbf{i}\xi(\beta\xi^2+\gamma|\xi|-\delta)}{1+\varepsilon\xi^2}t\right]\widehat{u_0}(\xi)\\
&- \frac{\mathbf{i}\xi}{1+\varepsilon\xi^2}\int_0^t \exp\left[-\frac{\alpha|\xi|^2 - \mathbf{i}\xi(\beta\xi^2+\gamma|\xi|-\delta)}{1+\varepsilon\xi^2}(t-\tau)\right]\widehat{u^3}(\xi,\tau)\mathrm{d}\tau,
\end{aligned}
$$

where C is a constant independent of time (depending only on the initial data u_0).

Sketch of Proof. The first four estimates in (I) and (II) follow from standard techniques in dynamical systems: multiplying the differential equation by $2u$ and $2u_{xx}$, respectively, integrating the results with respect to (x,τ) over $\mathbb{R}\times[0,t]$, using the boundary conditions $\lim_{x\to\pm\infty} u(x,t)=0$ and $\lim_{x\to\pm\infty} u_x(x,t)=0$, for all $t>0$, respectively, we obtain

$$
\begin{aligned}
&\int_{\mathbb{R}}\left[|u(x,t)|^2+\varepsilon|u_x(x,t)|^2\right]\mathrm{d}x + 2\alpha\int_0^t\left[\int_{\mathbb{R}}|u_x(x,\tau)|^2\mathrm{d}x\right]\mathrm{d}\tau\\
=\ &\int_{\mathbb{R}}\left[|u_0(x)|^2+\varepsilon|u_{0x}(x)|^2\right]\mathrm{d}x,\\
&\int_{\mathbb{R}}\left[|u_x(x,t)|^2+\varepsilon|u_{xx}(x,t)|^2\right]\mathrm{d}x + 2\alpha\int_0^t\int_{\mathbb{R}}|u_{xx}(x,\tau)|^2\mathrm{d}x\mathrm{d}\tau\\
=\ &\int_{\mathbb{R}}\left[|u_{0x}(x)|^2+\varepsilon|u_{0xx}(x)|^2\right]\mathrm{d}x\\
+\ &6\int_0^t\left[\int_{\mathbb{R}}u_{xx}(x,\tau)u^2(x,\tau)u_x(x,\tau)\mathrm{d}x\right]\mathrm{d}\tau.
\end{aligned}
$$

Applying the Cauchy-Schwartz inequality, we have

$$
\begin{aligned}
\left|6\int_{\mathbb{R}}u(x,t)^2u_x(x,t)u_{xx}(x,t)\mathrm{d}x\right| &\le \alpha\int_{\mathbb{R}}|u_{xx}(x,t)|^2\mathrm{d}x+\frac{9}{\alpha}\int_{\mathbb{R}}|u(x,t)|^4|u_x(x,t)|^2\mathrm{d}x\\
&\le \alpha\int_{\mathbb{R}}|u_{xx}(x,t)|^2\mathrm{d}x+\frac{9}{\alpha}\left[\int_{\mathbb{R}}|u(x,t)|^2\mathrm{d}x\right]\left[\int_{\mathbb{R}}|u_x(x,t)|^2\mathrm{d}x\right]^2.
\end{aligned}
$$

Then

$$
\begin{aligned}
&\int_{\mathbb{R}}\left[|u_x(x,t)|^2+\varepsilon|u_{xx}(x,t)|^2\right]\mathrm{d}x + \alpha\int_0^t\int_{\mathbb{R}}|u_{xx}(x,\tau)|^2\mathrm{d}x\mathrm{d}\tau\\
\le\ &\int_{\mathbb{R}}\left[|u_{0x}(x)|^2+\varepsilon|u_{0xx}(x)|^2\right]\mathrm{d}x\\
+\ &\frac{9}{\alpha}\int_{\mathbb{R}}\left[|u_0(x)|^2+\varepsilon|u_{0x}(x)|^2\right]\mathrm{d}x\int_0^t\left[\int_{\mathbb{R}}|u_x(x,\tau)|^2\mathrm{d}x\right]^2\mathrm{d}\tau.
\end{aligned}
$$

By using the Gronwall's inequality, we get the desired results. The technical details are standard and omitted. The next estimate follows from the Fourier splitting method and iteration technique. The results involving the nonlinearity $(u^3)_x$ follow from routine calculations. Taking the Fourier transform and using integrating factor idea yield the expression for $\widehat{u}(\xi,t)$. See [90], [92], [93], [97], [98], [100] for more details. #

6.2. The n-dimensional Magnetohydrodynamics Equations

Without loss of generality, suppose that the Reynolds number $RE = 1$ and the magnetic Reynolds number $RM = 1$.

Theorem 6.2 *Suppose that the initial data* $\mathbf{u}_0$ *and* $\mathbf{A}_0 \in L^1(\mathbb{R}^n) \cap H^2(\mathbb{R}^n)$, *such that* $\nabla \cdot \mathbf{u}_0 = \nabla \cdot \mathbf{A}_0 = 0$. *Then there exists at least a global weak solution* $\mathbf{u}, \mathbf{A} \in L^\infty(\mathbb{R}^+; L^2(\mathbb{R}^n)) \cap L^2_{\text{loc}}(\mathbb{R}^+; H^1(\mathbb{R}^n))$ *to the Cauchy problems* $\big(\mathbf{u}(\mathbf{x},0), \mathbf{A}(\mathbf{x},0)\big) = \big(\mathbf{u}_0(\mathbf{x}), \mathbf{A}_0(\mathbf{x})\big)$ *for the Magnetohydrodynamics equations, such that*

$$\lim_{|\mathbf{x}|\to\infty} \big(\mathbf{u}(\mathbf{x},t), \mathbf{A}(\mathbf{x},t)\big) = \lim_{|\mathbf{x}|\to\infty} \big(\nabla\mathbf{u}(\mathbf{x},t), \nabla\mathbf{A}(\mathbf{x},t)\big) = \mathbf{0},$$

for all $t > 0$. *Moreover there hold the following estimates:*

(I)

$$\begin{aligned}
\sup_{t\in\mathbb{R}^+} \int_{\mathbb{R}^n} \big|\big(\mathbf{u}(\mathbf{x},t), \mathbf{A}(\mathbf{x},t)\big)\big|^2 \mathrm{dx} &\leq \int_{\mathbb{R}^n} \big|\big(\mathbf{u}_0(\mathbf{x}), (\mathbf{A}_0(\mathbf{x})\big)\big|^2 \mathrm{dx}, \\
2\alpha \int_0^\infty \left[\int_{\mathbb{R}^n} \big|\big(\nabla\mathbf{u}(\mathbf{x},t), \nabla\mathbf{A}(\mathbf{x},t)\big)\big|^2 \mathrm{dx}\right] dt &\leq \int_{\mathbb{R}^n} \big|\big(\mathbf{u}_0(\mathbf{x}), \mathbf{A}_0(\mathbf{x})\big)\big|^2 \mathrm{dx}, \\
(1+t)^{n/2} \int_{\mathbb{R}^n} \big|\big(\mathbf{u}(\mathbf{x},t), \mathbf{A}(\mathbf{x},t)\big)\big|^2 \mathrm{dx} &\leq C(\mathbf{u}_0, \mathbf{A}_0),
\end{aligned}$$

for all spatial dimension $n \geq 2$.

(II) *If* $n = 2$, *then we have*

$$\begin{aligned}
&\sup_{t\in\mathbb{R}^+} \int_{\mathbb{R}^2} \big|\big(\nabla\mathbf{u}(\mathbf{x},t), \nabla\mathbf{A}(\mathbf{x},t)\big)\big|^2 \mathrm{dx} \\
&\leq \int_{\mathbb{R}^2} \big|\big(\nabla\mathbf{u}_0(\mathbf{x}), \nabla\mathbf{A}_0(\mathbf{x})\big)\big|^2 \mathrm{dx} \exp\left\{\frac{1}{2}\left[\int_{\mathbb{R}^2} \big|\big(\mathbf{u}_0(\mathbf{x}), \mathbf{A}_0(\mathbf{x})\big)\big|^2 \mathrm{dx}\right]^4\right\}, \\
&\alpha \int_0^\infty \left[\int_{\mathbb{R}^2} |(\Delta\mathbf{u}(\mathbf{x},t), \Delta\mathbf{A}(\mathbf{x},t))|^2 \mathrm{dx}\right] dt \\
&\leq \int_{\mathbb{R}^2} \big|\big(\nabla\mathbf{u}_0(\mathbf{x}), \nabla\mathbf{A}_0(\mathbf{x})\big)\big|^2 \mathrm{dx} \exp\left\{\frac{1}{2}\left[\int_{\mathbb{R}^2} \big|\big(\mathbf{u}_0(\mathbf{x}), \mathbf{A}_0(\mathbf{x})\big)\big|^2 \mathrm{dx}\right]^4\right\}.
\end{aligned}$$

(III)

$$\begin{aligned}
&\widehat{\mathbf{u}}(\xi,t) = \exp\big(-|\xi|^2 t\big)\widehat{\mathbf{u}}_0(\xi) \\
&- \int_0^t \exp\big[-|\xi|^2(t-\tau)\big]\left[\widehat{(\mathbf{u}\cdot\nabla)\mathbf{u}}(\xi,\tau) - \widehat{(\mathbf{A}\cdot\nabla)\mathbf{A}}(\xi,\tau) + \nabla\widehat{\left(p + \frac{\kappa}{2}|A|^2\right)}(\xi,\tau)\right] d\tau, \\
&\widehat{\mathbf{A}}(\xi,t) = \exp\big(-|\xi|^2 t\big)\widehat{\mathbf{A}}_0(\xi) \\
&- \int_0^t \exp\big[-|\xi|^2(t-\tau)\big]\left[\widehat{(\mathbf{u}\cdot\nabla)\mathbf{A}}(\xi,\tau) - \widehat{(\mathbf{A}\cdot\nabla)\mathbf{u}}(\xi,\tau)\right] d\tau,
\end{aligned}$$

(IV)

$$\begin{aligned}
&\int_{\mathbb{R}^n} \mathbf{u}\cdot[(\mathbf{u}\cdot\nabla)\mathbf{u}]\mathrm{dx} = 0, \\
&\left|\int_{\mathbb{R}^n} \mathbf{v}\cdot[(\mathbf{u}\cdot\nabla)\mathbf{u}]\mathrm{dx}\right| = \left|\int_{\mathbb{R}^n} \mathbf{u}\cdot[(\mathbf{u}\cdot\nabla)\mathbf{v}]\mathrm{dx}\right| \leq \|\nabla\mathbf{v}(\cdot,t)\|_{L^\infty(\mathbb{R}^n)} \int_{\mathbb{R}^n} |\mathbf{u}(\mathbf{x},t)|^2 \mathrm{dx}.
\end{aligned}$$

(V)

$$\sup_{t\in\mathbb{R}^+} \|\widehat{P}(\cdot,t)\|_{L^\infty(\mathbb{R}^n)} \leq \big\|(\mathbf{u}_0, \mathbf{A}_0)\big\|^2_{L^2(\mathbb{R}^n)},$$
$$2\int_0^\infty \big\|\widehat{\triangle P}(\cdot,t)\big\|_{L^\infty(\mathbb{R}^n)} dt \leq \big\|(\mathbf{u}_0, \mathbf{A}_0)\big\|^2_{L^2(\mathbb{R}^n)},$$
$$\big|\widehat{(\mathbf{u}\cdot\nabla)\mathbf{u}}(\xi,t)\big| \leq |\xi|\|\mathbf{u}(\cdot,t)\|^2_{L^2(\mathbb{R}^n)},$$
$$\big|\widehat{(\mathbf{A}\cdot\nabla)\mathbf{A}}(\xi,t)\big| \leq |\xi|\|\mathbf{A}(\cdot,t)\|^2_{L^2(\mathbb{R}^n)},$$
$$\big|\widehat{(\mathbf{u}\cdot\nabla)\mathbf{A}}(\xi,t)\big| \leq |\xi|\|\mathbf{u}(\cdot,t)\|_{L^2(\mathbb{R}^n)}\|\mathbf{A}(\cdot,t)\|_{L^2(\mathbb{R}^n)},$$
$$\big|\widehat{(\mathbf{A}\cdot\nabla)\mathbf{u}}(\xi,t)\big| \leq |\xi|\|\mathbf{u}(\cdot,t)\|_{L^2(\mathbb{R}^n)}\|\mathbf{A}(\cdot,t)\|_{L^2(\mathbb{R}^n)},$$
$$\big|\big(\widehat{\mathbf{u}}(\xi,t), \widehat{\mathbf{A}}(\xi,t)\big)\big| \leq \big\|(\mathbf{u}_0, \mathbf{A}_0)\big\|_{L^1(\mathbb{R}^n)} + 3|\xi|\int_0^t \|(\mathbf{u},\mathbf{A})(\cdot,\tau)\|^2 d\tau.$$

where C is a constant independent of time (depending only on the initial data $\mathbf{u}_0$).
Sketch of Proof. Multiplying the differential equations

$$\mathbf{u}_t + (\mathbf{u}\cdot\nabla)\mathbf{u} - (\mathbf{A}\cdot\nabla)\mathbf{A} - \triangle\mathbf{u} + \nabla\left(p + \frac{\kappa}{2}|\mathbf{A}|^2\right) = 0$$

and

$$\mathbf{A}_t + (\mathbf{u}\cdot\nabla)\mathbf{A} - (\mathbf{A}\cdot\nabla)\mathbf{u} - \triangle\mathbf{A} = 0$$

by $2\mathbf{u}$ and $2\mathbf{A}$, respectively, integrating the results with respect to $(\mathbf{x},\tau)$ over $\mathbb{R}^n \times [0,t]$, using the boundary conditions $\lim_{|\mathbf{x}|\to\infty}\big(\mathbf{u}(\mathbf{x},t), \mathbf{A}(\mathbf{x},t)\big) = (\mathbf{0},\mathbf{0})$, we have

$$\int_{\mathbb{R}^n}\left[|\mathbf{u}(\mathbf{x},t)|^2 + |\mathbf{A}(\mathbf{x},t)|^2\right] d\mathbf{x} + 2\int_0^t\int_{\mathbb{R}^n}\left[|\nabla\mathbf{u}(\mathbf{x},\tau)|^2 + |\nabla\mathbf{A}(\mathbf{x},\tau)|^2\right] d\mathbf{x}d\tau = \int_{\mathbb{R}^n}\left[|\mathbf{u}_0(\mathbf{x})|^2 + |\mathbf{A}_0(\mathbf{x})|^2\right] d\mathbf{x}.$$

Taking the Fourier transform and using integrating factor idea yield the expressions for $\widehat{\mathbf{u}}(\xi,t)$ and $\widehat{\mathbf{A}}(\xi,t)$:

$$\widehat{\mathbf{u}}(\xi,t) = \exp\big(-|\xi|^2 t\big)\widehat{\mathbf{u}}_0(\xi)$$
$$-\int_0^t \exp\left[-|\xi|^2(t-\tau)\right]\left[\widehat{(\mathbf{u}\cdot\nabla)\mathbf{u}}(\xi,\tau) - \widehat{(\mathbf{A}\cdot\nabla)\mathbf{A}}(\xi,\tau) + \nabla\widehat{\left(p + \frac{\kappa}{2}|\mathbf{A}|^2\right)}(\xi,\tau)\right] d\tau,$$
$$\widehat{\mathbf{A}}(\xi,t) = \exp\big(-|\xi|^2 t\big)\widehat{\mathbf{A}}_0(\xi)$$
$$-\int_0^t \exp\left[-|\xi|^2(t-\tau)\right]\left[\widehat{(\mathbf{u}\cdot\nabla)\mathbf{A}}(\xi,\tau) - \widehat{(\mathbf{A}\cdot\nabla)\mathbf{u}}(\xi,\tau)\right] d\tau.$$

The decay estimate follows from the Fourier splitting method and iteration technique. The results involving the nonlinearity $(\mathbf{u}\cdot\nabla)\mathbf{u}$, $(\mathbf{A}\cdot\nabla)\mathbf{A}$, $(\mathbf{u}\cdot\nabla)\mathbf{A}$ and $(\mathbf{A}\cdot\nabla)\mathbf{u}$ follow from routine calculations. #

We see that all of the assumptions mentioned in (3)-(4)-(5)-(6) are satisfied by the global solutions of these equations.

For more detailed results on existence, uniqueness, regularity and decay estimates of global solutions of various partial differential equations, please see [1]-[2], [4], [5]-[8], [9]-[12], [13], [14]-[16], [17]-[20], [21]-[23], [24]-[26], [27]-[32], [33]-[36], [40], [41]-[46], [48], [54], [57]-[59], [61]-[62], [72], [79]-[80], [81]-[83], and [109]-[111].

6.3. The Fourier Transform and Its Properties

Let $\phi \in C_0^\infty(\mathbb{R}^n)$, define the Fourier transform and the inverse Fourier transform of ϕ by

$$\widehat{\phi}(\xi) = \int_{\mathbb{R}^n} \exp(-\mathbf{i}\mathbf{x}\cdot\xi)\phi(\mathbf{x})\mathrm{d}\mathbf{x}, \qquad \check{\phi}(x) = \frac{1}{(2\pi)^n}\int_{\mathbb{R}^n} \exp(+\mathbf{i}\mathbf{x}\cdot\xi)\phi(\xi)\mathrm{d}\xi.$$

Note that $C_0^\infty(\mathbb{R}^n)$ is dense in $L^p(\mathbb{R})$, where $1 \le p \le \infty$, in the sense of $\|\cdot\|_{L^p(\mathbb{R}^n)}$-norm. The definitions of the Fourier transform and the inverse Fourier transform may be extended to $L^p(\mathbb{R}^n)$. Observe that there holds the estimate $\|\widehat{\phi}\|_{L^q(\mathbb{R}^n)} \le C(p)\|\phi\|_{L^p(\mathbb{R}^n)}$, where $1 \le p \le 2$ and $\frac{1}{p}+\frac{1}{q}=1$, and $C(p)>0$ is a constant independent of ϕ.

Example 1. Let $\phi(\mathbf{x}) = \delta\exp\left(-\gamma|\mathbf{x}|^2\right)$, with a positive constant $\gamma > 0$, then the Fourier transform

$$\widehat{\phi}(\xi) = \delta\left(\frac{\pi}{\gamma}\right)^{n/2}\exp\left(-\frac{1}{4\gamma}|\xi|^2\right) \qquad \text{in } \mathbb{R}^n.$$

Let us review several important properties of the Fourier transform.

If ϕ is even, then $\widehat{\phi}$ is also even. If ϕ is odd, then $\widehat{\phi}$ is also odd. Additionally

$$\widehat{\phi_x}(\xi) = \mathbf{i}\xi\widehat{\phi}(\xi), \qquad \widehat{\phi_{xx}}(\xi) = -\xi^2\widehat{\phi}(\xi) \qquad \text{in } \mathbb{R}.$$

In particular, $\widehat{\Delta\phi}(\xi) = -|\xi|^2\widehat{\phi}(\xi)$ in $\mathbb{R}^n$. Furthermore

$$(\widehat{\phi})\check{} = (\check{\phi})\widehat{} = \phi, \qquad \left|\widehat{\phi\psi}(\xi)\right| \le \|\phi\|_{L^2(\mathbb{R}^n)}\|\psi\|_{L^2(\mathbb{R}^n)}.$$

6.4. Several Well-Known Inequalities and an Identity

Some inequalities used in this book chapter are listed below. Denote partial differential operators by

$$\mathcal{D}^{\mathcal{A}} = \frac{\partial^{\alpha_1+\alpha_2+\cdots+\alpha_n}}{\partial x_1^{\alpha_1}\partial x_2^{\alpha_2}\cdots\partial x_n^{\alpha_n}},$$

where $\mathcal{A} = (\alpha_1, \alpha_2, \cdots, \alpha_n)^T$.

Lemma 6.1 (The Gagliardo-Nirenberg's interpolation inequality)
Let $p \ge 1$, $q \ge 1$ and $r \ge 1$. For any positive integers m and k with $m > k$, there exists a number $\alpha \in (k/m, 1)$ and a positive constant $C = C(p,q,r,m,k)$, such that for all functions $\mathbf{u} \in C_0^\infty(\mathbb{R}^n)$, there holds the estimate

$$\left[\sum_{|\mathcal{A}|=k}\|\mathcal{D}^{\mathcal{A}}\mathbf{u}\|_{L^p(\mathbb{R}^n)}^p\right]^{1/p} \le C\left[\sum_{|\mathcal{A}|=m}\|\mathcal{D}^{\mathcal{A}}\mathbf{u}\|_{L^r(\mathbb{R}^n)}^r\right]^{\alpha/r}\|\mathbf{u}\|_{L^q(\mathbb{R}^n)}^{1-\alpha},$$

where

$$\frac{n}{p} - k = \alpha\left(\frac{n}{r} - m\right) + (1-\alpha)\frac{n}{q}.$$

This estimate is also correct if $\mathbf{u} \in W^{m,r}(\mathbb{R}^n) \cap L^q(\mathbb{R}^n)$.

Lemma 6.2 (The Cauchy-Schwartz's inequality)
Let $a_1, a_2, \cdots, a_m$ *and* $b_1, b_2, \cdots, b_m$ *be complex numbers. Then*

$$\left|a_1b_1 + a_2b_2 + \cdots + a_mb_m\right|^2 \leq \left[|a_1|^2 + |a_2|^2 + \cdots + |a_m|^2\right]\left[|b_1|^2 + |b_2|^2 + \cdots + |b_m|^2\right].$$

Lemma 6.3 (The Hölder's inequality)
Let $\phi \in L^p(\mathbb{R}^n)$ *and* $\psi \in L^q(\mathbb{R}^n)$, *where* $p, q \in [1, \infty]$, *such that* $\frac{1}{p} + \frac{1}{q} = 1$. *Then*

$$\left|\int_{\mathbb{R}^n} \phi(\mathbf{x})\psi(\mathbf{x})\mathrm{d}\mathbf{x}\right| \leq \|\phi\|_{L^p(\mathbb{R}^n)}\|\psi\|_{L^q(\mathbb{R}^n)}.$$

Lemma 6.4 (The Gronwall's inequality)
Let the nonnegative, integrable functions g *and* h *satisfy the inequality*

$$g(t) \leq C + \int_0^t g(\tau)h(\tau)\mathrm{d}\tau,$$

for all $t > 0$, *and for some positive constant* $C > 0$. *Then we have*

$$g(t) \leq C\exp\left\{\int_0^t h(\tau)\mathrm{d}\tau\right\},$$

for all $t > 0$, *where the function* $h \in L^1(0, \infty)$.

Lemma 6.5 The Plancherel's identity: For any function $\phi \in L^2(\mathbb{R}^n)$, there holds

$$\int_{\mathbb{R}^n} |\phi(\mathbf{x})|^2\mathrm{d}\mathbf{x} = \frac{1}{(2\pi)^n}\int_{\mathbb{R}^n} \left|\widehat{\phi}(\xi)\right|^2 \mathrm{d}\xi.$$

Acknowledgments

Glory to the Lord Almighty, Jesus Christ!
The author gains valuable research insights and experiences while he was studying mathematical courses or doing research projects at the following institutions: Lehigh University (2002 - 2008, Tenure Track Assistant Professor); The University of Minnesota (1999 - 2002, Dunham Jackson Assistant Professor); The Ohio State University (1994 - 1999, Ph.D student, advisor: Professor David Terman); Beijing Institute of Applied Physics and Computational Mathematics (Phase I: 1986 - 1989, Master of Science Program, advisors: Professor Yulin Zhou and Professor Boling Guo, members of the Chinese Academy of Sciences; Phase II: 1989 - 1994, Research Assistant).

The author is very grateful to Professor Donald G. Aronson, Professor Yulin Zhou and Professor Boling Guo for their valuable comments and suggestions on the early version of the manuscript.

References

[1] John Albert, Dispersion of low-energy waves for the generalized Benjamin-Bona-Mahony equation. *Journal of Differential Equations*, **63**(1986), 117-134.

[2] John Albert, On the decay of solutions of the generalized Benjamin-Bona-Mahony equations. *Journal of Mathematical Analysis and Applications*, **141**(1989), 527-537.

[3] Charles J. Amick, On the asymptotic form of Navier-Stokes flow past a body in the plane. *Journal of Differential Equations*, **91**(1991), 149-167.

[4] Charles J. Amick, Jerry L. Bona and Maria E. Schonbek, Decay of solutions of some nonlinear wave equations. *Journal of Differential Equations*, **81**(1989), 1-49.

[5] Hugo Beirao da Veiga, Existence and asymptotic behavior for strong solutions of the Navier-Stokes equations in the whole space. *Indiana University Mathematics Journal*, **36**(1987), 149-166.

[6] Hugo Beirao da Veiga, On the smoothness of a class of weak solutions to the Navier-Stokes equations. *Journal of Mathematical Fluid Mechanics*, **2**(2000), 315-323.

[7] Hugo Beirao da Veiga, Time periodic solutions of the Navier-Stokes equations in unbounded cylindrical domains-Leray's problem for periodic flows. *Archive for Rational Mechanics and Analysis*, **178**(2005), 301-325.

[8] Hugo Beirao da Veiga and Paolo Secchi, L^p-stability for the strong solutions of the Navier-Stokes equations in the whole space. *Archive for Rational Mechanics and Analysis*, **98**(1987), 65-69.

[9] Jerry L. Bona and Laihan Luo, Decay of solutions to nonlinear, dispersive wave equations. *Differential and Integral Equations*, **6**(1993), 961-980.

[10] Jerry L. Bona and Laihan Luo, More results on the decay of solutions to nonlinear, dispersive wave equations. *Discrete and Continuous Dynamical Systems*, **1**(1995), 151-193.

[11] Jerry L. Bona and Laihan Luo, A generalized Korteweg-de Vries equation in a quarter plane. Applied analysis (Baton Rouge, LA, 1996), 59-125, Contemporary Mathematics, 221, American Mathematical Society, Providence, RI, 1999.

[12] Jerry L. Bona and Laihan Luo, Asymptotic decomposition of nonlinear, dispersive wave equations with dissipation. Advances in nonlinear mathematics and science. *Physica D*, **152/153**(2001), 363-383.

[13] Luis A. Caffarelli, Robert V. Kohn and Louis Nirenberg, Partial regularity of suitably weak solutions of the Navier-Stokes equations, *Communications on Pure and Applied Mathematics*, **35**(1982), 771-831.

[14] Ana Carpio, Asymptotic behavior for the vorticity equations in dimensions two and three. *Communications in Partial Differential Equations*, **19**(1994), 827-872.

[15] Ana Carpio, Existence of global solutions to some nonlinear dissipative wave equations. *Journal de Mathematiques Pures et Appliquees*, **73**(1994), 471-488.

[16] Ana Carpio, Large-time behavior in incompressible Navier-Stokes equations. *SIAM Journal on Mathematical Analysis*, **27**(1996), 449-475.

[17] Dongho Chae, On the regularity conditions for the dissipative quasi-geostrophic equations. *SIAM Journal on Mathematical Analysis*, **37**(2006), 1649-1656.

[18] Dongho Chae, On the continuation principles for the Euler equations and the quasi-geostrophic equation. *Journal of Differential Equations*, **227**(2006), 640-651.

[19] Alexandre J. Chorin and Ole H. Hald, Viscosity-dependent inertial spectra of the Burgers and Korteweg-de Vries-Burgers equations. *Proceedings of the National Academy of Sciences, USA*, **102**(2005), 3921-3923.

[20] Peter Constantin, Diego Cordoba and Jiahong Wu, On the critical dissipative quasi-geostrophic equation. Dedicated to Professors Ciprian Foias and Roger Temam (Bloomington, IN, 2000). *Indiana University Mathematics Journal*, **50** (2001), Special Issue, 97-107.

[21] Peter Constantin and Charles Fefferman, Direction of vorticity and the problem of global regularity for the Navier-Stokes equations. *Indiana University Mathematics Journal*, **42**(1993), 775-789.

[22] Peter Constantin and Ciprian Foias, "Navier-Stokes equations." University of Chicago Press, Chicago, 1988.

[23] Peter Constantin and Jiahong Wu, Behavior of solutions of 2D quasi-geostrophic equations. *SIAM Journal on Mathematical Analysis*, **30**(1999), 937-948.

[24] Daniel. B. Dix, Temporal asymptotic behavior of solutions of the Benjamin-Ono-Burgers equations. *Journal of Differential Equations*, **90**(1991), 238-287.

[25] Daniel. B. Dix, The dissipation of nonlinear dispersive waves: The case of asymptotically weak nonlinearity. *Communications in Partial Differential Equations*, **17**(1992), 1665-1693.

[26] Daniel. B. Dix, Large-time behaviour of solutions of Burgers' equation. *Proceedings of the Royal Society of Edinburgh, Section A*, **132**(2002), 843-878.

[27] Gema Duro and Enrique Zuazua, Large time behavior for convection-diffusion equations in $\mathbb{R}^n$ with asymptotically constant diffusion. *comptes rendus de l'Academie des sciences - series i - mathematics*, **321**(1995), 1419-1424.

[28] Gema Duro and Enrique Zuazua, Large time behavior for convection-diffusion equations in $\mathbb{R}^n$ with asymptotically constant diffusion. *Communications in Partial Differential Equations*, **24**(1999), 1283-1340.

[29] Gema Duro and Enrique Zuazua, Large time behavior for convection-diffusion equations in $\mathbb{R}^n$ with periodic coefficients. *Journal of Differential Equations*, **167**(2000), 275-315.

[30] Miguel Escobedo and Enrique Zuazua, Self-similar solutions for a convection-diffusion equation with absorption in $\mathbb{R}^n$. *Israel Journal of Mathematics*, **74**(1991), 47-64.

[31] Miguel Escobedo and Enrike Zuazua, Large time behavior for convection-diffusion equations in $\mathbb{R}^n$. *Journal of Functional Analysis*, **100**(1991), 119-161.

[32] Miguel Escobedo and Enrique Zuazua, Long-time behavior for a convection-diffusion equation in higher dimensions. *SIAM Journal on Mathematical Analysis*, **28**(1997), 570-594.

[33] Zhaosheng Feng, An exact solution to the Korteweg-de Vries-Burgers equation. *Applied Mathematics Letters*, **18** (2005), 733-737.

[34] Ciprian Foias, Oscar P. Manley, Roger Temam and Yvain M. Treve, Asymptotic analysis of the Navier-Stokes equations, *Physica D*, **9**(1983), 157-188.

[35] Ciprian Foias and Roger Temam, Some analytic and geometric properties of the solutions of the Navier-Stokes equations, *Journal de Mathematiques Pures et Appliquees*, **58**(1979), 339-368.

[36] Ciprian Foias and Roger Temam, Approximation of attractors by algebraic or analytic sets. *SIAM Journal on Mathematical Analysis*, **25**(1994), 1269-1302.

[37] A.S. Fokas and Laihan Luo, Global solutions and their asymptotic behavior for Benjamin-Ono-Burgers type equations. *Differential and Integral Equations*, **13**(2000), 115-124.

[38] Gerald B. Folland, Real Analysis, Modern Techniques and Their Applications. Second Edition. A Wiley-Interscience Publication John Wiley & Sons, Inc. New York-Chichester-Weinheim-Brisbane-Singapore-Toronto.

[39] Avner Friedman, "Foundations of Modern Analysis." Holt, Rinehart and Winston Inc. New York. 1970.

[40] Boling Guo and Linghai Zhang, Decay of solutions to magnetohydrodynamics equations in two space dimensions. *Proceedings of the Royal Society of London, Series A: Mathematical and Physical Sciences*, **449**(1995), 79-91.

[41] Nakao Hayashi, Elena I. Kaikina and Pavel I. Naumkin, Large-time behaviour of solutions to the dissipative nonlinear Schrödinger equation. *Proceedings of the Royal Society of Edinburgh, Section A*. **130**(2000), 1029-1043.

[42] Cheng He and Tetsuro Miyakawa, Nonstationary Navier-Stokes flows in a two-dimensional exterior domain with rotational symmetries. *Indiana University Mathematics Journal*, **55**(2006), 1483-1555.

[43] Cheng He and Zhouping Xin, On the decay properties of solutions to the non-stationary Navier-Stokes equations in $\mathbb{R}^3$. *Proceedings of the Royal Society of Edinburgh, Section A*, **131**(2001), 597-619.

[44] Cheng He and Zhouping Xin, On the regularity of weak solutions to the magnetohydrodynamic equations. *Journal of Differential Equations*, **213**(2005), 235-254.

[45] Cheng He and Zhouping Xin, Partial regularity of suitable weak solutions to the incompressible magnetohydrodynamic equations. *Journal of Functional Analysis*, **227**(2005), 113-152.

[46] John G. Heywood, Remarks on the possible global regularity of solutions of the three-dimensional Navier-Stokes equations. Progress in theoretical and computational fluid mechanics (Paseky, 1993), 1-32, Pitman Research Notes in Mathematics, Series **308**, Longman Scientific Technology, Harlow, 1994.

[47] Eberhard Hopf, The partial differential equation $u_t + uu_x = \mu u_{xx}$. *Communications on Pure and Applied Mathematics*, **3**(1950), 201-230.

[48] Chun-Hsiung Hsia and Xiaoming Wang, On a Burgers' type equation. *Discrete and Continuous Dynamical Systems, Series B*, **6** (2006), 1121-1139.

[49] Tosio Kato, "Perturbation Theory for Linear Operators." 1976. Second edition. Grundlehren der Mathematischen Wissenschaften, Band 132. Springer-Verlag, Berlin-New York.

[50] Tosio Kato, Strong L^p-solutions of the Navier-Stokes equations in $\mathbb{R}^m$, with applications to weak solutions. *Mathematische Zeitschrift*, **187**(1984), 471-480.

[51] Tosio Kato, Strong solutions of the Navier-Stokes equations in Morrey spaces, *Boletim da Sociedade Brasileira de Matematica*, **22**(1992), 127-155.

[52] Tosio Kato, The Navier-Stokes equations for an incompressible fluid in $\mathbb{R}^n$ with measure as the initial vorticity, *Differential and Integral Equations*, **7**(1994), 949-966.

[53] Tosio Kato, On the smoothness of trajectories in incompressible perfect fluids. Nonlinear wave equations (Providence, Rhode Island, 1998), 109-130, *Contemporay Mathematics,* **263**, American Mathematical Society, Providence, RI, 2000.

[54] Olga Alexandrovna Ladyzhenskaya, "The Mathematical Theory of Viscous Incompressible Flow." Second English edition. Mathematical Applications. Gordon and Breach, New York, 1969.

[55] Jean Leray, Sur le mouvement d'un liquide visqueux emplissant l'espace. *Acta Mathematica*, **63**(1934), 193-248.

[56] Fang-Hua Lin, A new proof of the Caffarelli-Kohn-Nirenberg theorem. *Communications on Pure and Applied Mathematics*, **51**(1998), 241-257.

[57] Andrew J. Majda, Mathematical fluid dynamics: the interaction of nonlinear analysis and modern applied mathematics. *American Mathematical Society centennial publications*, Vol. II (Providence, RI, 1988), 351-394.

[58] Ming Mei, Large-time behavior of solution for generalized Benjamin-Bona-Mahony-Burgers equations. *Nonlinear Analysis, Theory and Methods*, **33**(1998), 699-714.

[59] Ming Mei, L^q-decay rates of solutions for Benjamin-Bona-Mahony-Burgers equations. *Journal of Differential Equations*, **158**(1999), 314-340.

[60] Jindrich Necas, Michael Ruzicka and Vladimir Sverak, On Leray's self-similar solutions of the Navier-Stokes equations. *Acta Mathematica*, **176**(1996), 283-294.

[61] Harumi Ono, Algebraic solitary waves in stratified fluids. *Journal of the Physical Society of Japan*, **39**(1975), 1082-1091.

[62] Raul Prado and Enrique Zuazua, Asymptotic expansion for the generalized Benjamin-Bona-Mahony-Burgers equation. *Differential and Integral Equations*, **15**(2002), 1409-1434.

[63] Maria E. Schonbek, L^2 decay for weak solutions of the Navier-Stokes equations. *Archive for Rational Mechanics and Analysis*, **88**(1985), 209-222.

[64] Maria E. Schonbek, Large time behaviour of solutions of the Navier-Stokes equations. *Communications in Partial Differential Equations* , **11**(1986), 733-763.

[65] Maria E. Schonbek, Lower bounds of rates of decay for solutions to the Navier-Stokes equations. *Journal of the American Mathematical Society*, **4**(1991), 423-449.

[66] Maria E. Schonbek and Tomas P. Schonbek, Asymptotic behavior to dissipative quasi-geostrophic flows. *SIAM Journal on Mathematical Analysis*, **35**(2003), 357-375.

[67] Maria E. Schonbek and Tomas Schonbek, Moments and lower bounds in the far-field of solutions to quasi-geostrophic flows. *Discrete and Continuous Dynamical Systems*, **13**(2005), 1277-1304.

[68] Maria E. Schonbek and Michael Wiegner, On the decay of higher-order norms of the solutions of Navier-Stokes equations. *Proceedings of the Royal Society of Edinburgh, Section A*, **126**(1996), 677-685.

[69] George R. Sell, Global attractors for the three-dimensional Navier-Stokes equations. *Journal of Dynamics and Differential Equations* , **8**(1996), 1-33.

[70] James Serrin, On the interior regularity of weak solutions of the Navier-Stokes equations. *Archive for Rational Mechanics and Analysis*, **9**(1962), 187-195.

[71] James Serrin, The initial value problems for the Navier-Stokes equations, in "Nonlinear Problems" (Proceedings of Symposium, Madison, Wisconsin, R. Langer, ed.), University of Wisconsin Press, Madison, 1963, 69-98.

[72] Milena Stanislavova, Atanas Stefanov, and Bixiang Wang, Asymptotic smoothing and attractors for the generalized Benjamin-Bona-Mahony equation on $\mathbb{R}^3$. *Journal of Differential Equations*, **219**(2005), 451-483.

[73] Elias M. Stein, "Singular Integrals and Differentiability Properties of Functions." Princeton University Press, Princeton, New Jersey, 1970.

[74] Angus E. Taylor, "Introduction to Functional Analysis." John Wiley & Sons, Inc., New York; Chapman & Hall, Ltd., London 1958.

[75] Roger Temam, "Navier-Stokes Equations, Theory and Numerical Analysis." North-Holland, Amsterdam and New York, 1979.

[76] Roger Temam, Navier-Stokes equations and Nonlinear Functional Analysis, CBMS-NSF Regional Conference Series in Applied Mathematics, SIAM, Philadelphia, 1983.

[77] Roger Temam, Attractors for Navier-Stokes equations, in *Nonlinear Partial Differential Equations and Their Applications*, College de France Seminar, **7**, Haim Brezis, Jacques-Louis Lions (Eds.), Pitman, London, 1985.

[78] Roger Temam, Generalization of the Sobolev-Tieb-Thirring inequality and applications to the dimension of attractors. *Differential and Integral Equations*, **1**(1988), 1-21.

[79] Roger Temam, Infinite-dimensional dynamical systems in mechanics and physics. Applied Mathematical Sciences, 68. Springer-Verlag, New York, 1988.

[80] Xiaoming Wang, Infinite Prandtl number limit of Rayleigh-Benard convection. *Communications on Pure and Applied Mathematics*, **57** (2004), 1265-1282.

[81] Michael Wiegner, Decay results for weak solutions of the Navier-Stokes equations on $\mathbb{R}^n$. *Journal of London Mathematical Society*, **35**(1987), 303-313.

[82] Michael Wiegner, Decay and stability in L^p for strong solutions of the Cauchy problem for the Navier-Stokes equations, in "The Navier-Stokes Equations, Theory and Numerical Methods", edited by J. Heywood *et al*, Springer-Verlag, New York. *Lecture Notes in Mathematics*, **1431**(1990), 95-99.

[83] Michael Wiegner, Decay of the L^∞-norm of solutions of the Navier-Stokes equations in unbounded domains. *Acta Applied Mathematics*, **37**(1994), 215-219.

[84] Dao-Xing Xia, Zhuo-Ren Wu, Shao-Zong Yan and Wu-Chang Shu, "Real Analysis and Functional Analysis." Advanced Educational Publishing House, Beijing 1988. (In Chinese)

[85] Zhouping Xin and Ping Zhang, On the uniqueness and large time behavior of the weak solutions to a shallow water equation. *Communications in Partial Differential Equations*, **27**(2002), 1815-1844.

[86] Gong-Qing Zhang and Mao-Zheng Guo, "Introduction of Functional Analysis." Peking University Press. 1990. (In Chinese)

[87] Linghai Zhang, Decay of solutions to 2-dimensional Navier-Stokes equations. *Chinese Advances in Mathematics*, **22**(1993), 469-472.

[88] Linghai Zhang, Decay of solutions of the multidimensional generalized Kuramoto-Sivashinsky system. *IMA Journal of Applied Mathematics*, **50**(1993), 29-42.

[89] Linghai Zhang, Decay of solutions of a higher order multidimensional nonlinear Korteweg-de Vries-Burgers system. *Proceedings of the Royal Society of Edinburgh, Section A: Mathematics*, **124**(1994), 263-271.

[90] Linghai Zhang, Decay estimates for the solutions of some nonlinear evolution equations. *Journal of Differential Equations*, **116**(1995), 31-58.

[91] Linghai Zhang, Sharp rate of decay of solutions to 2-dimensional Navier-Stokes equations. *Communications in Partial Differential Equations*, **20**(1995), 119-127.

[92] Linghai Zhang, Decay of solution of generalized Benjamin-Bona-Mahony-Burgers equations in n-space dimensions. *Nonlinear Analysis, Theory and Methods*, **25**(1995), 1343-1369.

[93] Linghai Zhang, Decay estimates for solutions to initial value problems for the generalized nonlinear Korteweg-de Vries-Burgers equation. *Chinese Annals of Mathematics*, **16A**(1995), 22-32.

[94] Linghai Zhang, Uniform stability and asymptotic behavior of solutions of 2-dimensional Magnetohydrodynamics equations. *Chinese Annals of Mathematics*, **19B**(1998), 35-58.

[95] Linghai Zhang, Long time uniform stability for solutions of n-dimensional Navier-Stokes equations. *Quarterly of Applied Mathematics*, **57**(1999), 283-315.

[96] Linghai Zhang, Uniform stability for solutions of n-dimensional Navier-Stokes equations. *Bulletin of the Institute of Mathematics, Academia Sinica*, **27**(1999), 265-315.

[97] Linghai Zhang, Local Lipschitz continuity of a nonlinear bounded operator induced by a generalized Benjamin-Ono-Burgers equation. *Nonlinear Analysis, Series A: Theory and Methods*, **39**(2000), 379-402.

[98] Linghai Zhang, Long-time asymptotic behaviors of solutions of n-dimensional dissipative partial differential equations. *Discrete and Continuous Dynamical Systems*, **8**(2002), 1025-1042.

[99] Linghai Zhang, On the modified Navier-Stokes equations in n-dimensional spaces. *Bulletin of the Institute of Mathematics, Academia Sinica*, **32**(2004), 185-193.

[100] Linghai Zhang, Dissipation and decay estimates. *Acta Mathematicae Applicatae Sinica, English Series*, **20**(2004), 49-76.

[101] Huijiang Zhao, Decay estimates for the solutions of some multidimensional nonlinear evolution equations. *Communications in Partial Differential Equations* , **25**(2000), 377-422.

[102] Huijiang Zhao, Asymptotic behaviors of solutions of quasilinear hyperbolic equations with linear damping. II. *Journal of Differential Equations* , **167**(2000), 467-494.

[103] Huijiang Zhao, Asymptotics of solutions of nonlinear parabolic equations. *Journal of Differential Equations* , **191**(2003), 544-594.

[104] Huijiang Zhao and Shaoqiang Tang, Nonlinear stability and optimal decay rate for a multidimensional generalized Kuramoto-Sivashinsky system. *Journal of Mathematical Analysis and Applications* , **246**(2000), 423-445.

[105] Yulin Zhou, Initial value problems for nonlinear degenerate systems of filtration type. *Chinese Annals of Mathematics, Series B* , **5**(1984), 633-652.

[106] Yulin Zhou and Boling Guo, Periodic boundary value problems and initial value problems for a system of higher order equations of generalized Korteweg-de Vries type. *Acta Mathematica Sinica*, **27**(1984), 154-176. (in Chinese)

[107] Yulin Zhou and Boling Guo, Initial value problems for a nonlinear singular integral-differential equation of deep water. Partial differential equations (Tianjin, 1986). *Lecture Notes in Mathematics*, **1306**(1988), 278-290. Springer, Berlin.

[108] Yulin Zhou, Boling Guo and Shaobin Tan, Existence and uniqueness of smooth solution for system of ferro-magnetic chain. *Science in China, Series A*, **34**(1991), 257-266.

[109] Songmu Zheng, Asymptotic behavior for strong solutions of the Navier-Stokes equations with external forces. *Nonlinear Analysis, Series A: Theory and Methods* , **45**(2001), 435-446.

[110] Enrique Zuazua, Weakly nonlinear large time behavior in scalar convection-diffusion equations. *Differential and Integral Equations* , **6**(1993), 1481-1491.

[111] Enrique Zuazua, Some recent results on the large time behavior for scalar parabolic conservation laws. Elliptic and parabolic problems (Pont-a-Mousson, 1994), 251-263, *Pitman Res. Notes Math. Series,* **325**, Longman Sci. Tech., Harlow, 1995.

This book Chapter was reviewed by (1) Professor Donald G. Aronson, School of Mathematics, University of Minnesota. 206 Church Street Southeast, Minneapolis, Minnesota 55455 USA. Email: don@math.umn.edu; and by (2) Professor Yulin Zhou and Professor Boling Guo, Division of Applied Mathematics, Beijing Institute of Applied Physics and Computational Mathematics, Beijing 100088 People's Republic of China.

In: Progress in Nonlinear Analysis Research
Editor: Erik T. Hoffmann, pp. 137-181

ISBN 978-1-60456-359-7

Chapter 7

THE DEVELOPMENT OF LYAPUNOV'S DIRECT METHOD IN THE APPLICATION TO NEW TYPES OF PROBLEMS OF HYDRODYNAMIC STABILITY THEORY

Yu.G. Gubarev*
Lavrentyev Institute for Hydrodynamics,
Siberian Division of Russian Academy for Sciences,
Lavrentyev Avenue 15, Novosibirsk, 630090, Russia

Abstract

The problems of linear stability of steady axial–symmetric sheared jet flows of non–viscous ideally conducting incompressible fluid with free surface in the magnetic field are being investigated. The sufficient conditions for stability, the necessary and sufficient conditions for stability, or the sufficient conditions for instability of these flows regarding small axial–symmetric long–wave perturbations are gained by Lyapunov's direct method. The a priori upper and lower exponential estimates, which are significative of the possible time growth of the investigated small perturbations, are constructed for those stationary flows at issue which turned out to be unstable. The examples of the steady flows and their small perturbations evolving in time according to the constructed estimates are presented.

Key Words: sheared magnetohydrodynamic jet flows, long–wave approximation, stability, instability, Lyapunov's direct method

1. Introduction

Nowadays the fundamental analytical methods of investigations are the first (spectral) and the second (direct) Lyapunov's methods [1, 2] in hydrodynamic stability theory.

The spectral method showed a good performance for studying problems referring to linear stability of quiescent states and stationary flows of fluid when evolution of small

*E-mail address: gubarev@hydro.nsc.ru

perturbations is described by systems of linear partial differential equations with constant coefficients [3–5]. But if these coefficients are the functions of space coordinates, then the spectral method, generally speaking, allows to get only the results about instability, and, as a rule, for the simplest quiescent states and stationary flows mentioned above. It is connected with the fact, that so–called normal waves [6] do not only form a complete system of solutions for linear partial differential equations with variable coefficients, but may dissatisfy them.

As for Lyapunov's direct method, it is free from restrictions which are peculiar to the spectral method [7]. The gist of the method with reference to studying stability of quiescent states and stationary flows of fluid is in seeking some "trial" functionals (so–called Lyapunov's functionals), possessing specific properties, whereas neither these quiescent states and stationary flows nor their perturbations aren't found. Hence, Lyapunov's direct method in contrast to the spectral one enables to get results about stability not for special cases of quiescent states and stationary flows of fluid, but for extensive classes of them and, moreover, regarding perturbations of arbitrarily general form. For example, to demonstrate instability of any fluid stationary flows in the magnetic field regarding any perturbations, it is necessary to construct Lyapunov's functional which, on the one hand, grows by virtue of magnetohydrodynamics equations, and, on the other hand, depends on perturbations in such a way, that its growth results in their growth according to instability definition which is adequate to the studied problem. In its turn, the given properties of the desired functional allow to ascertain conditions, which permit growing with time perturbations of the considered flows, and to construct estimates which would give researchers quantitative information about growth character of these perturbations.

By the present moment, one managed to achieve impressive success in the field of studying stability of quiescent states and stationary flows of fluid by Lyapunov's direct method. It is corroborated in the papers [4, 6–12]. At the same time, a number of problems of hydrodynamic stability theory do not have a satisfactory solution.

One of these problems is undoubtedly the evidence of instability of quiescent states and stationary flows of fluid, as nowadays, generally speaking, Lyapunov's functionals constructing algorithms, possessing the property to grow with time by force of corresponding equations of motion, are unknown. There are only some examples of Lyapunov's increasing functionals which are constructed while considering problems of linear and/or really nonlinear instability of quiescent states [13–19] and stationary symmetric flows of fluid, reduced by transformations of motion equations to effective quiescent states [20–24]. The characteristic feature of the given growing functionals lays in their following from one and the same differential relation (so–called virial equality) with using one and the same functional which has the term "virial" in the world scientific literature [13–25]. It is astonishing but to present day this virial is the only known functional that enables to construct increasing with time Lyapunov's functionals according to any motion equations.

The given chapter presents the new approaches to constructing Lyapunov's increasing with time functionals by force of motion equations and their applying for problems of linear stability of steady magnetohydrodynamic flows which cannot be reduced to effective quiescent states in any case (with the purpose of getting either necessary and sufficient conditions for stability or sufficient conditions for instability of the latter ones with respect to interested small perturbations).

Generally speaking, the given chapter can be divided into three tightly bound parts. The first part is devoted to investigating linear stability of stationary axial–symmetric sheared jet flows of inviscid perfectly conducting incompressible fluid with free boundary in the azimuthal magnetic field, which is directly proportional to a radial coordinate, concerning small axial–symmetric long–wave perturbations (paragraph 2). The second part is devoted to studying linear stability of steady axial–symmetric sheared jet flows of non–viscous ideally conducting incompressible fluid with free surface in the azimuthal magnetic field, which is an arbitrary function of radius, concerning small axial–symmetric long–wave perturbations (paragraph 3). And the third part is devoted to investigating linear stability of stationary axial–symmetric sheared jet flows of inviscid perfectly conducting incompressible fluid with free boundary in the poloidal magnetic field concerning small axial–symmetric long–wave perturbations (paragraph 4).

2. Stability of Steady Axial–Symmetric Sheared Jet Flows of Non–viscous Ideally Conducting Incompressible Fluid with Free Surface in the Azimuthal Magnetic Field Being Directly Proportional to a Radial Coordinate

An unrestrictedly long cylindrical jet of homogeneous in density inviscid perfectly conducting incompressible fluid, which is in infinite open space, is being investigated. The azimuthal magnetic field is supposed to be "frozen" into jet matter, and longitudinal constant electrical current, generating the quasi–stationary azimuthal magnetic field in unbounded environment, flows on free surface of the jet. Besides, the considered magnetohydrodynamic flows of an ideal fluid are supposed to be axial–symmetric, in addition, its azimuthal component of velocity field is identically equal to zero. At last, it is considered that the action of surface tension forces on free boundary of the conducting jet may be disregarded.

According to the above assumptions, the system of equations of one–fluid ideal magnetohydrodynamics [26] will have the form

$$\rho\left(\frac{\partial v_1}{\partial t^*}+v_1\frac{\partial v_1}{\partial r^*}+v_3\frac{\partial v_1}{\partial z^*}\right)+\frac{H_2^2}{4\pi r^*}=-\frac{\partial P_*}{\partial r^*},$$
$$\rho\left(\frac{\partial v_3}{\partial t^*}+v_1\frac{\partial v_3}{\partial r^*}+v_3\frac{\partial v_3}{\partial z^*}\right)=-\frac{\partial P_*}{\partial z^*}, \tag{1}$$
$$\frac{\partial H_2}{\partial t^*}+v_1\frac{\partial H_2}{\partial r^*}+v_3\frac{\partial H_2}{\partial z^*}-\frac{v_1H_2}{r^*}=0,\quad \frac{1}{r^*}\frac{\partial(v_1r^*)}{\partial r^*}+\frac{\partial v_3}{\partial z^*}=0,$$

where $\rho\equiv\text{const}$ is density field; v_1, v_3 are radial and axial components of velocity field; H_2 is the azimuthal component of the magnetic field inside the investigated jet; P is pressure field; $P_*\equiv P+H_2^2/(8\pi)$ is modified pressure field; t^* is time; r^*, z^* are cylindrical coordinates; π is the known constant. The axis z^* of cylindrical coordinate system is supposed to coincide with the axis of symmetry of the conducting jet.

Disregarding displacement current, the azimuthal component H_2^* of the magnetic field from the outside the investigated jet is defined by the formula

$$H_2^*=\frac{2J}{cr^*} \tag{2}$$

(here $J \equiv \mathrm{const}$ is value of surface longitudinal constant electrical current, and c is the speed of light).

There are the following boundary conditions on the axis of symmetry of the conducting jet and its free surface:

$$v_1 = 0, \ \left| \frac{H_2}{r^*} \right| < +\infty \ \ (r^* = 0);$$
$$P_* = \frac{H_2^{*2}}{8\pi}, \ v_1 = \frac{\partial r_1}{\partial t^*} + v_3 \frac{\partial r_1}{\partial z^*} \ \ (r^* = r_1(t^*, z^*)). \tag{3}$$

The initial data for the first three ratios of the system (1) and the last of the boundary conditions (3) are given in the following form

$$v_1(0, r^*, z^*) = v_{10}(r^*, z^*), \ v_3(0, r^*, z^*) = v_{30}(r^*, z^*),$$
$$H_2(0, r^*, z^*) = H_{20}(r^*, z^*), \ r_1(0, z^*) = r_{10}(z^*), \tag{4}$$

where it is necessary that the functions v_{10}, v_{30}, H_{20}, and r_{10} satisfy the forth equation of the system (1) and the first three of the ratios (3).

Further, in the mixed problem (1)–(4) the reduction to dimensionless variables takes place and then transition to long–wave approximation is realized. At that, the following values: L, which is the representative distance scale of variation of hydrodynamic and magnetic fields along coordinate axis z^*, v_0, which is the characteristic fluid velocity, and r_0, which is the representative radius of the considered jet, — are chosen as parameters of dimension exclusion procedure. By means of the given parameters, the dimensionless values t, η, z, q, w, p_*, h, and κ are constructed in such a way, that the relations take place

$$t^* = \frac{tL}{v_0}, \ r^{*2} = \eta L^2 \delta^2, \ z^* = zL, \ 2v_1 r^* = q v_0 L \delta^2,$$
$$v_3 = w v_0, \ P_* = p_* \rho v_0^2, \ H_2 = \frac{h r^* \sqrt{4\pi\rho v_0^2}}{L\delta}, \ H_2^* r^* = \kappa \sqrt{4\pi\rho v_0^2}\, L\delta,$$

where $\delta \equiv r_0/L \ll 1$ is dimensionless characteristic radius of the conducting jet.

As a result of the dimension exclusion procedure with using the relations mentioned above, the equations system (1) will have the form

$$\frac{\delta^2}{2}\left(q_t + q q_\eta - \frac{q^2}{2\eta} + w q_z\right) + \eta h^2 = -\,2\eta p_{*\eta},$$
$$w_t + q w_\eta + w w_z = -\,p_{*z}, \tag{5}$$
$$h_t + q h_\eta + w h_z = 0, \ q_\eta + w_z = 0$$

(here and below independent variables written as indexes denote the corresponding partial derivatives). At the same time, taking into account the ratio (2), the boundary conditions (3) are transformed to the form

$$q = 0, \ |h| < +\infty \ \ (\eta = 0); \tag{6}$$

$$p_* = \frac{\kappa^2}{2\eta_1}, \; q = \eta_{1t} + w\eta_{1z} \quad (\eta = \eta_1 (t, \, z)) \, ,$$

where

$$\kappa \equiv \frac{J}{cr_0\sqrt{\pi\rho v_0^2}} = \text{const}.$$

At last, the initial data (4) will have the form

$$q(0, \, \eta, \, z) = q_0 (\eta, \, z) \, , \; w(0, \, \eta, \, z) = w_0 (\eta, \, z) \, ,$$
$$h(0, \, \eta, \, z) = h_0 (\eta, \, z) \, , \; \eta_1 (0, \, z) = \eta_{10} (z) \, . \tag{7}$$

And now if to omit the summands, which are proportional to the multiplier δ^2, in the first equation of the system (5), and to remove the expression for the function $q(0, \eta, z)$ from the ratios (7), then the initial boundary value problem (5)–(7) will reduce at once to the form which corresponds to long–wave approximation.

However, this long–wave presentation of the mixed problem (5)–(7) cannot be considered as the final one in any case, as it may be simplified yet more due to the change of Eulerian independent variables $(t, \, z, \, \eta)$ to the mixed Eulerian–Lagrangian independent variables $(t', \, z', \, \nu)$ [27]; the change is carried out by analogy with the paper [28] according to formulae

$$t = t', \; z = z', \; \eta = R(t', \, z', \, \nu) \, ; \; \nu \in [0, \, 1] \, .$$

Here the function R is considered to satisfy the equation

$$q = R_{t'} + wR_{z'} \tag{8}$$

and the boundary conditions

$$R(t', \, z', \, 0) = 0, \; R(t', \, z', \, 1) = \eta_1 (t', \, z') \, . \tag{9}$$

The main point of the given change of independent variables is in the fact, that trajectories of fluid particles motion in the jet may be numbered using Lagrangian variable ν. Besides, the boundary conditions (6) are met automatically for the function q by the definition of the function R (8), (9). At last, (and it is undoubtedly the most important thing) unknown free surface of the conducting jet $\eta = \eta_1$ will turn to the known fixed boundary $\nu = 1$ due to the realized change of independent variables.

Thus, (after disregarding summands having the multiplier δ^2) the ratios system (5) with new mixed Eulerian–Lagrangian independent variables will have the form

$$R_\nu h^2 = -\, 2p_{*\nu}, \; R_\nu (w_t + ww_z) = -\, R_\nu p_{*z} + R_z p_{*\nu},$$
$$h_t + wh_z = 0, \; q_\nu + R_\nu w_z - R_z w_\nu = 0, \tag{10}$$

where the variables t' and z' are written without primes for convenience of writing further formulae. The given equations are complemented by the following initial conditions:

$$w(0, \, z, \, \nu) = w_0 (z, \, \nu) \, , \; h(0, \, z, \, \nu) = h_0 (z, \, \nu) \, , \tag{11}$$

$$R(0,\ z,\ \nu) = R_0(z,\ \nu)\,.$$

Here the function $R_0(z,\ \nu)$ is supposed to be monotone increasing by the argument ν under the requirement of the one–to–one change of independent variables.

Further, to represent the system (10) more clearly, primarily its first ratio is integrated within the limits from ν to 1, the variable of integration is ν, and then, by means of the boundary conditions (6), the dimensionless modified pressure field p_* is excluded from this ratio and is substituted into the second equation of the same ratios system, and it allows to get the relation

$$w_t + w w_z = \frac{\kappa^2 R_{1z}}{2R_1^2} - \frac{\left(h_1^2 R_1\right)_z}{2} + \frac{\left(h^2\right)_z R}{2} + \frac{1}{2}\left(\int\limits_{\nu}^{1} R\left(h^2\right)_{\nu_1} d\nu_1\right)_z, \tag{12}$$

where the independent variable ν is designated as ν_1 (to distinguish it from the variable ν on the lower limit of the functional from the right part), and the values of functions h and R on free surface of the jet $\nu = 1$ are designated as h_1 and R_1 correspondingly, moreover, according to the second boundary condition of the system (9), $R_1(t,\ z) \equiv \eta_1(t,\ z)$.

In addition, the function q is replaced by the corresponding expression (8) in the last of the equations (10), thus the given equation has the form

$$R_{\nu t} + \left(w R_\nu\right)_z = 0. \tag{13}$$

Below, the azimuthal component of the magnetic field inside the conducting jet is supposed to be directly proportional to the radial coordinate: $h \equiv h_1 = \text{const}$ [28, 29]. This assumption implies, on the one hand, reducing of the third ratio of the system (10) to an identity, and, on the other hand, considerable simplifying of the equation (12) which now have the form

$$w_t + w w_z = \left[\left(\frac{\kappa}{R_1}\right)^2 - h_1^2\right]\frac{R_{1z}}{2}. \tag{14}$$

The relations (11) will be the initial conditions for the ratios (13), (14), if the data only for the functions R and w are kept in them, i.e.

$$R(0,\ z,\ \nu) = R_0(z,\ \nu)\,,\ w(0,\ z,\ \nu) = w_0(z,\ \nu)\,. \tag{15}$$

It should be noted that the equations similar to ratios (13), (14) can be obtained, when R is considered to be a monotone decreasing function by the argument ν, as well. At that, the difference, in comparison with above–said, will be in the only fact, that the straight line $\nu = 0$ will be free boundary of the investigated jet, while the straight line $\nu = 1$ will be its axis of symmetry.

The initial boundary value problem (13)–(15) has the energy integral

$$E_1 \equiv \frac{1}{2}\int\limits_{-\infty}^{+\infty}\left(\int\limits_{0}^{1} w^2 R_\nu d\nu + \kappa^2 \ln R_1 + \frac{h_1^2}{2} R_1^2\right) dz = \text{const} \tag{16}$$

assuming that the solutions of the given problem are either periodic along the axis z or localized on it.

It is easy to find out that the mixed problem (13)–(15) has one more motion integral. To do this, the equation (14) is necessary to be differentiated by the independent variable ν, it will lead to the ratio

$$w_{\nu t} + (ww_z)_\nu = 0. \tag{17}$$

Further, the important relation

$$C_t + wC_z = 0 \tag{18}$$

is implied from the equations (13) and (17) (here $C \equiv R_\nu / w_\nu$), the use of this relation together with the ratio (17) does allow to show that the functional

$$I \equiv \int_{-\infty}^{+\infty}\int_0^1 w_\nu F(C)\, d\nu dz, \tag{19}$$

where $F(C)$ is an arbitrary function of its argument, is the desired additional motion integral [28–30].

The exact stationary solutions of the initial boundary value problem (13)–(15) have the form

$$w = w^0(\nu),\ R = R^0(\nu),\ R_1 = R_1^0 \equiv 1. \tag{20}$$

Here w^0 is some function of the independent variable ν, and R^0 is the monotone increasing one of the independent variable ν; the steady radius of the conducting jet is equal to its representative radius r_0 in this case. It is easy to ascertain that the functions w^0, R^0, and R_1^0 (20) satisfy the equations (13), (14) identically.

The purpose of the further investigation is to find out the conditions under which the stationary flows (20) will be stable regarding small axial–symmetric long–wave perturbations $w'(t,\ z,\ \nu)$, $R'(t,\ z,\ \nu)$, and $R_1'(t,\ z)$.

To achieve this purpose, the mixed problem (13)–(15) and the ratios (17), (18) are linearized near exact stationary solutions (20), and this linearization results in the initial boundary value problem in the form

$$\begin{gathered}
w_t' + w^0 w_z' = \frac{1}{2}\left(\kappa^2 - h_1^2\right) R_{1z}',\ R_{\nu t}' + w^0 R_{\nu z}' + \frac{dR^0}{d\nu} w_z' = 0,\\
w_{\nu t}' + \frac{dw^0}{d\nu} w_z' + w^0 w_{z\nu}' = 0,\ C_t' + w^0 C_z' = 0;\\
C' \equiv \left(\frac{dw^0}{d\nu}\right)^{-1}\left[R_\nu' - C^0 w_\nu'\right],\ C^0 \equiv \frac{dR^0}{d\nu}\left(\frac{dw^0}{d\nu}\right)^{-1};\\
w'(0,\ z,\ \nu) = w_0'(z,\ \nu),\ R'(0,\ z,\ \nu) = R_0'(z,\ \nu).
\end{gathered} \tag{21}$$

The functional

$$E \equiv \int_{-\infty}^{+\infty}\int_0^1 \left[\frac{1}{2}\frac{dR^0}{d\nu} w'^2 + w^0 w' R_\nu' + \frac{1}{2}\frac{dw^0}{d\nu}\frac{d^2F}{dC^2}\left(C^0\right) C'^2\right] d\nu dz +$$

$$+\frac{h_1^2-\kappa^2}{4}\int\limits_{-\infty}^{+\infty} R_1'^2 dz = \text{const} \quad (22)$$

is conserved on the solutions of this problem.

One can check that the first variation δJ_1 of the integral $J_1 \equiv E_1 + I = \text{const}$ (16), (19) vanishes on the steady flows (20), if the functions w^0, R^0, and F reduce the equation

$$\frac{dF}{dC}\left(C^0\right) = -\frac{w^{02}}{2}$$

to identity, and its second variation $\delta^2 J_1$, rewritten in the corresponding designations, coincides with the functional E by form.

The exact stationary solutions (20) of the mixed problem (13)–(15) will be stable to small axial–symmetric long–wave perturbations (21) when and only when the integral E (22) is of fixed sign.

To find out, if the functional E is of fixed sign, the following form

$$E = \int\limits_{-\infty}^{+\infty}\int\limits_{0}^{1} (B\mathbf{u},\, \mathbf{u})\, d\nu dz;\ \mathbf{u} \equiv \begin{pmatrix} w' \\ R'_\nu \\ C' \\ R'_1 \end{pmatrix} \quad (23)$$

is convenient, where $B \equiv ||b_{ik}||$ is square matrix of order 4×4 with nonzero elements

$$b_{11} = \frac{1}{2}\frac{dR^0}{d\nu},\ b_{12} = b_{21} = \frac{w^0}{2},$$

$$b_{24} = b_{42} = \frac{h_1^2-\kappa^2}{8},\ b_{33} = \frac{1}{2}\frac{dw^0}{d\nu}\frac{d^2F}{dC^2}\left(C^0\right).$$

According to Silvester's criterion [31], the integrand of the functional E (23) will be positive (negative) definite when and only when the principal minors of the matrix B are positive (have the sign $(-1)^m$) (here m is an order of principal minor).

It is easy to conclude that the principal minors of the matrix B do not have the required definiteness in sign. So, for positive definiteness of the integrand of the functional E the inequalities

$$\frac{dR^0}{d\nu} > 0,\ -w^{02} > 0$$

must particularly be true. It is obvious that the second inequality is not valid in principle. At the same time, for the negative definiteness of the given integrand, the ratios

$$\frac{dR^0}{d\nu} < 0,\ -w^{02} > 0$$

are necessary to be true, but it is not realized either because of the character of the function R^0 monotonicity and falsity of the second inequality.

Thus, according to Silvester's criterion, the functional E (23) has neither positive nor negative definiteness. It means, in its turn, that the sufficient conditions for stability of the exact stationary solutions (20) of the initial boundary value problem (13)–(15) regarding small axial–symmetric long–wave perturbations $w'(t,\ z,\ \nu)$, $R'(t,\ z,\ \nu)$, and $R'_1(t,\ z)$ (21) are absent, if they are considered to be conditions for definiteness in sign of the energetic integral of motion E.

Below, the necessary and sufficient condition for stability of the subclass

$$\frac{d}{d\nu}\left(w^0 C^0\right) \leq 0 \tag{24}$$

of the steady flows (20) regarding those small axial–symmetric long–wave perturbations (21), which keep the function $C^0(\nu)$ value invariable in any fluid particle and meet a number of restrictions on the symmetry axis of the considered jet and its free surface, will be obtained by Lyapunov's direct method [1, 2].

To show the instability of some exact stationary solution (20), (24) of the mixed problem (13)–(15) regarding small axial–symmetric long–wave perturbations $w'(t, z, \nu)$, $R'(t, z, \nu)$, and $R'_1(t,\ z)$ (21), it is necessary to single out at least one rapidly (at the minimum, exponentially) growing with time perturbation among the given ones.

For this purpose, the partial class of axial–symmetric sheared jet flows of non–viscous ideally conducting incompressible fluid with free boundary in the azimuthal magnetic field is being investigated. The characteristic feature of this class is that the small perturbations $C'(t,\ z,\ \nu)$ (21) are equal to zero for its flows. In other words, the value of the function C^0 (21), (24) is supposed to be invariable during perturbations in any fluid particle. So, the given perturbations are deviations of trajectories of fluid particle motion from corresponding current lines of the steady flows (20), (24).

From the physical point of view, the above requirement, concerning small perturbations, is based on the fact, that velocity circulation on any fluid loop in the axial plane, which is set at the initial time, will be kept while developing perturbations, since, according to the initial boundary value problem (13)–(15), the value of the function C (18) is invariable in fluid particles.

These perturbations can be introduced in the most effective way by Lagrangian displacements field $\xi = \xi(t,\ z,\ \nu)$ [25] which is defined by the equation

$$\xi_t = w' - w^0 \xi_z. \tag{25}$$

Due to the ratio (25), the mixed problem (21) can be rewritten in the form

$$w'_t + w^0 w'_z = \frac{1}{2}\left(\kappa^2 - h_1^2\right) R'_{1z},\ R'_\nu = -\frac{dR^0}{d\nu}\xi_z,$$

$$w'_\nu = -\frac{dw^0}{d\nu}\xi_z,\ R'_\nu = C^0 w'_\nu; \tag{26}$$

$$\xi(0,\ z,\ \nu) = \xi_0(z,\ \nu),\ w'(0,\ z,\ \nu) = w'_0(z,\ \nu).$$

Direct calculations demonstrate that the functional E (22) can be reduced to the form

$$E = \frac{1}{2}\int\limits_{-\infty}^{+\infty}\left[\int\limits_0^1 \frac{d}{d\nu}\left(R^0 - w^0 C^0\right) w'^2 d\nu + \frac{h_1^2 - \kappa^2}{2} R_1'^2\right] dz \tag{27}$$

and will be a motion integral for the initial boundary value problem (25), (26) if the following equalities take place:

$$\int_{-\infty}^{+\infty} \left(w^0 C^0 w'^2\right)\Big|_{\nu=1} dz = \int_{-\infty}^{+\infty} \left(w^0 C^0 w'^2\right)\Big|_{\nu=0} dz,$$
$$\int_{-\infty}^{+\infty} \left(w^0 C^0 w'\right)\Big|_{\nu=1} R'_{1z} dz = \int_{-\infty}^{+\infty} \left(w^0 C^0 w'\right)\Big|_{\nu=0} R'_{1z} dz. \tag{28}$$

The important thing is that the ratios (28) are connected with each other: when one of them is true, the other is true automatically. Besides, since the function $w'(t,\ z,\ \nu)$ (25), (26) as a function of the independent variable ν has some arbitrariness on the symmetry axis $\nu = 0$ of the conducting jet and its free surface $\nu = 1$, the given equalities can be interpreted as additional boundary conditions of the mixed problem (25), (26).

The analysis of the functional E (27) shows that if the inequality

$$h_1^2 \geq \kappa^2 \tag{29}$$

is true, then, taking into consideration the properties of the function R^0 monotony and the integral E independence of time, the stability of the exact stationary solutions (20), (24) of the initial boundary value problem (13)–(15) regarding small axial–symmetric long–wave perturbations (25), (26), (28) will take place.

Assume that the inequality (29) is dissatisfied so, that the ratio

$$h_1^2 < \kappa^2 \tag{30}$$

is true. Then it turns out that the instability of the steady flows (20), (24) regarding small axial–symmetric long–wave perturbations (25), (26), (28) can be shown.

In fact, when the auxiliary functional

$$M \equiv \int_{-\infty}^{+\infty}\int_0^1 \frac{dR^0}{d\nu}\xi^2 d\nu dz \tag{31}$$

is doubly differentiated with respect to the independent variable t and when the relations (25)–(28) are applied, it is easy to come to the known virial equality [22, 24, 25]

$$\frac{d^2 M}{dt^2} = 4\,(T - \Pi)\,;$$
$$T \equiv \frac{1}{2}\int_{-\infty}^{+\infty}\int_0^1 \frac{dR^0}{d\nu} w'^2 d\nu dz, \quad \Pi \equiv \frac{h_1^2 - \kappa^2}{4}\int_{-\infty}^{+\infty} R_1'^2 dz. \tag{32}$$

Multiplying this equality by some constant λ and taking into consideration the ratio

$$E \equiv T + T_1 + \Pi = \text{const}; \tag{33}$$

$$T_1 \equiv -\frac{1}{2}\int\limits_{-\infty}^{+\infty}\int\limits_{0}^{1}\frac{d}{d\nu}\left(w^0C^0\right)w'^2 d\nu dz \geq 0,$$

it is possible to derive the equation

$$\frac{dE_\lambda}{dt} = 2\lambda E_\lambda - 4\lambda T_\lambda - 2\lambda T_1;$$

$$E_\lambda \equiv \Pi_\lambda + T_\lambda,\ 2\Pi_\lambda \equiv 2\left(\Pi + T_1\right) + \lambda^2 M, \tag{34}$$

$$2T_\lambda \equiv 2T - \lambda\frac{dM}{dt} + \lambda^2 M = \int\limits_{-\infty}^{+\infty}\int\limits_{0}^{1}\frac{dR^0}{d\nu}\left(w' - \lambda\xi\right)^2 d\nu dz \geq 0,$$

which is basic one for further statement.

As the values T_1 (33) and T_λ are non–negative, due to $\lambda > 0$ the following differential inequality is consequence of the ratio (34):

$$\frac{dE_\lambda}{dt} \leq 2\lambda E_\lambda.$$

It is easy to get the important estimate when this inequality is integrated:

$$E_\lambda(t) \leq E_\lambda(0)\exp(2\lambda t). \tag{35}$$

The ratio (35) is true for any solutions of the mixed problem (25), (26), (28) and for arbitrary positive values of the constant λ. Moreover, while obtaining the given inequality it was not necessary to impose any restriction on the functional Π (32) sign.

The ratio (35) allows to conclude that the integral E_λ may consider below as Lyapunov's functional [1, 2, 22, 24], as further, due to this inequality, double–sided exponential estimates of small axial–symmetric long–wave perturbations $\xi(t,\ z,\ \nu)$ (25), (26), (28) growth will be constructed, at that, the most rapidly growing small perturbations will be sorted out and described among the latter ones.

Taking into consideration the ratio (30), by choosing suitable initial field of Lagrangian displacements $\xi_0(z,\ \nu)$ and perturbations of velocity field $w_0'(z,\ \nu)$ (26) it is easy to provide inequalities validity: $\Pi(0) < 0,\ T(0) + T_1(0) \leq \left|\Pi(0)\right|$. As a result, the integral $E_\lambda(0)$ (as it follows from its definition (34)) will become a quadratic polynomial with the parameter λ, with the positive coefficient $M(0)$ (31) by λ^2 and the absolute term $E(0) \leq 0$ (27), viz:

$$E_\lambda(0) \equiv E(0) - \frac{\lambda}{2}\frac{dM}{dt}(0) + \lambda^2 M(0). \tag{36}$$

When the parameter λ values are taken from the interval

$$0 < \lambda < \Lambda \equiv A_1 + \sqrt{A_2};$$

$$A_1 \equiv \left[4M(0)\right]^{-1}\frac{dM}{dt}(0),\ A_2 \equiv A_1^2 - \frac{E(0)}{M(0)}, \tag{37}$$

the ratio (36) results in the estimate: $E_\lambda(0) < 0$. The given estimate and inequality (35) prove that small axial–symmetric long–wave perturbations (25), (26), (28) grow in time no slower than exponentially.

On condition that $\lambda = \Lambda - \delta_1$ (with any parameter δ_1 from the interval $]0,\ \Lambda[$), the ratio (35) has the form

$$E_{\Lambda-\delta_1}(t) \le E_{\Lambda-\delta_1}(0)\exp\left[2\left(\Lambda-\delta_1\right)t\right] \quad \left(E_{\Lambda-\delta_1}(0) < 0\right). \tag{38}$$

Since, due to the expression (34) for the functional E_λ, the inequality $E_\lambda(t) \ge \Pi(t)$ is true, the ratio (38) can be written in the form

$$-\Pi(t) \ge \left|E_{\Lambda-\delta_1}(0)\right|\exp\left[2\left(\Lambda-\delta_1\right)t\right]$$

or, finally,

$$\left(\kappa^2 - h_1^2\right)\int\limits_{-\infty}^{+\infty} R_1'^2 dz \ge 4\left|E_{\Lambda-\delta_1}(0)\right|\exp\left[2\left(\Lambda-\delta_1\right)t\right]. \tag{39}$$

The inequality (39) shows that the value $\Lambda - \delta_1$ (37), (38) is the lower bound for values of small axial–symmetric long–wave perturbations $\xi(t,\ z,\ \nu)$ (25), (26), (28) increments.

The estimate (39) can be greatly improved if the initial field of Lagrangian displacements $\xi_0(z,\ \nu)$ and perturbations of the velocity field $w_0'(z,\ \nu)$ (26) are brought under the additional condition

$$w_0'(z,\ \nu) = \lambda\xi_0(z,\ \nu). \tag{40}$$

In fact, then $T_\lambda(0) = 0$, $E_\lambda(0) = \Pi_\lambda(0)$ imply from the ratios (34), (36). In its turn, these equalities help to make sure that the estimate $\Pi_\lambda(0) < 0$ is true on the interval

$$0 < \lambda < \Lambda_1 \equiv \sqrt{-\frac{2\Pi(0)}{M(0) + A_3}};\ A_3 \equiv -\int\limits_{-\infty}^{+\infty}\int\limits_0^1 \frac{d}{d\nu}\left(w^0 C^0\right)\xi_0^2 d\nu dz. \tag{41}$$

It follows that, if $\lambda = \Lambda_1 - \delta_2$ (with an arbitrary parameter δ_2 from the interval $]0,\ \Lambda_1[$), the inequality (35) can be brought to the form

$$E_{\Lambda_1-\delta_2}(t) \le \Pi_{\Lambda_1-\delta_2}(0)\exp\left[2\left(\Lambda_1-\delta_2\right)t\right] \quad \left(\Pi_{\Lambda_1-\delta_2}(0) < 0\right). \tag{42}$$

While doing computations similar to the above ones during the estimate (39) validation, the inequality (42) can be rewritten in the form

$$-\Pi(t) \ge \left|\Pi_{\Lambda_1-\delta_2}(0)\right|\exp\left[2\left(\Lambda_1-\delta_2\right)t\right]$$

or, finally,

$$\left(\kappa^2 - h_1^2\right)\int\limits_{-\infty}^{+\infty} R_1'^2 dz \ge 4\left|\Pi_{\Lambda_1-\delta_2}(0)\right|\exp\left[2\left(\Lambda_1-\delta_2\right)t\right]. \tag{43}$$

According to the ratio (43), the value $\Lambda_1 - \delta_2$ (41), (42) is the lower estimate for values of small axial–symmetric long–wave perturbations (25), (26), (28), (40) increments.

The matching of the inequalities (39) and (43) with each other allows to note that small axial–symmetric long–wave perturbations $\xi(t,\ z,\ \nu)$ (25), (26), (28), when their initial data satisfy the restriction (40), grow in the fastest way than the other perturbations of the investigated subclass, and (as it will be demonstrated below) the increments of the fastest increasing ones are calculated by formula

$$\Lambda_1^+ \equiv \sup_{\xi_0(z,\ \nu)} \Lambda_1. \tag{44}$$

In fact, assume that $\lambda > \Lambda_1^+$. In this case, the ratio $\Pi_\lambda(0) > 0$ will be true for all possible initial fields of Lagrangian displacements $\xi_0(z, \nu)$ (26). So, the integral $E_\lambda(0)$ (34), (36) will be positively defined for all possible initial fields of Lagrangian displacements $\xi_0(z,\ \nu)$ and perturbations of velocity field $w_0'(z,\ \nu)$ (26) too.

As a result, with $\lambda = \Lambda_1^+ + \delta_3$, where $\delta_3 > 0$ is a parameter, the estimate

$$E_{\Lambda_1^+ + \delta_3}(t) \leq E_{\Lambda_1^+ + \delta_3}(0) \exp\left[2\left(\Lambda_1^+ + \delta_3\right) t\right] \tag{45}$$

follows from the inequality (35). By force of the given estimate, the value $\Lambda_1^+ + \delta_3$ is the upper bound for the values of small axial–symmetric long–wave perturbations (25), (26), (28) increments.

The matching of the inequalities (43) and (45) enables to conclude: the parameter Λ_1^+ (41), (44) indicates the growth rate ω of small perturbations (25), (26), (28) both on the low and upper bounds, i.e.

$$\Lambda_1^+ - \delta_2 \leq \omega \leq \Lambda_1^+ + \delta_3. \tag{46}$$

At that, the ratio (46) shows that the small axial–symmetric long–wave perturbations $\xi(t,\ z,\ \nu)$ (25), (26), (28) grow in the fastest way if their increment value is near to the parameter Λ_1^+.

Thus, if the condition (30) is true, then, after ascertaining the value of the parameter Λ_1^+ by means of the relations (41), (44) and estimating the growth rate ω (46) of the fastest growing small perturbations (25), (26), (28), (40), it is possible to answer to the question, for which representative times the small axial–symmetric long–wave perturbations (25), (26), (28) will be destroying the stationary axial–symmetric sheared jet flows (20), (24) of inviscid perfectly conducting incompressible fluid with free surface in the azimuthal magnetic field which is directly proportional to a radial coordinate?

Further, one builds an example of steady flows (20), (24) and the initial small axial–symmetric long–wave perturbations (25), (26), (28), superposed upon these flows, which, generally speaking, will evolve with time according to the obtained estimates (39), (45). This example should not be considered as a comparison with the physical phenomenon at issue, but it is an illustration to the analytical treatment above.

So, the stationary axial–symmetric sheared magnetohydrodynamic jet flows of an ideal fluid

$$w^0(\nu) = C_1 \exp\left(-C_2 \nu\right),\ R^0(\nu) = \nu,\ R_1^0 = 1 \tag{47}$$

are investigated (here $C_1,\ C_2$ are some positive constants) in the area which is an infinite band of the kind

$$\left[(z,\ \nu):\ -\infty < z < +\infty,\ 0 \leq \nu \leq 1\right]. \tag{48}$$

It is easy to verify that these flows are typical samples of the partial class (24) of the steady flows (20).

When the inequality (30) is true, the stationary flows (47), (48) will be unstable, for example, to small axial–symmetric long–wave perturbations $\xi(t,\ z,\ \nu)$ (25), (26), (28), for which the initial field of Lagrangian displacements $\xi_0(z,\ \nu)$ is given in the form

$$\xi_0(z,\ \nu) = (2\nu - 1)\exp(C_2\nu)\sin\frac{2\pi z}{l}, \tag{49}$$

where l is an arbitrary positive constant value. From the physical point of view, these perturbations are periodical (with the wave–length l) fluctuations of free boundary of the investigated jet and the axial velocity of fluid flowing inside the one.

Indeed, using the function $R_1(t,\ z)$ (9), (12) definition and the equations (26), it is easy to get the ratios

$$R'_{0\nu}(z,\ \nu) = -\frac{2\pi}{l}(2\nu - 1)\exp(C_2\nu)\cos\frac{2\pi z}{l},$$

$$R'_1(0,\ z) \equiv \int_0^1 R'_{0\nu}(z,\ \nu)d\nu = -\frac{2\pi}{lC_2}\left[\left(1 - \frac{2}{C_2}\right)\exp C_2 + \frac{2}{C_2} + 1\right]\cos\frac{2\pi z}{l},$$

$$w'_{0\nu}(z,\ \nu) = \frac{2\pi C_1 C_2}{l}(2\nu - 1)\cos\frac{2\pi z}{l},$$

$$w'_0(z,\ \nu) = \int_0^\nu w'_{0\nu_1}(z,\ \nu_1)d\nu_1 = \frac{2\pi C_1 C_2 \nu}{l}(\nu - 1)\cos\frac{2\pi z}{l}.$$

It is worth noting that, since here

$$w'_0(z,\ 0) = 0,\ w'_0(z,\ 1) = 0, \tag{50}$$

the boundary conditions (28) are satisfied identically, and therefore they are concordant with the initial conditions (26), (50) when $t = 0$.

Taking into consideration the periodicity of the field $\xi_0(z,\ \nu)$ (49) by the independent variable z and the expressions (32), (33) for the functionals T, T_1, and Π, it is possible to calculate values of the latter ones at the initial time:

$$T(0) \equiv \frac{1}{2}\int_0^l\int_0^1 \frac{dR^0}{d\nu}w_0'^2(z,\ \nu)d\nu dz = \frac{\pi^2 C_1^2 C_2^2}{30l},$$

$$T_1(0) \equiv -\frac{1}{2}\int_0^l\int_0^1 \frac{d}{d\nu}\left(w^0 C^0\right)w_0'^2(z,\ \nu)d\nu dz = 0,$$

$$\Pi(0) \equiv \frac{h_1^2 - \kappa^2}{4}\int_0^l R_1'^2(0,\ z)dz = \frac{\pi^2\left(h_1^2 - \kappa^2\right)}{2lC_2^2}\left[\left(1 - \frac{2}{C_2}\right)\exp C_2 + \frac{2}{C_2} + 1\right]^2.$$

This implies that the inequality $\Pi(0) < 0$ is true. The ratio $T(0) + T_1(0) \leq |\Pi(0)|$ will be correct if the constants C_1 and C_2 are chosen properly. For example,

$$0 < C_1 \leq (3 - e)\sqrt{15\left(\kappa^2 - h_1^2\right)},\ C_2 = 1,$$

where e is the known constant.

In total, for the steady flows (47), (48) it is possible to write out the estimates of the low and upper bounds (39), (45) in the evident form (the second one with parameter Λ_1 instead of Λ_1^+), which characterize a growth of the small axial–symmetric long–wave perturbations (25), (26), (28), (49), and it is the evidence of these flows instability. It is important to note that here the value Λ_1 (41), but not Λ_1^+ (44), evaluates the growth rate ω (46) of the small perturbations (25), (26), (28), (49) at the low and upper bounds.

Finally, the fastest growing small axial–symmetric long–wave perturbations of the stationary flows (47), (48) will be the ones which have the initial field of Lagrangian displacements in the form $\xi_0(z,\ \nu) = f\left(w^0 - \lambda z\right)$ (by force of the equations (26) and the equality (40)), where the function f is required to be either periodical or localized by the coordinate z. Then one can judge about properties of the given perturbations growth, settling on the low (43) and upper (45) estimates, and their growth rate ω (46) can be calculated due to the parameter Λ_1^+ (41), (44).

It will be demonstrated below by Lyapunov's direct method that the inequality (30) is the sufficient condition for instability of the exact stationary solutions (20) of the initial boundary value problem (13)–(15) regarding the small axial–symmetric long–wave perturbations $w'(t,\ z,\ \nu)$, $R'(t,\ z,\ \nu)$, and $R_1'(t,\ z)$ (21), at that an growing (at the minimum, exponentially) with time perturbation will be looked for among representatives of the subclass (25), (26) of the latter ones. It means that the steady flows (20) are considered further as free from the restriction (24), and the small perturbations (25), (26) — from the requirements (28).

According to the assumptions above, the integral E (22), conserving, naturally, not only on the solutions of the mixed problem (21), but on the solutions of the initial boundary value problem (25), (26) as well, will have the form (33) with the only exception, that another functional

$$T_1 \equiv \int_{-\infty}^{+\infty}\int_0^1 w^0 w' R_\nu' d\nu dz \tag{51}$$

will be designated by the symbol T_1 (essentially, that the integral T_1 in the form (51) is not already of fixed sign). In given case, the virial equality (32) and the basic equation (34) will be true as well (the latter one — accurate to the form of the functional T_1).

Taking into consideration the circumstances above and assuming, that the condition (30) is true, it is possible to obtain the original differential inequality

$$\frac{d^2M}{dt^2} - 2\lambda\frac{dM}{dt} + 2\lambda^2 M \geq 0 \tag{52}$$

from the ratio (34), where λ is some positive constant. Under certain conditions [32], this inequality gives the following lower exponential estimate:

$$M(t) \geq C_3 \exp(\lambda t) \tag{53}$$

(here C_3 is known positive constant).

Indeed, the relation (52) can be formally integrated over the half–open intervals $2\pi n/\lambda \le t < \pi/(2\lambda) + 2\pi n/\lambda$ $(n = 0,\ 1,\ 2,\ ...)$ with help of the following changes of the functional M:

1) $$M_1(t) \equiv \exp(-\,\lambda t)\, M(t),$$

$$\frac{d^2 M_1}{dt^2} + \lambda^2 M_1 \ge 0;$$

2) $$M_2(t) \equiv M_1(t)\cos^{-1}(\lambda t),$$

$$\frac{d}{dt}\left[\frac{dM_2}{dt}\cos(\lambda t)\right] - \lambda\frac{dM_2}{dt}\sin(\lambda t) \ge 0;$$

3) $$M_3(t) \equiv (dM_2/dt)\cos^2(\lambda t),$$

$$\frac{dM_3}{dt} \ge 0.$$

Integrating the last inequality and turning back to M, it isn't difficult to arrive at the relation

$$M(t) \ge [C_{1n}\cos(\lambda t) + C_{2n}\sin(\lambda t)]\exp(\lambda t), \tag{54}$$

where C_{1n} and C_{2n} are arbitrary constants.

Since the inequality (54) is non–strict, and the intervals $2\pi n/\lambda \le t < \pi/(2\lambda) + 2\pi n/\lambda$ $(n = 0,\ 1,\ 2,\ ...)$ have the empty intersection, one can express the constants C_{1n} and C_{2n} through the values of the functional M (31) and its first derivative dM/dt at the time moments $t = 2\pi n/\lambda$. As a result, the relation (54) takes the form

$$M(t) \ge g(t); \tag{55}$$

$$g(t) \equiv \left[M\left(\frac{2\pi n}{\lambda}\right)\cos(\lambda t) + \left[\frac{1}{\lambda}\frac{dM}{dt}\left(\frac{2\pi n}{\lambda}\right) - M\left(\frac{2\pi n}{\lambda}\right)\right]\sin(\lambda t)\right]\exp(\lambda t - 2\pi n).$$

To justify formal integration of the inequality (52) over the half–open intervals $2\pi n/\lambda \le t < \pi/(2\lambda) + 2\pi n/\lambda$ $(n = 0,\ 1,\ 2,\ ...)$, which finally results in the lower estimate (55), it is needed to calculate the first order derivative of the function $g(t)$:

$$\frac{dg}{dt} = \left[\frac{dM}{dt}\left(\frac{2\pi n}{\lambda}\right)\cos(\lambda t) + \left[\frac{dM}{dt}\left(\frac{2\pi n}{\lambda}\right) - 2\lambda M\left(\frac{2\pi n}{\lambda}\right)\right]\sin(\lambda t)\right] \times$$
$$\times \exp(\lambda t - 2\pi n). \tag{56}$$

With the relations (55) and (56) in hands, one can conclude that the function $g(t)$ is positive and strictly increasing on the intervals $2\pi n/\lambda \le t < \pi/(2\lambda) + 2\pi n/\lambda$ $(n = 0,\ 1,\ 2,\ ...)$ if the following inequalities are valid [33]:

$$M\left(\frac{2\pi n}{\lambda}\right) > 0,\ \frac{dM}{dt}\left(\frac{2\pi n}{\lambda}\right) \ge 2\lambda M\left(\frac{2\pi n}{\lambda}\right). \tag{57}$$

These inequalities are the sought–for conditions which make the formal integration of the relation (52) true.

Values of the functional M and its first derivative dM/dt at the left ends of the half–open intervals $2\pi n/\lambda \le t < \pi/(2\lambda) + 2\pi n/\lambda$ $(n = 0,\ 1,\ 2,\ \ldots)$ may be arbitrary; in particular, they can be chosen to satisfy the conditions

$$M\left(\frac{2\pi n}{\lambda}\right) \equiv M(0)\exp(2\pi n),\ \frac{dM}{dt}\left(\frac{2\pi n}{\lambda}\right) \equiv \frac{dM}{dt}(0)\exp(2\pi n).$$

Then the inequalities (57) are fulfilled identically if

$$M(0) > 0,\ \frac{dM}{dt}(0) \ge 2\lambda M(0).$$

The function g (55) can be rewritten as follows:

$$g(t) = \left[M(0)\cos(\lambda t) + \left[\frac{1}{\lambda}\frac{dM}{dt}(0) - M(0)\right]\sin(\lambda t)\right]\exp(\lambda t).$$

It is possible to reason in the same way while integrating the relation (52) over the rest intervals $\pi/(2\lambda) + 2\pi n/\lambda \le t < \pi/\lambda + 2\pi n/\lambda$, $\pi/\lambda + 2\pi n/\lambda \le t < 3\pi/(2\lambda) + 2\pi n/\lambda$, and $3\pi/(2\lambda) + 2\pi n/\lambda \le t < 2\pi/\lambda + 2\pi n/\lambda$ $(n = 0,\ 1,\ 2,\ \ldots)$. So, the results of such integration are placed below as auxiliary formulae without details:

i) $t \in [\pi/(2\lambda) + 2\pi n/\lambda,\ \pi/\lambda + 2\pi n/\lambda),\ n = 0,\ 1,\ 2,\ \ldots$:

1)
$$M_1(t) \equiv \exp(-\lambda t)\, M(t),$$
$$\frac{d^2M_1}{dt^2} + \lambda^2 M_1 \ge 0;$$

2)
$$M_2(t) \equiv M_1(t)\cos^{-1}(\lambda t),$$
$$\frac{d}{dt}\left[\frac{dM_2}{dt}\cos(\lambda t)\right] - \lambda\frac{dM_2}{dt}\sin(\lambda t) \ge 0;$$

3)
$$M_3(t) \equiv (dM_2/dt)\cos^2(\lambda t),$$
$$\frac{dM_3}{dt} \le 0;$$

4)
$$M(t) \ge [C_{3n}\cos(\lambda t) + C_{4n}\sin(\lambda t)]\exp(\lambda t);\ C_{3n},\ C_{4n} \text{ — const};$$

5)
$$M(t) \ge g_1(t);$$

$$g_1(t) \equiv \left[M\left(\frac{\pi}{2\lambda} + \frac{2\pi n}{\lambda}\right)\sin(\lambda t) - \left[\frac{1}{\lambda}\frac{dM}{dt}\left(\frac{\pi}{2\lambda} + \frac{2\pi n}{\lambda}\right) - M\left(\frac{\pi}{2\lambda} + \frac{2\pi n}{\lambda}\right)\right] \times\right.$$
$$\left. \times \cos(\lambda t)\right]\exp(\lambda t - \frac{\pi}{2} - 2\pi n),$$

$$\frac{dg_1}{dt} = \left[\frac{dM}{dt}\left(\frac{\pi}{2\lambda} + \frac{2\pi n}{\lambda}\right)\sin(\lambda t) - \left[\frac{dM}{dt}\left(\frac{\pi}{2\lambda} + \frac{2\pi n}{\lambda}\right) - 2\lambda M\left(\frac{\pi}{2\lambda} + \frac{2\pi n}{\lambda}\right)\right] \times\right.$$

$$\times\ \cos(\lambda t)]\exp(\lambda t - \frac{\pi}{2} - 2\pi n);$$

6) $$M\left(\pi/(2\lambda) + 2\pi n/\lambda\right) > 0,$$

$$\frac{dM}{dt}\left(\frac{\pi}{2\lambda} + \frac{2\pi n}{\lambda}\right) \geq 2\lambda M\left(\frac{\pi}{2\lambda} + \frac{2\pi n}{\lambda}\right);$$

7) $$M\left(\pi/(2\lambda) + 2\pi n/\lambda\right) \equiv M(0)\exp(\pi/2 + 2\pi n),$$

$$\frac{dM}{dt}\left(\frac{\pi}{2\lambda} + \frac{2\pi n}{\lambda}\right) \equiv \frac{dM}{dt}(0)\exp\left(\frac{\pi}{2} + 2\pi n\right);$$

$$M(0) > 0,\ \frac{dM}{dt}(0) \geq 2\lambda M(0);$$

$$g_1(t) = \left[M(0)\sin(\lambda t) - \left[\frac{1}{\lambda}\frac{dM}{dt}(0) - M(0)\right]\cos(\lambda t)\right]\exp(\lambda t);$$

ii) $t \in [\pi/\lambda + 2\pi n/\lambda,\ 3\pi/(2\lambda) + 2\pi n/\lambda)$, $n = 0,\ 1,\ 2,\ \ldots$:

1) $$M_1(t) \equiv \exp(-\lambda t)\, M(t),$$

$$\frac{d^2 M_1}{dt^2} + \lambda^2 M_1 \geq 0;$$

2) $$M_2(t) \equiv M_1(t)\cos^{-1}(\lambda t),$$

$$\frac{d}{dt}\left[\frac{dM_2}{dt}\cos(\lambda t)\right] - \lambda\frac{dM_2}{dt}\sin(\lambda t) \geq 0;$$

3) $$M_3(t) \equiv (dM_2/dt)\cos^2(\lambda t),$$

$$\frac{dM_3}{dt} \leq 0;$$

4) $$M(t) \geq [C_{5n}\cos(\lambda t) + C_{6n}\sin(\lambda t)]\exp(\lambda t);\ C_{5n},\ C_{6n} \text{ — const};$$

5) $$M(t) \geq g_2(t);$$

$$g_2(t) \equiv -\left[M\left(\frac{\pi}{\lambda} + \frac{2\pi n}{\lambda}\right)\cos(\lambda t) + \left[\frac{1}{\lambda}\frac{dM}{dt}\left(\frac{\pi}{\lambda} + \frac{2\pi n}{\lambda}\right) - M\left(\frac{\pi}{\lambda} + \frac{2\pi n}{\lambda}\right)\right]\times\right.$$

$$\times\ \sin(\lambda t)]\exp(\lambda t - \pi - 2\pi n),$$

$$\frac{dg_2}{dt} = -\left[\frac{dM}{dt}\left(\frac{\pi}{\lambda} + \frac{2\pi n}{\lambda}\right)\cos(\lambda t) + \left[\frac{dM}{dt}\left(\frac{\pi}{\lambda} + \frac{2\pi n}{\lambda}\right) - 2\lambda M\left(\frac{\pi}{\lambda} + \frac{2\pi n}{\lambda}\right)\right]\times\right.$$

$$\times\ \sin(\lambda t)]\exp(\lambda t - \pi - 2\pi n);$$

6) $$M\left(\pi/\lambda + 2\pi n/\lambda\right) > 0,$$

$$\frac{dM}{dt}\left(\frac{\pi}{\lambda} + \frac{2\pi n}{\lambda}\right) \geq 2\lambda M\left(\frac{\pi}{\lambda} + \frac{2\pi n}{\lambda}\right);$$

7) $$M\left(\pi/\lambda+2\pi n/\lambda\right)\equiv M(0)\exp(\pi+2\pi n),$$

$$\frac{dM}{dt}\left(\frac{\pi}{\lambda}+\frac{2\pi n}{\lambda}\right)\equiv\frac{dM}{dt}(0)\exp\left(\pi+2\pi n\right);$$

$$M(0)>0,\ \frac{dM}{dt}(0)\geq 2\lambda M(0);$$

$$g_2(t)=-\left[M(0)\cos(\lambda t)+\left[\frac{1}{\lambda}\frac{dM}{dt}(0)-M(0)\right]\sin(\lambda t)\right]\exp(\lambda t);$$

iii) $t\in[3\pi/(2\lambda)+2\pi n/\lambda,\ 2\pi/\lambda+2\pi n/\lambda),\ n=0,\ 1,\ 2,\ \ldots$:

1) $$M_1(t)\equiv\exp(-\lambda t)\,M(t),$$

$$\frac{d^2M_1}{dt^2}+\lambda^2M_1\geq 0;$$

2) $$M_2(t)\equiv M_1(t)\cos^{-1}(\lambda t),$$

$$\frac{d}{dt}\left[\frac{dM_2}{dt}\cos(\lambda t)\right]-\lambda\frac{dM_2}{dt}\sin(\lambda t)\geq 0;$$

3) $$M_3(t)\equiv(dM_2/dt)\cos^2(\lambda t),$$

$$\frac{dM_3}{dt}\geq 0;$$

4) $$M(t)\geq[C_{7n}\cos(\lambda t)+C_{8n}\sin(\lambda t)]\exp(\lambda t);\ C_{7n},\ C_{8n}\text{ — const};$$

5) $$M(t)\geq g_3(t);$$

$$g_3(t)\equiv\left[-M\left(\frac{3\pi}{2\lambda}+\frac{2\pi n}{\lambda}\right)\sin(\lambda t)+\left[\frac{1}{\lambda}\frac{dM}{dt}\left(\frac{3\pi}{2\lambda}+\frac{2\pi n}{\lambda}\right)-M\left(\frac{3\pi}{2\lambda}+\frac{2\pi n}{\lambda}\right)\right]\times\right.$$

$$\left.\times\cos(\lambda t)\right]\exp(\lambda t-\frac{3\pi}{2}-2\pi n),$$

$$\frac{dg_3}{dt}=\left[-\frac{dM}{dt}\left(\frac{3\pi}{2\lambda}+\frac{2\pi n}{\lambda}\right)\sin(\lambda t)+\left[\frac{dM}{dt}\left(\frac{3\pi}{2\lambda}+\frac{2\pi n}{\lambda}\right)-2\lambda M\left(\frac{3\pi}{2\lambda}+\frac{2\pi n}{\lambda}\right)\right]\times\right.$$

$$\left.\times\cos(\lambda t)\right]\exp(\lambda t-\frac{3\pi}{2}-2\pi n);$$

6) $$M\left(3\pi/(2\lambda)+2\pi n/\lambda\right)>0,$$

$$\frac{dM}{dt}\left(\frac{3\pi}{2\lambda}+\frac{2\pi n}{\lambda}\right)\geq 2\lambda M\left(\frac{3\pi}{2\lambda}+\frac{2\pi n}{\lambda}\right);$$

7) $$M\left(3\pi/(2\lambda)+2\pi n/\lambda\right)\equiv M(0)\exp(3\pi/2+2\pi n),$$

$$\frac{dM}{dt}\left(\frac{3\pi}{2\lambda}+\frac{2\pi n}{\lambda}\right)\equiv\frac{dM}{dt}(0)\exp\left(\frac{3\pi}{2}+2\pi n\right);$$

$$M(0) > 0, \ \frac{dM}{dt}(0) \geq 2\lambda M(0);$$

$$g_3(t) = \left[-M(0)\sin(\lambda t) + \left[\frac{1}{\lambda}\frac{dM}{dt}(0) - M(0) \right] \cos(\lambda t) \right] \exp(\lambda t).$$

From the final expressions for the functions $g(t),\ g_i(t)\ (i = \overline{1,\ 3})$, it is easily seen that the graphs of the functions are curves which lie across an exponential half–strip; moreover, their left–hand ends lean on the lower bound of this half–strip

$$f_1(t) \equiv M(0)\exp(\lambda t),$$

and the right–hand ends adjoin to its upper bound

$$g_4(t) \equiv \left[\frac{1}{\lambda}\frac{dM}{dt}(0) - M(0) \right] \exp(\lambda t).$$

The exception is the case when

$$\frac{dM}{dt}(0) = 2\lambda M(0).$$

Then the exponential half–strip degenerates into its former lower bound $f_1(t)$, and both ends of curves which are the graphs of the functions $g(t),\ g_i(t)\ (i = \overline{1,\ 3})$ on the corresponding time half–open intervals lie on this line.

From such geometric properties of the functions $g(t),\ g_i(t)\ (i = \overline{1,\ 3})$, one can conclude that the time growth of the integral M (31) is definitely no slower than exponential. Thus, it is showed that, indeed, under the conditions

$$M\left(\frac{\pi n}{2\lambda}\right) > 0, \ \frac{dM}{dt}\left(\frac{\pi n}{2\lambda}\right) \geq 2\lambda M\left(\frac{\pi n}{2\lambda}\right) \ (n = 0,\ 1,\ 2,\ \ldots)$$
$$M\left(\frac{\pi n}{2\lambda}\right) \equiv M(0)\exp\left(\frac{\pi n}{2}\right), \ \frac{dM}{dt}\left(\frac{\pi n}{2\lambda}\right) \equiv \frac{dM}{dt}(0)\exp\left(\frac{\pi n}{2}\right) \tag{58}$$

the relation (52) implies the lower exponential estimate (53).

So, the inequality (53) demonstrates that among small axial–symmetric long–wave perturbations (25), (26) with initial data

$$M(0) > 0, \ \frac{dM}{dt}(0) \geq 2\lambda M(0) \tag{59}$$

of the exact stationary solutions (20) to the mixed problem (13)–(15) there can be growing (at the minimum, exponentially) ones in time. Moreover, of basic importance is that the additional conditions (58) are automatically fulfilled for the exponentially growing in time small axial–symmetric long–wave perturbations (25), (26), (59).

As a characteristic feature of the given growth, it is necessary to note the absence of any constraints on the positive parameter λ value in the exponent which is in the estimate (53). In this aspect, the discovered instability can be explained as the specific manifestation of small–scale perturbations, excluded from the consideration by using long–wave approximation, in the field of investigated large–scale fluid motions.

Due to the comparison of the ratio (53) and the inequality (35), it follows that just the integral M (31) and nothing else performs the role of Lyapunov's functional here.

Hence, it is showed that, really, the relation (30) is the sufficient condition for instability of the steady flows (20) against the small axial–symmetric long–wave perturbations (21), (25), (26), (59).

Moreover, the above described process for obtaining the lower estimate (53) clearly demonstrates that the inequality (29) is the necessary and sufficient condition for stability with respect to the small axial–symmetric long–wave perturbations (25), (26), (28) both for the exact stationary solutions (20), (24) to the initial boundary value problem (13)–(15) and for the steady flows (20),

$$\frac{d}{d\nu}\left(R^0 - w^0 C^0\right) \geq 0.$$

At last, the minutely expounded procedure for integrating the relation (52) visually shows that one can extract the information about initial data of the increasing small perturbations while carrying out researches in class of piecewise continuous functions too.

Further, an illustrative example of the exact stationary solutions (20) to the mixed problem (13)–(15) and the small axial–symmetric long–wave perturbations (25), (26), (59), superposed on them, is constructed, where they develop with time according to the found lower estimate (53) when the ratio (30) is valid.

Specifically, the steady axial–symmetric sheared magnetohydrodynamic jet flows of an ideal fluid

$$w^0(\nu) = b - \nu,\ R^0(\nu) = \nu,\ R_1^0 = 1 \tag{60}$$

are being investigated, where $b > 1$ is a constant, in the infinite band (48). It is clear that the given flows belong to the class of the stationary flows (20).

If the inequality (30) is true then the steady flows (60), (48) will be unstable regarding the small axial–symmetric long–wave perturbations $\xi(t,\ z,\ \nu)$ (25), (26) of the form

$$\xi(t,\ z,\ \nu) = \frac{\alpha \exp(\sigma\beta t)}{\left[\sigma^2 + \left(\frac{1}{2} - \nu\right)^2\right]^2}\left(\left[\sigma^2 - \left(\frac{1}{2} - \nu\right)^2\right][\cos(\gamma\beta t)\cos(\beta z) - \sin(\gamma\beta t) \times\right.$$

$$\left.\times \sin(\beta z)] + 2\sigma\left(\frac{1}{2} - \nu\right)[\cos(\gamma\beta t)\sin(\beta z) + \sin(\gamma\beta t)\cos(\beta z)]\right). \tag{61}$$

Here α is an arbitrary constant, and β is a positive one, whereas

$$\sigma \equiv \sqrt{\frac{\kappa^2 - h_1^2}{2} - \frac{1}{4}},\ \gamma \equiv \frac{1}{2} - b.$$

In fact, it is easy to make sure by means of the direct check that the function $\xi(t, z, \nu)$ (61) is the solution of the initial boundary value problem (25), (26), and it also satisfies the ratios (32), (33) (with the integral T_1 in the form (51)), (53), and (59). Besides, this function can be interpreted as Hadamard's example [34], since the constant β in the exponent from the right part of the expression (61) conserves some uncertainty.

3. The Instability of the Steady Axial–Symmetric Sheared Jet Flows of Non–viscous Ideally Conducting Incompressible Fluid with Free Boundary in the Azimuthal Magnetic Field which is the Preset Radius Function

The mixed problem, which is formed by the third equation of the system (10) and the ratios (11)–(13), is considered below in more general formulation, when h is an arbitrary function of the independent variable ν: $h = h(\nu)$ [35]. It means that during the process of the fluid flow the dimensionless ratio of the azimuthal component of the magnetic field in the studied jet to the radial coordinate is the value, which is unchangeable on any trajectory of the fluid particle motion. It is important that the given restriction does not contradict the third equation of the system (10), since if we take $h = h_0(\nu)$ under the condition $t = 0$ then, according to this equation, such kind of dependence of the function h on the variable ν will not change at any further moment $t > 0$.

As a result, the initial boundary value problem, which includes the third ratio of the system (10) and the relations (11)–(13) as well, can be reformulated in the form

$$w_t + ww_z = \frac{1}{2}\left[\left(\frac{\kappa}{R_1}\right)^2 - h_1^2\right] R_{1z} + \int\limits_{\nu}^{1}\left[hR_z\frac{dh}{d\nu_1}\right] d\nu_1,$$

$$R_{\nu t} + (wR_\nu)_z = 0; \tag{62}$$

$$w(0,\, z,\, \nu) = w_0(z,\, \nu),\ R(0,\, z,\, \nu) = R_0(z,\, \nu).$$

It is easy to demonstrate that the functional

$$E_1 \equiv \frac{1}{2}\int\limits_{-\infty}^{+\infty}\left(\int\limits_0^1 \left(w^2 + h^2 R\right) R_\nu d\nu + \kappa^2 \ln R_1\right) dz \tag{63}$$

is the energy integral of the given problem, and the functional

$$I \equiv \int\limits_{-\infty}^{+\infty}\int\limits_0^1 \left[w_\nu F_1(h)\right] d\nu dz \tag{64}$$

is its additional motion integral, where $F_1(h)$ is a function of its argument [28–30].

The functions w^0, R^0, and R_1^0 (20) are the exact stationary solutions of the mixed problem (62). It is not difficult to ascertain that these functions reduce the first two equations from the ratios of the initial boundary value problem (62) to identities indeed.

The purpose of the further research is in finding out if the stationary solutions (20) can be stable to the small axial–symmetric long–wave perturbations $w'(t,\, z,\, \nu)$, $R'(t,\, z,\, \nu)$, and $R_1'(t,\, z)$, the evolution of which is described by the mixed problem in the form

$$w_t' + w^0 w_z' = \frac{\kappa^2 - h_1^2}{2} R_{1z}' + \int\limits_{\nu}^{1}\left[hR_z'\frac{dh}{d\nu_1}\right] d\nu_1,$$

$$R'_{\nu t} + w^0 R'_{\nu z} + \frac{dR^0}{d\nu} w'_z = 0; \tag{65}$$

$$w'(0,\, z,\, \nu) = w'_0(z,\, \nu),\ R'(0,\, z,\, \nu) = R'_0(z,\, \nu).$$

On the analogy of the above paragraph, the given problem is obtained by linearization of the initial boundary value problem (62) near its exact solutions (20).

As the direct calculations show, the functional in the form

$$E \equiv \frac{1}{2} \int\limits_{-\infty}^{+\infty} \left(\int\limits_0^1 \left[\frac{dR^0}{d\nu} w'^2 + 2w^0 w' R'_\nu + h^2 R' R'_\nu \right] d\nu - \frac{\kappa^2}{2} R_1'^2 \right) dz \tag{66}$$

is the energy integral for the mixed problem (65). Besides, it is possible to check that the first variation δJ_1 of the functional $J_1 \equiv E_1 + I$ (63), (64) is made vanishing on the stationary solutions (20) in the case, when the functions w^0, R^0, h, and F_1 are connected by the equalities

$$w^0 \frac{dw^0}{d\nu} + hR^0 \frac{dh}{d\nu} = 0,\ w^0 \frac{dR^0}{d\nu} = \frac{dF_1}{dh} \frac{dh}{d\nu}. \tag{67}$$

At that, the second variation $\delta^2 J_1$ of the given integral, rewritten in the appropriate designations, coincides with the functional E (66) by form.

The exact stationary solutions (20), (67) of the initial boundary value problem (62) will be stable regarding the small axial–symmetric long–wave perturbations (65) if and only if the integral E is of fixed sign.

To define if the functional E (66) has the property of fixed sign definiteness, it is convenient to write it in the form (23) with the vector

$$\mathbf{u} \equiv \begin{pmatrix} w' \\ R' \\ R'_\nu \\ R'_1 \end{pmatrix} \tag{68}$$

and the nonzero constituents

$$b_{11} = \frac{dR^0}{d\nu},\ b_{13} = b_{31} = w^0,\ b_{23} = b_{32} = \frac{h^2}{2},\ b_{34} = b_{43} = -\frac{\kappa^2}{4} \tag{69}$$

of the quadratic matrix B of order 4.

It is not difficult to make the conclusion that the principal minors of the matrix B do not have the fixed sign definiteness which is required according to Silvester's criterion [31]. Indeed, the calculation of only its first two principal minors gives

$$\Delta_1 = \frac{dR^0}{d\nu} > 0,\ \Delta_2 = 0.$$

This implies that, since the principal minor Δ_2 of the matrix B is equal to zero, the integral E (23), (68), (69) is not of fixed sign by force of Silvester's criterion.

Hence, the sufficient conditions for stability of the exact stationary solutions (20), (67) of the mixed problem (62) regarding the small axial–symmetric long–wave perturbations

$w'(t,\ z,\ \nu)$, $R'(t,\ z,\ \nu)$, and $R_1'(t,\ z)$ (65), which are interpreted as the conditions for the fixed sign definiteness of the functional E (66), which is the energetic one by its nature, do not exist with this method.

Further, the sufficient conditions for the instability of the stationary solutions (20), (67) to the small axial–symmetric long–wave perturbations (65) will be found out by Lyapunov's direct method [1, 2], and also the lower estimate will be constructed, which will demonstrate that these perturbations may grow with time no slower than exponentially.

As in the above paragraph, the subclass of the axial–symmetric sheared jet flows of an ideal perfectly conducting incompressible fluid with free boundary in the azimuthal magnetic field is considered below, where the small axial–symmetric long–wave perturbations $w'(t,\ z,\ \nu)$, $R'(t,\ z,\ \nu)$, and $R_1'(t,\ z)$ (65) are deviations of the trajectories of the fluid particles motion from the corresponding current lines of the steady flows (20), (67). The given perturbations can be described in the most obvious way by means of the field of Lagrangian displacements $\xi = \xi(t,\ z,\ \nu)$ [25] which is represented by the ratio (25).

By means of the equation (25), the initial boundary value problem (65) can be rewritten in the form

$$w_t' + w^0 w_z' = \frac{\kappa^2 - h_1^2}{2} R_{1z}' + \int\limits_{\nu}^{1} \left[h R_z' \frac{dh}{d\nu_1} \right] d\nu_1,$$

$$R_\nu' = -\frac{dR^0}{d\nu}\xi_z; \tag{70}$$

$$\xi(0,\ z,\ \nu) = \xi_0(z,\ \nu),\ w'(0,\ z,\ \nu) = w_0'(z,\ \nu).$$

By the direct calculations, it is not difficult to show that the energy integral of the mixed problem (25), (70) will be the functional E (66) and no other one.

It is expedient to use the auxiliary integral

$$\Pi \equiv \frac{1}{2} \int\limits_{-\infty}^{+\infty} \int\limits_{0}^{1} \left[h^2 R' R_\nu' \right] d\nu dz - \frac{\kappa^2}{4} \int\limits_{-\infty}^{+\infty} R_1'^2 dz \tag{71}$$

for the sake of the further statement. It allows, for example, to represent the functional E (66) in the form

$$E \equiv T + \Pi + T_1 = \text{const} \tag{72}$$

(here the values T and T_1 are taken from the ratios (32) and (51) accordingly).

The double differentiation of the integral M (31) by the independent variable t and the realization of the number of the obtained functional transformations with applying the relations (25), (70), (71) result in the virial equality [22, 24, 25], which already appeared above, in the form

$$\frac{d^2 M}{dt^2} = 4(T - \Pi).$$

Now, multiplying the given equality by the arbitrary constant λ and taking into consideration the expression (72), it is possible to derive the key equation of the kind (34) with the integrals $M,\ T,\ T_1$, and Π in the form (31), (32), (51), and (71).

If the inequalities $\lambda > 0$ and

$$h\frac{dh}{d\nu} > 0,\ h_1^2 < \kappa^2 \tag{73}$$

are valid then it is easy to make sure that this key equation will imply the differential ratio (52), and hence, the lower estimate (53), as it is demonstrated above.

The presence of the given lower estimate is evidence of the fact, that the small axial–symmetric long–wave perturbations $\xi(t,\ z,\ \nu)$ (25), (59), (70) of the exact stationary solutions (20), (67) really may grow with time no slower than exponentially; therefore, the inequalities (73) are the very desired sufficient conditions for the linear instability.

4. The Stability of the Steady Axial–Symmetric Sheared Jet Flows of Non–viscous Ideally Conducting Incompressible Fluid with Free Boundary in the Poloidal Magnetic Field

The axial–symmetric perfectly conducting fluid jet of an infinite length in the magnetic field is being investigated, at that, the longitudinal direct electrical current J is considered to flow on its free surface. Besides, the given jet is supposed to be placed along the axis of the ideally conducting cylindrical shell of the radius r_*, osculating nowhere with the latter due to the vacuum gap between them. The cylindrical system of coordinates $(r^*,\ \varphi,\ z^*)$ is set so, that its axis z^* coincides with the axis of symmetry of the conducting jet. Besides the used ones, some new designations are applied: H_1, H_3 are the radial and axial components of the magnetic field inside the jet; H_1^*, H_3^* are also the radial and axial components of the magnetic field, but outside the conducting jet. It is assumed that $v_2 \equiv 0,\ H_2 \equiv 0$ when there are fluid motions in the jet (v_2 is the azimuthal component of velocity field). Moreover, these motions are supposed to be axial–symmetric, and fluid is supposed to be ideal, incompressible, homogeneous in density. At last, the action of surface tension on free boundary of the conducting jet is not taken into consideration.

By force of the enumerated assumptions, the equations of one–fluid non–dissipative magnetohydrodynamics [26, 36] will have the form

$$\begin{gathered}
\rho\left(\frac{\partial v_1}{\partial t^*} + v_1\frac{\partial v_1}{\partial r^*} + v_3\frac{\partial v_1}{\partial z^*}\right) = -\frac{\partial P_*}{\partial r^*} + \frac{H_1}{4\pi}\frac{\partial H_1}{\partial r^*} + \frac{H_3}{4\pi}\frac{\partial H_1}{\partial z^*}, \\
\rho\left(\frac{\partial v_3}{\partial t^*} + v_1\frac{\partial v_3}{\partial r^*} + v_3\frac{\partial v_3}{\partial z^*}\right) = -\frac{\partial P_*}{\partial z^*} + \frac{H_1}{4\pi}\frac{\partial H_3}{\partial r^*} + \frac{H_3}{4\pi}\frac{\partial H_3}{\partial z^*}, \\
\frac{1}{r^*}\frac{\partial (v_1 r^*)}{\partial r^*} + \frac{\partial v_3}{\partial z^*} = 0,\quad \frac{\partial (A r^*)}{\partial t^*} + v_1\frac{\partial (A r^*)}{\partial r^*} + v_3\frac{\partial (A r^*)}{\partial z^*} = 0; \\
P_* \equiv P + \frac{H_1^2 + H_3^2}{8\pi},\quad H_1 \equiv -\frac{\partial A}{\partial z^*},\quad H_3 \equiv \frac{1}{r^*}\frac{\partial (A r^*)}{\partial r^*},
\end{gathered} \tag{74}$$

where A is the azimuthal component of the magnetic field vector potential in the studied jet.

If to disregard the displacement current then the ratios for the magnetic field components in the layer between the interior surface of the cylindrical shell and free boundary of the jet [37] will be written in the form

$$\frac{\partial H_2^*}{\partial z^*} = 0, \quad \frac{\partial H_1^*}{\partial z^*} - \frac{\partial H_3^*}{\partial r^*} = 0, \quad \frac{1}{r^*}\frac{\partial (H_2^* r^*)}{\partial r^*} = 0,$$
$$\frac{1}{r^*}\frac{\partial (H_1^* r^*)}{\partial r^*} + \frac{\partial H_3^*}{\partial z^*} = 0. \tag{75}$$

The following boundary conditions are set: a) on the symmetry axis of the conducting jet (with $r^* = 0$)

$$v_1 = 0, \; H_1 = 0; \tag{76}$$

b) on its free surface (with $r^* = r_1(t^*, z^*)$)

$$P_* = \frac{H_1^{*2} + H_2^{*2} + H_3^{*2}}{8\pi}, \; v_1 = \frac{\partial r_1}{\partial t^*} + v_3 \frac{\partial r_1}{\partial z^*},$$
$$H_1 - H_3 \frac{\partial r_1}{\partial z^*} = 0, \; H_1^* - H_3^* \frac{\partial r_1}{\partial z^*} = 0; \tag{77}$$

c) on the interior boundary of the cylindrical shell (with $r^* = r_*$)

$$H_1^* = 0. \tag{78}$$

The initial data for the equations system (74) and the second ratio from the system (77) are set in the form

$$v_1(0, r^*, z^*) = v_{10}(r^*, z^*), \; v_3(0, r^*, z^*) = v_{30}(r^*, z^*),$$
$$A(0, r^*, z^*) = A_0(r^*, z^*), \; r_1(0, z^*) = r_{10}(z^*), \tag{79}$$

at that, the functions v_{10}, v_{30}, A_0, and r_{10} are required to be in no contradiction with the third equation of the system (74), the conditions (76), and also the first, third and forth relations from the ratios system (77).

Further, the transition to long–wave approximation is realized in the initial boundary value problem (74)–(79), which is anticipated by the reduction to dimensionless variables, at that, the values L, v_0, and r_0 are chosen as dimensionless parameters, similarly to paragraph 2 of the present chapter. Due to the given parameters, the dimensionless values t, η, z, q, w, p_*, h, H, a, h^*, κ, and H^* are constructed so that the expressions

$$t^* = \frac{tL}{v_0}, \; r^{*2} = \eta L^2 \delta^2, \; z^* = zL, \; 2v_1 r^* = q v_0 L \delta^2,$$
$$v_3 = w v_0, \; P_* = p_* \rho v_0^2, \; H_1 r^* = h v_0 \sqrt{\pi\rho}\, L\delta^2, \; H_3 = 2H v_0 \sqrt{\pi\rho}, \tag{80}$$
$$Ar^* = a v_0 \sqrt{\pi\rho}\, L^2 \delta^2, \; H_1^* r^* = h^* v_0 \sqrt{\pi\rho}\, L\delta^2, \; H_2^* r^* = 2\kappa v_0 \sqrt{\pi\rho}\, L\delta, \; H_3^* = 2H^* v_0 \sqrt{\pi\rho}$$

take place (here $\delta \equiv r_0/L \ll 1$ is the dimensionless characteristic radius of the considered jet, as before).

The result of the reduction to dimensionless variables of the mixed problem (74)–(79) with using the relations (80) will be, above all, the equations system

$$\left[q_t + qq_\eta - \frac{q^2}{2\eta} + wq_z\right]\delta^2 = -\,4\eta p_{*\eta} + \left[hh_\eta - \frac{h^2}{2\eta} + Hh_z\right]\delta^2,$$

$$w_t + qw_\eta + ww_z = -\,p_{*z} + hH_\eta + HH_z,\ q_\eta + w_z = 0, \tag{81}$$

$$a_t + qa_\eta + wa_z = 0;\ h \equiv -\,a_z,\ H \equiv a_\eta,$$

which arises from the system (74). In its turn, the ratios system (75) is converted to the form

$$h_z^*\delta^2 - 4\eta H_\eta^* = 0,\ H_z^* + h_\eta^* = 0,\ \kappa = \text{const}. \tag{82}$$

Along with this, taking into consideration the latter equality of the system (82), the boundary conditions (76)–(78) can be represented in the form

$$q = 0,\ h = 0\quad(\eta = 0)\,;$$

$$q = \eta_{1t} + w\eta_{1z},\ p_* = \frac{1}{2}\left[\frac{(h^*\delta)^2}{4\eta_1} + \frac{\kappa^2}{\eta_1} + H^{*2}\right]\quad(\eta = \eta_1\,(t,\,z))\,;$$

$$h - H\eta_{1z} = 0,\ h^* - H^*\eta_{1z} = 0\quad(\eta = \eta_1\,(t,\,z))\,; \tag{83}$$

$$h^* = 0\quad(\eta = \eta_*)\,,$$

where the function $\eta_1(t,\,z)$ describes the change of the form of free surface of the jet with time, and the value η_* corresponds to the radius of the enveloping cylindrical shell. At last, the initial data (79) will have the form

$$q(0,\,\eta,\,z) = q_0(\eta,\,z),\ w(0,\,\eta,\,z) = w_0(\eta,\,z),$$

$$a(0,\,\eta,\,z) = a_0(\eta,\,z),\ \eta_1(0,\,z) = \eta_{10}(z). \tag{84}$$

This directly implies that if to remove the summands, which are proportional to the multiplier δ^2, and the expression for the function $q(0,\,\eta,\,z)$ in the ratios of the initial boundary value problem (81)–(84), then it will have the form, which does correspond to the long–wave approximation.

It is remarkable that the long–wave image of the equations system (82) will have the solution, reducing to identities the sixth and the seventh boundary conditions of the system (83):

$$h^*(t,\,\eta,\,z) = (\eta_* - \eta)\,H_z^*,\ H^*(t,\,z) = \frac{\Phi}{\eta_* - \eta_1}.$$

Here $\Phi = \Phi(t)$ is an arbitrary time function, which is, from the physics perspective, the dimensionless axial flux of the magnetic field through the vacuum gap between free boundary of the conducting jet and the interior surface of the cylindrical shell.

It is considered below that the axial flux of the magnetic field through the vacuum layer between the jet and the shell is fixed, i.e. $\Phi(t) \equiv \Phi_0 = \text{const}$. This assumption implies the absence of any external to the investigated mechanical system sources, generating the change of the given flux value with time, and completely goes with the ratios which are obtained from the mixed problem (81)–(84) as a result of the long–wave approximation realization. Hence, the forth equality of the system of the boundary conditions (83) will turn into the relation

$$p_* = \frac{1}{2}\left[\frac{\kappa^2}{\eta_1} + \frac{\Phi_0^2}{(\eta_* - \eta_1)^2}\right] \quad (\eta = \eta_1(t,\, z))\,. \tag{85}$$

The further research of the long–wave modification of the initial boundary value problem (81)–(84) can be significantly simplified if to realize the replacement of the Eulerian independent variables $(t,\, \eta,\, z)$ by the mixed Eulerian–Lagrangian ones $(t',\, z',\, \nu)$ in it; the replacement is defined by the ratios (8), (9) in paragraph 2 of the present chapter [27, 28, 38].

In fact, disregarding the summands, which have factor δ^2, and omitting the variables t' and z' primes, it is not difficult to rewrite the equations system (81) by means of the new mixed Eulerian–Lagrangian independent variables in the form

$$\begin{gathered} p_{*\nu} = 0,\; R_\nu\,(w_t + w w_z) = -\,R_\nu p_{*z} + hH_\nu + R_\nu H H_z - H R_z H_\nu, \\ q_\nu + R_\nu w_z - R_z w_\nu = 0,\; a_t + w a_z = 0; \\ h \equiv -\,a_z + \frac{a_\nu R_z}{R_\nu},\; H \equiv \frac{a_\nu}{R_\nu}. \end{gathered} \tag{86}$$

The boundary conditions for the ratios (86) will be the equality (85) and the relation

$$a_z = 0 \quad (\nu = 0,\, 1)\,, \tag{87}$$

which is followed from the second and the fifth expressions (83), and two latter ratios of the equations system (86) as well. The initial data are added to the ratios (8), (9), (85)–(87) in the form

$$\begin{gathered} w\,(0,\, z,\, \nu) = w_0\,(z,\, \nu)\,,\; R\,(0,\, z,\, \nu) = R_0\,(z,\, \nu)\,, \\ a\,(0,\, z,\, \nu) = a_0\,(z,\, \nu)\,, \end{gathered} \tag{88}$$

where the function $R_0\,(z,\, \nu)$, on the analogy of paragraphs 2, 3 of the present chapter, is considered to be monotone growing one by the argument ν.

On further examination of the initial boundary value problem (8), (9), (85)–(88) the value a is supposed to be the known function of the variable ν, namely $a = a_*(\nu)$. It means that the dimensionless product of the azimuthal component of the magnetic field vector potential by the radial coordinate is constant on each line $\nu = \text{const}$ in the process of fluid motions. It is essential that the given restriction does not contradict the forth equation of the system (86), since if $a_0 \equiv a_*(\nu)$ at the initial time $t = 0$ then by force of this equation such kind of the value a dependence on the independent variable ν will not change at any further moment $t > 0$. Moreover, the boundary condition (87) is satisfied identically for the function a in the form $a_*(\nu)$.

To put the mixed problem (8), (9), (85)–(88) into the maximally obvious form, by means of the ratio (85) and the first equation of the system (86) the relation

$$p_{*z} = \left[-\frac{\kappa^2}{2\eta_1^2} + \frac{\Phi_0^2}{(\eta_* - \eta_1)^3} \right] \eta_{1z} \tag{89}$$

is found out. Now, replacing the partial derivative p_{*z} by its expression (89), and the function q by its presentation (8) in the ratios system (86) and taking into consideration the above assumption about the form of the functional dependence of the value a on the variable ν, it is easy to write the initial boundary value problem (8), (9), (85)–(88) in the final form

$$\begin{gathered} w_t + w w_z = \left[\frac{\kappa^2}{2R_1^2} - \frac{\Phi_0^2}{(R_* - R_1)^3} \right] R_{1z} - \frac{R_{\nu z}}{R_\nu^3} \left(\frac{da_*}{d\nu} \right)^2, \\ R_{\nu t} + (w R_\nu)_z = 0; \; h \equiv \frac{R_z}{R_\nu} \frac{da_*}{d\nu}, \; H \equiv R_\nu^{-1} \frac{da_*}{d\nu}; \\ w(0, z, \nu) = w_0(z, \nu), \; R(0, z, \nu) = R_0(z, \nu). \end{gathered} \tag{90}$$

Here, as before, R_1 is the function R value on free boundary of the conducting jet $\nu = 1$, and the relation $R_1(t, z) \equiv \eta_1(t, z)$ is true for R_1 according to the second equality of the system (9), but the designation R_* for the radius of the cylindrical shell, which envelopes the studied jet, is set instead of η_* for the sake of uniformity of the further statement.

The mixed problem (90) has the energy integral in the form

$$E_1 \equiv \frac{1}{2} \int_{-\infty}^{+\infty} \left(\int_0^1 \left[w^2 R_\nu + R_\nu^{-1} \left(\frac{da_*}{d\nu} \right)^2 \right] d\nu + \kappa^2 \ln R_1 + \right.$$

$$\left. + \frac{\Phi_0^2}{R_* - R_1} \right) dz = \text{const} \tag{91}$$

with the same restrictions on the behavior of its solutions as functions of the independent variable z, as in the previous two paragraphs of the present chapter. Besides, it is not difficult to show that the functional

$$I \equiv \int_{-\infty}^{+\infty} \int_0^1 w_\nu F_2(a_*) \, d\nu dz, \tag{92}$$

where F_2 is a function of its argument, is one more motion integral of the initial boundary value problem (90) [28–30, 38].

The mixed problem (90) has the exact stationary solutions in the form

$$\begin{gathered} w = w^0(\nu), \; R = R^0(\nu), \; R_1 = R_1^0 \equiv 1, \\ h = h^0 \equiv 0, \; H = H^0(\nu) \equiv \left(\frac{dR^0}{d\nu} \right)^{-1} \frac{da_*}{d\nu}. \end{gathered} \tag{93}$$

Here w^0 is an arbitrary function of the variable ν, and R^0 is monotone growing one; the characteristic radius r_0 (80) is taken again as the radius of the unperturbed conducting jet, as in paragraph 2 of the present chapter.

The goal of the further research is to find out the sufficient conditions for the linear stability of the stationary solutions (93) regarding the small axial–symmetric long–wave perturbations $w'(t,\ z,\ \nu)$ and $R'(t,\ z,\ \nu)$.

This purpose can be achieved by linearization of the initial boundary value problem (90) near exact stationary solutions (93). As a result, the mixed problem

$$w'_t + w^0 w'_z = \left[\frac{\kappa^2}{2} - \frac{\Phi_0^2}{(R_* - 1)^3}\right] R'_{1z} - \left(\frac{dR^0}{d\nu}\right)^{-3} \left(\frac{da_*}{d\nu}\right)^2 R'_{\nu z},$$

$$R'_{\nu t} + w^0 R'_{\nu z} + \frac{dR^0}{d\nu} w'_z = 0;\ R'_1(t,\ z) \equiv R'(t,\ z,\ 1),$$

$$h' \equiv \left(\frac{dR^0}{d\nu}\right)^{-1} \frac{da_*}{d\nu} R'_z,\ H' \equiv -\left(\frac{dR^0}{d\nu}\right)^{-2} \frac{da_*}{d\nu} R'_\nu; \tag{94}$$

$$w'\left(0,\ z,\ \nu\right) = w'_0\left(z,\ \nu\right),\ R'\left(0,\ z,\ \nu\right) = R'_0\left(z,\ \nu\right)$$

arises. The functional

$$E \equiv \frac{1}{2} \int_{-\infty}^{+\infty} \left(\int_0^1 \left[\frac{dR^0}{d\nu} w'^2 + 2w^0 R'_\nu w' + \left(\frac{dR^0}{d\nu}\right)^{-3} \left(\frac{da_*}{d\nu}\right)^2 R'^2_\nu \right] d\nu + \right.$$

$$\left. + \left[\frac{\Phi_0^2}{(R_* - 1)^3} - \frac{\kappa^2}{2}\right] R'^2_1 \right) dz \tag{95}$$

is conserved on the solutions of the given initial boundary value problem with time.

It is easy to ascertain that the first variation δJ_1 of the integral $J_1 \equiv E_1 + I$ (91), (92) is becoming equal to zero for the stationary solutions (93) if the functions $w^0,\ R^0,\ a_*$, and F_2 reduce the equations

$$w^0 \frac{dR^0}{d\nu} = \frac{dF_2}{da_*} \frac{da_*}{d\nu},$$

$$\left(\frac{dR^0}{d\nu}\right)^{-1} \frac{da_*}{d\nu} \frac{d}{d\nu} \left[\left(\frac{dR^0}{d\nu}\right)^{-1} \frac{da_*}{d\nu} \right] = w^0 \frac{dw^0}{d\nu} \tag{96}$$

to identities. At that, the second variation $\delta^2 J_1$ of the functional J_1, written out in the proper designations, coincides with the integral E (95) by form.

The exact stationary solutions (93), (96) of the mixed problem (90) will be stable to the small axial–symmetric long–wave perturbations (94) if and only if the functional E is of fixed sign.

To find out, if the integral E (95) has the desired fixed sign definiteness, it is convenient to represent it in the form (23) with the vector $\mathbf{u}$ and nonzero elements b_{ik} of the quadratic matrix B of order 3 in the form

$$\mathbf{u} \equiv \begin{pmatrix} w' \\ R'_\nu \\ R'_1 \end{pmatrix}; \; b_{11} = \frac{dR^0}{d\nu}, \; b_{12} = b_{21} = w^0, \tag{97}$$

$$b_{22} = \left(\frac{dR^0}{d\nu}\right)^{-3} \left(\frac{da_*}{d\nu}\right)^2, \; b_{23} = b_{32} = \frac{1}{2}\left[\frac{\Phi_0^2}{(R_* - 1)^3} - \frac{\kappa^2}{2}\right].$$

It is not difficult to conclude that by force of Silvester's criterion [31] the principal minors of the matrix B do not have the required fixed sign definiteness. In fact, the direct calculation of the given minors leads to the ratios

$$\Delta_1 = \frac{dR^0}{d\nu} > 0, \; \Delta_2 = \left(\frac{dR^0}{d\nu}\right)^{-2} \left(\frac{da_*}{d\nu}\right)^2 - w^{02},$$

$$\Delta_3 = -\frac{1}{4}\left[\frac{\Phi_0^2}{(R_* - 1)^3} - \frac{\kappa^2}{2}\right]^2 \frac{dR^0}{d\nu} < 0.$$

The comparison of signs of the principal minors Δ_1 and Δ_3 enables to state unambiguously that according to Silvester's criterion the functional E (23), (97) does not have the property of fixed sign definiteness.

However, if we use the method of perfect squares separation [39] then the integral E (95) can be transformed to the shape

$$E = \frac{1}{2}\int_{-\infty}^{+\infty}\left(\int_0^1 \frac{dR^0}{d\nu}\left[\left(w' + w^0\left[\frac{dR^0}{d\nu}\right]^{-1} R'_\nu\right)^2 + \left(\frac{dR^0}{d\nu}\right)^{-4}\left(\left[\frac{da_*}{d\nu}\right]^2 - \right.\right.\right.$$

$$\left.\left.\left. - w^{02}\left[\frac{dR^0}{d\nu}\right]^2\right) R'^2_\nu\right] d\nu + \left[\frac{\Phi_0^2}{(R_* - 1)^3} - \frac{\kappa^2}{2}\right] R'^2_1\right) dz, \tag{98}$$

from which immediately follows that the functional E (98) is non–negative one if and only if such inequalities

$$\left(\frac{da_*}{d\nu}\right)^2 \geq w^{02}\left(\frac{dR^0}{d\nu}\right)^2, \; \frac{\Phi_0^2}{(R_* - 1)^3} \geq \frac{\kappa^2}{2} \tag{99}$$

are valid.

As a result, the relations (99) are the desired sufficient conditions for the linear stability of the exact stationary solutions (93), (96) of the initial boundary value problem (90) against the small axial–symmetric long–wave perturbations $w'(t, z, \nu)$ and $R'(t, z, \nu)$ (94), which one should regard as the conditions for the fixed sign definiteness of the energetic (in essence) functional E (95), (98).

The consideration of the linear stability of the stationary solutions (93) of the mixed problem (90) is done below in the class of such fluid motions, that

$$R(t, z, \nu) \equiv R_1(t, z) R^0(\nu). \tag{100}$$

This ratio allows to lead the second equation of the initial boundary value problem (90) to the form

$$R_{1t} + (wR_1)_z = 0. \tag{101}$$

If to differentiate the relation (101) by the independent variable ν now then the equation

$$u_{z\nu} = 0;\ u \equiv wR_1, \tag{102}$$

which is the compatibility condition for the class (100) of fluid motions, is obtained.

The functions R (100) and u (102) substitution in the first, third and forth ratios of the mixed problem (90) and in the equation (101) as well enables to write them in the form

$$u_t + \left(\frac{u^2}{R_1}\right)_z = \left[\frac{\kappa^2}{2R_1} - \frac{\Phi_0^2 R_1}{(R_* - R_1)^3} - \left(\frac{dR^0}{d\nu}R_1\right)^{-2}\left(\frac{da_*}{d\nu}\right)^2\right] R_{1z},$$

$$R_{1t} + u_z = 0; \tag{103}$$

$$h = \left(\frac{dR^0}{d\nu}\right)^{-1} \frac{R_0 R_{1z}}{R_1}\frac{da_*}{d\nu},\ H = \left(\frac{dR^0}{d\nu}R_1\right)^{-1}\frac{da_*}{d\nu}.$$

In its turn, differentiating the first ratio of the system (103) primarily, by the variable ν, then, by the independent variable z, and finally, by the variable ν again, it is easy to get the second compatibility condition for the fluid motions in the form (100):

$$\left[\left(\frac{da_*}{d\nu}\frac{d}{d\nu}\left[\left(\frac{dR^0}{d\nu}\right)^{-1}\frac{da_*}{d\nu}\right]\right)^{-1}\frac{dR^0}{d\nu}\left(\frac{u^2}{R_1}\right)_{zz\nu}\right]_\nu = 0. \tag{104}$$

The equations (102)–(104) can be supplemented with the following initial data

$$u(0,\ z,\ \nu) = u_0(z,\ \nu),\ R_1(0,\ z) = R_{10}(z). \tag{105}$$

As a result, the initial boundary value problem (102)–(105), which describes the particular class (100), (102), (104) of the unsteady axial–symmetric sheared magnetohydrodynamic jet flows of non–viscous perfectly conducting incompressible fluid with free surface, is formulated.

The integrals E_1 (91) and I (92) will be conserved with time on the solutions of the mixed problem (102)–(105) as well, however, their form will become rather different by force of the functions R (100) and u (102) definitions above:

$$E_1 = \frac{1}{2}\int\limits_{-\infty}^{+\infty}\left(\int\limits_0^1\left[R_1^{-1}\frac{dR^0}{d\nu}\left(u^2 + \left[\frac{dR^0}{d\nu}\right]^{-2}\left[\frac{da_*}{d\nu}\right]^2\right)\right] d\nu + \kappa^2 \ln R_1 + \right.$$

$$\left. + \frac{\Phi_0^2}{R_* - R_1}\right) dz, \tag{106}$$

$$I = \int_{-\infty}^{+\infty} \int_{0}^{1} \frac{u_\nu}{R_1} F\left(a_*\right) d\nu dz.$$

The exact stationary solutions of the initial boundary value problem (102)–(105) are the functions

$$u = w^0(\nu),\ R_1 = R_1^0 \equiv 1,$$
$$h = h^0 \equiv 0,\ H = H^0(\nu) \equiv \left(\frac{dR^0}{d\nu}\right)^{-1} \frac{da_*}{d\nu}. \tag{107}$$

The purpose of the further investigation is to find the sufficient conditions for the linear stability of the stationary solutions (107) regarding the small axial–symmetric long–wave perturbations $u'(t,\ z,\ \nu)$ and $R_1'(t,\ z)$. It is clear that these stability conditions will be the sufficient conditions for the linear stability of the exact stationary solutions (93) of the mixed problem (90) regarding the same perturbations simultaneously.

To achieve the given purpose, the linearization of the initial boundary value problem (102)–(105) is realized near the stationary solutions (107). It results in the mixed problem which is written out below:

$$u_t' + 2w^0 u_z' = \left[\frac{\kappa^2}{2} + w^{02} - \frac{\Phi_0^2}{(R_* - 1)^3} - \left(\frac{dR^0}{d\nu}\right)^{-2} \left(\frac{da_*}{d\nu}\right)^2\right] R_{1z}',$$
$$R_{1t}' + u_z' = 0;\ u_{z\nu}' = 0,$$
$$\frac{d}{d\nu}\left[w^0 - \left(\frac{dR^0}{d\nu}\frac{dw^0}{d\nu}\right)^{-1} \frac{da_*}{d\nu}\frac{d}{d\nu}\left(\left[\frac{dR^0}{d\nu}\right]^{-1}\frac{da_*}{d\nu}\right)\right] = 0; \tag{108}$$
$$h' = \left(\frac{dR^0}{d\nu}\right)^{-1} \frac{da_*}{d\nu} R^0 R_{1z}',\ H' = -\left(\frac{dR^0}{d\nu}\right)^{-1} \frac{da_*}{d\nu} R_1';$$
$$u'(0,\ z,\ \nu) = u_0'(z,\ \nu),\ R_1'(0,\ z) = R_{10}'(z).$$

This problem has the functional, which is conserved with time, in the form

$$E_2 \equiv \frac{1}{2}\int_{-\infty}^{+\infty}\int_{0}^{1} \left(\frac{dR^0}{d\nu}\left[u'^2 + \left(\left[\frac{dR^0}{d\nu}\right]^{-2}\left[\frac{da_*}{d\nu}\right]^2 - \frac{\kappa^2}{2} + \right.\right.\right.$$
$$\left.\left.\left. + \frac{\Phi_0^2}{[R_* - 1]^3} - w^{02}\right) R_1'^2\right]\right) d\nu dz. \tag{109}$$

It is not difficult to check that the first variation δJ_1 of the integral $J_1 = E_1 + I$ (106) is equal to zero on the exact stationary solutions (107), and its second variation $\delta^2 J_1$ is identical to the functional E_2 in the proper designations if the ratios

$$w^0 \frac{dR^0}{d\nu} = \frac{dF_2}{da_*}\frac{da_*}{d\nu},\ w^0(1) F_2\left[a_*(1)\right] = w^0(0) F_2\left[a_*(0)\right] \tag{110}$$

are true.

The stationary solutions (107) of the initial boundary value problem (102)–(105) (hence, the exact stationary solutions (93) of the mixed problem (90)) will be stable regarding the small axial–symmetric long–wave perturbations (108) if and only if everywhere inside the jet

$$\left(\frac{dR^0}{d\nu}\right)^{-2}\left(\frac{da_*}{d\nu}\right)^2-\frac{\kappa^2}{2}+\frac{\Phi_0^2}{(R_*-1)^3}-w^{02}\geq 0, \tag{111}$$

since the given ratio, at least, provides nonnegativity of the integral E_2, as it is followed from the expression (109).

It should be noted that the inequality (111) is the very sufficient condition for the linear stability of the stationary solutions (107) (or (93)), which are required to find out.

It is considered further that the ratio (111) is not valid, at least, somewhere within the conducting jet. Then one can hope to demonstrate the linear instability of the exact stationary solutions (107) of the initial boundary value problem (102)–(105) (and, naturally, the stationary solutions (93) of the mixed problem (90) as well) regarding the small axial–symmetric long–wave perturbations $u'(t,\ z,\ \nu)$, $R_1'(t,\ z)$ (108). If one manages to achieve the given purpose then one will display that the condition (111) for the linear stability of the stationary solutions (107) (or (93)) is not only sufficient but necessary too.

To put the planned thing into practice, it is necessary to single out only one quickly (at the minimum, exponentially) growing with time perturbation among the small axial–symmetric long–wave ones (108). It can be realized in the most effective way when the research is concentrated on the small perturbations, which are deviations of the trajectories of fluid particles motion from the corresponding current lines of the steady flows (107).

The given perturbations can be characterized by means of the field of Lagrangian displacements $\xi=\xi(t,\ z,\ \nu)$ [25], which will satisfy here the equation

$$\xi_t=u', \tag{112}$$

in much more obvious and simple way, than in any other ones. In fact, applying the ratio (112), it is easy to show that the initial boundary value problem (108) can be brought to the mixed problem, which includes the only evolutionary differential equation, i.e.

$$\begin{gathered}\xi_{tt}+2w^0\xi_{tz}=\left[\left(\frac{dR^0}{d\nu}\right)^{-2}\left(\frac{da_*}{d\nu}\right)^2-\frac{\kappa^2}{2}+\frac{\Phi_0^2}{(R_*-1)^3}-w^{02}\right]\xi_{zz},\\ R_1'=-\xi_z;\ \xi_{z\nu}=0,\\ \frac{d}{d\nu}\left[w^0-\left(\frac{dR^0}{d\nu}\frac{dw^0}{d\nu}\right)^{-1}\frac{da_*}{d\nu}\frac{d}{d\nu}\left(\left[\frac{dR^0}{d\nu}\right]^{-1}\frac{da_*}{d\nu}\right)\right]=0;\\ h'=-\left(\frac{dR^0}{d\nu}\right)^{-1}\frac{da_*}{d\nu}R^0\xi_{zz},\ H'=\left(\frac{dR^0}{d\nu}\right)^{-1}\frac{da_*}{d\nu}\xi_z;\\ \xi(0,\ z,\ \nu)=\xi_0(z,\ \nu),\ \xi_t(0,\ z,\ \nu)=u'(0,\ z,\ \nu)=u_0'(z,\ \nu).\end{gathered} \tag{113}$$

It is clear that the initial boundary value problem (112), (113) is overspecified one because of the presence of the third ratio in the system (113). Besides, this ratio testifies that the function $\xi(t,\ z,\ \nu)$, being the solution of the mixed problem (112), (113), has the form

$$\xi(t,\ z,\ \nu) = f_2(t,\ z) + f_3(t,\ \nu), \tag{114}$$

where $f_2(t,\ z)$ and $f_3(t,\ \nu)$ are some functions of their arguments, and nothing else.

Hence, the initial data $\xi_0(z,\ \nu)$ and $u_0'(z,\ \nu)$ for the initial boundary value problem (112), (113) must be given in the form

$$\xi_0(z,\ \nu) = f_4(z) + f_5(\nu),\ u_0'(z,\ \nu) = f_6(z) + f_7(\nu) \tag{115}$$

(here f_4 and f_6 are arbitrary functions of the independent variable z, and f_5 and f_7 are some functions of the variable ν).

Therefore, the number of questions arises: if the mixed problem (112), (113) has the solutions in the form (114), if the overdetermination of the initial boundary value problem (112), (113) imposes additional (besides (115)) constraints on the initial perturbations $\xi_0(z,\ \nu)$, $u_0'(z,\ \nu)$ choice.

One can give convincing answers to the given questions if one reformulates the mixed problem (112), (113) in the form

$$\begin{gathered}
R'_{1tt} + 2w^0 R'_{1tz} = \left[\left(\frac{dR^0}{d\nu}\right)^{-2}\left(\frac{da_*}{d\nu}\right)^2 - \frac{\kappa^2}{2} + \frac{\Phi_0^2}{(R_*-1)^3} - w^{02}\right] R'_{1zz}; \\
\frac{d}{d\nu}\left[w^0 - \left(\frac{dR^0}{d\nu}\frac{dw^0}{d\nu}\right)^{-1}\frac{da_*}{d\nu}\frac{d}{d\nu}\left(\left[\frac{dR^0}{d\nu}\right]^{-1}\frac{da_*}{d\nu}\right)\right] = 0; \\
h' = \left(\frac{dR^0}{d\nu}\right)^{-1}\frac{da_*}{d\nu} R^0 R'_{1z},\ H' = -\left(\frac{dR^0}{d\nu}\right)^{-1}\frac{da_*}{d\nu} R'_1; \\
R'_1(0,\ z) = R'_{10}(z)\left(= -\frac{df_4}{dz}\right),\ R'_{1t}(0,\ z) = \left(R'_{1t}\right)_0(z)\left(= -\frac{df_6}{dz}\right).
\end{gathered} \tag{116}$$

It is not difficult to note that the initial boundary value problem (116) contains one evolutionary differential equation to determine the only desired function $R_1'(t,\ z)$, at that, it is very important that this equation is homogeneous. Thus, the initial data $R'_{10}(z)$ and $\left(R'_{1t}\right)_0(z)$ can be taken arbitrarily, without any restrictions.

As a result, if the solution $R_1'(t,\ z)$ of the mixed problem (116) is found then one can calculate the function $\xi(t,\ z,\ \nu)$ as the solution of the initial boundary value problem (112), (113), which completely corresponds to the representation (114), by means of the second ratio of the system (113).

So, being based on the above reasons, it is easy to conclude that the mixed problem (112), (113) can have the solutions in the form (114), at that, its overdetermination does not have any additional (except (115)) restrictions on the initial perturbations $\xi_0(z,\ \nu)$, $u_0'(z,\ \nu)$.

Below, the additional functionals

$$T \equiv \frac{1}{2}\int_{-\infty}^{+\infty}\int_0^1 \left[\frac{dR^0}{d\nu}\left(u' - w^0 R_1'\right)^2\right] d\nu dz,$$

$$\Pi \equiv \frac{1}{2}\int\limits_{-\infty}^{+\infty}\int\limits_{0}^{1}\left(\frac{dR^0}{d\nu}\left[\left(\frac{dR^0}{d\nu}\right)^{-2}\left(\frac{da_*}{d\nu}\right)^2+\frac{\Phi_0^2}{(R_*-1)^3}-\frac{\kappa^2}{2}\right]R_1'^2\right)d\nu dz, \tag{117}$$

$$T_1 \equiv \int\limits_{-\infty}^{+\infty}\int\limits_{0}^{1}\left[\frac{dR^0}{d\nu}w^0\left(u'-w^0R_1'\right)R_1'\right]d\nu dz,\ T_2 \equiv \frac{1}{2}\int\limits_{-\infty}^{+\infty}\int\limits_{0}^{1}\frac{dR^0}{d\nu}u'^2 d\nu dz,$$

$$\Pi_1 \equiv \frac{1}{2}\int\limits_{-\infty}^{+\infty}\int\limits_{0}^{1}\left(\frac{dR^0}{d\nu}\left[\left(\frac{dR^0}{d\nu}\right)^{-2}\left(\frac{da_*}{d\nu}\right)^2-\frac{\kappa^2}{2}+\frac{\Phi_0^2}{(R_*-1)^3}-w^{02}\right]R_1'^2\right)d\nu dz$$

are introduced for the sake of the further consideration so that

$$E_2 \equiv T+\Pi+T_1 \equiv T_2+\Pi_1 = \text{const}. \tag{118}$$

The double differentiation of the integral M (31) with time and the transformations of the obtained functional with the use of the relations (108), (112), (113), and (117) allow to get the important equation [22, 24, 28, 29, 38]

$$\frac{d^2M}{dt^2} = 4(T-\Pi),$$

which is called a virial equality in the world scientific literature [25]. Now, multiplying the given equality by some constant factor λ, taking into consideration the ratio (118) and applying the equation

$$\frac{dT_1}{dt} = 0,$$

which directly implies from the initial boundary value problems (108) and (112), (113), the key ratio

$$\frac{dE_\lambda}{dt} = 2\lambda E_\lambda - 4\lambda T_\lambda \tag{119}$$

can be derived, where

$$E_\lambda \equiv \Pi_\lambda + T_\lambda,\ 2\Pi_\lambda \equiv 2\Pi + \lambda^2 M,$$

$$2T_\lambda \equiv 2T + \lambda^2 M - \lambda\frac{dM}{dt} = \int\limits_{-\infty}^{+\infty}\int\limits_{0}^{1}\frac{dR^0}{d\nu}\left(u'-w^0R_1'-\lambda\xi\right)^2 d\nu dz \geq 0.$$

With the strictly positive constant λ, taking into consideration nonnegativity of the integral T_λ, the equation (119) can be brought to the differential inequality

$$\frac{dE_\lambda}{dt} \leq 2\lambda E_\lambda.$$

After integrating this inequality, the basic estimate

$$E_\lambda(t) \leq E_\lambda(0)\exp(2\lambda t) \tag{120}$$

will take place again, as in paragraph 2 of the present chapter.

This estimate is also true both for any solutions of the mixed problem (112), (113) and for arbitrary positive values of the constant λ. Moreover (what is essential), while getting it, it was not necessary to impose constraints on the functional Π sign either. At last, it enables to conclude that integral E_λ can be taken further as Lyapunov's functional [1, 2, 28, 29, 38].

Repeating in many respects the corresponding statements in paragraph 2 of the present chapter, one can construct double–sided exponential estimates of the growth of the small axial–symmetric long–wave perturbations (112)–(115) of the stationary solutions (107) of the initial boundary value problem (102)–(105) by means of the basic integral inequality (120) and proper fitting of the functions $\xi_0(z,\,\nu),\; u_0'(z,\,\nu)$ under the conditions, when the ratio

$$\left(\frac{dR^0}{d\nu}\right)^{-2}\left(\frac{da_*}{d\nu}\right)^2-\frac{\kappa^2}{2}+\frac{\Phi_0^2}{(R_*-1)^3}-w^{02}<0 \tag{121}$$

is valid either anywhere inside the jet or in some part of the jet. Therefore, without any additional vindications, they are written out directly in the complete form below:

a) the lower estimate

$$\int\limits_{-\infty}^{+\infty}\int\limits_0^1\frac{dR^0}{d\nu}\left[\frac{\kappa^2}{2}-\left(\frac{dR^0}{d\nu}\right)^{-2}\left(\frac{da_*}{d\nu}\right)^2-\frac{\Phi_0^2}{(R_*-1)^3}+w^{02}\right]R_1'^2d\nu dz\ \geq$$

$$\geq\ 2\left|E_{\Lambda-\delta_1}(0)\right|\exp\left[2\left(\Lambda-\delta_1\right)t\right];$$
$$\lambda\equiv\Lambda-\delta_1,\ \delta_1\in\,]0,\ \Lambda[,\ E_{\Lambda-\delta_1}(0)<0, \tag{122}$$
$$\Lambda\equiv[4M(0)]^{-1}\frac{dM}{dt}(0)+\sqrt{[4M(0)]^{-2}\left[\frac{dM}{dt}(0)\right]^2-\frac{E_2(0)-T_1(0)}{M(0)}};$$
$$\Pi_1(0)<0,\ T_2(0)\leq\left|\Pi_1(0)\right|,\ T_2(0)-T_1(0)\leq\left|\Pi_1(0)\right|;$$

b) the upper estimate

$$E_{\Lambda^++\delta_2}(t)\leq E_{\Lambda^++\delta_2}(0)\exp\left[2\left(\Lambda^++\delta_2\right)t\right];$$
$$\lambda\equiv\Lambda^++\delta_2,\ \delta_2>0,\ \Lambda^+\equiv\sup_{\xi_0(z,\,\nu),\,u_0'(z,\,\nu)}\Lambda,\ E_{\Lambda^++\delta_2}(0)>0; \tag{123}$$
$$\Pi_1(0)<0,\ T_2(0)\leq\left|\Pi_1(0)\right|,\ T_2(0)-T_1(0)\leq\left|\Pi_1(0)\right|.$$

The comparison of the inequalities (122) and (123) implies that the parameter Λ^+ estimates the growth rate ω of the small perturbations (112)–(115) both at the lower and upper bounds, namely

$$\Lambda^+-\delta_1\leq\omega\leq\Lambda^++\delta_2. \tag{124}$$

The given ratio testifies that the small axial–symmetric long–wave perturbations (112)–(115) with the increments, which are close to the parameter Λ^+ value, will grow in the fastest way.

Thus, having calculated the value of the parameter Λ^+, which characterizes the growth rate ω (124) of the fastest growing small perturbations (112)–(115), by means of the expressions (122) and (123), if the inequality (121) is true, one can formulate the answer to the question: for which time will the small axial–symmetric long–wave perturbations (112)–(115) lead the steady axial–symmetric sheared magnetohydrodynamic jet flows (107) (or (93), which is equal) of inviscid ideally conducting incompressible fluid with free surface to destruction?

It is worth noting that the presence of the estimates (122), (123) does prove that the condition (111) for the linear stability of the exact stationary solutions (107) (or (93)) is simultaneously sufficient and necessary one. Besides, this condition is the partial converse of sufficient conditions (99) for the linear stability of the stationary solutions (93), (96) of the initial boundary value problem (90).

Further, the examples of the steady flows (93), (107) and their initial small perturbations (113), (115) are constructed, and they illustrate the stated results of this paragraph.

So, the stationary axial–symmetric sheared magnetohydrodynamic jet flow

$$w^0(\nu) = 2 - \nu,\ R^0(\nu) = \nu,\ R_1^0 = 1,\ h^0 = 0,\ \kappa = 2, \tag{125}$$

$$a_*(\nu) = \frac{\nu^2}{2} + 2\nu + 1,\ H^0(\nu) = \nu + 2,\ \Phi_0 = \sqrt{2},\ R_* = 2$$

of non–viscous perfectly conducting incompressible fluid in the domain, which is an infinite band of the kind (48), is being studied. It is clear that the given flow is the typical sample of the stationary solutions (93) of the mixed problem (90) (hence, the one of the exact stationary solutions (107) of the initial boundary value problem (102)–(105)).

Here, for the flow (125), (48) the function $F_2\,(a_*)$ (92) must satisfy the relations (110) which get the form

$$\frac{dF_2}{d\nu} = 2 - \nu,\ 2F_2(0) = F_2(1),$$

from which follows that

$$F_2(\nu) = -\frac{\nu^2}{2} + 2\nu + \frac{3}{2}.$$

Now, the steady flow (125), (48) is the one, for which the inequality (111) is valid. In fact, the direct calculations show that

$$\left(\frac{dR^0}{d\nu}\right)^{-2}\left(\frac{da_*}{d\nu}\right)^2 - \frac{\kappa^2}{2} + \frac{\Phi_0^2}{(R_* - 1)^3} - w^{02} = 8\nu \geq 0$$

on the variation interval (48) of the independent variable ν.

It testifies that the stationary axial–symmetric sheared magnetohydrodynamic jet flow (125), (48) will be stable by the linear approximation to the small axial–symmetric long–wave perturbations (108) and, all the more, (112)–(115).

The steady axial–symmetric sheared magnetohydrodynamic jet flow

$$w^0(\nu) = 4 - \nu,\ R^0(\nu) = \nu,\ R_1^0 = 1,\ h^0 = 0,$$

$$a_*(\nu) = \frac{\nu+1}{2}\sqrt{\nu^2 + 2\nu + 2} + \frac{1}{2}\ln\left(\nu + 1 + \sqrt{\nu^2 + 2\nu + 2}\right) + 1, \tag{126}$$

$$H^0(\nu) = \sqrt{\nu^2 + 2\nu + 2},\ \kappa = 8,\ \Phi_0 = 4,\ R_* = 3$$

of inviscid ideally conducting incompressible fluid within the same infinite band (48) is being investigated below.

The given flow is also one of the representatives of the stationary solutions (93) (and, naturally, along with them — the exact stationary solutions (107)), at that, the function $F_2(a_*)$ has the following form for it:

$$F_2(\nu) = -\frac{\nu^2}{2} + 4\nu + \frac{21}{2}.$$

It lets this function to transform the ratios

$$\frac{dF_2}{d\nu} = 4 - \nu, \; 4F_2(0) = 3F_2(1),$$

which follow from the relations (110), into identities.

Besides, the inequality (121) is true for the steady flow (126), (48). In fact, the direct calculations show that

$$\left(\frac{dR^0}{d\nu}\right)^{-2}\left(\frac{da_*}{d\nu}\right)^2 - \frac{\kappa^2}{2} + \frac{\Phi_0^2}{(R_* - 1)^3} - w^{02} = 2(5\nu - 22) < 0$$

with any values of the variable ν from the interval $[0, \; 1]$.

As a result, the stationary axial–symmetric sheared magnetohydrodynamic jet flow (126), (48) will be unstable, for example, regarding the small axial–symmetric long–wave perturbations (112)–(115) with the initial data $\xi_0(z, \; \nu)$ and $u_0'(z, \; \nu)$ in the form

$$\xi_0(z, \; \nu) = \sin\frac{2\pi z}{l_1} + \nu, \; u_0'(z, \; \nu) = 0, \tag{127}$$

where l_1 is an arbitrary positive constant.

Actually, using the ratios (113), (117), (118), and (122), and also taking into consideration periodicity of the function $\xi_0(z, \; \nu)$ (127) by the independent variable z, it is not difficult to find out that

$$\Pi_1(0) = \frac{4\pi^2}{l_1^2}\int_0^{l_1}\int_0^1 \left[(5\nu - 22)\cos^2\frac{2\pi z}{l_1}\right] d\nu dz = -\frac{39\pi^2}{l_1} < 0,$$

$$T_2(0) + \Pi_1(0) = \Pi_1(0) = -\frac{39\pi^2}{l_1} < 0, \tag{128}$$

$$T_2(0) - T_1(0) + \Pi_1(0) = \frac{4\pi^2}{l_1^2}\int_0^{l_1}\int_0^1 \left[\left(\nu^2 - 3\nu - 6\right)\cos^2\frac{2\pi z}{l_1}\right] d\nu dz = -\frac{86\pi^2}{3l_1} < 0.$$

As a result, the expressions (128) imply truth of two latter inequalities (122). In its turn, it enables to conclude that the stationary flow (126), (48) is, in fact, unstable regarding the small perturbations (112)–(115), (127). The given perturbations will be developed with time according to the estimates (122) and (123) (but it is necessary to substitute the value Λ (122) for the parameter Λ^+ in the second one), whereas their growth rate ω (124) will be defined by means of the parameter Λ.

Further consideration is concentrated on the linear stability of the exact stationary solutions (93), (96) of the initial boundary value problem (90) regarding the small axial–symmetric long–wave perturbations (94) again.

It will be demonstrated below that the sufficient conditions (99) for the linear stability of the steady flows (93), (96) regarding the small perturbations (94) can be converted partially. Moreover, a priori lower exponential estimate for growth of studied perturbations will be constructed for unstable stationary flows (93), (96).

Alias, these results can be obtained by investigation of such perturbed fluid motions which serve as deviations of fluid particles motion paths from the corresponding stream lines of steady flows (93), (96) and describe with the use of the Lagrangian displacements field $\xi = \xi(t,\ z,\ \nu)$ [25] in the form

$$\xi_t = w' - w^0\xi_z. \tag{129}$$

With the help of the equation (129), we are able to put the mixed problem (94) into the shape

$$w'_t + w^0 w'_z = \left[\frac{\kappa^2}{2} - \frac{\Phi_0^2}{(R_* - 1)^3}\right] R'_{1t} + \left(\frac{dR^0}{d\nu}\right)^{-2}\left(\frac{da_*}{d\nu}\right)^2 \xi_{zz},$$

$$R'_\nu = -\frac{dR^0}{d\nu}\xi_z;\ h' = \left(\frac{dR^0}{d\nu}\right)^{-1}\frac{da_*}{d\nu}R'_z,\ H' = \left(\frac{dR^0}{d\nu}\right)^{-1}\frac{da_*}{d\nu}\xi_z; \tag{130}$$

$$\xi(0,\ z,\ \nu) = \xi_0(z,\ \nu), w'(0,\ z,\ \nu) = w'_0(z,\ \nu).$$

Further, with the use of relations of the initial boundary value problem (129), (130), we differentiate the functional M (31) by its argument t twice:

$$\frac{dM}{dt} = 2\int_{-\infty}^{+\infty}\int_0^1 \frac{dR^0}{d\nu}\xi w' d\nu dz, \tag{131}$$

$$\frac{d^2M}{dt^2} = 2\int_{-\infty}^{+\infty}\int_0^1 \left[\frac{dR^0}{d\nu}w'^2 - \left(\frac{dR^0}{d\nu}\right)^{-3}\left(\frac{da_*}{d\nu}\right)^2 R'^2_\nu\right] d\nu dz + 2\left[\frac{\kappa^2}{2} - \frac{\Phi_0^2}{(R_* - 1)^3}\right] \times$$

$$\times \int_{-\infty}^{+\infty} R'^2_1 dz.$$

After that, by the ratios (31), (131), we make up the equality in the form

$$\frac{d^2M}{dt^2} - 2\lambda\frac{dM}{dt} + 2\lambda^2 M = 2\int_{-\infty}^{+\infty}\int_0^1 \frac{dR^0}{d\nu}(w' - \lambda\xi)^2 d\nu dz - 2\int_{-\infty}^{+\infty}\int_0^1 \left[\left(\frac{dR^0}{d\nu}\right)^{-3} \times\right.$$

$$\left.\times \left(\frac{da_*}{d\nu}\right)^2 R'^2_\nu\right] d\nu dz - 2\left[\frac{\Phi_0^2}{(R_* - 1)^3} - \frac{\kappa^2}{2}\right]\int_{-\infty}^{+\infty} R'^2_1 dz$$

(here λ is a certain positive constant).

For one's part, it isn't difficult to convert the last equality in the next series of inequalities:

$$\frac{d^2M}{dt^2}-2\lambda\frac{dM}{dt}+2\lambda^2 M\geq -2\int\limits_{-\infty}^{+\infty}\int\limits_{0}^{1}\left(\frac{dR^0}{d\nu}\right)^{-3}\left(\frac{da_*}{d\nu}\right)^2 R_\nu'^2 d\nu dz-2\left[\frac{\Phi_0^2}{(R_*-1)^3}-\right.$$

$$\left.-\frac{\kappa^2}{2}\right]\int\limits_{-\infty}^{+\infty}\int\limits_{0}^{1}R_1' R_\nu' d\nu dz\geq -2\int\limits_{-\infty}^{+\infty}\int\limits_{0}^{1}\left(\frac{dR^0}{d\nu}\right)^{-3}\left(\frac{da_*}{d\nu}\right)^2 R_\nu'^2 d\nu dz-\left[\frac{\Phi_0^2}{(R_*-1)^3}-\right.$$

$$\left.-\frac{\kappa^2}{2}\right]\int\limits_{-\infty}^{+\infty}\int\limits_{0}^{1}\left(R_1'^2+R_\nu'^2\right)d\nu dz=\left[\frac{\kappa^2}{2}-\frac{\Phi_0^2}{(R_*-1)^3}\right]\int\limits_{-\infty}^{+\infty}R_1'^2 dz+\int\limits_{-\infty}^{+\infty}\int\limits_{0}^{1}\left[\frac{\kappa^2}{2}-\right.$$

$$\left.-\frac{\Phi_0^2}{(R_*-1)^3}-2\left(\frac{dR^0}{d\nu}\right)^{-3}\left(\frac{da_*}{d\nu}\right)^2\right]R_\nu'^2 d\nu dz.$$

If

$$\frac{\kappa^2}{2}-\frac{\Phi_0^2}{(R_*-1)^3}-2\left(\frac{dR^0}{d\nu}\right)^{-3}\left(\frac{da_*}{d\nu}\right)^2>0 \tag{132}$$

then one can break the given series and take, in the upshot, the differential inequality in the shape

$$\frac{d^2M}{dt^2}-2\lambda\frac{dM}{dt}+2\lambda^2 M\geq 0 \tag{133}$$

which is of principal for subsequent consideration.

Carried out the procedure of the relation (133) integration (by the way, this procedure is entirely identical one to the integration procedure realized earlier for the inequality (52); therefore, the given procedure of integration isn't reproduced here), it isn't hard to reveal the counting collection of conditions

$$M\left(\frac{\pi n}{2\lambda}\right)>0,\ \frac{dM}{dt}\left(\frac{\pi n}{2\lambda}\right)\geq 2\lambda M\left(\frac{\pi n}{2\lambda}\right)\ (n=0,\ 1,\ 2,\ ...),$$
$$M\left(\frac{\pi n}{2\lambda}\right)\equiv M(0)\exp\left(\frac{\pi n}{2}\right),\ \frac{dM}{dt}\left(\frac{\pi n}{2\lambda}\right)\equiv\frac{dM}{dt}(0)\exp\left(\frac{\pi n}{2}\right), \tag{134}$$

under which a priori lower exponential estimate

$$M(t)\geq C_4\exp(\lambda t), \tag{135}$$

where C_4 is the known positive constant quantity, follows from this relation with necessity.

Hence, the conditions (134) and inequality (135) make it clear that growing (at the minimum, exponentially) in time perturbations may be among the small axial–symmetric long–wave perturbations (129), (130) of the exact stationary solutions (93), (96) of the mixed problem (90), for which the initial data are given in the form

$$M(0)>0,\ \frac{dM}{dt}(0)\geq 2\lambda M(0). \tag{136}$$

At that, as before, the counting collection of conditions (134) is met for the exponentially growing in time small perturbations (129), (130), (136) identically and automatically.

Thus, it is shown that the ratio (132) is the sufficient condition for the linear instability of steady flows (93), (96) regarding the small axial–symmetric long–wave perturbations (94), (129), (130), (136). Furthermore, as expected, this ratio represents the partial converse of sufficient conditions (99) for the linear stability of stationary flows (93), (96) against the small perturbations (94).

5. Conclusion

In the present chapter one has considered the problems on the linear stability of the steady axial–symmetric sheared jet flows of non–viscous perfectly conducting incompressible fluid with free surface either in the azimuthal magnetic field, which is directly proportional to the radial coordinate, or in the azimuthal or poloidal magnetic fields, which are the prescribed functions of the radius. Both the sufficient or the necessary and sufficient conditions for stability and the sufficient conditions for instability of these flows regarding the small axial–symmetric long–wave perturbations have been obtained by Lyapunov's direct method. The estimates, which demonstrate that the studied small perturbations may grow with time no slower than exponentially, have been constructed for unstable stationary flows. One has given the examples of the steady flows and the small perturbations superposed upon them, which illustrate the obtained conditions for stability/instability and the constructed estimates of growth.

It is worth emphasizing that, from mathematics perspective, the results of the present chapter are, in most cases, of a priori character, as existence theorems regarding solutions of the investigated mixed problems for systems of differential equations with partial derivatives are not proved.

Finally, concerning the connection between the results in this chapter and results of other authors, special attention should be paid to the following important circumstances:

1) since the differential inequalities (52) and (133) contain no information about the exact stationary solutions (20) to the mixed problem (13)–(15); (20), (67) to the initial boundary value problem (62), and (93), (96) to the mixed problem (90), we can expect that these inequalities will appear in one or another form while considering another hydrodynamical models;

2) we can determine the character of instability (absolute or conditional) in dependence on whether requirements on the considered steady–state hydrodynamical flows are imposed while deriving the relations of the form (52), (133); and if yes, these requirements are sufficient conditions for instability of the flows, if no, the flows are absolutely unstable ones;

3) without specifying the solutions to initial boundary value problems on perturbations of steady–state flows, the integration of differential inequalities of the form (52), (133) allows us to find initial data such that small perturbations of these flows may become growing;

4) existence of relations in the form (52), (133) immediately converts the known sufficient conditions for linear stability of steady–state hydrodynamical flows; in other words, if the differential inequalities of the form (52), (133) are fulfilled, then only necessary and sufficient conditions for linear stability can take place; and

5) the differential inequalities of the form (52), (133) can serve as formal relations if analogs of energy functionals of corresponding linearized mixed problems are definite or constant ones in sign.

Based on the above mentioned circumstances, we can surely conclude that the described new methods to construct the growing in time Lyapunov's functionals for the studied equations of motion is undoubtedly a good support while considering various theoretical linear problems on hydrodynamic stability.

Acknowledgements

This chapter was supported by the Russian Foundation for Basic Research (project No. 07–01–00585) and by the Integration Program No. 13 of the Russian Academy for Sciences.

References

[1] A.M. Lyapunov, *The General Problem of the Stability of Motion*. London: Taylor & Francis (1992) 270 pp.

[2] N.G. Chetaev, *Stability of Motion*. Moscow: Nauka (1990) 176 pp. (in Russian).

[3] S. Chandrasekhar, *Hydrodynamic and Hydromagnetic Stability*. Oxford: Clarendon Press (1961) 652 pp.

[4] N.N. Moiseev, V.V. Rumyantsev, *Dynamics of a Body with Cavities Containing a Liquid*. Moscow: Nauka (1965) 439 pp. (in Russian).

[5] A.D. Myshkis, V.G. Babskii, N.D. Kopachevskii et al., *Low–Gravity Fluid Mechanics. Mathematical Theory of Capillary Phenomena*. Berlin, Heidelberg: Springer–Verlag (1987) 584 pp.

[6] L.A. Dikii, *Hydrodynamic Stability and Dynamics of Atmosphere*. Leningrad: Gidrometeoizdat (1976) 108 pp. (in Russian).

[7] Yu.G. Gubarev, *Lyapunov's Direct Method in Magnetohydrodynamics*. Novosibirsk: Novosibirsk State University Press (2002) 170 pp. (in Russian).

[8] M.A. Grinfel'd, Stability of two–dimensional curvilinear flows of an ideal barotropic fluid. *Fluid Dyn.* **16** (1982) 653–658.

[9] D.D. Holm, J.E. Marsden, T. Ratiu, A. Weinstein, Nonlinear stability of fluid and plasma equilibria. *Phys. Rev.* **123**(1/2) (1985) 1–116.

[10] V.A. Vladimirov, On nonlinear stability of incompressible fluid flows. *Arch. Mech.* **38**(5/6) (1986) 689–696.

[11] V.A. Vladimirov, Yu.G. Gubarev, The conditions for the nonlinear stability of plane and helical MHD flows. *J. Appl. Math. Mech.* **59**(3) (1995) 415–423.

[12] V.A. Vladimirov, H.K. Moffatt, K.I. Ilin, On general transformations and variational principles for the magnetohydrodynamics of ideal fluids. IV. Generalized isovorticity principle for three–dimensional flows. *J. Fluid Mech.* **390** (1999) 127–150.

[13] S.Ya. Belov, V.A. Vladimirov, An example of inversion of the Lagrange's theorem in hydrodynamics of two–layered liquids. *Dinamika Sploshn. Sredy, Lavrentyev Institute for Hydrodynamics of SD USSR, Novosibirsk*. **84** (1988) 21–27 (in Russian).

[14] V.A. Vladimirov, V.V. Rumyantsev, Inversion of Lagrange's theorem for a rigid body with a cavity containing an ideal liquid. *J. Appl. Math. Mech.* **53**(4) (1989) 474–477.

[15] V.A. Vladimirov, On the instability of equilibrium in fluids. *J. Appl. Mech. Tech. Phys.* **30**(2) (1989) 269–276.

[16] V.A. Vladimirov, V.V. Rumyantsev, Inversion of Lagrange's theorem for a rigid body with a cavity containing a viscous liquid. *J. Appl. Math. Mech.* **54**(2) (1990) 154–163.

[17] Yu.G. Gubarev, On the converse Lagrange's theorem in magnetohydrodynamics. *Ibid*. **54**(6) (1990) 812–815.

[18] Idem, Instability of a self–gravitating compressible medium. *J. Appl. Mech. Tech. Phys.* **35**(4) (1994) 546–555.

[19] Yu.G. Gubarev, M.–B.A. Negmatov, The instability of the states of rest of a viscous ideally conducting compressible medium containing magnetic field. *J. Appl. Math. Mech.* **64**(4) (2000) 619–625.

[20] K.I. Ilin, Stability of a baroclinic vortex. *Izv. Acad. Sci. USSR Atmospher. Ocean. Phys.* **27**(5) (1991) 403–406.

[21] H.–R. Cho, T.G. Shepherd, V.A. Vladimirov, Application of the Lyapunov's direct method to the problem of symmetric stability in the atmosphere. *J. Atmos. Sci.* **50**(6) (1993) 822–836.

[22] Yu.G. Gubarev, Instability of axially symmetric MHD flows. *Fluid Dyn.* **30**(6) (1995) 814–820.

[23] P.N. Svirkunov, On conditions for the symmetric instability of rotational motions of an ideal stratified fluid. *J. Appl. Math. Mech.* **62**(6) (1998) 996–1001.

[24] Yu.G. Gubarev, Instability of helical magnetohydrodynamic flows. *Fluid Dyn.* **34**(1) (1999) 126–132.

[25] S. Chandrasekhar, *Ellipsoidal Figures of Equilibrium*. New Haven: Yale University Press (1969) 252 pp.

[26] L.D. Landau, E.M. Lifshitz, *Course of Theoretical Physics. Vol. 8. Electrodynamics of Continuous Media*. Oxford: Pergamon Press (1984) 460 pp.

[27] V.E. Zakharov, Benney's equations and quasiclassical approximation in the inverse problem method. *Funktsional. Anal. i Prilozhen*. **14**(2) (1980) 15–24 (in Russian).

[28] Yu.G. Gubarev, V.V. Nikulin, Linear long–wave instability of a single class of steady–state jet flows of an ideal fluid in the field of a self–electric current. *Fluid Dyn.* **36**(2) (2001) 225–235.

[29] Yu.G. Gubarev, Stability of steady–state shear jet flows of an ideal fluid with a free boundary in an azimuthal magnetic field against small long–wave perturbations. *J. Appl. Mech. Tech. Phys.* **45**(2) (2004) 239–248.

[30] Idem, On an analogy between the Benney's equations and the Vlasov–Poisson's equations. *Dinamika Sploshn. Sredy, Lavrentyev Institute for Hydrodynamics of SB RAS, Novosibirsk*. **110** (1995) 78–90 (in Russian).

[31] V.A. Yakubovich, V.M. Starzhinskii, *Linear Differential Equations with Periodical Coefficients and its Applications*. Moscow: Nauka (1972) 718 pp. (in Russian).

[32] S.A. Chaplygin, *New Method for Approximate Integration of Differential Equations* . Moscow, Leningrad: State Press of Technical and Theoretical Literature (1950) 102 pp. (in Russian).

[33] V.A. Ilin, V.A. Sadovnichii, Bl. Kh. Sendov, *Mathematical Analysis*. Moscow: Nauka (1979) 720 pp. (in Russian).

[34] S.K. Godunov, *Equations of Mathematical Physics*. Moscow: Nauka (1979) 392 pp. (in Russian).

[35] Yu.G. Gubarev, Sufficient conditions for linear long–wave instability of steady shear jet flows of an ideal fluid with a free boundary in an azimuthal magnetic field. In: G.A. Shvetsov (ed.), *Proceedings of International Workshop on High Energy Density Hydrodynamics*, Novosibirsk, Russia, 11–15 August 2003. Novosibirsk: Lavrentyev Institute for Hydrodynamics of SB RAS Press (2003) pp. 94–103 (in Russian).

[36] R.V. Polovin, V.P. Demutskii, *Fundamentals of Magnetohydrodynamics*. Moscow: Energoatomizdat (1987) 206 pp. (in Russian).

[37] L.D. Landau, E.M. Lifshitz, *Theoretical Physics. Vol. 2. Theory of Field*. Moscow: Nauka (1988) 512 pp. (in Russian).

[38] Yu.G. Gubarev, V.V. Nikulin, Criterion for linear long–wave stability of stationary jet magnetohydrodynamic flows of an ideal fluid. *Prikl. Matem. Mekh.* **67**(5) (2003) 849–863 (in Russian).

[39] V.S. Vladimirov (ed.), *Collection of problems on the equations of mathematical physics*. Moscow: Nauka (1974) 272 pp. (in Russian).

In: Progress in Nonlinear Analysis Research
Editor: Erik T. Hoffmann, pp. 183-208
ISBN 978-1-60456-359-7

Chapter 8

ADAPTIVE CONTROL WITH STABILITY AND ROBUSTNESS ANALYSIS FOR NONLINEAR PLANT WIDE SYSTEMS BY MEANS OF NEURAL NETWORKS

Dimitri Lefebvre, Salem Zerkaoui, Fabrice Druaux, and Edouard Leclercq
GREAH - University LE HAVRE - 25 rue Philippe Lebon - 76068 LE HAVRE

Abstract

Adaptive control by means of neural networks for nonlinear plant wide dynamical systems is an open but promising issue. For real world applications, practitioners have to paid attention to external disturbances, parameters uncertainty and measurement noise, as long as these factors will influence the stability and robustness of the closed loop system. This chapter presents some of the most popular control schemes based on behavioural models and adaptive control with neural networks. Stability and robustness are discussed and the main difficulties are mentioned: the initialization and pre training phases, the determination of the networks size, and the arbitrary value of the adaptive rate. Then an indirect adaptive control scheme is detailed. This scheme is based on fully connected neural networks and is inspired from the standard real time recurrent learning. It is characterized by a small number of neurons that depends only on the number of system inputs and outputs and by a permanent updating of all parameters. The stability analysis is concerned by combining Lyapunov approach and linearization around the nominal parameters to establish analytical sufficient conditions for the global robust stability of the closed loop system. The scheme is applied to control the Tennessee Eastman Challenge Process. Performance evaluation such as set point stabilization, processing modes changes and disturbances rejection are pointed out, and results are discussed according to the Down and Vogel control objectives.

1. Introduction

Among the nonlinear complex processes the control engineers are confronted with, plant wide nonlinear chemical or biological systems are frequently described. These processes consist of a large number of different units (reactors, separators, strippers, cooling systems, and so on).

Such plants are strongly nonlinear, multivariable and delayed systems. Moreover, numerous installations are open-loop unstable. Plant wide control means the control of the overall system and it generally implies the decomposition of the process into some sub-processes or units and a selection of pertinent measured and controlled variables. The global control scheme usually accepted consists of a cascade control strategy divided into two or three successive stages according to figure 1. Such global strategies were introduced 30 years ago [15; 25; 26; 28; 42]. They are developed by scientists and engineers in numerous journals (Biochemical Engineering Journal, Chemical Engineering Science, Computers and Chemical Engineering, Control Engineering Practice, Journal of Process Control, and so on) and conferences (American Control Conference, International Control Conference, European Control Conference, and so on) concerned with chemical and automatic control. The different stages have decreasing dynamics from stage 1 to stage 3.

At the low level are the stages 1 and 2. The objective of the first stage is the steady state stabilization. Consequently, it is essential to perfectly analyze steady state conditions before developing any particular strategy. The first difficulty to be overcame for the multivariable processes is a judicious choice of some measured and manipulated variables. Generally, the measured variables are the levels in different units of the plant such as chemical reactors, liquid/gas separators and the pressure and temperature in these units. The controller associated to the measurements acts on manipulated variables, principally feed flow valves, product flow valves, cooling water equipment valves, purge valves, and so on. This stage consists on single input – single output loops providing set points trough a stabilizing algorithm thanks to local controllers like PI or PID regulators (base control).

The second stage is implemented for disturbances rejection. Controllers tuned for stage 1 only reject some disturbances like cooling water temperature variations and cooling water valves sticking. Numerous other disturbances that act on the feed component ratio, the feed component temperature or flow, and the changes in kinetics can not be compensated by low level controllers and require additive loops. The stage 2 generally consists on a decentralized control in which the plant is divided into sub-units. The first step is the selection of a set of reliable measured and manipulated variables sensitive to the disturbances. The control strategies associate a single feedback loop to each sub-unit. Finally, tuning these loops, the engineers focus their attention on the stability of the model in the presence of disturbances.

The higher control level (i.e. stage 3) is implemented to ensure the global control of the plant. Several problems can be considered: safety and/or environmental issues; production rate optimization; physical constraints on the process. This stage essentially results on composition and flow control considering some economic aspects.

Whatever the stage considered the development of a controller is strongly linked to the model that describes the dynamics from the controlled variables towards the measured ones. With regard to the diagram on figure 1, it is obvious that the model of a particular stage includes all-or-part of the precedent stages. Since the last two decades, researchers and engineers have proved that for plant wide control of industrial processes, nonlinear models associated with adaptive control algorithms are suitable. This choice is a direct consequence of the necessity to get stable and robust performance for systems whose dynamics are not perfectly known and disturbed by many unpredictable perturbations. Among the different methods for adaptive modelling and control design, neural networks are more and more frequently encountered in recent literature [1; 2; 7; 11; 13; 45; 46]. The first issue of this chapter is to motivate and explain this growing interest for neural networks. The most popular control schemes based on behavioural models and adaptive control with neural networks are presented. Then, an indirect control scheme with fully connected neural networks is detailed. This control scheme is shown to be stable and robust with regard to disturbances and parameters uncertainties. It adapts itself thanks to a Real Time Recurrent Learning (RTRL) algorithm and is based on a self tuning of all parameters including the adaptive and time parameters. Usual problems as network dimension, pre-training and parameters initialization are overcame.

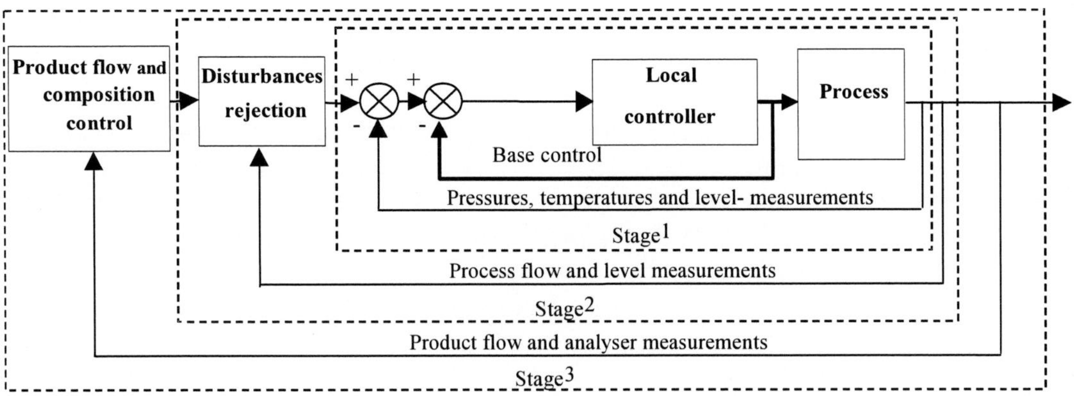

Figure 1. Usual cascaded control strategy for plant-wide chemical or biological processes: stage 1 is defined for stabilization, it is based upon few experimentally selected measurements and manipulated variables; stage 2 is developed for rejection of disturbances, it is based upon the most reliable measurements; stage 3 is an optimization control strategy based upon a global objective for product flow and composition.

The chapter is organized into five sections. In section two, adaptive control schemes are described and neural networks are introduced as a suitable tool to model and control the systems. In section three, a particular attention is paid to fully connected neural networks with stable and robust autonomous adapting algorithms. Several recent contributions are presented in a framework. An application to the Tennessee Eastman Challenge Process is presented in section four in order to illustrate the efficiency of the indirect control scheme with fully connected neural networks. Stabilization, disturbances rejection and processing mode changes are investigated and compared with other works. The conclusion sums up the main characteristic of adaptive control with fully connected neural networks and introduces one of the most challenge for adaptive control in the future research.

2. Adaptive Control Schemes for Nonlinear Systems

Control design for complex systems is a difficult task as long as non linearities such as saturations, dead zones, or hysteresis phenomena result in unsuitable behaviours like instabilities, oscillatory trajectories or limit cycles. This section is about some linear and nonlinear control schemes that have been investigated for stages 1 and 2 of the above diagram (figure 1).

2.1. Control Design Based on Physical Model

Control design based on physical models (i.e. models resulting from the knowledge of the physical phenomena that drive the considered systems) can be worked out according to a local linear approximation of the system behaviour or according to the global nonlinear model.

2.1.1. Control Design Based on Linear Approximations

For some applications, a linear approximation of the nonlinear behaviour is worked out [6; 23]. In that case, linear controllers can be designed that are efficient as long as the system remains in a neighbourhood of the nominal trajectory. When it leaves such a neighbourhood, the controller becomes inefficient and practitioners must look at other solutions like nonlinear control schemes. Thanks to the huge progress in computer science and to the development of innovative technologies like DSP processors, it is possible today to implement nonlinear controller for actual applications. In comparison with the methods based on linear approximations, nonlinear controllers cover a wide region in the state space.

The gain scheduling control is probably one of the most popular approaches for nonlinear control design. The basic idea is to consider several nominal positions in state space and to attach to each position a linear model with its linear controller. Each local controller is suitable in a small area of the state space and a global control signal can be processed with an interpolation of the local controllers outputs [21; 39]. The gain scheduling control requires a state estimator in order to evaluate the interpolation function and leads sometimes to instabilities in case of frequent and abrupt changes of the state. Other drawbacks are: the complexity of the method increases quickly with the number of linear models that are considered; the performance and robustness regarding the model uncertainties are not warranted.

H∞ robust control uses a frequential approach to design the controller. The basic idea is to work out the closed loop transfers of the outputs with respect to the perturbations and uncertainties and to compute the minimal magnitude of these transfers according to the control signal [5; 33]. The problem is usually formulated with a Riccati equation. Major drawbacks of the method are that it needs a good knowledge of the perturbations influence and is only applicable when uncertainties and perturbations are upper bounded.

2.1.2. Control Design Based on Lyapunov Approach

Designing nonlinear controllers directly from the nonlinear models is an alternative solution for the control of complex systems. The use of Lyapunov functions makes efficient such an approach. The basic idea is to define a candidate function that depends on the control system parameters. The control is designed so that the function will decrease asymptotically. Global asymptotic stability of the closed loop system is obtained as a consequence. Backstepping and sliding modes control are two popular methods based on the Lyapunov approach.

The aim of sliding modes control is to drive the trajectory in a specific region of the state space, named the sliding surface, according to the desired dynamics [16]. The tracking is obtained with a two stages strategy. The first stage consists to drive the system near the sliding surface. The second stage is to drive the system along this surface. Chattering due to abrupt variations in the control signal may occur and physical models of the system are required in order to avoid chattering [32; 59].

Adaptive backstepping is another solution for high order systems [17]. The basic idea is to consider some state space variables as additive inputs and intermediate control signals are designed with a recursive strategy. The method is efficient as long as the derivatives of the state space variables satisfy specific conditions (strict – feedback systems) [14]. The complexity and the presence of uncertainties are two major drawbacks for this method.

The design of control schemes based on physical nonlinear models is a difficult challenge for multivariable problems because of the cross-coupling between inputs and outputs. Few contributions have been proposed with adaptive observers [38] and backstepping [20]. But numerous results are restricted to affine systems for which uncertainties are upper bounded. As a conclusion let us point out that the previous methods need a partial or complete physical knowledge of the process dynamics. They are not efficient for plant wide control problems. In that case, adaptive control based on behavioural models must be investigated and among the existing methods, neural networks are often preferred.

2.2. Control Design Based on Behaviour Models

Due to the increasing complexity of dynamical systems, it becomes difficult to design controllers using standard control techniques. This difficulty is enhanced when it is impossible to obtain a physical model of the plant, due to its complexity and to the interactions with the environment. Furthermore, the environment may be unknown and time dependent that makes the model inexact and the resulting controller is inefficient. Whatever the suggested method and the considered stage of control, the control strategy for plant wide systems is generally based upon model predictive control and adaptive methods that mainly require a behavioural model of the plant. With adaptive control, the controller adapts itself to changing system parameters and to the environment influence. In this context, neural networks are often used to approximate the system behaviour and lead to efficient predictors (section 2.2.1). As a consequence neural networks can be combined with standard control techniques (section 2.2.2) or used simultaneously as predictors and controllers in adaptive control schemes (section 2.2.3).

2.2.1. Plant Wide Systems Modelling with Neural Networks

Neural networks, thanks to their approximation properties, make possible to build behavioral models. As a consequence, the control of the chemical or biological systems mainly include neural networks for process modeling. The most popular structures are RBF (Radial Bases Functions) networks or MLP (Multi-Layer Perceptron) networks and sometimes DRNN (Differential Recurrent Neural Networks) networks. [18] proposes a RBF neural model adapted trough a multi-resolution adaptation algorithm. A RBF structure is also used in [46] associated with a self organizing learning algorithm for off-line training of the model. A multi-model method is proposed in [53] that associates three multi-input / single-output RBF networks for prediction. Concerning MLP structures [51; 52] use neural predictors that are adapted on-line using an extended Kalman Filter algorithm. Off-line training of an MLP is also presented in [1; 11]. [7] combines the off-line training with on-line adaptation of the predictor. A DRNN tuned with the Levenberg-Marquardt algorithm is developed in [2]. This last network is defined according to a non-linear state-space equation in order to predict the process dynamic. All off-line training methods need a training data set. Consequently open-loop systems can easily be modeled while unstable-open loop systems have to be previously stabilized.

2.2.2 Combining Neural Networks with Standard Techniques

The neural controllers can be combined with other conventional control strategies such as back stepping and sliding modes control [14; 36; 59] but the most popular strategy consists in combining neural networks predictors with PID controllers. As example, in [7] the neural network model is linearized for tuning a PID through a General Minimum Variance control law. [52] tune the PID using a Lyapunov method with a quadratic tracking error. Neural networks are also combined with reference model control and self tuning regulators. The main characteristic of the control with reference model (MRAC) is the adaptation of the controller parameters in order to minimize the closed loop tracking error resulting from the reference model [37]. Backpropagation of the error gradient is usually used to drive the adaptive algorithm. In comparison with MRAC control, self tuning regulators include an identification of the parameters of the process [37]. This approach is very flexible and can be associated with numerous strategies to design the controller but analysis of the closed loop system is difficult due to the nonlinear relationships from process identification to adaptive controller.

Combining neural networks with standard techniques leads to some difficulties. Training must be reiterated when the system or environment change. Nonlinear control laws are difficult to formulate and the complexity is growing up according to the number of unknown parameters. As a consequence the above mentioned methods present strong limitations for real-time applications. For all these reasons, adaptive control schemes based on neural predictors and controllers have been investigated.

2.2.3. Adaptive Control Schemes with Neural Networks

Neural network controllers made of RBF or MLP using self organizing algorithm, Kalman filter optimization method, Back Propagation Through Time, automatic differentiation techniques or multi-resolution adaptation algorithm are developed in [1; 2; 18; 46; 53]. The design of efficient adaptation algorithms is the main issue in the development of controllers based on neural networks for unknown systems. The most popular algorithms used for parameters adaptation are the gradient back propagation learning algorithm and numerous variants. By evaluating the gradient of a cost function (depending on the actual and the desired output of the plant) with respect to the weights of the networks, it is possible to adjust iteratively the value of the weights in the direction opposite to the gradient. As a consequence the cost function is expected to decrease according to the number of iterations. Another important issue is the estimation of the Jacobian matrix of the plant that is required by the controller and cannot be evaluated analytically unless the process dynamics are known [30]. Three architectures have been proposed [3; 30] in order to solve that problem: direct inverse control, direct control and indirect control.

The inverse system model is cascaded with the plant so that the composed system results in an identity mapping between the actual and the desired outputs of the plant. Thus, the inverse model acts as a controller [27] (figure 2). Obviously, this approach is only applicable when the process behaves monotonically as a forward function at steady state (minimum phase and causality are required) [31]. The implementation of this structure requires two stages. The first stage is the inverse model training. A cost function defined as the quadratic error between the estimated and the desired control signal is minimized. The second stage is to use the inverse model as a controller. Throughout this stage, the inverse model parameters are considered as constant and lead to an open-loop controller. As a consequence, this control technique is not robust to parametric variations and external disturbances. For example, [51] have proposed a controller resulting of the inversion of the neural model for multivariable systems.

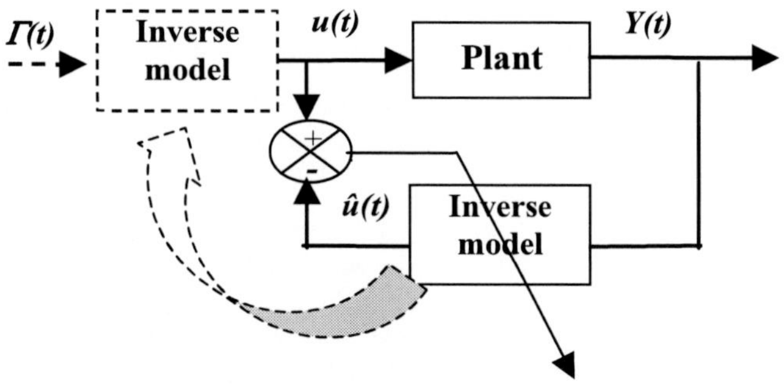

Figure 2. Direct inverse control scheme: training of the inverse model (continuous lines) and control with the obtained inverse model (dotted lines).

The second structure is the direct adaptive control [14; 36; 47; 48] (Figure 3). The main idea is to learn directly the controller parameters using the tracking error. For that purpose, the Jacobian matrix of the plant is estimated with a first-order approximation [8], or simply by the sign of the plant Jacobian [35]. The main drawback of this approach comes from the plant Jacobian approximation. When disturbances and measurement noises affect the plant outputs, these outputs quickly vary compared with the plant inputs. These variations lead to uncertainties on the control parameters. Consequently, this structure is very sensitive to the measurement noises and has poor robustness properties.

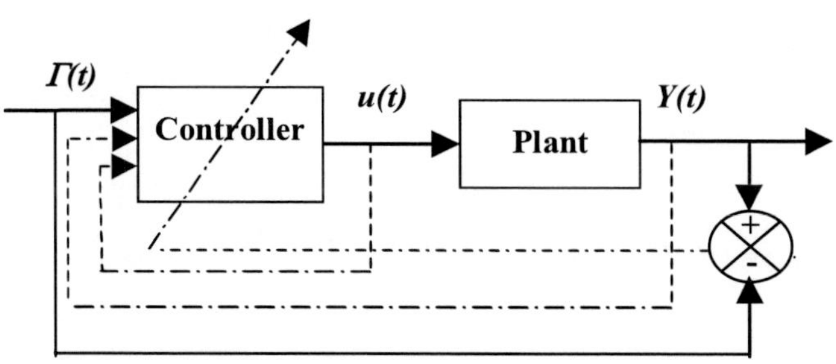

Figure 3. Direct adaptive control scheme.

The third structure is based on the estimation of the plant parameters and on the adaptation of the controller parameters (figure 4). It is supposed that the estimated parameters of the process remain close to the exact values. If the process parameters change, the model and the controller adapt themselves to the new set point. The global approach is called "indirect adaptive control" since the controller adapts itself indirectly according to the identification of the plant parameters [10; 12; 29; 43]. The evaluation of the Jacobian matrix is obtained through a pre-training phase. This last structure has more advantages than the direct and inverse structures: less sensitivity to the noise and better disturbance rejection. The main drawback results from the separation of the training and operational phases. In fact an offline training is required. In case of non-stationary environment, the uncertainties affect the model parameters and then the controller parameters. This problem must be overcame by using on-line learning.

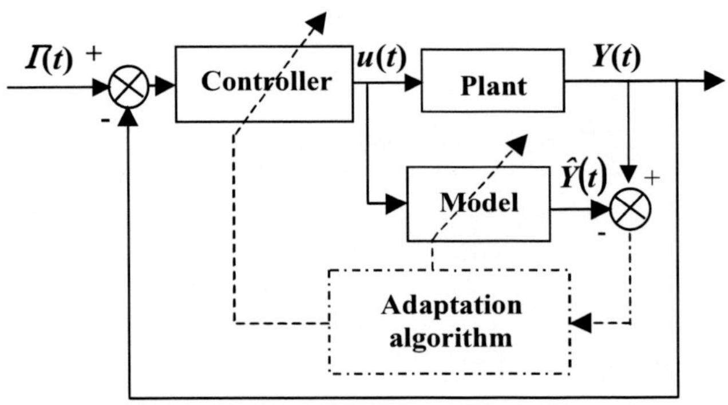

Figure 4. Indirect adaptive control scheme.

As a conclusion, the indirect adaptive control scheme appears as a smart solution to control nonlinear plant wide systems as long as disturbances are rejected and the Jacobian matrix is estimated thanks to off-line training. However, for practical uses, this scheme must be implemented with on-line training and some difficulties must also be overcame: the initialization phase, the presence of local minima, the size of the networks and the arbitrary value of the adaptive rate. In the next section, an indirect control scheme with fully connected neural networks is proposed that is stable and robust with regard to disturbances and parameters uncertainties. This structure adapts itself thanks to a real-time recurrent learning (RTRL) based algorithm. It is based on a self tuning of the adaptive rate and time parameter and an initialization to zero with a small numbers of neurons that only depends on the considered inputs and outputs.

3. Robust Control with Fully Connected Neural Networks

3.1. Adaptive NN Control

The proposed control scheme is an InDirect Neural network Controller (*IDNC*) composed of two separate neural networks: the Neural Controller (*NC*) and the Adaptive instantaneous

Monitoring Network (*AMN*) [10]. The aim of *AMN* is to provide an estimation of the process output(s) during a short time window in order to drive *NC*. The subscripts *m* and *c* are used to distinguish the *AMN* and *NC* respectively. The updating of *NC* and *AMN* is synchronous (figure 5, the dash lines show the RTRL paths to update the parameters of *AMN* and *NC*). Let us define N_{IN} and N_{OUT} respectively as the number of plant inputs and outputs, $IN = \{1,2, .. N_{IN}\}$ and $OUT = \{N_{IN}+1, N_{IN}+2, .. N_{IN}+N_{OUT}\}$ as the set of input and output indexes.

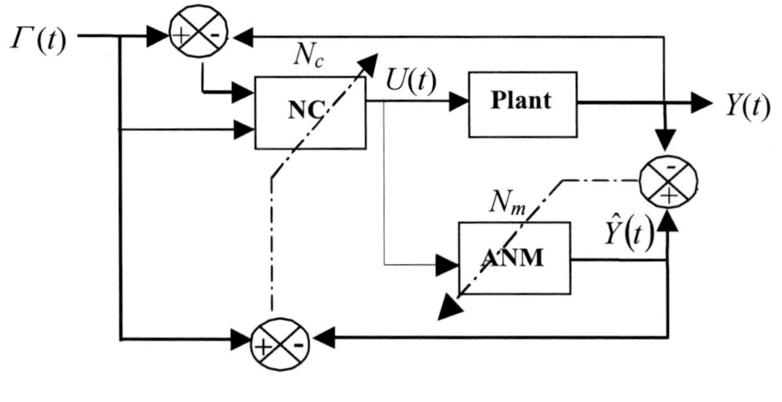

Figure 5. Structure of the indirect neural network control, the dimension of $Y(t)$, $\hat{Y}(t)$ and $\Gamma(t)$ is N_{OUT}, the dimension of $U(t)$ is N_{IN}.

When the process is running, the neural networks continuously adapt. Comparison between physical measurements and neural network outputs is processed at each sample time. All measured and manipulated variables X correspond to physical signals (temperatures, pressures, levels, compositions, valve setting points, etc) and have positive values constrained within an interval $]X_{min} ; X_{max}[$. The activation functions of the neurons are hyperbolic functions, therefore the network outputs evolve within the interval]-1 ; +1[. To avoid negative output values and zero crossing during adaptation, it is necessary to transform all measurements X from range $] X_{min} ; X_{max} [$ into]0 ; 1[according to the nominal value X_m (the setting point) and to the admissible variance (acceptable limit) of X around X_m.

The autonomous evolution of *AMN* and *NC* starts from zero values [22]. It results in a compact structure with a small number of nodes. For *AMN*, the total number of neurons N_m is chosen equal to $N_{IN}+N_{OUT}$, so that any node is either an input node or an output node but not both at same time, in order to avoid to perturb any output signal with input ones. Additional nodes are useless [22]. For *NC*, the total number of neurons N_c is chosen equal to $2.N_{OUT} + N_{IN}$. The network *NC* has $2.N_{OUT}$ inputs: the N_{OUT} plant desired outputs and the N_{OUT} output tracking error functions. There is no pre-training or post-training phase but only an on-line updating of weights, time parameters and adaptation parameters.

To overcome the problem of persistent excitation and to provide an instantaneous model that adapts itself when plant or environment changes, *AMN* parameters are updated in real time. The idea is to compute an instantaneous behavioural model from input and output data of the plant. This instantaneous model is used to automatically update the controller parameters in order to track the process variations. The real time adaptation provides an efficient compensation of the unpredictable process disturbances and sensor noises.

3.1.1. Adaptive Instantaneous Monitoring Network

The Adaptive instantaneous Monitoring Network (*AMN*) is developed with fully connected recurrent neural networks [10; 22]. The dynamics of the N_m neurons are given according to equation (1):

$$\frac{1}{|\tau_m(t)|}\frac{d\hat{Y}_i(t)}{dt} = -\hat{Y}_i(t) + \tanh\left(\sum_{j=1}^{N_m} W_{ij}(t)\hat{Y}_j(t) + X_i(t)\right) \tag{1}$$

Let us define $Y(t) = (Y_i(t))$, $i \in \{1, 2, ...N_{OUT}\}$, as the output vector of the plant at time t and $\hat{Y}(t) = (\hat{Y}_i(t))$, $i \in OUT$, as the estimated output vector of the plant at time t. A matrix representation of (1) is given by (2):

$$\frac{1}{|\tau_m(t)|}\frac{d\hat{Y}(t)}{dt} = -\hat{Y}(t) + B(t) \tag{2}$$

with $B(t) = (B_i(t)) \in IR^{Nm}$ and :

$$B_i(t) = \tanh\left(\sum_{j=1}^{N_m} W_{ij}(t)\hat{Y}_j(t) + X_i(t)\right) \tag{3}$$

where $X_i(t) = U_i(t)$ if $i \in IN$ and $X_i(t) = 0$ if $i \in OUT$. $\hat{Y}_i(t)$, W_{ij}, $1/\tau_m(t)$ and $U_i(t)$ represent respectively the ith neuron state, the ith estimated output vector of the plant, the *ANM* weight from j^{th} neuron to i^{th} neuron, the *ANM* adaptive time parameter and the ith *ANM* input (figure 6).

The autonomous adaptation algorithm is inspired from the RTRL algorithm [49; 60]. The *ANM* weights are updated according to equation (4) :

$$\frac{dW_{ij}(t)}{dt} = -|\eta_m(t)|\frac{\partial E_m(t)}{\partial W_{ij}(t)} \tag{4}$$

where $\eta_m(t)$ stands for the updating rate and $E_m(t)$ is the model error between the plant outputs and the *ANM* outputs, which is defined as:

$$E_m(t) = \frac{1}{2}\sum_{i \in OUT}\left(\hat{Y}_i(t) - Y_{i-N_{IN}}(t)\right)^2 \tag{5}$$

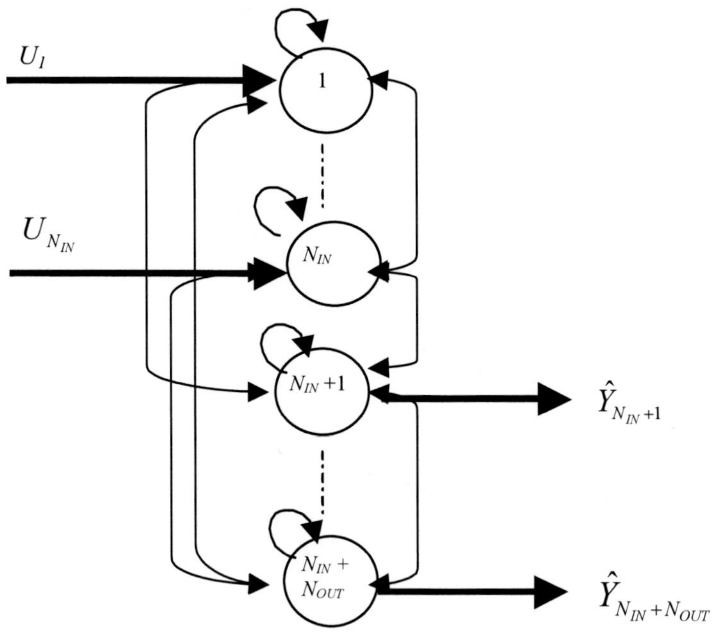

Figure 6. *ANM* fully connected structure with $N_{IN} + N_{OUT}$ neurons, N_{IN} inputs U_i and N_{OUT} outputs $\hat{Y}_j$.

3.1.2. Neural Controller

The Neural Controller (*NC*) is also a fully connected recurrent neural network based on autonomous RTRL algorithm. The $2N_{OUT}$ *NC* inputs are both the N_{OUT} plant desired outputs $\Gamma_i(t)$ and the N_{OUT} output error functions $\Gamma_i(t)-Y_i(t)$. The *NC* outputs are the N_{IN} control inputs $U_i(t)$. Control inputs are calculated by comparing the *AMN* outputs with the desired system responses and according to the dynamic activation of neurons given by equation (6):

$$\frac{1}{|\tau_c(t)|}\frac{dU(t)}{dt} = -U(t) + D(t) \tag{6}$$

with $D(t) = (D_i(t)) \in IR^{Nc}$ and :

$$D_i(t) = \tanh\left(\sum_{j=1}^{N_c} \Phi_{ij}(t)U_j(t) + Z_i(t)\right) \tag{7}$$

and $Z_i(t) = \Gamma_i(t)$ if $i \in [1, N_{OUT}]$, $Z_i(t) = \Gamma_{i\text{-}Nout}(t)\text{-}Y_{i\text{-}Nout}(t)$ if $i \in [N_{OUT}+1, 2\,N_{OUT}]$ and $Z_i(t) = 0$ if $i \in [2\,N_{OUT}+1, 2\,N_{OUT} + N_{IN}]$. $\Gamma_i(t)$ corresponds to the ith plant desired output to be tracked and $\Gamma_i(t)\text{-}Y_i(t)$ stands for the ith output error function. $\Phi_{ij}(t)$ and $1/\tau_c(t)$ represent respectively the *NC* weight values from j^{th} neuron to i^{th} neuron and the *NC* adaptive time parameter (figure 7).

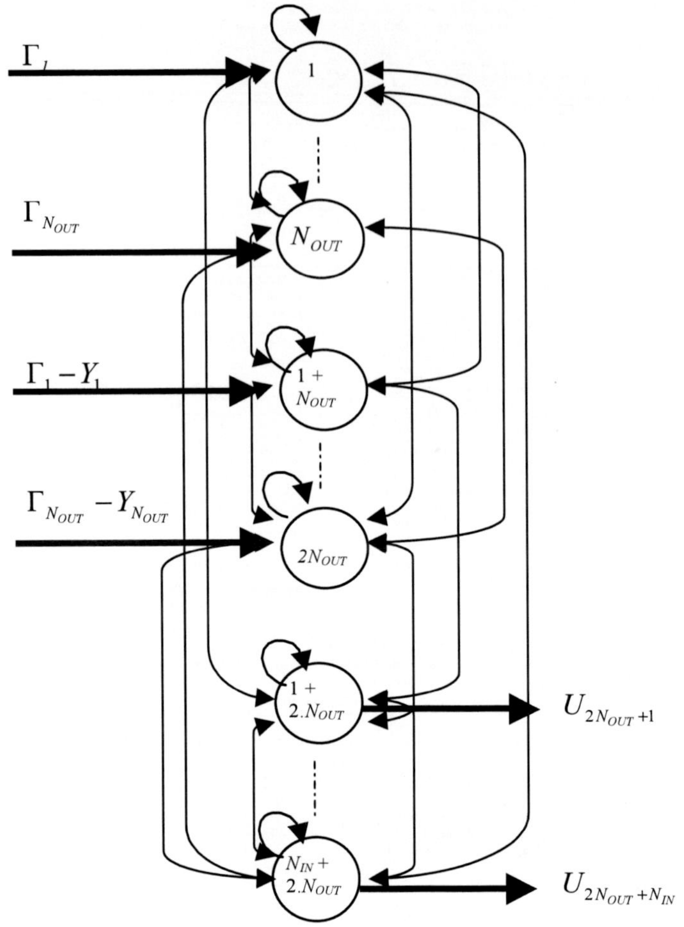

Figure 7. *NC* fully connected structure, with $2.N_{OUT}+N_{IN}$ neurons, N_{OUT} inputs Γ_i, N_{OUT} inputs $\Gamma_i - Y_i$, and N_{IN} outputs U_j.

The vector *D(t)* is considered constant. The weights are then updated with equation (8) [54]:

$$\frac{d\Phi_{ij}(t)}{dt} = -|\eta_c(t)| \frac{\partial E_c(t)}{\partial \Phi_{ij}(t)} \tag{8}$$

where $\eta_c(t)$ stands for the updating rate and $E_c(t)$ is the tracking error between the plant desired outputs and the plant outputs, which is defined as:

$$E_c(t) = \frac{1}{2} \sum_{i=1}^{N_{OUT}} \left(\Gamma_i(t) - Y_i(t)\right)^2 \tag{9}$$

The N_{IN} outputs U_j for $j \in [2N_{OUT}+1,\ 2N_{OUT}+N_{IN}]$ are the N_{IN} inputs of *ANM*.

3.1.3. Automatic Adaptation of the Parameters τ and η

In order to obtain an autonomous control design able to adapt itself to a large variety of uncertain or unknown processes with few initial constraints, we consider updating rates $\eta_m(t)$, $\eta_c(t)$ and time parameters $1/\tau_m(k)$, $1/\tau_c(k)$ as time functions. The dynamics of $\eta_c(t)$ and $\tau_c(t)$ are worked out in order to get the dynamics of the *NC* close to the *AMN* ones.

$$\begin{aligned}\tau_c(t) &= \tau_m(t) = \tau(t) \\ \eta_c(t) &= \eta_m(t) = \eta(t)\end{aligned} \tag{10}$$

Initial values are $\tau(0)= \eta(0)= 0$. Parameters $\tau(t)$ and $\eta(t)$ are updated according to (11) [54]:

$$\frac{d\eta(t)}{dt} = -\frac{\partial E_m(t)}{\partial \eta(t)}$$

$$\frac{d\tau(t)}{dt} = -|\eta(t)|\frac{\partial E_m(t)}{\partial \eta(t)} \tag{11}$$

In case of stable open loop systems, the number of neurons for *NC* can be reduced to $Nc = N_{OUT} + N_{IN}$ [10; 57].

3.2. Stability and Robustness Analysis

Stability and robustness properties of the closed loop system are important issues to be addressed. Indeed, small parameter uncertainties and external disturbances can have an unfavourable impact on performance as well as stability. In addition, the dynamic behaviour of the networks can lead to instability of the plant.

3.2.1. Stability Analysis

We propose to study the stability of the *IDNC* with the well-known Lyapunov approach [19]. This approach is based on the consideration of a generalized energy content of the system. The method consists in finding a function *E(t),* positive definite with a negative semi-definite derivative and a unique global minimum at equilibrium.

The purpose of the control design is to force the output of the controlled process to track a desired trajectory. Stability is investigated according to the Lyapunov candidate function defined by equation (12):

$$E(t) = E_c(t) + E_m(t) \tag{12}$$

The theorem 1 provides sufficient conditions to ensure *IDNC* stability.

Theorem 1 [22; 54; 56]:

Let $|\eta|$ the adaptive updating rate of both the model and control networks, suppose the Lyapunov function is defined as in (12). Then the variation of the Lyapunov function could be expressed as $\Delta E(k) = a|\eta|^2 - 2b|\eta| - c \leq 0$, where the three parameters a, b and c depend on the process dynamics.

The sufficient stability condition for the *IDNC* in the sense of Lyapunov should satisfy the following updating rate bounds:

$$0 \leq \frac{b - \sqrt{\Delta'}}{a} \leq |\eta| \leq \frac{b + \sqrt{\Delta'}}{a} \tag{13}$$

where $\Delta' = b^2 + a.c$ depends on the updating procedure and must be non-negative.

Let us mention that theorem 1 implicitly leads to a stabilization algorithm that acts as a disturbance in the updating rules of parameters. In order to minimize this disturbance when parameter η does not satisfy (13), the nearest bound of admissible range has to be chosen to correct the parameter $|\eta|$. Then the algorithm ensures an evolution of the absolute value of the updating rate as smooth as possible with respect to asymptotic stability.

3.2.2. Sensor Noise Rejection

In order to apply neural networks to industrial problems, it is useful to have an understanding of such issues as robustness and parameter sensitivity of the adaptive algorithm. Robustness means that *IDNC* achieves stability and maintains similar performance even if the system to be controlled differs from the model used due to measurement noise or disturbances on the system. . Particularly, in closed loop, sensor noise should be rejected and structural uncertainties should be compensated. However, sensor noises typically have a high frequency range, while structural uncertainties usually lie towards low frequencies [56]. Hence, it is possible to consider separately the disturbance rejection and the sensor noise attenuation. On the one hand, sensor noise disturbs explicitly the *AMN* and influences the process indirectly via the feedback. Thus, their effects can be attenuated while acting on *AMN*, by restrictions of the updating rate parameter. On the other hand, the uncertainties are compensated thanks to the sensitivity functions which are used in the adaptation procedure of the network parameters [54].

Considering sensor noise, our aim is to provide a suitable indicator in order to evaluate the noise rejection and to decrease such disturbances. Let us consider the case of sensor noise ε_m that quickly varies compared to the process signal. The sensor noise ε_m acts additively on the plant output Y, so that the nominal plant output will be denoted by $Y_n = Y - \varepsilon_m$. In order to reject the measurement noise, the *AMN* must follow the nominal plant output Y_n rather than the disturbed signal Y.

Theorem 2 [54; 56; 57].

Let $|\eta|$ the adaptive learning rate of both the monitoring and control networks. Suppose that the sensor noise quickly varies compared to the process dynamics. The *IDNC* compensates the sensor noise, if the following sufficient condition holds:

$$|\eta| \le \varphi \tag{14}$$

where φ depends on the tolerate noise ratio and on the process dynamics.

This robustness criterion could be considered during the updating algorithm.

3.2.3. Robustness Stability Analysis

Objectives of the robustness analysis are to establish the conditions which ensure the asymptotic stability of disturbed equilibrium, under the influence of disturbances affecting the model of the controlled system and to deduce a confidence band for the disturbed parameters (ΔW, $\Delta\tau$) of the AMN such that the disturbed system always has an asymptotically stable equilibrium.

Lyapunov approach and the linearization around the values of the nominal parameters of *ANM*, leads to the following sufficient conditions for robust stability:

Theorem 3 [55; 56; 57]

A sufficient condition on the uncertainty parameters (ΔW, $\Delta\tau$) for robust stability is given by:

$$\max\left(\|\Delta W\|_\infty, \|\Delta W^T\|_\infty, |\Delta\tau|\right) \le K \tag{15}$$

where $\|.\|_\infty$ is the infinite norm of a matrix and K a confidence band of uncertainties which depends on the nominal parameters of *ANM*.

The figure 8 sums up the different conditions on the networks parameters that perform the robustness and the stability of the control system. Theorems 1 and 2 constrain explicitly the evolution of the adaptation parameter η and indirectly the evolution of other parameters. Theorem 3 only informs on the robustness stability performance.

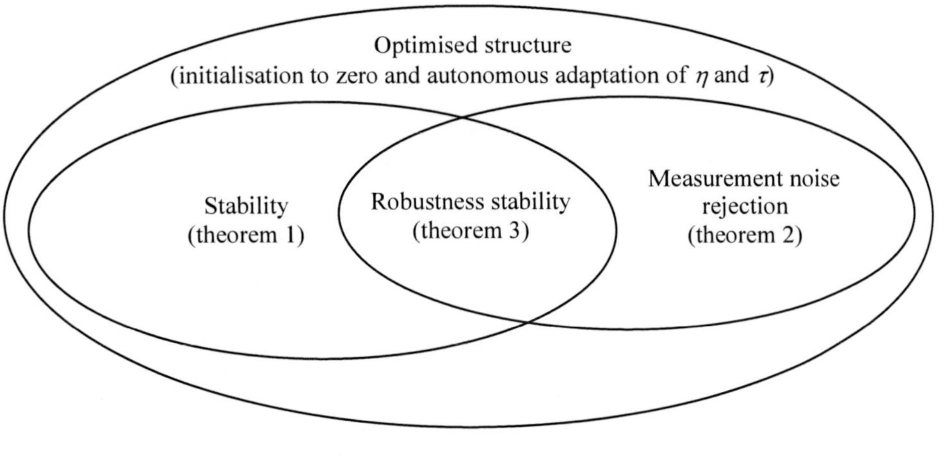

Figure 8. Updating including the different conditions in the network parameters space evolution.

4. Control Design for the Tennessee Eastman Challenge Process

The TECP [9] is a simulator of a multivariable non-linear, high dimensionality, unstable open-loop chemical plant provided by the Eastman Company. This simulator illustrates the main issues concerned with the control of plant wide systems. The TECP requires coordination of four unit operations: an exothermic 2–phase reactor, a flash separator, a compressor and a re-boiled stripper. The process results in final products *G* and *H* from four reactants *A, C, D* and *E* (figure 9). The plant has 7 operating modes: one base mode, 3 modes to minimize the production cost depending on the *G*/*H* ratio and 3 modes to maximize the production flow depending on the *G*/*H* ratio (table 2), 41 measured variables, 12 manipulated variables.

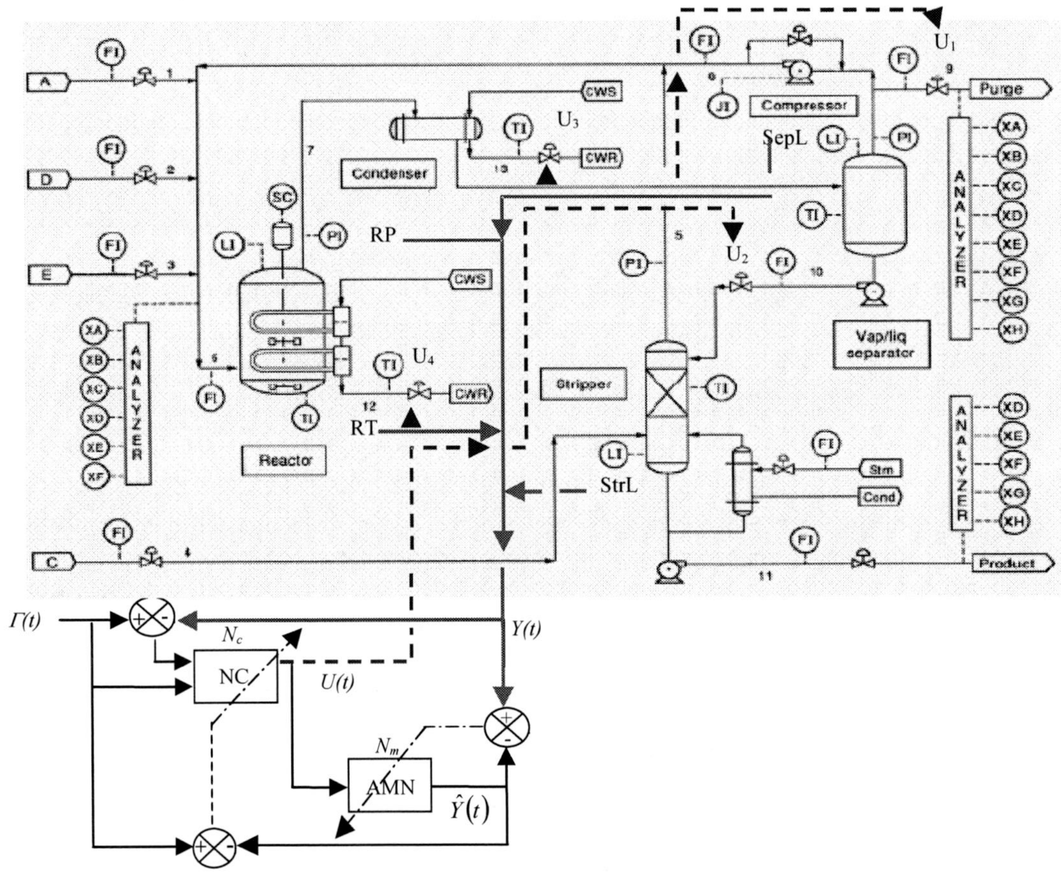

Figure 9. Tennessee Eastman Challenge Process connected with *IDNC*.

The process can be disturbed by 20 disturbances named IDV(1) to IDV(20) [9; 40]. The main control objective is to maintain measurement of product rate and composition at set points while keeping other variables within specified shutdown limits.

The TECP is controlled with the *IDNC* described in section 3. Four inputs (manipulated variables) and four outputs (measured variables) have been selected. The choice of these variables results from a study of the more commonly found in the literature [24; 34].

The controller regulates the temperature (*RT*) and pressure (*RP*) in reactor, and the levels in separator (*SepL*) and stripper (*StrL*). For this purpose, the manipulated variables are the purge valve U_1, the stripper input valve (separator liquid flow) U_2, the condenser cooling water valve U_3, and reactor cooling water valve U_4 (figure 9, full lines represent measurements and dashed line represent manipulated variables updating). The sampling period for measurements and for the purge (U_1) and reactor cooling water valve (U_4) is 72 seconds and the sampling period for stripper input (U_2) and condenser cooling water valve (U_3) is 180 seconds. The proposed *IDNC* has 20 nodes: N_m = 8 nodes and N_c = 12 nodes according to the number of inputs and outputs that are considered and the schemes in figures 6 and 7 [56; 58].

The TECP offers numerous opportunities to discuss and evaluate control strategies. The next sections discuss stabilization of the reactor in base mode, modes switching and disturbances rejection.

4.1. Stabilization in Base Mode

Let us first illustrate the performance of *IDNC* in the base mode without perturbation. In the table 1 are gathered the mean value and standard deviation for the most significant variables for 600 hours simulations. Let us define *StrF* as the product flow in stripper and *XD, XE, XF, XG, XH* the composition of products. Our results show that performance is very close to the ones found in literature [9; 40; 50].

In comparison with other approaches reported in the literature [40; 50], our control scheme includes simultaneously stages 1 and 2. That is not the case for [40; 50] where the control scheme only includes stage 1 and requires 12 measured variables and 11 manipulated ones in [40] and 9 manipulated ones in [50]. As a consequence, we obtain better performance with a small number of measured and manipulated variables.

The results in table 1 prove the capacity of *IDNC* to compensate slow decays in TECP with only 4 measured variables and 4 manipulated variables. Looking at the standard deviation, one can also notice the weak dispersion of measurements around the mean values. All measured variables are near the reference values in order to satisfy the safety requirements of the plant.

Table 1. Control design for TECP base mode

	IDNC		J.J. Downs [9]
	Mean value	**Standard Deviation**	**Mean value**
RP (kPa)	2704.9	1.3748	2705.0
RT (°C)	120.40	0.0694	120.4
SepL (%)	50.010	1.0524	50.0
StrL (%)	50.009	1.0559	50.0
StrF (m^3/h)	22.950	0.1156	22.949
XD (Mole%)	0.0179	0.0098	0.0179
XE (Mole%)	0.8329	0.0112	0.8357
XF (Mole%)	0.0986	0.0010	0.0986
XG (Mole%)	53.724	0.4867	53.724
XH (Mole%)	43.831	0.5047	43.828

Similar conclusions hold by considering the production cost J (equation (16)) that includes the cost for compressor and stripper J_1, the cost due to the loss of product J_2 and reactants J_3 [4; 34; 44]:

$$J = J_1 + J_2 + J_3$$

$$J_1 = 0.0536x_{20} + 0.0318x_{19}$$

$$J_2 = 0.4479x_{10}\left(2.206x_{29} + 6.177x_{31} + 22.06x_{32} + 14.56x_{33} + 17.89x_{34} + 30.44x_{35} + 22.94x_{36}\right)$$

$$J_3 = 0.0921x_{17}\left(22.06x_{37} + 14.56x_{38} + 17.89x_{39}\right) \qquad (16)$$

where x_{19} stands for the stripper steam flow (kg/h); x_{20} stands for the compressor power (kW); x_{10} stands for the purge rate (kscmh); x_{17} stands for the stripper underflow (m3/h); x_{29} (resp. x_{31} to x_{36}) stands for the purge A % (mol%) (resp. purge C % to purge H %); x_{37} (resp. x_{38} and x_{39}) stands for the product D % (mol%) (resp. product E % and product F %). The cost with IDNC control (170.42 $/h) is very similar with the cost obtained by Downs and Vogel (170.62 $/h) and other authors in base mode.

4.2. Modes Switching

The TECP has 7 operating modes (table 2), and this section illustrates the performance of the *IDNC* in steady state for different modes and also during the transients from one mode to the other.

Table 2. TECP operating modes

Mode	G/H Mass ratio	Objective	Production rate
0	50/50	Stabilization	Base mode
1	50/50	Production cost minimization	7038 kg/hr G and 7038 kg/hr H (base mode)
2	10/90	Production cost minimization	1408 kg/hr G and 12669 kg/hr H
3	90/10	Production cost minimization	10000 kg/hr G and 1111 kg/hr H
4	50/50	Maximum production rate	
5	10/90	Maximum production rate	
6	90/10	Maximum production rate	

The *IDNC* is suitable to drive the plant from one mode to another according a first order trajectory of parameter 0.2 h^{-1}. Let us consider the following scenario:

- From t = 0 to t = 3 hours, TECP is in base mode (mode 0) with desired values: RP = 2705 kPa; RT = 120.4 °C; $SepL$ = 50 %; $StrL$ = 50 %.
- At t = 3h TECP switches from mode 0 to mode 1.

- From t = 3 to t = 20 hours, TECP is in mode 1 with desired values: *RP* = 2800 kPa; *RT* = 122.9 °C; *SepL* = 50 %; *StrL* = 50 %.
- At t = 20h TECP switches from mode 1 to mode 2.
- From t = 20 hours to the end of simulation (60 hours), TECP is in mode 2 with desired values: *RP* = 2800 kPa; *RT* = 124.2 °C; *SepL* = 50 %; *StrL* =50 %

In the modes 0 and 1, the product ratio is *G* / *H* = 50 /50 whereas in mode 2, *G* / *H* = 10 /90. The figures 10-a to 10-d illustrate the evolution of measured variables for the previous scenario.

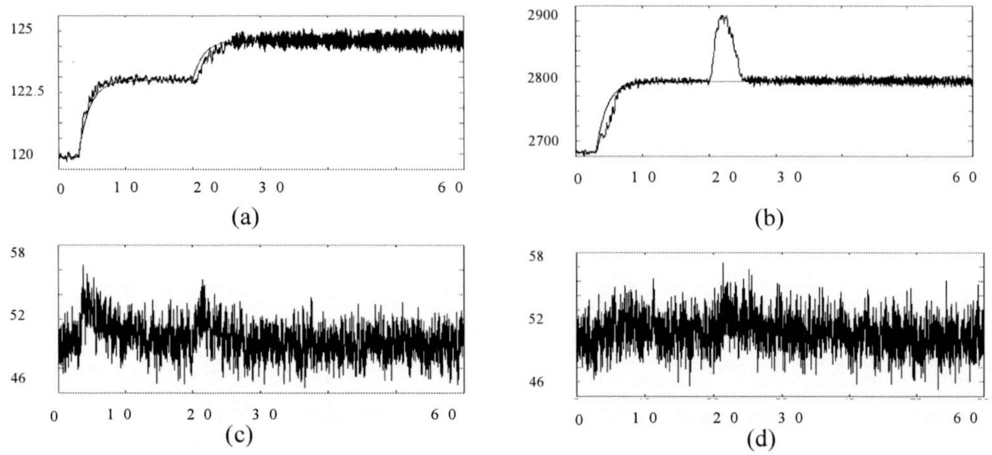

Figure 10. Modes switching: Measured variables in function of time (hours) a) Reactor pressure (kPa), b) Reactor temperature (°C), c) Separator level (%), d) Stripper level (%).

Mean values and variances are worked out during the steady state (for mode 0 the steady state holds from t =1 to t =3 hours; for mode 1 the steady state holds from t = 10 to t = 20 hours and data are processed from t = 15 to t = 20 hours; for mode 2 the steady state holds from t = 30 to t = 60 hours and data are processed from t = 50 to t = 60 hours) and reported in table 3.

Table 3. Mean values and variances of the measured variable during modes switching

	Mode 0 (from t = 1 to t = 3 hours)		Mode 1 (from t = 15 to t = 20 hours)		Mode 2 (from t = 50 to t = 60 hours)	
	Mean value	**Variance**	**Mean value**	**Variance**	**Mean value**	**Variance**
RP	2705	2.9	2800	5.4	2800	13.9
RT	120.41	0.01	122.92	0.01	124.20	0.04
SepL	50.34	0.97	50.24	1.52	49.79	1.38
StrL	49.84	1.25	50.44	1.21	49.99	1.23

The table 4 presents the quality of the production and compares the results with the ones obtained by [34; 44].

Table 4. Mean values and variances of the production in case of mode switching

Product	IDNC and Z. Tian [44]	Mode 0	Mode 1	Mode 2
D	Mean value IDNC (Mole %)	0.0185	0.0096	0.0011
	Variance IDNC	1.07e-004	9.00e-005	8.96e-005
	Mean value [44] (Mole %)	0.02	0.01	0.00
E	Mean value IDNC (mol. mass %)	0.8297	0.5907	0.9357
	Variance IDNC	1.00e-004	1.64e-004	2.49e-004
	Mean value [44] (Mole %)	0.84	0.58	0.92
F	Mean value IDNC (Mole %)	0.1062	0.1322	0.2720
	Variance IDNC	9.34e-005	1.14e-004	9.75e-005
	Mean value [44] (Mole %)	0.10	0.19	0.29
G	Mean value IDNC (Mole %)	53.80	54.03	11.59
	Variance IDNC	0.29	0.20	0.22
	Mean value [44] (Mole %)	53.72	53.83	11.66
H	Mean value IDNC (Mole %)	43.77	43.81	85.82
	Variance IDNC	0.20	0.25	0.29
	Mean value [44] (Mole %)	43.83	43.91	85.64

The results in modes 0, 1 and 2 are very similar to the ones proposed by [34]. In comparison to stage 1 local controllers, the *IDNC* provides good results not only during the steady states regarding each mode, but also during the transients. It is suitable to drive the process from one mode to another within an acceptable time interval. The controller adapts itself in the different modes and there is no need to develop a multi-model controller or to tune the controller parameters depending on the new setting point.

4.3. Disturbance Rejection

In order to test the capacity of *IDNC* to reject disturbances, and to investigate the robustness of the control scheme, various simulations are performed with different perturbations (table 5). These perturbations have been proposed by [9] for performance evaluation and comparative studies purposes.

For example, let us consider a random perturbation in *A*, *B*, and *C* feed composition from time t = 10 to t = 58 hours. The figure 11 illustrates the process behaviour in presence of disturbance IDV(8). Our results can be compared with those obtained by [41] with 12 local PID control loops. The reactor pressure varies within the interval]2640 kPa; 2790 kPa[for *IDNC* and within]2600 kPa; 2810 kPa[for Sozio controller, and the reactor temperature varies within]119,7°C; 121°C[for *IDNC* and]119°C; 122°C[for Sozio controller. The variability of controlled variables is smaller with *IDNC*. Moreover *IDNC* manipulates only 4 variables instead of 12, and it is more robust because all manipulated variables are always far from saturation (this is not the case with PID controller as long as the valve V_1 that controls the input flow of reactant *A* is saturated [41]).

The table 5 sums up the results obtained for each perturbation IDV(1) to IDV(20) with *IDNC* [57; 58] in comparison with the performance obtained by [40] that uses a controller similar to the one introduced by [24]. Only 3 perturbations are not rejected by the *IDNC*:

(IDV(1), IDV(6) and IDV(7)), and lead to the process shutdown. The reason for this problem is that the *IDNC* does not control any reactant input flow valve. In order to reject these three perturbations, the *IDNC* has to be completed with at least one more input-output couple.

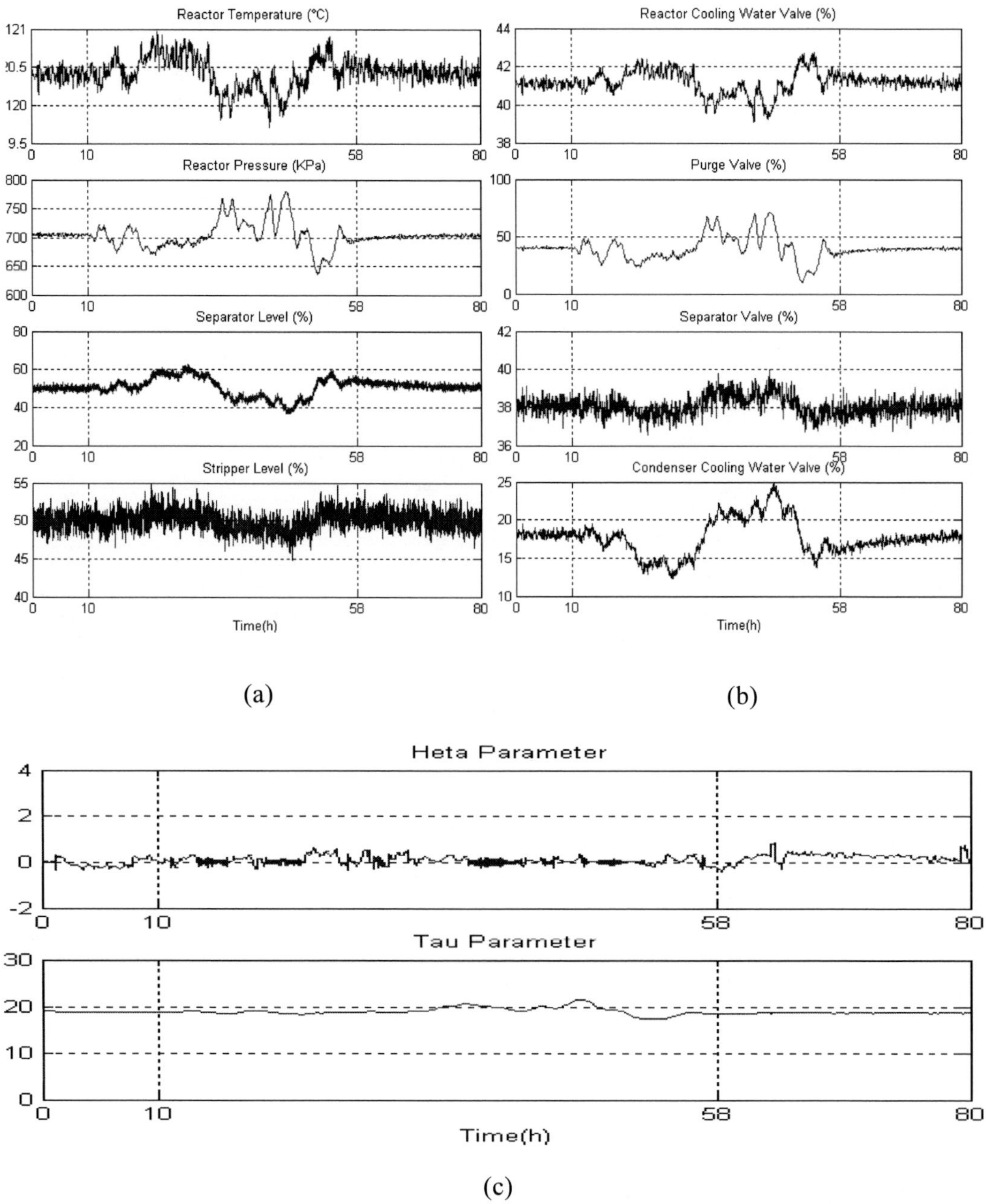

Figure 11. Simulation in presence of IDV (8) during 48 hours from t = 10 h to t = 58 h. a) Controlled variables b) Manipulated variables c) Adaptive parameters.

Table 5. Disturbance scenarii for the TECP

Disturbance	Variable under disturbance	Type of disturbance	IDNC	[40]
IDV(1)	A/C Feed ratio B composition	Step	Unstable	Stable
IDV(2)	B composition A/C ratio constant	Step	Stable	Stable
IDV(3)	D feed temperature	Step	Stable	Stable
IDV(4)	Reactor cooling water inlet temperature	Step	Stable	Stable
IDV(5)	Condenser cooling water inlet temperature	Step	Stable	Stable
IDV(6)	Loss of A feed	Step	Unstable	Unstable
IDV(7)	Loss of C header pressure – reduced availability	Step	Unstable	Stable
IDV(8)	A,B, C feed composition	Random variation	Stable	Stable
IDV(9)	D feed temperature	Random variation	Stable	Stable
IDV(10)	C feed temperature	Random variation	Stable	Stable
IDV(11)	Reactor cooling water inlet temperature	Random variation	Stable	Stable
IDV(12)	Condenser cooling water inlet temperature	Random variation	Stable	Stable
IDV(13)	Reaction kinetics	Slow drift	Stable	Unstable
IDV(14)	Reactor cooling water inlet valve	Sticking	Stable	Stable
IDV(15)	Condenser cooling water inlet valve	Sticking	Stable	Stable
IDV(16)	Unknown		Stable	Unstable
IDV(17)	Unknown		Stable	Unstable
IDV(18)	Unknown		Stable	Unstable
IDV(19)	Unknown		Stable	Stable
IDV(20)	Unknown		Stable	Unstable

5. Conclusion

Due to the large number of variables and dynamics and also to the interactions with a moving environment, the control design for plant wide systems appears as a complicated task. Behavioural models are often preferred because physical models are not available or reliable. In this context, neural networks have proved their efficiency to approximate the system behaviours as long as data can be used to train the controllers. In order to start the controller or to use it for changing environment, without any pre-training phase, an adaptation algorithm has been developed to update all parameters including the adaptive parameters and time constants that drive the dynamics of the controller. The proposed structure is efficient to control uncertain or unknown systems. The number of neurons only depends on the plant structure. No training phase is required because the autonomous adaptation of all networks parameters evolves from initial conditions set to zero. Stability and robustness analysis for closed loop systems are performed based on Lyapunov theory and sensitivity functions. These analysis lead to additive constraints that are included to the updating algorithm so that the control system keeps the variation of adaptation rate in acceptable range. Several simulations with the Tennessee Eastman Challenge Process illustrate the good performance of the proposed controller. But control design for plant wide systems remains an open issue and the necessity to integrate high level objectives like economical or ecological constraints in a framework will motivate our future works. The use of adaptive control for multi-level optimization with numerous environment constraints and several time horizons is probably one of most challenge in the near future.

References

[1] B.M. Åkesson, H.T. Toivonen, "A neural network model predictive controller", *Journal of Process Control*, 2006, vol. 16, pp 937 – 946.

[2] R.K. Al Seyab, Y. Cao, "Differential recurrent neural network based predictive control", *Computers & Chemical Engineering*, 2007.

[3] A.G. Barto, "Connectionist Learning for Control: An Overview", *Technical Report*, Dept. of Computer and Information Science, University of Massachusetts, Amherst, MA, 1989.

[4] V.M. Becerra, Z.H. Abu-el-zeet, P.D. Roberts, "Integrating predictive control and economic optimization", *Computing and control engineering journal*, 1999, pp.198-208.

[5] J. Bernussou, "Commande Robuste : Développements Et Applications", *Hermes Science Publications,* ISBN-13: 978-2866015008, 1996.

[6] Charara, J.D. Miras, B. Caron, "Nonlinear Control of a Magnetic Levitation System without Premagnetisation", *IEEE Transactions on Control Systems Technology, Special Issue on Magnetic Bearing Control,* 1996, vol. 5 (4), pp. 513-523.

[7] J. Chen, T-C. Huang, "Applying neural networks to on-line updated PID controllers for nonlinear process control", *Journal of Process Control*, 2004, vol. 14, pp 211 – 230.

[8] X. Cui, K.G. Shin, "Design of an industrial process controller using neural networks", *Proceedings of American Control Conference* 1992, vol. l, pp.6508-513.

[9] J.J. Downs, E.F. Vogel E.F., "A plant-wide industrial process control problem", *Computers and Chemical Engineering,* 1993, vol. 17 (3), pp.245-255.

[10] F. Druaux, E. Leclercq, D. Lefebvre, "Adaptive Neural Network Control for Uncertain or Unknown Non-linear Systems", *Proceedings of IEEE International Conference on Methods and Models in Automation and Robotics,* 2004, Miedzyzdroje, Poland, pp. 1309-1314.

[11] L. Ender, R. M. Filho, "Design of multivariable controller based on neural networks", *Computers and Chemical Engineering*, 2000, vol. 24, 937 – 943.

[12] S.S. Ge, C. Wang, Y.H. Tan, "Adaptive Control of Partially Known Nonlinear Multivariable Systems Using Neural Networks", *Proceedings of the IEEE International Symposium on Intelligent Control*, 2001, Mexico City, Mexico, pp. 292 – 297.

[13] S.S. Ge, C.C. Hang, T.H. Lee, T. Zhang, "Stable Adaptive Neural Network Control", *Springer*, ISBN: 0-7923-7597-1, 2002.

[14] S.S. Ge, C. Wang, "Direct Adaptive NN Control of a Class of Nonlinear Systems", *IEEE Transactions on Neural Networks,* 2002, vol. 13 (1), pp. 214-221.

[15] D. Grant Fisher, "Computer control of decentralized process systems", *Computers and Chemical Engineering*, 1983, vol. 7, pp. 395-421.

[16] Hamzaoui, N. Essounbouli, J. Zaytoon, "Commande par mode glissant flou d'un système non linéaire incertain, " *Revue électronique des Sciences et Technologies de l'Automatique, e-STA,* 2003, pp.88-95.

[17] Kanellakopoulos, P.V.Kokotovic, R. Morse, "Systematic design of adaptive controllers for feedback linearisable systems", *IEEE Transactions on Automatic Control*, 1991, vol. 36 (11), pp. 1241-1253.

[18] M. Kavchak, H. Budman, "Adaptive neural network structures for non-linear process estimation and control", *Computers and Chemical Engineering*, 1999, vol. 23, pp. 1209 - 1228.

[19] H.K. Khalil, "Lyapunov stability. In: Control Handbook", *William S. Levine,* 1996, pp. 889–895.

[20] M. Krstic, I. Kanellakopoulos, P. Kokotovic, "Nonlinear and Adaptive Control Design", *Wiley-Interscience*, ISBN-13 : 978-0471127321, 1995.

[21] D.A. Lawrence, W.J. Rugh, "Gain Scheduling Dynamic Linear Controllers for a Nonlinear Plant", *Automatica*, 1995, vol. 31, pp. 381-390.

[22] F. Leclercq, F. Druaux, D. Lefebvre, S. Zerkaoui, "Autonomous learning algorithm for fully connected recurrent networks", *Neurocomputing*, 2005, vol. 63, pp. 25-44.

[23] J. Lévine, J. Lottin, J.C. Ponsart, "A Nonlinear Approach to the Control of Magnetic Bearings", *IEEE Transactions on Control Systems Technology, Special Issue on Magnetic Bearing Control*, 1996, Vol. 5(4), pp. 524-544.

[24] T.J. McAvoy, N. Ye, "Base control for the Tennessee Eastman Problem", *Computers and Chemical Engineering,* 1994, vol. 18, pp. 383 – 413.

[25] M. Morari, A.K.W. Fung, "Nonlinear inferential control", *Computers and Chemical Engineering*, 1982, vol. 6, pp. 271-281.

[26] M. Morari, "Design of resilient processing plant – III: A general framework for the assessment of dynamic resilience", *Chemical Engineering Science*, 1983, vol. 3, pp. 1881-1891.

[27] W.T. Miller, F.H. Glanz, L.G. Kraft, "Real-Time Dynamic Control of an Industrial Manipulator Using a Neural-Network- Based Learning Controller", *IEEE Transactions on Robotics and Automation*, 1990, vol. 6 (1), pp. 2-9.

[28] M. Morari, G. Stephanopoulos, A. Rutherford, "Stability analysis via decomposition", *Chemical Engineering Science*, 1979, vol. 34, pp. 11-15.

[29] K.S. Narendra, K. Parthasarathy, "Identification and Control of Dynamical Systems Using Neural Networks", *IEEE Transactions on Neural Networks,* 1991, vol. 1 (1), pp. 4 – 27.

[30] M. Nørgaard O. Ravn, N.K. Poulsen, L.K. Hansen, "Neural Networks for Modeling and Control of Dynamic Systems", *Springer-Verlag*, ISBN-13: 978-1852332273, 2000.

[31] O. Omidvar and D.L., Elliott, "neural systems for control," *Academic Press*, ISBN: O-12-526430-5 1997..

[32] R. Ouiguini, R. Bouzid, Y. Sellami, "Une commande robuste par mode glissant flou Appliquée à la poursuite de trajectoire d'un robot mobile non holonome", *Conférence Internationale sur les Systèmes de Télécommunications, d'Electronique Médicale et d'Automatique,* Université de Tlemcen, 2003.

[33] Oustaloup, "Robustesse : analyse et synthèse de commandes robustes", *Hermes*, ISBN 2-86601-442-1, 1994.

[34] N.L. Ricker, "Optimal steady-state operation of the Tennessee Eastman Challenge Process", *Computers and Chemical Engineering*, 1995, vol. 19 (9), pp. 949-959.

[35] M. Saerens, A. Soquet, "Neural Controller Based on Back-Propagation Algorithm", *IEE Proceedings of Radar and Signal Processing,* 1991, vol. 138 (1), pp. 55-62.

[36] R.M. Sanner, J-J.E. Slotine, "Gaussian networks for direct adaptive control", *IEEE Transactions on Neural Networks,* 1992, vol. 3, pp.837-863.

[37] S. Sastry, M. Bodson, "Adaptive Control: Stability, Convergence, and Robustness", *Prentice-Hall*, ISBN 0-13-004326-5, 1994.

[38] M. Sekher, M. Farza, M. M'Saad, "Synthèse d'observateurs pour une classe de systèmes non linéaires MIMO", *Journées Doctorales et Nationales du GDR MACS*, 2005, Lyon, France.

[39] J.S. Shamma, M. Athans, "Analysis of Gain Scheduled Control for Nonlinear Plants", *IEEE Transactions on Automatic Control,* 1990, vol. 35 (8), pp. 898-907.

[40] Singhal, "Tennessee Eastman Plant Simulation with Base Control System of McAvoy and Ye", *Research report,* 2000, Dep. of Chemical Engineering, University of California, Santa Barbara, USA.

[41] J.C. Sozio, "Intelligent parameter adaptation for chemical processes", 143 p. *Thèse de Master of Science*, 1999, Faculty of the Virginia Polytechnic Institute and State University.

[42] G. Stephanopoulos, Y. Arkun, M.Morari, "A unified approach to the synthesis of control structures for complex chemical plants", Comp*uter and Chemical Engineering*, 1979, vol. 3, p. 573.

[43] L. Tian, C. Collins, "A dynamic recurrent neural network-based controller for a rigid–flexible manipulator system", *Mechatronics*, 2004, vol. 14, pp. 471–490.

[44] Z. Tian, K. A. Hoo, Multiple model-based control of the Tenessee – Eastman Process, *Ind. Eng. Chem Res*., 2005, vol. 44, pp. 3187 –3202.

[45] V. Uraikul, C.W. Chan, P. Tontiwachwuthikul, "Artificial intelligence for monitoring and supervisory control of process systems", *Engineering Applications of Artificial Intelligence*, 2007, vol. 20, pp. 115 – 131.

[46] Ch. Venkateswarlu, K. Venkat Rao, "Dynamic recurrent radial basis function network model predictive control of unstable nonlinear processes", *Chemical Engineering Science*, 2005, vol. 60, pp. 6718 – 6732.

[47] D. Wang, P. Bao, "Enhancing the estimation of plant Jacobian for adaptive neural inverse control", *Neurocomputing*, 2000, vol. 34, pp. 99-115.

[48] J. Wang, "Sensitivity identification enhanced control strategy for nonlinear process systems", *Computers and Chemical Engineering*, 2003, vol. 27, pp.1631-1640.

[49] R.J. Williams, D. Zipser, "A learning algorithm for continually running fully recurrent neural networks", *Neural Computation*, 1989, vol. 1(2), pp. 270–280.

[50] T.M. Yeh, M.C. Huang, C.T. Huang, "Estimate of process compositions and plantwide control from multiple secondary measurements using artificial neural networks", *Computers and Chemical Engineering*, 2003, vol. 27, pp. 55-72.

[51] D.L. Yu, T.K. Chang, D.W. Yu, "Adaptive neural model-based fault tolerant control for multi-variable processes", *Engineering Application of Artificial Intelligence*, 2005, vol. 18, pp. 393 – 411.

[52] D.L. Yu, T.K. Chang, D.W. Yu, "A stable self-learning PID control for multivariable time varying systems", *Control Engineering Practice*, 2007.

[53] D.W. Yu, D.L. Yu, "Multi-rate model predictive control of a chemical reactor based on three neural models", *Biochemical Engineering Journal*, 2007.

[54] S. Zerkaoui, F. Druaux, E. Leclercq, D. Lefebvre, "Stable adaptive control with recurrent neural networks for square MIMO non linear systems", *IFAC WC, International Federation of Automatic Control,* 2005, Prague, Tchèque.

[55] S. Zerkaoui, F. Druaux, E. Leclercq, D. Lefebvre, "Robust stability analysis for indirect neural Adaptive control", *Control 06, International Control Conference*, 2006, Glasgow, Scotland.

[56] S. Zerkaoui, "Adaptive neural controller for non linear systems", 2007, PhD Dissertation, University Le Havre, France.

[57] S. Zerkaoui, F. Druaux, E. Leclercq, D. Lefebvre, "Robust stability analysis of adaptive control based on recurrent ANN", *International Journal of Adaptive Infrastructures*, to appear 2007.

[58] S. Zerkaoui, F. Druaux, E. Leclercq, D. Lefebvre, "Multivariable adaptive control for non-linear systems : application to the Tennessee Eastman Challenge Process", *ECC,* 2007, Kos, Greece.

[59] M. Zhihong, H. R. Wu, M. Palaniswami, "An Adaptive Tracking Controller Using Neural Networks for a Class of Nonlinear Systems", *IEEE Transactions on Neural Networks*, 1998, vol. 9(5), pp. 947-955.

[60] D. Zipser, "A subgrouping strategy that reduces complexity and seeds up learning in recurrent networks", *Neural Computation,* 1989, vol. 1, pp. 552-558.

In: Progress in Nonlinear Analysis Research
Editor: Erik T. Hoffmann, pp. 209-241
ISBN 978-1-60456-359-7

Chapter 9

NONLINEAR DIFFUSION EQUATIONS WITH DISCONTINUOUS COEFFICIENTS IN POROUS MEDIA

Gabriela Marinoschi*
Institute of Mathematical Statistics and Applied Mathematics,
Calea 13 Septembrie 13, 050711 Bucharest, Romania

Abstract

In this work some mathematical aspects induced by the strongly nonlinearity of diffusion equations with convective terms modeling flows in porous media are investigated. Specifically, the interest lies in studying the properties of the solutions to some types of diffusion equations in which the diffusion coefficient and the convective term are nonlinear discontinuous functions of the solution. Particularly, this kind of equations can arise in soils science, describing the water infiltration in nonhomogeneous saturated-unsaturated soils characterized by strongly nonlinear hydraulic properties, but generally, they may be adequate for modeling the dynamics of various fluids in porous materials, as well as other physical diffusion processes, such as those arising in biology.

The mathematical approach is illustrated in the case of fast diffusion equations with flux and Robin boundary conditions and is developed in the framework of the theory of evolution equations with m-accretive nonlinear multivalued operators in Hilbert spaces. First, the study of the existence of the solution to an appropriate abstract approximating problem involving a quasi m-accretive operator will be done. Next, compactness results and a passing to the limit technique will prove the existence of the solution to the original problem. Additional properties of the solutions to some other models will be discussed. The theoretical results will be illustrated at the end by numerical applications to a real problem of water infiltration in nonlinear soils.

Mathematics Subject Classification (2000). Primary 35K55 - 47H06 - 76S05

Key Words: nonlinear parabolic PDE - m-accretive operators - flows in porous media

*E-mail address: gabimarinoschi@yahoo.com, gmarino@acad.ro

1. Introduction

The aim of this paper is to present a functional study of some mathematical models that can apply to physical processes of fluid diffusion in porous materials, whose typical representative is water infiltration in soils. Other applications concern more complex flows developing in biological materials or environments and the nonlinear heat propagation.

Apart from other diffusion processes, the diffusion of a fluid in a porous medium, consisting in a solid matrix and a void part, has a specific behavior in relation with the medium structure, specifically with the pore geometry and the volume of voids. Under certain conditions depending on the soil structure, the rate at which the fluid may be supplied on a part of the domain boundary, the initial distribution of the fluid in the soil, the presence of underground sources and the boundary permeability, a part or more of the flow domain starts to saturate. The pores in these regions can become at a certain time completely filled with the fluid. At the interface between the saturated and unsaturated parts a free boundary is formed and it begins to advance towards the unsaturated part. When all the pores in a part of the domain are completely filled with the fluid, its concentration in the pores attaints the maximum value, or the *saturation value,* which will be denoted by θ_s. This is a positive number equal to the value of the medium porosity and it is specific to each material.

Especially in the neighborhood of the saturation value the functions characterizing the process exhibit a particular behavior which imprints to the flow a different character from a more linear one up to a highly nonlinear one. On the other hand the possible inhomogeneities of the medium can induce some discontinuities in the behavior of the specific parameters. These characteristics will be considered in the models discussed in this paper.

Let us consider Ω an open bounded subset of $\mathbf{R}^N$, with a sufficient smooth boundary, $\Gamma = \partial\Omega$, e.g., of class C^1. We denote by $x = (x_1, ..., x_N)$ the space variable and by t the time running in the finite interval $(0, T)$. The models we shall discuss rely on the diffusion equation with a convective term, written for the unknown $\theta(x, t)$ as follows

$$\frac{\partial\theta}{\partial t} - \Delta\beta^*(\theta) + \nabla\cdot\mathbf{K}(\theta) \ni f \text{ for } (x,t) \in \Omega\times(0,T). \tag{1.1}$$

The unknown function θ can represent the volumetric content of the fluid in the pores or the temperature in a heat process. Roughly speaking, β^* is provided by the integral of the *diffusion coefficient*, denoted in the subsequent part by β, and $\mathbf{K}$ describes the contribution of the convection. The function f, depending on x and t, is a known source (or sink) in the domain. This equation models, among others, the diffusion with convection in a porous medium with a constant porosity. The particular influence that the porous medium has upon the diffusion is hidden in the properties of the functions β, β^* and $\mathbf{K}$ which will be presented in detail a little further. Concerning these properties we shall focus on the behavior of the functions around $\theta = \theta_s$ and on the possible discontinuities they may have.

First we refer to the vector $\mathbf{K} = \{K_j\}_{j=1,...,N}$, and state that its components are continuous on $\mathbf{R}$, except possibly at a finite number of points,

$$K_j \in C^0(\mathbf{R}\backslash\{r_1, ..., r_p\}),\ \{r_1, ..., r_p\} \subset (0, \theta_s), \tag{1.2}$$

at which they display a first species discontinuity, i.e., the lateral limits are finite but different.

Moreover, we consider that K_j are nonnegative functions, $K_j(r) \geq 0$ for $r \in \mathbf{R}$, which satisfy

$$K_j(r) = 0 \text{ for } r \leq 0 \text{ and } K_j(r) = K_s \text{ for } r \geq \theta_s,\ j = 1, ..., N, \tag{1.3}$$

where K_s is a positive number, $K_s = \max_{r \in \mathbf{R}} K_j(r) = K_j(\theta_s)$, for all $j = 1, ..., N$.

The assumptions (1.3) follow from the extensions of the functions K_j to the left of $r = 0$ and the right of $r = \theta_s$, by their values at $r = 0$ and $r = \theta_s$, respectively.

Finally we assume that each K_j is Lipschitz with constant M on the subsets $[0, r_1) \cup ...(r_k, r_{k+1}) \cup ...(r_p, \theta_s]$, for all $j = 1, ..., N$.

Since all components of $\mathbf{K}$ behave similarly and the fact that they might be different would not determine a major change in the proofs, we can set with no loss of generality

$$K_j = K \text{ for all } j = 1, ..., N \tag{1.4}$$

and denote the lateral limits at the discontinuity points by

$$K^l(r_j) = \lim_{r \nearrow r_j} K(r) \text{ and } K^r(r_j) = \lim_{r \searrow r_j} K(r),\ j = 1, ..., p. \tag{1.5}$$

By (1.3) $K^l(r_j)$ and $K^r(r_j)$ are positive numbers less than K_s.

The nonlinear function $\beta^* : (-\infty, \theta_s] \to \mathbf{R}$ and we assume it to be multivalued and defined by

$$\beta^*(\theta) = \begin{cases} \int_0^r \beta(\xi) d\xi \text{ if } r < \theta_s \\ [K_s^*, +\infty) \text{ if } r = \theta_s, \end{cases} \tag{1.6}$$

continuous on $(-\infty, \theta_s)$ including at the left of θ_s, where K_s^* is a known positive real number.

The function $\beta : (-\infty, \theta_s) \to \mathbf{R}$ is continuous except at the same points $\{r_1, ..., r_p\}$ considered for K_j, at which it has finite lateral limits. Moreover we require that

$$\beta(r) \geq \rho \text{ for any } r \in (-\infty, \theta_s),\ \beta(r) = \rho \text{ for } r \in (-\infty, 0] \tag{1.7}$$

and each piece is monotonically increasing. Also we denote

$$\begin{aligned} \beta^l(r_j) &= \lim_{r \nearrow r_j} \beta(r) \text{ and } \beta^r(r_j) = \lim_{r \searrow r_j} \beta(r), \\ \text{with } \beta^r(r_j) &\geq \beta^l(r_j),\ j = 1, ..., p. \end{aligned} \tag{1.8}$$

Concerning the behavior of β at $r = \theta_s$ we assert that this imprints the particular character of the diffusion equation (1.1) either of being closer to a linear one or having a highly nonlinear feature. To explain this we recall the classification made for diffusion equations by D. Aronson in [3] and adapt it in the case of diffusion in porous media. Thus, we notice that there are three classes of diffusion equations depending on the form of the functions β around θ_s,

$$\beta(r) = \frac{1}{(\theta_s - r)^{1-m}} \tag{1.9}$$

and on the parameter m. If $m \geq 1$ the diffusion is slow and we are led to the *porous media equation,* characterizing for example the diffusion of a gas in a porous medium.

If $m \in (0,1)$ the diffusion is fast and for $m \leq 0$ it is superfast. In soil science, β has more complicate expressions but which exhibit the properties of β given by (1.9) in various cases. The most used, which fit the data in the best way are those corresponding to the fast diffusion. For this reason we shall mainly focus on this case. However, we shall also describe what happens in other situations.

Thus we shall assume that β blows up at the saturation point

$$\lim_{r \nearrow \theta_s} \beta(r) = +\infty$$

and has the left limit of its integral finite

$$\lim_{r \nearrow \theta_s} \int_0^{\theta_s} \beta(r) = K_s^*,$$

situations which corresponds to (1.9) on (r_p, θ_s) with $m \in (0,1)$.

The situation in which $\lim_{r \to \theta_s} \beta(r)$ is finite is assigned to $m \geq 1$, namely to a slow diffusion.

The superdiffusion with $m \leq 0$ is characterized by a blowing up of both β and β^* at $r = \theta_s$.

In the monograph [24] the way in which these diffusion models were deduced for describing water infiltration in porous media is presented in detail.

Previous existence and uniqueness results for solutions to the elliptic-parabolic problems leading to equations of type (1.1) with continuous coefficients have been obtained using a technique inspired by the method of entropy solutions introduced by S. N. Krushkov in [18]. Studies following Krushkov's method are due, e.g., to H. W. Alt and S. Luckhaus (see [1]), Ph. Bénilan et al., (see [6], [7]), J. Carillo (see [11], [12]), F. Otto (see [27]). We also recall the papers [2], [8], [14], [15], [16], [26] dealing with various mathematical aspects of the models with continuous coefficients.

Nonlinear problems of type (1.1) with β^* blowing up at θ_s have been studied for various boundary conditions in [20] and [5] for the superdiffusion model with continuous coefficients, and in [21], [23] for β^* multivalued and K Lipschitz. The monograph [24] gathers basic results concerning especially the fast diffusion equation with a convective Lipschitz term and their applications to water diffusion in soils.

To end this brief presentation of the state of the art we have to mention that a special attention has been devoted to various aspects in diffusion problems by J. L. Vázquez and we cite here some titles from his extended work on this topic: [13], [29], [30] and the monographs [32], dealing especially with the study of porous media equation, and [31] which focuses on the large time behavior of solutions to nonlinear diffusion equations.

In this chapter the mathematical treatment is developed in the framework of the theory of evolution equations with m-accretive nonlinear multivalued operators in Hilbert spaces and will mainly refer to the model of fast diffusion. After setting the appropriate functional framework, the original boundary value problem is described by an equivalent abstract Cauchy problem with a multivalued nonlinear operator. Its presence imposes the study of the well-posedness of the solution to an appropriate abstract approximating problem involving a quasi m-accretive operator. Next, compactness results and a passage to the limit technique will prove the existence and uniqueness of the solution to the original problem. Some possible proof differences occurring in other models will be briefly discussed. Further, results on the solution asymptotic behavior at large time will be developed. The theoretical results will be illustrated at the end by numerical applications to a real problem of water infiltration in nonlinear soils.

In this perspective this chapter is organized in 6 sections including, besides this introduction:

2. Statement of the problem
3. Main results for the approximating problem
4. Main results for the original problem
5. Comments on other models
6. Numerical results.

2. Statement of the Problem

Assume that the boundary $\Gamma = \partial\Omega$ of the flow domain Ω is piecewise smooth and formed by the disjoint parts Γ_u and Γ_α, such that

$$\Gamma = \Gamma_u \cup \overline{\Gamma_\alpha},\ \Gamma_u \cap \Gamma_\alpha = \emptyset.$$

The theory will be illustrated by considering boundary conditions of flux and Robin type, the latter characterizing a variable permeability of the boundary Γ_α, described by the function α.

Noticing that each term in the sum $\nabla \cdot \mathbf{K}(\theta) = \sum\limits_{j=1}^{N} \frac{\partial K(\theta)}{\partial x_j}$ brings the same contribution from the mathematical point of view, we shall consider, for simplicity, that $\mathbf{K} = (0, 0, ..., K)$ and specify at the end how the result changes in the case of more nonzero components.

Therefore, the diffusion model with convection we shall deal with, reads

$$\frac{\partial\theta}{\partial t} - \Delta\beta^*(\theta) + \frac{\partial K(\theta)}{\partial x_N} \ni f \text{ in } Q = \Omega \times (0, T), \tag{2.1}$$

$$\theta(x, 0) = \theta_0(x) \text{ in } \Omega, \tag{2.2}$$

$$(K(\theta)i_N - \nabla\beta^*(\theta)) \cdot \nu \ni u \text{ on } \Sigma_u = \Gamma_u \times (0, T), \tag{2.3}$$

$$(K(\theta)i_N - \nabla\beta^*(\theta)) \cdot \nu - \alpha\beta^*(\theta) \ni f_0 \text{ on } \Sigma_\alpha = \Gamma_\alpha \times (0, T). \tag{2.4}$$

Here ν is the outward normal to Γ, i_N is the unit vector along Ox_N and f_0 and u are known on Σ_α and Σ_u, respectively.

Equation (2.3) expresses the continuity of the normal component of the inflow flux and (2.4) describes that the outflow through the semipermeable boundary is directly proportional to the water diffusivity and to the permeability of the boundary.

The properties assumed in this work for these functions are explained below.

2.1. Basic Assumptions

We can gather the assumptions concerning the functions β, β^* and K, made at the beginning in the following mathematical relationships:

(i_β) $\beta(r) \geq \rho > 0$, for any $r \in (-\infty, \theta_s)$, $\beta(r) = \rho$ for $r \in (-\infty, 0]$;

(i_β) $\lim_{r \nearrow \theta_s} \beta(r) = +\infty$;

(iii_β) $K_s^* = \lim_{r \nearrow \theta_s} \int_0^r \beta(\xi) d\xi,\ 0 < K_s^* < \infty$,

implying

(i) $(\beta^*(r) - \beta^*(\overline{r}))(r - \overline{r}) \geq \rho(r - \overline{r})^2$, for any $r, \overline{r} \in (-\infty, \theta_s]$;

(ii) $\lim\limits_{r \to -\infty} \beta^*(r) = -\infty$;

(iii) $\lim_{r \nearrow \theta_s} \beta^*(r) = K_s^*$.

For K we consider (1.3) and that there exists a positive constant M such that

(i_K) $|K(r) - K(\overline{r})| \leq M |r - \overline{r}|$, for any $r, \overline{r}$ belonging to each separate interval $[0, r_1), ...(r_k, r_{k+1}), ...(r_p, \theta_s]$.

Finally, we assume that $\alpha : \Gamma_\alpha \to [\alpha_m, \alpha_M]$ is positive and continuous

$$0 < \alpha_m \leq \alpha(x) \leq \alpha_M. \tag{2.5}$$

2.2. Functional Framework

For the sake of simplicity we shall denote the scalar product and the norm in $L^2(\Omega)$ by $(\cdot, \cdot)$ and $\|\cdot\|$, respectively. Also, sometimes we shall no longer write in the integrands those function arguments which represent the integration variables.

The problem will be treated within the functional framework represented by $V = H^1(\Omega)$ with the norm defined by

$$\|\psi\|_V = \left(\int_\Omega |\nabla \psi|^2 \, dx + \int_{\Gamma_\alpha} \alpha(x) \, |\psi|^2 \, d\sigma \right)^{1/2}, \tag{2.6}$$

and its dual, V'. It can be easily checked that the norm (2.6) is equivalent to the standard Hilbertian norm on $H^1(\Omega)$. Indeed, on the one hand, we have via the trace theorem that

$$\|\psi\|_V \leq c_{VH} \|\psi\|_{H^1(\Omega)}, \text{ with } c_{VH}^2 = 1 + \alpha_M c_{tr}^2, \tag{2.7}$$

where c_{tr} is the constant occurring in the trace theorem.

On the other hand, since $\Gamma_\alpha \subset \Gamma$ and $\text{meas}(\Gamma_\alpha) \neq 0$, we have from the Poincaré inequality that

$$\|\psi\|_{H^1(\Omega)}^2 \leq c_P(\|\psi\|_{L^2(\Gamma_\alpha)}^2 + \|\nabla\psi\|^2) \leq c_P \left(\frac{1}{\alpha_m} \|\psi\|_V^2 + \|\nabla\psi\|^2 \right) \leq c_H^2 \|\psi\|_V^2,$$

where $c_H^2 = c_P(1 + \frac{1}{\alpha_m})$. Hence

$$\|\psi\|_{H^1(\Omega)} \leq c_H \|\psi\|_V \tag{2.8}$$

which together with (2.7) implies that the norms in $H^1(\Omega)$ and V are equivalent. We also retain that

$$\|\psi\|_{L^2(\Gamma_\alpha)} \le c_{\Gamma_\alpha} \|\psi\|_V, \text{ with } c_{\Gamma_\alpha} = \frac{1}{\sqrt{\alpha_m}}, \tag{2.9}$$

implied by (2.6) and

$$\|\psi\|_{L^2(\Gamma_u)} \le \|\psi\|_{L^2(\Gamma)} \le c_{tr} \|\psi\|_{H^1(\Omega)} \le c_{\Gamma_u} \|\psi\|_V, \text{ with } c_{\Gamma_u} = c_H c_{tr}, \tag{2.10}$$

which follows by the trace theorem.

We must underline that c_H^2, $c_{\Gamma_\alpha}^2$ and $c_{\Gamma_u}^2$ depend on $\frac{1}{\alpha_m}$ and recall that we have assumed that $\alpha_m > 0$.

We endow the dual V' with the scalar product

$$(\theta, \overline{\theta})_{V'} = \theta(\psi), \ \forall \theta, \overline{\theta} \in V', \tag{2.11}$$

where $\psi \in V$ is the solution to the boundary value problem

$$-\Delta\psi = \overline{\theta}, \ \frac{\partial \psi}{\partial \nu} + \alpha\psi = 0 \text{ on } \Gamma_\alpha, \ \frac{\partial \psi}{\partial \nu} = 0 \text{ on } \Gamma_u. \tag{2.12}$$

Here, $\frac{\partial}{\partial \nu}$ denotes the normal derivative and $\theta(\psi)$ represents the value of $\theta \in V'$ at $\psi \in V$, or the pairing between V' and V, denoted also by $\langle \cdot, \cdot \rangle_{V',V}$. Further, we shall introduce the definition of the solution to (2.1)-(2.4) in a generalized sense, defining first a new multivalued function

$$\mathcal{K}(r) = \begin{cases} K(r), & r < r_1 \\ [K^l(r_1), K^r(r_1)], & r = r_1 \\ K(r), & r_1 < r < r_2 \\ \dots & \\ [K^l(r_p), K^r(r_p)], & r = r_p \\ K(r), & r_p < r. \end{cases} \tag{2.13}$$

Definition 2.1. Let

$$\theta_0 \in L^2(\Omega), \ \theta_0 \le \theta_s \text{ a.e. } x \in \Omega,$$

$$f \in L^2(0,T; V'), \ u \in L^2(0,T; L^2(\Gamma_u)), \ f_0 \in L^2(0,T; L^2(\Gamma_\alpha)).$$

We mean by a *solution* to (2.1)-(2.4) a function $\theta \in C([0,T]; L^2(\Omega)) \cap L^2(0,T;V)$, such that

$$\frac{d\theta}{dt} \in L^2(0,T;V'), \tag{2.14}$$

$$\theta(x,t) \le \theta_s \text{ a.e. } (x,t) \in Q, \tag{2.15}$$

which satisfies the initial condition

$$\theta(x,0) = \theta_0 \text{ in } \Omega, \tag{2.16}$$

and the equation

$$\begin{aligned}
&\left\langle \frac{d\theta}{dt}(t), \psi \right\rangle_{V',V} + \int_\Omega \left(\nabla\eta(t)\cdot\nabla\psi - \kappa(t)\frac{\partial\psi}{\partial x_N} \right) dx \qquad (2.17)\\
&= \langle f(t), \psi\rangle_{V',V} - \int_{\Gamma_\alpha} (\alpha\eta(t) + f_0(t))\psi d\sigma - \int_{\Gamma_u} u(t)\psi d\sigma,\\
\text{a.e. } t &\in (0,T), \text{ for any } \psi \in V,\\
\text{where } \eta &\in L^2(0,T;V) \text{ and } \kappa \in L^2(0,T;L^2(\Omega)) \text{ are such that}\\
\eta(x,t) &\in \beta^*(\theta(x,t)) \text{ a.e. } (x,t)\in Q \text{ and } \kappa \in \mathcal{K}(\theta(x,t)) \text{ a.e. } (x,t) \in Q.
\end{aligned}$$

By $\frac{d\theta}{dt}$ we mean the strong derivative of $\theta(t)$ in V' (equivalently the derivative in the sense of the V'-valued distributions on $(0,T)$). An equivalent form of (2.17) is

$$\begin{aligned}
&\int_0^T \left\langle \frac{d\theta}{dt}(t), \phi \right\rangle_{V',V} dt + \int_Q \left(\nabla\eta\cdot\nabla\phi - \kappa\frac{\partial\phi}{\partial x_N} \right) dxdt \qquad (2.18)\\
&= \int_0^T \langle f(t), \phi\rangle_{V',V}\, dt - \int_{\Sigma_\alpha} (\alpha\eta + f_0)\phi d\sigma dt - \int_{\Sigma_u} u\phi d\sigma dt,
\end{aligned}$$

for any $\phi \in L^2(0,T;V)$ and for some sections $\eta(x,t) \in \beta^*(\theta(x,t))$ and $\kappa(x,t) \in \mathcal{K}(\theta(x,t))$ a.e. on Q.

The solution introduced by the previous definition turns out to be a solution in the sense of distributions to (2.1) which satisfies the boundary conditions (2.3)-(2.4) in the sense of the trace theory (see [19]).

The arguments for the last assertions can be seen in [24].

Remark 2.2. By (i), the assumption $\zeta \in V$ where $\zeta(x) \in \beta^*(\theta(x))$ implies $\theta \in V$. This follows from the fact that the inverse of β^* is a Lipschitz function with the constant $\frac{1}{\rho}$.

It is also obvious that the boundedness of K implies that $\mathcal{K}(\theta) \in L^2(\Omega)$.

We shall write now problem (2.1)-(2.4) as an abstract Cauchy problem. To this end we define the multivalued operator $A : D(A) \subset V' \to V'$ by

$$\begin{aligned}
\langle A\theta, \psi\rangle_{V',V} &= \int_\Omega \left(\nabla\eta\cdot\nabla\psi - K(\theta)\frac{\partial\psi}{\partial x_N} \right) dx + \int_{\Gamma_\alpha} \alpha\eta\psi d\sigma, \qquad (2.19)\\
\text{for any } \psi &\in V \text{ and some } \eta \in \beta^*(\theta),
\end{aligned}$$

where

$$D(A) = \{\theta \in L^2(\Omega);\ \exists\eta \in V,\ \eta(x) \in \beta^*(\theta(x)),\ \text{a.e. } x \in \Omega\}. \qquad (2.20)$$

Moreover, we define the operator $B \in L(L^2(\Gamma_u); V')$ and the function $f_\Gamma \in L^2(0,T;V')$ by

$$Bu(\psi) = -\int_{\Gamma_u} u\psi d\sigma,\ \forall\psi \in V, \qquad (2.21)$$

$$f_\Gamma(t)(\psi) = -\int_{\Gamma_\alpha} f_0\psi d\sigma,\ \forall\psi \in V. \qquad (2.22)$$

With this notation we are led to the Cauchy problem

$$\frac{d\theta}{dt} + A\theta \ni f + Bu + f_\Gamma, \text{ a.e. } t \in (0,T), \tag{2.23}$$

$$\theta(0) = \theta_0. \tag{2.24}$$

Since the operator A is multivalued we have to consider an auxiliary problem by replacing β^* and K by more regular functions. This can be done by a standard regularization, using mollifiers, but in the proofs and in the perspective of numerical computations it is sufficient to use the functions presented below.

3. Main Results for the Approximating Problem

We approximate the multivalued function β^* by the continuous function defined for each $\varepsilon > 0$ by

$$\beta_\varepsilon^*(r) = \begin{cases} \beta^*(r), & r < \theta_s - \varepsilon \\ \beta^*(\theta_s - \varepsilon) + \frac{K_s^* - \beta^*(\theta_s - \varepsilon)}{\varepsilon}[r - (\theta_s - \varepsilon)], & r \geq \theta_s - \varepsilon, \end{cases} \tag{3.1}$$

so that, besides the properties (i) (for $r, \overline{r} \in \mathbf{R}$) and (ii), $\beta_\varepsilon^*(r)$ satisfies

(iv) $\lim_{r \to \infty} \beta_\varepsilon^*(r) = +\infty.$

This function is continuous on $\mathbf{R}$, differentiable on $\mathbf{R}\backslash\{r_1, ..., r_p, \theta_s - \varepsilon\}$ but at the left and right of these points the derivatives exist and are finite. On the open intervals between these points its derivative denoted by β_ε reads

$$\beta_\varepsilon(r) = \begin{cases} \beta(r), & r < \theta_s - \varepsilon \\ \frac{K_s^* - \beta^*(\theta_s - \varepsilon)}{\varepsilon}, & r > \theta_s - \varepsilon, \end{cases}$$

and it is monotonically increasing, such that

$$\max_{r \in \mathbf{R}} \beta_\varepsilon(r) = \frac{K_s^* - \beta^*(\theta_s - \varepsilon)}{\varepsilon} \tag{3.2}$$

for each ε positive.

Assume that all pieces of K are monotonically increasing and $K^r(r_j) \geq K^l(r_j)$. The function K will be replaced by a Lipschitz function K_ε as follows

$$K_\varepsilon(r) = \begin{cases} K(r), & r < r_1 - \varepsilon \\ K(r_1 - \varepsilon) + \frac{K^r(r_1) - K(r_1 - \varepsilon)}{\varepsilon}[r - (r_1 - \varepsilon)], & r_1 - \varepsilon \leq r \leq r_1 \\ K(r), & r_1 \leq r \leq r_2 - \varepsilon \\ K(r_2 - \varepsilon) + \frac{K^r(r_2) - K(r_2 - \varepsilon)}{\varepsilon}[r - (r_2 - \varepsilon)], & r_2 - \varepsilon \leq r \leq r_3 \\ \ldots & \\ K(r_p - \varepsilon) + \frac{K^r(r_p) - K(r_p - \varepsilon)}{\varepsilon}[r - (r_p - \varepsilon)], & r_p - \varepsilon \leq r \leq r_p \\ K_s, & r > r_p. \end{cases} \tag{3.3}$$

It follows that for each $\varepsilon > 0$, K_ε is bounded

$$0 \leq K_\varepsilon(r) \leq K_s \tag{3.4}$$

and has the derivative bounded by

$$M_\varepsilon = \max_{j=1,\dots,p}\left\{\left|\frac{K^r(r_j)-K(r_j-\varepsilon)}{\varepsilon}\right|\right\}. \tag{3.5}$$

Therefore, the Lipschitz function K_ε satisfies

(ii$_K$) $|K_\varepsilon(r)-K_\varepsilon(\overline{r})| \leq M_\varepsilon |r-\overline{r}|$, for any $r,\overline{r}\in \mathbf{R}$ and any ε positive.

In other situations, as for example when K is monotonically increasing on (r_{j-1},r_j) but $K^l(r_j)\geq K^r(r_j)$, we may replace K by a segment connecting the points $(r_j, K^l(r_j)$ and $(r_j+\varepsilon, K(r_j+\varepsilon))$. A similar replacement will be done if K is monotonically decreasing on (r_{j-1},r_j) with $K^l(r_j)\leq K^r(r_j)$.

With these considerations we can introduce the approximating problem

$$\frac{d\theta_\varepsilon}{dt}+A_\varepsilon\theta_\varepsilon = f+Bu+f_\Gamma, \text{ a.e. } t\in(0,T), \tag{3.6}$$

$$\theta_\varepsilon(0)=\theta_0, \tag{3.7}$$

where $A_\varepsilon : D(A_\varepsilon)\subset V'\to V'$ is the single-valued operator defined by

$$\langle A_\varepsilon\theta,\psi\rangle_{V',V} = \int_\Omega\left(\nabla\beta_\varepsilon^*(\theta)\cdot\nabla\psi - K_\varepsilon(\theta)\frac{\partial\psi}{\partial x_N}\right)dx+\int_{\Gamma_\alpha}\alpha\beta_\varepsilon^*(\theta)\psi d\sigma,\ \forall\psi\in V, \tag{3.8}$$

with the domain

$$D(A_\varepsilon)=\{\theta\in L^2(\Omega);\ \beta_\varepsilon^*(\theta)\in V\}.$$

Immediately we notice that $D(A_\varepsilon)=V$ for each $\varepsilon>0$.

Obviously, the strong solution to (3.6)-(3.7) is the solution in the generalized sense (similar to that of Definition 1.1) to the boundary value problem

$$\frac{\partial\theta_\varepsilon}{\partial t}-\Delta\beta_\varepsilon^*(\theta_\varepsilon)+\frac{\partial K_\varepsilon(\theta_\varepsilon)}{\partial x_N} = f \text{ in } Q, \tag{3.9}$$

$$\theta_\varepsilon(x,0)=\theta_0(x) \text{ in } \Omega, \tag{3.10}$$

$$(K_\varepsilon(\theta_\varepsilon)i_N-\nabla\beta_\varepsilon^*(\theta_\varepsilon))\cdot\nu = u \text{ on } \Sigma_u, \tag{3.11}$$

$$(K_\varepsilon(\theta_\varepsilon)i_N-\nabla\beta_\varepsilon^*(\theta_\varepsilon))\cdot\nu = \alpha\beta_\varepsilon^*(\theta_\varepsilon)+f_0 \text{ on } \Sigma_\alpha. \tag{3.12}$$

3.1. Existence and Properties of the Approximating Solution

The existence of the solution to the approximating problem relies on the quasi m-accretivity of the operator A_ε, proved below.

Proposition 3.1. *Let $\varepsilon>0$ be fixed. Under the hypotheses* (i)-(ii),(iv) *and* (ii$_K$) *the operator A_ε is quasi m-accretive on V'.*

Proof. Let λ be a positive real number. We have to prove that

$$\left((\lambda I+A_\varepsilon)\theta-(\lambda I+A_\varepsilon)\overline{\theta},\theta-\overline{\theta}\right)_{V'}\geq 0 \tag{3.13}$$

and

$$R(\lambda I+A_\varepsilon)=V', \tag{3.14}$$

for λ large enough. We have

$$\begin{aligned}
&((\lambda I + A_\varepsilon)\theta - (\lambda I + A_\varepsilon)\overline{\theta}, \theta - \overline{\theta})_{V'} \\
= \ & \lambda \left\|\theta - \overline{\theta}\right\|_{V'}^2 + \int_\Omega \nabla(\beta_\varepsilon^*(\theta) - \beta_\varepsilon^*(\overline{\theta})) \cdot \nabla\psi dx \\
& - \int_\Omega (K_\varepsilon(\theta) - K_\varepsilon(\overline{\theta}))\frac{\partial\psi}{\partial x_N}dx + \int_{\Gamma_\alpha} \alpha(\beta_\varepsilon^*(\theta) - \beta_\varepsilon^*(\overline{\theta}))\psi d\sigma,
\end{aligned}$$

where

$$-\Delta\psi = \theta - \overline{\theta},\ \frac{\partial\psi}{\partial\nu} + \alpha\psi = 0 \text{ on } \Gamma_\alpha \text{ and } \frac{\partial\psi}{\partial\nu} = 0 \text{ on } \Gamma_u.$$

Using Green's formula, (i) and (ii$_K$) we obtain

$$\begin{aligned}
&((\lambda I + A_\varepsilon)\theta - (\lambda I + A_\varepsilon)\overline{\theta}, \theta - \overline{\theta})_{V'} \\
= \ & \lambda \left\|\theta - \overline{\theta}\right\|_{V'}^2 + \int_\Omega (\beta_\varepsilon^*(\theta) - \beta_\varepsilon^*(\overline{\theta}))(\theta - \overline{\theta})dx - \int_\Omega (K_\varepsilon(\theta) - K_\varepsilon(\overline{\theta}))\frac{\partial\psi}{\partial x_N}dx \\
\geq \ & \lambda \left\|\theta - \overline{\theta}\right\|_{V'}^2 + \rho \left\|\theta - \overline{\theta}\right\|^2 - M_\varepsilon \left\|\theta - \overline{\theta}\right\| \left\|\theta - \overline{\theta}\right\|_{V'},
\end{aligned}$$

whence

$$\begin{aligned}
&((\lambda I + A_\varepsilon)\theta - (\lambda I + A_\varepsilon)\overline{\theta}, \theta - \overline{\theta})_{V'} \\
\geq \ & \left(\lambda - \frac{M_\varepsilon^2}{2\rho}\right) \left\|\theta - \overline{\theta}\right\|_{V'}^2 + \frac{\rho}{2} \left\|\theta - \overline{\theta}\right\|^2 \geq 0,
\end{aligned} \tag{3.15}$$

for λ large enough, $\lambda \geq \frac{M_\varepsilon^2}{2\rho}$.

To prove the m-accretivity we must show that for every $g \in V'$ there exists $\theta \in D(A_\varepsilon)$ solution to

$$\lambda\theta + A_\varepsilon\theta = g. \tag{3.16}$$

Let us denote $\zeta = \beta_\varepsilon^*(\theta) \in V$. Because β_ε^* is continuous and monotonically increasing on $(-\infty, \infty)$ and $Range(\beta_\varepsilon^*) = (-\infty, \infty)$, it follows that its inverse $G(\zeta) = (\beta_\varepsilon^*)^{-1}(\zeta)$ is Lipschitz, by (i), hence it is continuous from V to $L^2(\Omega)$, i.e.,

$$\left\|G(\zeta) - G(\overline{\zeta})\right\| \leq \frac{1}{\rho} \left\|\zeta - \overline{\zeta}\right\| \leq \frac{c_H}{\rho} \left\|\zeta - \overline{\zeta}\right\|_V. \tag{3.17}$$

Therefore, (3.16) can be rewritten as

$$\lambda G(\zeta) + A_G\zeta = g \tag{3.18}$$

with $A_G : V \to V'$ defined by

$$\begin{aligned}
&\langle A_G\zeta, \psi\rangle_{V',V} \\
= \ & \int_\Omega \nabla\zeta \cdot \nabla\psi dx - \int_\Omega K_\varepsilon(G(\zeta))\frac{\partial\psi}{\partial x_N}dx + \int_{\Gamma_\alpha} \alpha\zeta\psi d\sigma,\ \forall\psi \in V.
\end{aligned} \tag{3.19}$$

We are going to prove that $\lambda G + A_G$ is surjective. We have, in virtue of assumptions (i_β), (i) and (ii_K), that

$$\begin{aligned}
&\langle(\lambda G + A_G)\zeta - (\lambda G + A_G)\overline{\zeta}, \zeta - \overline{\zeta}\rangle_{V',V} \\
= &\ \lambda \int_\Omega (G(\zeta) - G(\overline{\zeta}))(\zeta - \overline{\zeta})dx + \int_\Omega \left|\nabla(\zeta - \overline{\zeta})\right|^2 dx \\
&- \int_\Omega (K_\varepsilon(G(\zeta)) - K_\varepsilon(G(\overline{\zeta})))\frac{\partial(\zeta - \overline{\zeta})}{\partial x_N}dx + \int_{\Gamma_\alpha} \alpha(\zeta - \overline{\zeta})^2 d\sigma \\
\geq &\ \lambda\rho \int_\Omega (G(\zeta) - G(\overline{\zeta}))^2 dx + \left\|\zeta - \overline{\zeta}\right\|_V^2 - M_\varepsilon \left\|G(\zeta) - G(\overline{\zeta})\right\| \left\|\zeta - \overline{\zeta}\right\|_V \\
\geq &\ \left(\lambda\rho - \frac{M_\varepsilon^2}{2}\right)\left\|G(\zeta) - G(\overline{\zeta})\right\|^2 + \frac{1}{2}\left\|\zeta - \overline{\zeta}\right\|_V^2 \geq 0,
\end{aligned}$$

for λ large enough (still for $\lambda \geq \frac{M_\varepsilon^2}{2\rho}$), so that $\lambda G + A_G$ is strongly monotone. This implies immediately that it is coercive, too. By (3.17) it follows that the function $\zeta \to K_\varepsilon(G(\zeta))$ is continuous on $L^2(\Omega)$, because

$$\left\|K_\varepsilon(G(\zeta)) - K_\varepsilon(G(\overline{\zeta}))\right\| \leq M_\varepsilon \left\|G(\zeta) - G(\overline{\zeta})\right\| \leq \frac{M_\varepsilon}{\rho}\left\|\zeta - \overline{\zeta}\right\|. \tag{3.20}$$

Thus, the operator $\lambda G + A_G$ is continuous from V to V', monotone and coercive and on the basis of Minty's theorem it is surjective, proving that (3.18) has a unique solution. This ends the proof of the quasi m-accretivity of A_ε.

Let us denote

$$j_\varepsilon(r) = \int_0^r \beta_\varepsilon^*(\xi)d\xi, \text{ for any } r \in \mathbf{R}. \tag{3.21}$$

It follows that j_ε is a proper, convex and (semi)continuous function and

$$\partial j_\varepsilon(r) = \beta_\varepsilon^*(r), \text{ for any } r \in \mathbf{R}. \tag{3.22}$$

Theorem 3.2 (existence and uniqueness in the approximating problem) *Let*

$$f \in W^{1,1}(0,T;V'),\ f_0 \in W^{1,1}(0,T;L^2(\Gamma_\alpha)),\ u \in W^{1,1}(0,T;L^2(\Gamma_u)), \tag{3.23}$$

$$\theta_0 \in D(A_\varepsilon) \tag{3.24}$$

hold and let us assume the hypotheses (i)-(ii),(iv), (1.3) *and* (ii_K). *Then, for each* $\varepsilon > 0$, *there exists a unique strong solution* $\theta_\varepsilon \in C([0,T];V')$ *to problem* (3.6)-(3.7) *such that*

$$\theta_\varepsilon \in W^{1,\infty}(0,T;V') \cap L^\infty(0,T;D(A_\varepsilon)) \cap L^\infty(0,T;V), \tag{3.25}$$

$$\beta_\varepsilon^*(\theta_\varepsilon) \in L^\infty(0,T;V), \tag{3.26}$$

$$j_\varepsilon(\theta_\varepsilon) \in L^\infty(0,T;L^1(\Omega)). \tag{3.27}$$

Moreover, the solution satisfies the estimate

$$\int_\Omega j_\varepsilon(\theta_\varepsilon(x,t))dx + \int_0^t \left\| \frac{d\theta_\varepsilon}{d\tau}(\tau)\right\|_{V'}^2 d\tau + \int_0^t \|\beta_\varepsilon^*(\theta_\varepsilon(\tau))\|_V^2\, d\tau \tag{3.28}$$

$$\leq \gamma_0(\alpha_m)\left(\int_\Omega j_\varepsilon(\theta_0(x))dx + \int_0^T \|f(\tau)\|_{V'}^2\, d\tau \right.$$

$$\left. + \int_0^T \|u(\tau)\|_{L^2(\Gamma_u)}^2\, d\tau + \int_0^T \|f_0(\tau)\|_{L^2(\Gamma_\alpha)}^2\, d\tau + 1\right),$$

which holds for each $\varepsilon > 0$ *and any* $t \in [0,T]$, *with* γ_0 *independent of* ε.

Finally, if θ_ε *and* $\overline{\theta_\varepsilon}$ *are two solutions to problem* (3.6)-(3.7) *corresponding to the pairs of data* $\{\theta_0, f, f_0, f_\Gamma, u\}$ *and* $\{\overline{\theta_0}, \overline{f}, \overline{f_0}, \overline{f_\Gamma}, \overline{u}\}$, *respectively, we have the estimate*

$$\left\|\theta_\varepsilon(t) - \overline{\theta_\varepsilon}(t)\right\|_{V'}^2 + \int_0^t \left\|\theta_\varepsilon(\tau) - \overline{\theta_\varepsilon}(\tau)\right\|^2 d\tau \tag{3.29}$$

$$\leq \gamma_1(\varepsilon,\alpha_m)\left(\left\|\theta_0 - \overline{\theta_0}\right\|_{V'}^2 + \int_0^T \left\|f(\tau) - \overline{f}(\tau)\right\|_{V'}^2 d\tau \right.$$

$$\left. + \int_0^T \|u(\tau) - \overline{u}(\tau)\|_{L^2(\Gamma_u)}^2\, d\tau + \int_0^T \left\|f_0(\tau) - \overline{f_0}(\tau)\right\|_{L^2(\Gamma_\alpha)}^2 d\tau\right),$$

for $\varepsilon > 0$ *and any* $t \in [0,T]$, *where* γ_1 *depends also on* ε.

Proof. By the trace theorem we notice first that $Bu + f_\Gamma + f \in W^{1,1}(0,T;V')$ and $Bu \in W^{1,1}(0,T;V')$.

Since A_ε is quasi m-accretive, $\theta_0 \in D(A_\varepsilon)$ and $Bu + f_\Gamma + f \in W^{1,1}(0,T;V')$, the existence of the solution follows from the general results concerning the evolution equations with m-accretive operators in Hilbert spaces (see [4], [10]). From Remark 2.2 we get that $\theta_\varepsilon \in L^\infty(0,T;V)$.

Concerning the estimate (3.28) we multiply equation (3.6) scalarly in V' by $\frac{d\theta_\varepsilon}{d\tau}$ and integrate over $(0,t)$, with $t \in [0,T]$, T finite. We obtain that

$$\int_0^t \left\|\frac{d\theta_\varepsilon}{d\tau}(\tau)\right\|_{V'}^2 d\tau + \int_0^t\int_\Omega \nabla\beta_\varepsilon^*(\theta_\varepsilon)\cdot\nabla\psi dx d\tau + \int_0^t\int_{\Gamma_a} \alpha\beta_\varepsilon^*(\theta_\varepsilon)\psi d\sigma d\tau$$

$$= \int_0^t\int_\Omega f\psi dx d\tau - \int_0^t\int_{\Gamma_\alpha} f_0\psi d\sigma d\tau - \int_0^t\int_{\Gamma_u} u\psi d\sigma d\tau$$

$$+ \int_0^t\int_\Omega K_\varepsilon(\theta_\varepsilon)\frac{\partial\psi}{\partial x_N} dx d\tau,$$

with ψ satisfying the boundary value problem

$$-\Delta\psi = \frac{d\theta_\varepsilon}{d\tau}(\tau),\ \frac{\partial\psi}{\partial\nu} = 0 \text{ on } \Gamma_u,\ \frac{\partial\psi}{\partial\nu} + \alpha\psi = 0 \text{ on } \Gamma_\alpha.$$

After an integration with respect to τ and using the boundedness of K_ε we obtain

$$\int_\Omega j_\varepsilon(\theta_\varepsilon(x,t))dx + \int_0^t \left\|\frac{d\theta_\varepsilon}{d\tau}(\tau)\right\|_{V'}^2 d\tau$$

$$\le \int_\Omega j_\varepsilon(\theta_0)dx + \frac{1}{2}\int_0^t \left\|\frac{d\theta_\varepsilon}{d\tau}(\tau)\right\|_{V'}^2 d\tau + 2\int_0^t \|f(\tau)\|_{V'}^2 d\tau$$
$$+2\int_0^t \left(c_{\Gamma_\alpha}^2 \|f_0(\tau)\|_{L^2(\Gamma_\alpha)}^2 + c_{\Gamma_u}^2 \|u(\tau)\|_{L^2(\Gamma_u)}^2\right) d\tau + 2\int_\Omega K_s^2 T dx.$$

This implies that

$$\int_\Omega j_\varepsilon(\theta_\varepsilon(x,t))dx + \int_0^t \left\|\frac{d\theta_\varepsilon}{d\tau}(\tau)\right\|_{V'}^2 d\tau \le 4S_\varepsilon, \tag{3.30}$$

where

$$S_\varepsilon = \int_\Omega j_\varepsilon(\theta_0)dx \tag{3.31}$$
$$+\int_0^T \left(\|f(\tau)\|_{V'}^2 + c_{\Gamma_u}^2 \|u(\tau)\|_{L^2(\Gamma_u)}^2 + c_{\Gamma_\alpha}^2 \|f_0(\tau)\|_{L^2(\Gamma_\alpha)}^2\right) d\tau + \overline{K}^2 T,$$

with

$$\overline{K} = K_s(\text{meas}(\Omega))^{1/2}. \tag{3.32}$$

Also we notice that

$$\frac{\rho}{2}\|\theta_\varepsilon(t)\|^2 \le \int_\Omega j_\varepsilon(\theta_\varepsilon(x,t))dx \tag{3.33}$$

and so we get

$$\|\theta_\varepsilon(t)\|^2 \le \frac{8}{\rho}S_\varepsilon, \quad \text{for any } t \in [0,T]. \tag{3.34}$$

Then we apply (3.8) with $\phi = \beta_\varepsilon^*(\theta_\varepsilon)$, integrate over $(0,t)$ and deduce by standard computations that

$$\int_\Omega j_\varepsilon(\theta_\varepsilon(x,t))dx + \int_0^t \|\beta_\varepsilon^*(\theta_\varepsilon(\tau))\|_V^2 d\tau \le 4S_\varepsilon. \tag{3.35}$$

Adding (3.30) and (3.35) we obtain (3.28) as claimed, with

$$\gamma_0(\alpha_m) = 8\max\{1, c_{\Gamma_u^2}, c_{\Gamma_\alpha}^2, \overline{K}^2 T\}. \tag{3.36}$$

By (3.28) we get that $j_\varepsilon(\theta_\varepsilon) \in L^\infty(0,T;L^1(\Omega))$, for each $\varepsilon > 0$.

To prove (3.29) we multiply the difference of equations (3.6), written for the two different pairs of data, scalarly in V' by $(\theta_\varepsilon - \overline{\theta_\varepsilon})$ and integrate over $(0,t)$ with $t \in (0,T)$. We get after some computations that

$$\left\|\theta_\varepsilon(t) - \overline{\theta_\varepsilon}(t)\right\|_{V'}^2 + \rho\int_0^t \left\|\theta_\varepsilon(\tau) - \overline{\theta_\varepsilon}(\tau)\right\|^2 d\tau$$
$$\le \left\|\theta_0 - \overline{\theta_0}\right\|_{V'}^2 + \left(1 + c_{\Gamma_\alpha}^2 + c_{\Gamma_u}^2 + \frac{M_\varepsilon^2}{\rho}\right)\int_0^t \left\|\theta_\varepsilon(\tau) - \overline{\theta_\varepsilon}(\tau)\right\|_{V'}^2 d\tau$$
$$+\int_0^t \left\|f(\tau) - \overline{f}(\tau)\right\|_{V'}^2 d\tau + \int_0^t \left\|f_0(\tau) - \overline{f_0}(\tau)\right\|_{L^2(\Gamma_\alpha)}^2 d\tau$$
$$+\int_0^t \|u(\tau) - \overline{u}(\tau)\|_{L^2(\Gamma_u)}^2 d\tau$$

and by Gronwall's lemma we deduce that

$$\begin{aligned}
&\left\|\theta_\varepsilon(t)-\overline{\theta_\varepsilon}(t)\right\|_{V'}^2+\rho\int_0^t\left\|\theta_\varepsilon(\tau)-\overline{\theta_\varepsilon}(\tau)\right\|^2 d\tau \\
\leq\ & \gamma_1(\alpha_m,\varepsilon)\left(\left\|\theta_0-\overline{\theta_0}\right\|_{V'}^2+\int_0^T\left\|f(\tau)-\overline{f}(\tau)\right\|_{V'}^2 d\tau\right. \\
&\left.+\int_0^T\left\|u(\tau)-\overline{u}(\tau)\right\|_{L^2(\Gamma_u)}^2 d\tau+\int_0^T\left\|f_0(\tau)-\overline{f_0}(\tau)\right\|_{L^2(\Gamma_\alpha)}^2 d\tau\right),
\end{aligned}$$

with

$$\gamma_1(\alpha_m,\varepsilon)=\exp\left[\left(1+c_{\Gamma_\alpha}^2+c_{\Gamma_u}^2+\frac{M_\varepsilon^2}{\rho}\right)T\right]. \tag{3.37}$$

The last estimate implies the uniqueness of the solution θ_ε for each ε positive. Obviously

$$\gamma_0(\alpha_m)=O\left(\frac{1}{\alpha_m}\right) \text{ and } \gamma_1(\alpha_m,\varepsilon)=O\left(\frac{1}{\alpha_m}\right) \text{ as } \alpha_m\to 0, \tag{3.38}$$

so that in the previous estimates we cannot consider the limit $\alpha_m\to 0$. ∎

In Theorem 3.3 below, we shall see that the above existence result remains true under a weaker regularity of f, u and f_0.

Theorem 3.3. *Let*

$$\begin{aligned}
f &\in L^2(0,T;V'),\ u\in L^2(0,T;L^2(\Gamma_u)),\ f_0\in L^2(0,T;L^2(\Gamma_\alpha)), \\
\theta_0 &\in L^2(\Omega).
\end{aligned} \tag{3.39}$$

Then, problem (3.6)-(3.7) *has, for each* $\varepsilon>0$, *a unique solution*

$$\begin{aligned}
\theta_\varepsilon &\in C([0,T];L^2(\Omega))\cap W^{1,2}(0,T;V')\cap L^2(0,T;V), \\
\beta_\varepsilon^*(\theta) &\in L^2(0,T;V),
\end{aligned} \tag{3.40}$$

that satisfies the estimates (3.28) and (3.29).

Moreover, if $\theta_0\leq\theta_s$ *a.e. on* Ω, *then* (3.28) *is independent of* ε.

Proof. Due to density arguments, let $\{f_n\}_{n\geq 1}$, $\{u_n\}_{n\geq 1}$ and $\{f_n^0\}_{n\geq 1}$ be such that

$$\begin{aligned}
f_n\in W^{1,1}(0,T;V'),\ f_n\to f \text{ strongly in } L^2(0,T;V') \\
u_n\in W^{1,1}(0,T;L^2(\Gamma_u)),\ u_n\to u \text{ strongly in } L^2(0,T;L^2(\Gamma_u)) \\
f_0^n\in W^{1,1}(0,T;L^2(\Gamma_\alpha)),\ f_0^n\to f_0 \text{ strongly in } L^2(0,T;L^2(\Gamma_\alpha)),
\end{aligned}$$

and let $\theta_0\in L^2(\Omega)$. As we have seen $D(A_\varepsilon)=V$ which is dense in $L^2(\Omega)$. Therefore, there exists $\{\theta_0^n\}_{n\geq 1}\subset D(A_\varepsilon)$ such that $\theta_0^n\to\theta_0$ strongly in $L^2(\Omega)$ and consequently strongly in V', too. Recall that ε is fixed. Then, for each $\varepsilon>0$ there exists a unique solution $\theta_{\varepsilon,n}$ to the second approximating problem

$$\begin{aligned}
\frac{d\theta_{\varepsilon,n}}{dt}+A_\varepsilon\theta_{\varepsilon,n} &= f_n+Bu_n+f_n^\Gamma,\ \text{a.e. } t\in(0,T), \\
\theta_{\varepsilon,n}(0) &= \theta_0^n,
\end{aligned} \tag{3.41}$$

which satisfies the conclusions of Theorem 3.2, with the estimate

$$\int_\Omega j_\varepsilon(\theta_{\varepsilon,n}(x,t))dx + \int_0^t \left\| \frac{d\theta_{\varepsilon,n}}{d\tau}(\tau)\right\|_{V'}^2 d\tau \tag{3.42}$$

$$+\int_0^t \|\beta_\varepsilon^*(\theta_{\varepsilon,n}(\tau))\|_V^2\, d\tau$$

$$\leq \gamma_0(\alpha_m)\left(\int_\Omega j_\varepsilon(\theta_0^n(x))dx + \int_0^T \|f_n(\tau)\|_{V'}^2\, d\tau\right.$$

$$\left. + \int_0^T \|u_n(\tau)\|_{L^2(\Gamma_u)}^2\, d\tau + \int_0^T \|f_0^n(\tau)\|_{L^2(\Gamma_\alpha)}^2\, d\tau + 1\right)$$

independently of n because the sequences on the right-hand side are bounded and

$$\int_\Omega j_\varepsilon(\theta_n^0(x))dx \leq \frac{K_s^* - \beta^*(\theta_s - \varepsilon)}{2\varepsilon}\, \|\theta_0^n\|^2 .$$

Estimate (3.42) implies the boundedness of the sequences $\{\frac{d\theta_{\varepsilon,n}}{dt}\}$ in $L^2(0,T;V')$, $\{\beta_\varepsilon^*(\theta_{\varepsilon,n})\}_{n\geq 1}$ in $L^2(0,T;V)$, the latter leading also to the boundedness of $\{\theta_{\varepsilon,n}\}_{n\geq 1}$ in $L^2(0,T;V)$. Therefore, selecting successive subsequences, if necessary, we obtain

$$\beta_\varepsilon^*(\theta_{\varepsilon,n}) \to \zeta_\varepsilon \text{ weakly in } L^2(0,T;V), \text{ as } n\to\infty,$$

$$\theta_{\varepsilon,n} \to \theta_\varepsilon \text{ weakly in } L^2(0,T;V), \text{ as } n\to\infty,$$

$$\frac{d\theta_{\varepsilon,n}}{dt} \to \frac{d\theta_\varepsilon}{dt} \text{ weakly in } L^2(0,T;V'), \text{ as } n\to\infty.$$

The latter two imply by Lions-Aubin theorem that

$$\theta_{\varepsilon,n} \to \theta_\varepsilon \text{ strongly in } L^2(Q), \text{ as } n\to\infty. \tag{3.43}$$

Also we deduce that $\zeta_\varepsilon = \beta_\varepsilon^*(\theta_\varepsilon)$ because of the continuity of β_ε^* in $\mathbf{R}$ and to the strong convergence of θ_ε to θ.

Moreover, since K_ε and j_ε are Lipschitz it follows that

$$K_\varepsilon(\theta_{\varepsilon,n}) \to K_\varepsilon(\theta_\varepsilon) \text{ strongly in } L^2(Q)$$

and

$$j_\varepsilon(\theta_{\varepsilon,n}) \to j_\varepsilon(\theta_\varepsilon) \text{ strongly in } L^2(Q).$$

Since the operator A_ε is quasi m-accretive on V', it follows that its realization on $L^2(0,T;V')$ is quasi m-accretive too, hence it is demiclosed, meaning that

$$A_\varepsilon\theta_{\varepsilon,n} \to A_\varepsilon\theta_\varepsilon \text{ weakly in } L^2(0,T;V), \text{ as } n\to\infty.$$

Eventually, we can pass to limit as $n\to\infty$ in (3.41) and obtain

$$\begin{aligned}\frac{d\theta_\varepsilon}{dt} + A_\varepsilon\theta_\varepsilon &= f + Bu + f_\Gamma, \text{ a.e. } t\in(0,T),\\ \theta_\varepsilon(0) &= \theta_0,\end{aligned}$$

which shows that θ_ε is a solution to (3.6)-(3.7).

By passing to the limit with $n \to \infty$ in (3.42) and using lower semicontinuity arguments it follows that θ_ε satisfies (3.28).

Now, if $\theta_0 \in L^2(\Omega)$ and $\theta_0 \leq \theta_s$, a.e. $x \in \Omega$, we have

$$j_\varepsilon(\theta_0) = \int_0^{\theta_0} \beta_\varepsilon^*(\xi)d\xi \leq \int_0^{\theta_s} \beta_\varepsilon^*(\xi)d\xi \leq K_s^*\theta_s < +\infty. \tag{3.44}$$

Thus, we emphasize that the right-hand side in (3.28) becomes independent of ε, being equal to $\gamma_0(\alpha_m)S_0$, where

$$\begin{aligned} S_0 &= K_s^*\theta_s \text{meas}(\Omega) + \int_0^T \|f(\tau)\|_{V'}^2\, d\tau \qquad (3.45)\\ &+ \int_0^T \|u(\tau)\|_{L^2(\Gamma_u)}^2\, d\tau + \int_0^T \|f_0(\tau)\|_{L^2(\Gamma_\alpha)}^2\, d\tau + 1. \end{aligned}$$

Finally, by (3.34) and (3.44), we obtain

$$\|\theta_\varepsilon(t)\| \leq C_S < +\infty, \; \forall t \in [0, T], \tag{3.46}$$

where

$$C_S = S_0\gamma_0(\alpha_m) \tag{3.47}$$

does not depend on ε. ■

3.2. Comparison Results for the Approximating Solution

Let us consider two time dependent functions $\theta_M \in C^1[0, T]$ and $\theta_m \in C^1[0, T]$ such that

$$\theta_m(t) < \theta_M(t) \text{ and } \theta_m'(t) \leq \theta_M'(t), \text{ for any } t \in [0, T].$$

Assume also that $\theta_m(0)$ and $\theta_M(0)$ do not vanish simultaneously and the same property is true for $\theta_m'(0)$ and $\theta_M'(0)$. Then, let us denote

$$f_M(t) = \theta_M'(t), \; u_{M\varepsilon}(t) = K_\varepsilon(\theta_M(t))i_N \cdot \nu, \tag{3.48}$$

$$f_{0\varepsilon}^M(x, t) = K_\varepsilon(\theta_M(t))i_N \cdot \nu - \alpha(x)\beta_\varepsilon^*(\theta_M(t)) \tag{3.49}$$

and

$$f_m(t) = \theta_m'(t), \; u_{m\varepsilon}(t) = K_\varepsilon(\theta_m(t))i_N \cdot \nu,$$

$$f_{0\varepsilon}^m(x, t) = K_\varepsilon(\theta_m(t))i_N \cdot \nu - \alpha(x)\beta_\varepsilon^*(\theta_m(t)).$$

It is obvious that $\theta_m(t)$ is the solution to the problem

$$\frac{\partial \theta_m}{\partial t} - \Delta\beta_\varepsilon^*(\theta_m) + \frac{\partial K_\varepsilon(\theta_m)}{\partial x_N} = f_m(t) \text{ in } Q,$$

$$\theta_m(x, 0) = \theta_m(0) \text{ in } \Omega,$$

$$(K_\varepsilon(\theta_m)i_N - \nabla\beta_\varepsilon^*(\theta_m)) \cdot \nu = u_{m\varepsilon}(t) \text{ on } \Sigma_u,$$

$$(K_\varepsilon(\theta_m)i_N - \nabla\beta^*_\varepsilon(\theta_m)) \cdot \nu = \alpha\beta^*_\varepsilon(\theta_m) + f^m_{0\varepsilon}(x,t) \text{ on } \Sigma_\alpha.$$

Analogously, $\theta_M(t)$ is the solution to the previous problem corresponding to f_M, $u_{M\varepsilon}$, $f^M_{0\varepsilon}$ instead of f_m, $u_{m\varepsilon}$, $f^m_{0\varepsilon}$.

Lemma 3.4. *Let*

$$f \in L^\infty(Q),\ u \in L^\infty(\Sigma_u),\ f_0 \in L^\infty(\Sigma_\alpha), \tag{3.50}$$

$$\theta_0 \in L^2(\Omega) \tag{3.51}$$

hold and assume still that

$$\theta_m(0) \leq \theta_0(x) \leq \theta_M(0) \text{ a.e. in } \Omega, \tag{3.52}$$

$$\theta'_m(t) \leq f(x,t) \leq \theta'_M(t) \text{ a.e. in } Q, \tag{3.53}$$

$$u_{M\varepsilon}(t) \leq u(x,t) \leq u_{m\varepsilon}(t) \text{ a.e. on } \Sigma_u, \tag{3.54}$$

$$f^M_{0\varepsilon}(x,t) \leq f_0(x,t) \leq f^m_{0\varepsilon}(x,t) \text{ a.e. on } \Sigma_\alpha. \tag{3.55}$$

Then, for each $\varepsilon > 0$, *we have*

$$\theta_m(t) \leq \theta_\varepsilon(x,t) \leq \theta_M(t) \text{ a.e. in } \Omega,\ \text{for each } t \in [0,T]. \tag{3.56}$$

Proof. By Theorem 3.3, problem (3.6)-(3.7) has a unique solution

$$\theta_\varepsilon \in C([0,T];L^2(\Omega)) \cap W^{1,2}(0,T;V') \cap L^2(0,T;V).$$

We multiply the equation

$$\frac{\partial(\theta_\varepsilon - \theta_m)}{\partial t} - \Delta(\beta^*_\varepsilon(\theta_\varepsilon) - \beta^*_\varepsilon(\theta_m)) + \frac{\partial K_\varepsilon(\theta_\varepsilon)}{\partial x_N} - \frac{\partial K_\varepsilon(\theta_m)}{\partial x_N} = f - f_m \tag{3.57}$$

by $(\theta_\varepsilon(x,t) - \theta_m(t))^-$ and then we integrate it over $\Omega \times (0,t)$. We get

$$\begin{aligned}
&\int_0^t \int_\Omega \left\{ -\frac{1}{2}\frac{\partial}{\partial \tau}[(\theta_\varepsilon - \theta_m)^-]^2 + \nabla(\beta^*_\varepsilon(\theta_\varepsilon) - \beta^*_\varepsilon(\theta_m)) \cdot \nabla(\theta_\varepsilon - \theta_m)^- \right\} dx d\tau \\
= &\int_0^t \int_\Omega (K_\varepsilon(\theta_\varepsilon) - K_\varepsilon(\theta_m)) \frac{\partial(\theta_\varepsilon - \theta_m)^-}{\partial x_N} dx d\tau \\
&+ \int_0^t \int_\Omega (f - f_m)(\theta_\varepsilon - \theta_m)^- dx d\tau \\
&- \int_0^t \int_{\Gamma_\alpha} (f_0 - f^m_{0\varepsilon})(\theta_\varepsilon - \theta_m)^- d\sigma d\tau - \int_0^t \int_{\Gamma_u} (u - u_{m\varepsilon})(\theta_\varepsilon - \theta_m)^- d\sigma d\tau \\
&- \int_0^t \int_{\Gamma_\alpha} \alpha(\beta^*_\varepsilon(\theta_\varepsilon) - \beta^*_\varepsilon(\theta_m))(\theta_\varepsilon - \theta_m)^- d\sigma d\tau.
\end{aligned}$$

We notice that

$$\alpha(\beta^*_\varepsilon(\theta_\varepsilon) - \beta^*_\varepsilon(\theta_m))(\theta_\varepsilon - \theta_m)^- \leq -\alpha\rho((\theta_\varepsilon - \theta_m)^-)^2,$$

and

$$\nabla(\beta_\varepsilon^*(\theta_\varepsilon)-\beta_\varepsilon^*(\theta_m))\cdot\nabla(\theta_\varepsilon-\theta_m)^- = -\beta_\varepsilon(\theta_\varepsilon)\nabla(\theta_\varepsilon-\theta_m)\cdot\nabla(\theta_\varepsilon-\theta_m)^-$$

so that we get that

$$\begin{aligned}
&\frac{1}{2}\int_\Omega[(\theta_\varepsilon(t)-\theta_m(t))^-]^2dx+\rho\int_0^t\left\|(\theta_\varepsilon(\tau)-\theta_m(\tau))^-\right\|_V^2d\tau\\
\le\ &\frac{1}{2}\int_\Omega[(\theta_0-\theta_m(0))^-]^2dx+M_\varepsilon\int_0^t\|\theta_\varepsilon(\tau)-\theta_m(\tau)\|\left\|(\theta_\varepsilon(\tau)-\theta_m(\tau))^-\right\|_V d\tau\\
&-\int_0^t\int_\Omega(f-f_m)(\theta_\varepsilon-\theta_m)^-dxd\tau+\int_0^t\int_{\Gamma_\alpha}(f_0-f_{0\varepsilon}^m)(\theta_\varepsilon-\theta_m)^-d\sigma d\tau\\
&+\int_0^t\int_{\Gamma_u}(u-u_{m\varepsilon})(\theta_\varepsilon-\theta_m)^-d\sigma d\tau-\int_0^t\int_{\Gamma_\alpha}\alpha\rho((\theta_\varepsilon-\theta_m)^-)^2d\sigma d\tau.
\end{aligned}$$

In the computations above we have used many times Stampacchia's lemma. Taking into account the hypotheses we obtain

$$\begin{aligned}
&\left\|(\theta_\varepsilon(t)-\theta_m(t))^-\right\|^2+\rho\int_0^t\left\|(\theta_\varepsilon(\tau)-\theta_m(\tau))^-\right\|_V^2d\tau\\
\le\ &\frac{M_\varepsilon^2}{\rho}\int_0^t\left\|(\theta_\varepsilon(\tau)-\theta_m(\tau))^-\right\|^2d\tau.
\end{aligned}$$

By Gronwall's lemma, we deduce that $\left\|(\theta_\varepsilon(t)-\theta_m(t))^-\right\|^2=0$, which implies that $\theta_\varepsilon(x,t)\ge\theta_m(t)$ a.e. on Ω, for each $t\in[0,T]$.

Similarly, we deduce $\left\|(\theta_\varepsilon(t)-\theta_M(t))^+\right\|^2=0$, i.e., $\theta_\varepsilon(x,t)\le\theta_M(t)$ a.e. on Ω, for each $t\in[0,T]$. ∎

The previous lemma has hypotheses depending on ε but for a particular choice of θ_m and θ_M we can impose sufficient conditions that do not depend on ε.

Corollary 3.5. *Let $\theta_m,\theta_M\in C^1([0,T])$ be such that*

$$\theta_m(t)<r_1<\theta_s\le\theta_M(t)\ \textit{for any}\ t\in[0,T]\ \textit{and}\ \theta_M(0)\equiv\theta_s. \tag{3.58}$$

Assume (3.50)-(3.53),

$$K(\theta_M(t))i_N\cdot\nu\le u(x,t)\le K(\theta_m(t))i_N\cdot\nu,\ \textit{a.e. on}\ \Sigma_u, \tag{3.59}$$

and

$$K_si_N\cdot\nu-\alpha K_s^*\le f_0(x,t)\le K(\theta_m(t))i_N\cdot\nu-\alpha\beta^*(\theta_m(t)),\ \textit{a.e. on}\ \Sigma_\alpha. \tag{3.60}$$

Then

$$\theta_m(t)\le\theta_\varepsilon(x,t)\le\theta_M(t),\ \textit{a.e. in}\ \Omega,\ \textit{for each}\ t\in[0,T]. \tag{3.61}$$

In particular, if $\theta_m(t)=0$ for any $t\in[0,T]$, we get

$$0\le\theta_\varepsilon(x,t)\le\theta_M(t),\ \textit{a.e. in}\ \Omega,\ \textit{for each}\ t\in[0,T]. \tag{3.62}$$

Proof. The hypothesis $\theta_m(t) < r_1 < \theta_s$ implies that $f_{0\varepsilon}^m = K_\varepsilon(\theta_m)i_N \cdot \nu - \alpha\beta_\varepsilon^*(\theta_m)$ can be replaced by $K(\theta_m)i_N \cdot \nu - \alpha\beta^*(\theta_m)$.

Now, for $\theta_M(t) \geq \theta_s$ we have $K_\varepsilon(\theta_M(t)) = K(\theta_M(t))$ and $\beta_\varepsilon^*(\theta_M) \geq K_s^*$, so that

$$f_{0\varepsilon}^M = K_\varepsilon(\theta_M)i_N \cdot \nu - \alpha\beta_\varepsilon^*(\theta_M) \leq K_\varepsilon(\theta_M)i_N \cdot \nu - \alpha K_s^* \leq K_s i_N \cdot \nu - \alpha K_s^*.$$

This and the assumption (3.60) imply

$$f_{0\varepsilon}^M \leq K_s i_N \cdot \nu - \alpha K_s^* \leq f_0(x,t) \leq K(\theta_m)i_N \cdot \nu - \alpha\beta^*(\theta_m) = f_{0\varepsilon}^m \text{ a.e. on } \Sigma_\alpha.$$

The latter inequalities, together with the other hypotheses represent the assumptions made in Lemma 3.4. Therefore we get $\theta_m(t) \leq \theta_\varepsilon(x,t) \leq \theta_M(t)$, a.e. in Ω, for each $t \in [0,T]$, as claimed. ■

4. Main Results for the Original Problem

Let us define the function $j : \mathbf{R} \to (-\infty, \infty]$ by

$$j(r) = \begin{cases} \int_0^r \beta^*(\xi)d\xi, \text{ if } r \leq \theta_s \\ +\infty, \text{ if } r > \theta_s, \end{cases} \tag{4.1}$$

where $j(\theta_s)$ should be understood as

$$j(\theta_s) = \lim_{r \nearrow \theta_s} \int_0^r \beta^*(\xi)d\xi. \tag{4.2}$$

It follows that j is a proper, convex, lower semicontinuous function and

$$\partial j(r) = \begin{cases} \beta^*(r), & r < \theta_s \\ [K_s^*, +\infty), & r = \theta_s \\ \emptyset, & r > \theta_s. \end{cases} \tag{4.3}$$

(The proof is done in [24], Chapter 5).

4.1. Existence and Properties of the Original Solution

We are going now to prove the existence of the solution to the original problem, by passing to limit in the approximating problem (3.6)-(3.7). In Theorem 3.3, we let ε tend to 0 and we obtain the following existence result:

Theorem 4.1. *Let f, u, f_0 and θ_0 satisfy*

$$f \in L^2(0,T;V'), \; u \in L^2(0,T;L^2(\Gamma_u)), \; f_0 \in L^2(0,T;L^2(\Gamma_\alpha)), \tag{4.4}$$

$$\theta_0 \in L^2(\Omega), \; \theta_0 \leq \theta_s, \; a.e. \; x \in \Omega. \tag{4.5}$$

Then, there exists a solution θ to the original problem (2.23)-(2.24),

$$\begin{aligned} \theta &\in C([0,T];L^2(\Omega)) \cap W^{1,2}(0,T;V') \cap L^2(0,T;V), \\ \beta^*(\theta) &\in L^2(0,T;V), \\ j(\theta) &\in L^1(Q). \end{aligned} \tag{4.6}$$

Moreover, the solution satisfies the estimate

$$\begin{aligned}
&\int_\Omega j(\theta(x,t))dx + \int_0^t \left\| \frac{d\theta}{d\tau}(\tau) \right\|_{V'}^2 d\tau \\
&+ \int_0^t \|\beta^*(\theta(\tau))\|_V^2 \, d\tau \\
\leq\ & \gamma_0(\alpha_m) \left(\int_\Omega j(\theta_0(x))dx + \int_0^T \|f(\tau)\|_{V'}^2 \, d\tau \right. \\
&\left. + \int_0^T \|u(\tau)\|_{L^2(\Gamma_u)}^2 \, d\tau + \int_0^T \|f_0(\tau)\|_{L^2(\Gamma_\alpha)}^2 \, d\tau + 1 \right), \\
\text{for any } t\ \in\ & [0,T].
\end{aligned} \tag{4.7}$$

Proof. Assume that (4.4) and (4.5) hold. Then the approximating problem (3.6)-(3.7) has a strong solution θ_ε, satisfying conclusions of Theorem 3.3. Since here we have imposed $\theta_0 \leq \theta_s$ we have by (3.44) that $j_\varepsilon(\theta_0) \leq K_s^* \theta_s$, so that the right-hand side term in (3.28) turns out to be bounded by $\gamma_0(\alpha_m)S_0$, independently of ε.

By (3.46) and (3.28) we deduce that $\{\theta_\varepsilon\}_{\varepsilon>0}$ lies in a bounded subset of $L^\infty(0,T;L^2(\Omega))$, $\{\frac{d\theta_\varepsilon}{dt}\}_{\varepsilon>0}$ is in a bounded subset of $L^2(0,T;V')$ and $\{\beta_\varepsilon^*(\theta_\varepsilon)\}_{\varepsilon>0}$ is included in a bounded subset of $L^2(0,T;V)$. Using (i) we get that $\{\theta_\varepsilon\}_{\varepsilon>0}$ lies in a bounded subset of $L^2(0,T;V)$, too.

Therefore, we can select some subsequences (denoted by θ_ε, too) such that

$$\begin{aligned}
&\theta_\varepsilon \to \theta \text{ weakly in } L^2(0,T;V), \\
&\frac{d\theta_\varepsilon}{dt} \to \frac{d\theta}{dt} \text{ weakly in } L^2(0,T;V'), \\
&\beta_\varepsilon^*(\theta_\varepsilon) \to \eta \text{ weakly in } L^2(0,T;V).
\end{aligned} \tag{4.8}$$

We also conclude that $\{\theta_\varepsilon\}_{\varepsilon>0}$ is compact in $L^2(0,T;L^2(\Omega))$, so that

$$\theta_\varepsilon \to \theta \text{ strongly in } L^2(0,T;L^2(\Omega)) \text{ as } \varepsilon \to 0\,. \tag{4.9}$$

By the trace theorem, the trace operator (denoted in the same way) is linear and continuous and we have

$$\|\beta_\varepsilon^*(\theta_\varepsilon)\|_{L^2(0,T;L^2(\Gamma))} \leq C\, \|\beta_\varepsilon^*(\theta_\varepsilon)\|_{L^2(0,T;H^1(\Omega))} \leq \text{ constant},$$

whence the trace of $\beta_\varepsilon^*(\theta_\varepsilon)$ converges to the trace of η,

$$\beta_\varepsilon^*(\theta_\varepsilon) \to \eta \text{ weakly in } L^2(0,T;L^2(\Gamma)).$$

Next we shall prove that

$$\eta \in \beta^*(\theta) \text{ a.e. on } Q, \tag{4.10}$$

using the inequalities

$$\int_Q j(\theta)dxdt \leq \liminf_{\varepsilon\to 0} \int_Q j_\varepsilon(\theta_\varepsilon)dxdt, \tag{4.11}$$

and

$$j_\varepsilon(z) \to j(z), \text{ as } \varepsilon \to 0, \ \forall z \in \mathbf{R}. \tag{4.12}$$

The latter convergence is obvious for $z < \theta_s$, where $j_\varepsilon(z) \equiv j(z)$ and j is continuous at the left of $z = \theta_s$,

$$\lim_{\varepsilon\to 0} j_\varepsilon(\theta_s) = \lim_{\varepsilon \to 0} \int_0^{\theta_s} \beta_\varepsilon^*(\xi)d\xi = K_s^*\theta_s = j(\theta_s),$$

due to (4.2). For $z > \theta_s$ we have

$$j_\varepsilon(z) = \int_0^{\theta_s} \beta_\varepsilon^*(\xi)d\xi + \int_{\theta_s}^{z} \left\{\beta^*(\theta_s - \varepsilon) + \frac{K_s^* - \beta^*(\theta_s-\varepsilon)}{\varepsilon}[\xi - (\theta_s - \varepsilon)]\right\} d\xi$$

which implies

$$\lim_{\varepsilon\to 0} j_\varepsilon(z) = +\infty = j(z) \text{ for } z > \theta_s.$$

Now, we are going to show (4.11). This is based on the fact that we can select a subsequence of $\{\theta_\varepsilon\}_\varepsilon$, denoted in the same way, such that

$$\liminf_{\varepsilon\to 0} j_\varepsilon(\theta_\varepsilon) \geq j(\theta). \tag{4.13}$$

The details can be found in [24], Chapter 5, Theorem 3.1, pp. 169.

Because $j_\varepsilon(\theta_\varepsilon) > 0$ we have by Fatou's lemma and (4.13) that

$$\liminf_{\varepsilon\to 0} \int_Q j_\varepsilon(\theta_\varepsilon)dxdt \geq \int_Q \liminf_{\varepsilon\to 0} j_\varepsilon(\theta_\varepsilon)dxdt \geq \int_Q j(\theta)dxdt.$$

From here and (3.28) we see also that $j(\theta) \in L^1(Q)$, which implies, in particular, that

$$\theta(x,t) \leq \theta_s \text{ a.e. } (x,t) \in Q. \tag{4.14}$$

We recall that j_ε is the subdifferential of β_ε^*, and then

$$j_\varepsilon(\theta_\varepsilon) \leq j_\varepsilon(z) + \beta_\varepsilon^*(\theta_\varepsilon)(\theta_\varepsilon - z), \text{ for any } z \in \mathbf{R},$$

which we apply in particular to $z : \Omega \times (0,T) \to \mathbf{R}$, i.e.,

$$\int_Q j_\varepsilon(\theta_\varepsilon)dxdt \leq \int_Q j_\varepsilon(z)dxdt + \int_Q \beta_\varepsilon^*(\theta_\varepsilon)(\theta_\varepsilon - z)dxdt, \ \forall\, z \in L^2(Q). \tag{4.15}$$

For $z \leq \theta_s$ we have $j_\varepsilon(z) \leq K_s^*\theta_s$ and using (4.12), we deduce by the Lebesgue dominated convergence theorem that

$$\lim_{\varepsilon\to 0} \int_Q j_\varepsilon(z)dxdt = \int_Q j(z)dxdt.$$

We remind that $\beta_\varepsilon^*(\theta_\varepsilon) \to \eta$ weakly in $L^2(0,T;V)$ and $\theta_\varepsilon \to \theta$ strongly in $L^2(0,T;L^2(\Omega))$ so that passing to the limit as $\varepsilon \to 0$ in (4.15) and taking into account (4.11) we obtain

$$\int_Q j(\theta)dxdt \leq \int_Q j(z)dxdt + \int_Q \eta(\theta - z)dxdt, \ \forall z \in L^2(Q), \ z \leq \theta_s. \tag{4.16}$$

This inequality allows us to conclude that $\partial j(\theta) \ni \eta$ a.e. on Q and by (4.3) it follows that $\eta \in \beta^*(\theta)$ a.e. on Q (see the argument in [24], Chapter 5, Theorem 3.1).

The last step of the proof refers to the convergence of $K_\varepsilon(\theta_\varepsilon)$. Recalling that $K_\varepsilon(\theta_\varepsilon)$ is bounded in $L^2(0,T;L^2(\Omega))$ we have that

$$K_\varepsilon(\theta_\varepsilon) \to \kappa \text{ weakly in } L^2(0,T;L^2(\Omega)) \text{ as } \varepsilon \to 0$$

and proceed to prove that $\kappa \in \mathcal{K}(\theta)$ defined in (2.13).

Let δ be arbitrary but fixed, $\delta > \varepsilon$ and denote

$$\Omega_\delta = [r_j - \delta, r_j] \text{ for } j \text{ fixed, } j = 1, ..., p.$$

For $\theta_\varepsilon \in (r_{j-1}, r_j - \delta)$ or $\theta_\varepsilon \in (r_p, \theta_s]$ the function $K_\varepsilon(\theta_\varepsilon) = K(\theta_\varepsilon)$ and K is Lipschitz on these intervals, so that $K_\varepsilon(\theta_\varepsilon) \to K(\theta)$ strongly as $\varepsilon \to 0$.

Let $\theta_\varepsilon \in \Omega_\delta$ and assume that K_ε is monotonically increasing on (r_{j-1}, r_j) and $K^l(r_j) \le K^r(r_j)$ such that

$$K_\varepsilon(r_{j-}\delta) \le K_\varepsilon(\theta_\varepsilon) \le K^l(r_j) \le K^r(r_j).$$

Recall that by the definition of K_ε we have $K_\varepsilon(r_{j-}\delta) = K(r_{j-}\delta)$. We denote

$$D_\delta = \{v \in L^2(0,T;L^2(\Omega));\ v(x,t) \in [K(r_j - \delta), K^r(r_j)] \text{ a.e. } (x,t) \in Q\}$$

which is a convex closed set. If $\theta_\varepsilon \in \Omega_\delta$ we have $K_\varepsilon(\theta_\varepsilon) \in D_\delta$ for all $\delta > 0$ and because D_δ is weakly closed too, it follows that the weakly limit of $K_\varepsilon(\theta_\varepsilon)$ belongs to D_δ, i.e.,

$$\kappa \in D_\delta \text{ for any } \delta > 0.$$

Therefore we get

$$\kappa \in \bigcap_{\delta>0} [K^l(r_j - \delta), K^r(r_j)] = [K^l(r_j), K^r(r_j)], \text{ a.e. } (x,t) \in Q. \tag{4.17}$$

The results shown above allow us to conclude that $\kappa \in \mathcal{K}(\theta)$ a.e. on Q, as claimed.

Finally we show that θ is the solution to the original problem. Since θ_ε is a solution to (3.6)-(3.7) we have for any $\phi \in L^2(0,T;V)$ that

$$\int_0^T \left\langle \frac{d\theta_\varepsilon}{dt}(t), \phi \right\rangle_{V',V} dt + \int_Q \left(\nabla \beta_\varepsilon^*(\theta_\varepsilon) \cdot \nabla \phi - K_\varepsilon(\theta_\varepsilon) \frac{\partial \phi}{\partial x_N} \right) dxdt$$
$$= \int_0^T \langle f(t), \phi \rangle_{V',V} dt - \int_{\Sigma_\alpha} (\alpha \beta_\varepsilon^*(\theta_\varepsilon) + f_0)\phi d\sigma dt - \int_{\Sigma_u} u\phi d\sigma dt.$$

Passing to the limit as $\varepsilon \to 0$ we obtain

$$\int_0^T \left\langle \frac{d\theta}{dt}(t), \phi \right\rangle_{V',V} dt + \int_Q \left(\nabla \eta \cdot \nabla \phi - \kappa \frac{\partial \phi}{\partial x_N} \right) dxdt$$
$$= \int_0^T \langle f(t), \phi \rangle_{V',V} dt - \int_{\Sigma_\alpha} (\alpha \eta + f_0)\phi d\sigma dt - \int_{\Sigma_u} u\phi d\sigma dt,$$

for any $\phi \in L^2(0,T;V)$,

where $\eta(x,t) \in \beta^*(\theta(x,t))$ and $\kappa(x,t) \in \mathcal{K}(\theta(x,t))$ a.e. $(x,t) \in Q$.

Moreover, by Ascoli-Arzelà theorem we deduce that

$$\theta_\varepsilon(t) \to \theta(t) \text{ strongly in } V' \text{ uniformly on bounded time intervals}$$

implying that $\theta_0 = \theta_\varepsilon(0) \to \theta(0)$ strongly in V' as $\varepsilon \to 0$.

This ends the proof that θ is a solution to (2.23)-(2.24).

Finally we notice that by weakly lower semicontinuity, we can pass to the limit in (3.28) and obtain (4.7) as claimed.

We make the remark that under the discontinuity property of K we cannot obtain the uniqueness of the solution. ■

4.2. Comparison Results for the Original Solution

We recall the notation made in the subsection 3.3, involving the functions $\theta_m \in C^1[0,T]$ and $\theta_M \in C^1[0,T]$.

Proposition 4.2. *Assume*

$$f \in L^\infty(Q),\ u \in L^\infty(\Sigma_u),\ f_0 \in L^\infty(\Sigma_\alpha),\ \theta_0 \in L^2(\Omega),$$

$$\theta_m(t) < r_1 < \theta_s,\ \forall t \in [0,T],$$

$$\theta_m(0) \le \theta_0(x) \le \theta_s \ a.e.\ in\ \Omega,$$

$$\theta_m'(t) \le f(x,t)\ a.e.\ in\ Q,$$

$$u(x,t) \le K(\theta_m(t)) i_N \cdot \nu\ a.e.\ on\ \Sigma_u,$$

$$f_0(x,t) \le K(\theta_m(t)) i_N \cdot \nu - \alpha\beta^*(\theta_m(t))\ a.e.\ on\ \Sigma_\alpha.$$

Then

$$\theta_m(t) \le \theta(x,t) \le \theta_s\ a.e.\ in\ \Omega,\ for\ each\ t \in [0,T]. \tag{4.18}$$

In particular, if $\theta_m(t) = 0$ *we have*

$$0 \le \theta(x,t) \le \theta_s\ a.e.\ in\ \Omega,\ for\ each\ t \in [0,T]. \tag{4.19}$$

Proof. The proof follows immediately from Theorem 4.1 and Corollary 3.5. By the hypotheses of Theorem 4.1 we obtain a solution θ to the Cauchy problem (2.23)-(2.24) such that $\theta(x,t) \le \theta_s$ a.e. $(x,t) \in Q$. Also we get an approximating solution θ_ε which tends strongly in $L^2(Q)$ to θ. By the hypotheses of Corollary 3.5 we have that $\theta_\varepsilon(x,t) \ge \theta_m(t)$ for all $t \in [0,T]$. By the strongly convergence of θ_ε to θ this inequality is preserved. We notice that the smaller θ_m, the larger the interval of boundedness for θ_ε and θ. Finally, if $\theta_m(t) = 0$, we have (4.19). ■

Remark 4.3. By this last result the solution to the diffusion model is placed in the domain corresponding to the physical admissible interval related to a real problem. We recall that θ can signify the concentration of the fluid in a porous medium, so that it should be positive and less than a maximal value θ_s.

5. Comments on Other Models

1. If $\mathbf{K} = (K_1, ..., K_N)$ with K_j bounded by K_s and having the points of discontinuity $\{r_1, ..., r_p\}$ the proofs remain the same with minor changes. Everywhere, $Ki_N \cdot \nu$ is replaced by $\mathbf{K} \cdot \nu$, M_ε from (3.5) becomes

$$M_\varepsilon = \max_{n=1,...,N} \max_{j=1,...,p} \left\{ \left| \frac{K_n^r(r_j) - K_n(r_j - \varepsilon)}{\varepsilon} \right| \right\}$$

and $\lambda > N_{\max} \frac{M_\varepsilon^2}{2\rho}$, where $N_{\max}$ is the number of non constant components of $\mathbf{K}$. Then, $\overline{K}$ given by (3.32) is in this case

$$\overline{K} = N_{\max} K_s (\text{meas}(\Omega))^{1/2}.$$

2. The case with K Lipschitz on $\mathbf{R}$ with the Lipschitz constant M was treated in the previous works of the author for various boundary conditions. In this case we have worked under the assumption (i_K) for $r, \overline{r} \in \mathbf{R}$ so that $K_\varepsilon \equiv K$. The main change is in (3.29) where γ_1 becomes independent of ε, because in (3.37) M_ε is replaced by M. Thus, we can deduce an estimate for the difference of two solutions to the original problem corresponding to two pairs of data

$$\begin{aligned} &\left\|\theta(t) - \overline{\theta}(t)\right\|_{V'}^2 + \int_0^t \left\|\theta(\tau) - \overline{\theta}(\tau)\right\|^2 d\tau \\ \leq\ & \gamma_1(\alpha_m) \left(\left\|\theta_0 - \overline{\theta_0}\right\|_{V'}^2 + \int_0^T \left\|f(\tau) - \overline{f}(\tau)\right\|_{V'}^2 d\tau \right. \\ &\left. + \int_0^T \|u(\tau) - \overline{u}(\tau)\|_{L^2(\Gamma_u)}^2 d\tau + \int_0^T \left\|f_0(\tau) - \overline{f_0}(\tau)\right\|_{L^2(\Gamma_\alpha)}^2 d\tau \right), \end{aligned}$$

multiplying directly the difference of the equations (2.23) written for θ and $\overline{\theta}$ by $(\theta - \overline{\theta})$ scalarly in V' and proceeding exactly like for the computation of (3.29). The estimate leads immediately to the uniqueness of the solution. This case was treated in detail in [24]. There, supplementary regularity properties of the solution and the free boundary existence have been proved. The case with a Lipschitz convective term and flux and Robin boundary conditions was studied in [5] for the superdiffusion model, too.

3. The case with homogeneous boundary Dirichlet conditions does not differ essentially from this one. Here, the space V previously used is replaced by $V = H_0^1(\Omega)$ and this case was treated in [24], including a degenerate situation when $\beta(0) = 0$, i.e., $\rho = 0$. In the latter case the solution does no longer belong to $L^2(0, T; V)$.

4. The boundary conditions which include a nonhomogeneous Dirichlet boundary condition of the form $\theta(x, t) = \theta_g(x)$ on a part Γ_g of the boundary, with $\theta_g(x) < \theta_s$ a.e. on Σ_α, require a change of variable $\widetilde{\theta} = \theta - \theta_g$ and the choice of the space $V = \{v \in H^1(\Omega); v(x) = 0 \text{ a.e. on } \Gamma_g\}$, with the norm of $H_0^1(\Omega)$, as showed in [22]. If θ_g is time dependent too, some difficulties occur because this leads to a non-autonomous Cauchy problem. This case was treated in [20] and [24] for the superdiffusion model. Another boundary condition that can be used with the fast diffusion model is $\theta(x, t) = \theta_s$ a.e. on a part of the boundary, Γ_g. Used together with a Robin boundary condition like (2.4) or

with a Neumann boundary condition like (2.3) on the other part of the boundary, it imposes the choice of the space $V = \{v \in H^1(\Omega); v(x) = 0 \text{ a.e. on } \Gamma_g\}$, with the gradient norm. However, a certain modification in the definition and proof of the solution should be done.

5. The model of slow diffusion in which β^* is the multivalued function (1.6) and β behaves around θ_s like (1.9) with $m \geq 1$ can be treated in a similar way to the fast diffusion case.

6. The superdiffusion model with β^* single valued and β of the form (1.9) with $m \leq 0$ involves some changes with respect to the previous cases. Examples developed for this case can be found in [24].

7. If the problem data f, u and f_0 are periodic with the period T, the problem (2.1), (2.3)-(2.4) with $\theta(x,0) = \theta(x,T)$ has at least a T-periodic solution θ, even if K_j is not Lipschitz. The periodic solution turns out to be unique for K_j Lipschitz (for all j) and under the condition $\rho > N_{\max} M c_H$. The proof may be led in a similar way to that in [25] where the case with a continuous K whose derivative blows up at $\theta = \theta_s$ was studied.

5.1. Considerations on the Longtime Behavior of the Solution

In the final part of these comments we shall refer to the longtime behavior of the solutions in particular cases.

First, we assume that $\mathbf{K} = (K_1, ..., K_N)$ with K_j Lipschitz for all $j = 1, ..., N$, with the constant M satisfying the relationship

$$\rho > N_{\max} M c_H, \tag{5.1}$$

where c_H is the constant given in (2.8) and $N_{\max}$ is the number of non constant components of $\mathbf{K}$. This relationship expresses the fact that the convective term is slowly varying with respect to θ and it is dominated by the diffusivity.

Let f_∞, f_∞^0, u_∞ be some constants. We consider problem (2.23)-(2.24)

$$\begin{aligned} \frac{d\theta}{dt} + A\theta &\ni f + Bu + f_\Gamma, \text{ a.e. } t \in (0,T), \\ \theta(0) &= \theta_0, \end{aligned} \tag{5.2}$$

and we take $T = +\infty$.

Theorem 5.1. *Assume that K_j are Lipschitz for all j, and (5.1) holds. Let*

$$\theta_0 \in L^2(\Omega),\ \theta_0 \leq \theta_s \text{ a.e. on } \Omega,$$

$$f \in L^2_{\mathrm{loc}}([0,+\infty); V'),\ f_0 \in L^2_{\mathrm{loc}}([0,+\infty); L^2(\Gamma_\alpha)),\ u \in L^2_{\mathrm{loc}}([0,+\infty); L^2(\Gamma_u)),$$

such that

$$\begin{aligned} \underset{s\in(t,+\infty)}{ess\ \sup} \|f(s) - f_\infty\|_{V'} &\to 0, \text{ as } t \to +\infty, \\ \underset{s\in(t,+\infty)}{ess\ \sup} \left\|f_0(s) - f_\infty^0\right\|_{L^2(\Gamma_\alpha)} &\to 0, \text{ as } t \to +\infty, \\ \underset{s\in(t,+\infty)}{ess\ \sup} \|u(s) - u_\infty\|_{L^2(\Gamma_u)} &\to 0, \text{ as } t \to +\infty. \end{aligned} \tag{5.3}$$

Let θ be the solution to (2.23)-(2.24). *Then there exists $\theta_\infty \in D(A)$ such that*

$$\|\theta(t) - \theta_\infty\| \to 0, \ as\ t \to +\infty. \tag{5.4}$$

Proof. We consider again the Cauchy problem (2.23)-(2.24) which has by Theorem 4.1 a solution θ which is unique as specified in the previous comment at point 2.

We introduce now the stationary problem

$$\begin{aligned} -\Delta\beta^*(\theta_\infty) + \nabla\cdot\mathbf{K}(\theta) &\ni f_\infty \text{ in } Q, \\ (\mathbf{K}(\theta_\infty) - \nabla\beta^*(\theta_\infty))\cdot\nu &\ni u_\infty \text{ on } \Sigma_u, \\ (\mathbf{K}(\theta_\infty) - \nabla\beta^*(\theta_\infty))\cdot\nu - \alpha\beta^*(\theta_\infty) &\ni f^0_\infty \text{ on } \Sigma_\alpha \end{aligned} \tag{5.5}$$

and the associated abstract problem $A\theta_\infty = f_\infty + f^\Gamma_\infty + Bu_\infty$, where $f^\Gamma_\infty(\psi) = -\int_{\Gamma_\alpha} f^0_\infty \psi d\sigma$ and $Bu(\psi) = -\int_{\Gamma_\alpha} u_\infty \psi d\sigma$.

We consider the approximating problem

$$\begin{aligned} -\Delta\beta^*_\varepsilon(\theta^\varepsilon_\infty) + \nabla\cdot\mathbf{K}(\theta^\varepsilon_\infty) &= f_\infty \text{ in } Q, \\ (\mathbf{K}_\varepsilon(\theta^\varepsilon_\infty) - \nabla\beta^*_\varepsilon(\theta^\varepsilon_\infty))\cdot\nu &= u_\infty \text{ on } \Sigma_u, \\ (\mathbf{K}_\varepsilon(\theta^\varepsilon_\infty) - \nabla\beta^*_\varepsilon(\theta^\varepsilon_\infty))\cdot\nu &= \alpha\beta^*_\varepsilon(\theta^\varepsilon_\infty) + f^0_\infty \text{ on } \Sigma_\alpha. \end{aligned} \tag{5.6}$$

In this case the corresponding operator A_ε defined by (3.8) is m-accretive and coercive, due to (5.1), because

$$\begin{aligned} (A_\varepsilon\theta - A_\varepsilon\overline{\theta}, \theta - \overline{\theta})_{V'} &\geq \rho\left\|\theta - \overline{\theta}\right\|^2 - \sum_{i=1}^{N_{\max}} M\left\|\theta - \overline{\theta}\right\|\left\|\frac{\partial\psi}{\partial x_i}\right\| \\ &\geq \rho\left\|\theta - \overline{\theta}\right\|^2 - N_{\max}M\left\|\theta - \overline{\theta}\right\|\left\|\theta - \overline{\theta}\right\|_{V'} \\ &\geq c_H^2(\rho - N_{\max}Mc_H)\left\|\theta - \overline{\theta}\right\|^2_{V'}, \end{aligned} \tag{5.7}$$

where we used the inequality $\left\|\theta - \overline{\theta}\right\|_{V'} \leq c_H\left\|\theta - \overline{\theta}\right\|$. Therefore, A_ε is surjective and (5.6) has a unique solution $\theta^\varepsilon_\infty = A_\varepsilon^{-1}(f_\infty) \in D(A_\varepsilon)$. Exactly like in Theorem 4.1 we prove that the sequence $\{\theta^\varepsilon_\infty\}_{\varepsilon>0}$ converges as $\varepsilon \to 0$ to θ_∞ which is the generalized solution to the stationary problem (5.5).

Next we shall show that $\lim_{t\to\infty}\|\theta(t) - \theta_\infty\| = 0$. To come to this end we multiply the equation

$$\frac{\partial(\theta - \theta_\infty)}{\partial t} - \Delta(\eta - \eta_\infty) + \nabla\cdot(\mathbf{K}(\theta) - \mathbf{K}(\theta_\infty)) = f - f_\infty \tag{5.8}$$

by $(\theta - \theta_\infty)$ and integrate it over Ω. Here $\eta \in \beta^*(\theta)$ a.e. on Q and $\eta_\infty \in \beta^*(\theta_\infty)$ a.e. on Ω. We have

$$\int_\Omega \left\{\frac{1}{2}\frac{\partial}{\partial\tau}(\theta - \theta_\infty)^2 + \nabla(\eta - \eta_\infty)\cdot\nabla(\theta - \theta_\infty)\right\} dx$$

$$+\int_{\Gamma_\alpha}\alpha(\beta^*(\theta)-\beta^*(\theta_\infty)(\theta-\theta_\infty)d\sigma$$
$$=\int_\Omega(\mathbf{K}(\theta)-\mathbf{K}(\theta_\infty))\cdot\nabla(\theta-\theta_\infty)dx+\int_\Omega(f-f_\infty)(\theta-\theta_\infty)dx$$
$$-\int_{\Gamma_u}(u(t)-u_\infty)(\theta-\theta_\infty)d\sigma-\int_{\Gamma_\alpha}(f_0(t)-f_\infty^0)(\theta-\theta_\infty)d\sigma.$$

Following standard computations we deduce that

$$\begin{aligned}&\frac{1}{2}\frac{d}{dt}\left\|\theta(t)-\theta_\infty\right\|^2+\rho\left\|\theta(t)-\theta_\infty\right\|_V^2\\ \leq\ & N_{\max}Mc_H\left\|\theta(t)-\theta_\infty\right\|_V^2+\left\|f(t)-f_\infty\right\|_{V'}\left\|\theta(t)-\theta_\infty\right\|_V+\\ &+\left\|f_0(t)-f_\infty^0\right\|_{L^2(\Gamma_\alpha)}\left\|\theta(t)-\theta_\infty\right\|+\left\|u(t)-u_\infty\right\|_{L^2(\Gamma_u)}\left\|\theta(t)-\theta_\infty\right\|.\end{aligned}$$

This yields further

$$\frac{d}{dt}\left\|\theta(t)-\theta_\infty\right\|^2+c_M\left\|\theta(t)-\theta_\infty\right\|^2\leq S(t),\tag{5.9}$$

where $c_M=\rho-N_{\max}Mc_H$ is positive by hypothesis and

$$S(t)=\frac{3}{c_M}\left(\left\|f(t)-f_\infty\right\|_{V'}^2+c_{\Gamma_\alpha}^2\left\|f_0(t)-f_\infty^0\right\|_{L^2(\Gamma_\alpha)}^2+c_{\Gamma_u}^2\left\|u(t)-u_\infty\right\|_{L^2(\Gamma_u)}^2\right).$$

We denote by $\mathcal{P}(t)$ the antiderivative of the function $S(t)\exp(c_Mt)$ and by (5.9) we deduce that the function

$$R(t)=e^{c_Mt}\left\|\theta_\varepsilon(t)-\theta_\infty^\varepsilon\right\|^2-\mathcal{P}(t)$$

is monotonically decreasing, so that

$$\left\|\theta(t)-\theta_\infty\right\|^2\leq e^{-c_Mt}\left\|\theta_0-\theta_\infty\right\|^2+\int_0^t e^{-c_M(t-\sigma)}S(\sigma)d\sigma.\tag{5.10}$$

By (5.3) we have that for any ε there exists $T(\varepsilon)$ such that for any $\sigma\geq T(\varepsilon)$ it follows that $\left\|f(\sigma)-f_\infty\right\|_{V'}<\varepsilon$. Similar relationships take place for the other two functions, u and f_0. Therefore, if t is large enough, $t\geq T(\varepsilon)$, we can write (5.10) in the following way

$$\begin{aligned}\left\|\theta(t)-\theta_\infty\right\|^2\ \leq\ & e^{-c_Mt}\left\|\theta_0-\theta_\infty\right\|^2+\int_0^{T(\varepsilon)}e^{-c_M(t-\sigma)}S(\sigma)d\sigma\\ &+\int_{T(\varepsilon)}^t e^{-c_M(t-\sigma)}S(\sigma)d\sigma\end{aligned}$$

which implies that

$$\begin{aligned}\left\|\theta(t)-\theta_\infty\right\|^2\ \leq\ & e^{-c_Mt}\left\|\theta_0-\theta_\infty\right\|^2+c_1\left(1-e^{-c_M(t-T(\varepsilon))}\right)\varepsilon^2\\ &+c_2e^{-c_M(t-T(\varepsilon))}\left(\left\|f(\sigma)\right\|_{L^2_{\text{loc}}([0,\infty;V')}^2+\left|f_\infty\right|^2\right.\\ &+\left\|f_0(\sigma)\right\|_{L^2_{\text{loc}}([0,\infty;L^2(\Gamma_\alpha))}^2+\left|f_\infty^0\right|^2\\ &\left.+\left\|u(\sigma)\right\|_{L^2_{\text{loc}}([0,\infty;L^2(\Gamma_u))}^2+\left|u_\infty\right|^2\right),\end{aligned}$$

with c_1 and c_2 constants depending on c_M, $T(\varepsilon)$ and the domain. Passing now to the limit we obtain that

$$\lim_{s\in(t,+\infty)} ||\theta(s)-\theta_\infty|| = 0, \text{ as } t\to\infty, \tag{5.11}$$

as claimed. ■

Remark 5.2. If condition (5.1) does not hold, the longtime behavior of the trajectory $\theta(t)$ might be more complex. However, since the trajectory $\{\theta(t); t\geq 0\}$ is bounded in $L^2(\Omega)$, (because $\theta(t)\leq\theta_s$ a.e. in Ω, for each $t\geq 0$), it is compact in V'. Then, the ω-limit set

$$\Gamma_\omega = \{\lim_{t_n\to\infty}\theta(t_n) \text{ in } V'\}$$

which contains the stationary solutions to (5.2) is nonempty and the general theory of infinite dimensional attractors can be applied in order to investigate the structure of Γ_ω (see [28]).

Remark 5.3. If the problem data f, u and f_0 are periodic with the period T, then the solution to (2.1)-(2.4), with any θ_0, satisfies

$$\lim_{t\to+\infty} ||\theta(t)-\omega(t)||_{V'} = 0, \tag{5.12}$$

where ω is a periodic solution to (2.1), (2.3)-(2.4) with $\theta(x,0) = \theta(x,T)$. As specified before, ω is unique if the convection is Lipschitz and dominated by the diffusivity, by (5.1). (The proof can be extended to a vector $\mathbf{K}$ with K_j discontinuous using the arguments from [25] and [17])). For $K=0$, the solution θ to (2.1)-(2.4) satisfies (5.12), even if the operator is degenerate ($\rho=0$), but in this case ω is no longer unique (see [25]).

6. Numerical Results

In this last part we shall make a numerical application to the saturated - unsaturated flow of water in nonhomogeneous soils illustrating the theoretical conclusions by numerical results obtained for some realistic scenarios. The diffusion model which better applies in this case is the fast diffusion one and the expressions of β and K are taken from the soil science literature.

We specify that in water infiltration models the function $\theta(x,t)$ represents the volumetric water content, or soil moisture, defined as the ratio between the volume of pores filled with water and a reference volume around the point x at time t.

We have chosen the hydraulic model of P. Broadbridge and I. White (see [9]) in which the diffusivity β and conductivity K are given by nonlinear expressions of the form

$$\beta(\theta) = \frac{c(c-1)}{(c-\theta)^2},\ K(\theta) = \frac{(c-1)\theta^2}{c-\theta}, \text{ with the parameter } c\in(1,+\infty), \tag{6.1}$$

where the increase of c from 1 to $+\infty$ indicates the decrease of the medium nonlinearity. We stress that we consider the dimensionless model which has the same form as (2.1)-(2.4).

We consider a 1-D dimensionless domain $\Omega=(0,L)$, $L=2$, $T=10$, $f_0(x,t)=0$, $\alpha=1$. In all computations β is continuous, i.e., is defined by only one value of c. We envisage the comparison of the results obtained for a continuous K with those determined for a discontinuous K of the form

$$K(\theta) = \begin{cases} K_1(\theta), & 0 \leq \theta < r \\ K_2(\theta), & r \leq \theta < 1 \end{cases}$$

where K_i have the expressions (6.1) corresponding to $c = c_i$ (for K_i) and to two values of r.

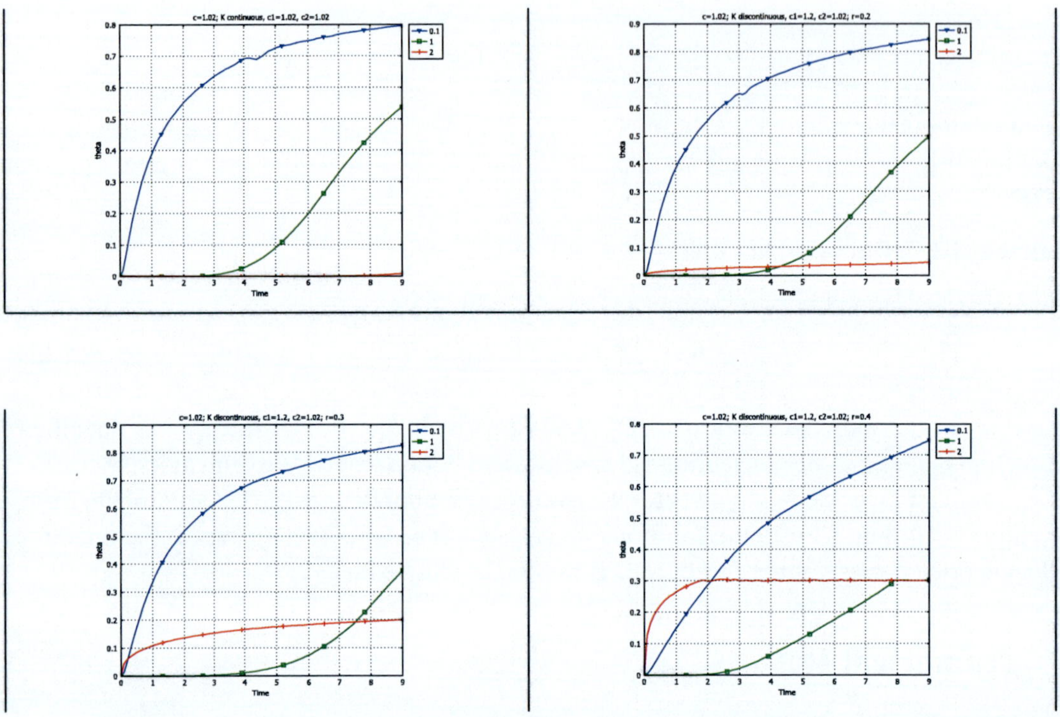

Figure 1.

Some scenarios using various combinations of the parameters c, c_i and r have been run and the results are in the Figures 1 and 2, where the graphics of the moisture at three levels are shown, respectively at $x = 0.1$ (marked by $\triangledown$), $x = 1$ (marked by $\square$) and $x = 2$ (marked by $|$). Figures 1a, b, c, d correspond to a highly nonlinear diffusion with $c = 1.02$ (in the expression of β) and show the different moisture evolutions determined by the change of K and r. Thus, In Fig. 1a we have a picture of the moisture evolution in time for a medium with a continuous conductivity ($c_1 = c_2 = 1.02$) while in Fig. 1b the effect due to the jump of K at $r = 0.2$ from a more linear behavior ($c_1 = 1.2$) to a certain nonlinear one ($c_2 = 1.02$) can be seen. Fig. 1c and 1d illustrate the differences occurred in the case 1b if the discontinuity point is translated to $r = 0.3$ and $r = 0.4$, respectively.

The other triplet of figures display what happens in a more linear soil (β corresponding to $c = 1.2$), starting from a continuous K (with $c_1 = c_2 = 1.2$) in Fig. 2a, Fig. 2b, 2c and 2d illustrate the results corresponding to a discontinuous K at $r = 0.2$, 0.3 and 0.4, respectively.

The discontinuity influence on the flow is better observed in the case when the diffusion coefficient is nonlinear ($c = 1.02$). Also one can notice that K_i corresponding to a c_i much larger than 1 has a stronger influence on the flow, determining a faster transport.

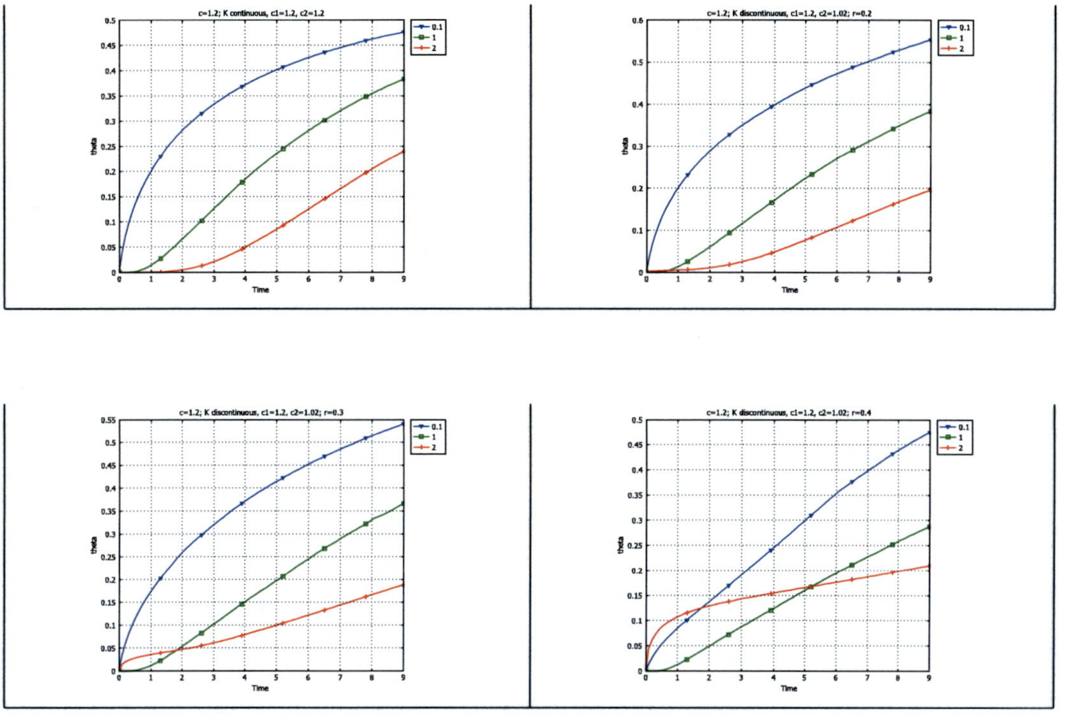

Figure 2.

Acknowledgements

The author would like to thank to the reviewer for the kindness of reading the manuscript with much care and for the observations he made. This work was supported by the project CEEX-D11-36/05 financed by the Romanian Ministry of Education, Research and Youth - National Authority for Scientific Research.

References

[1] Alt H.W.; Luckhaus, S. Quasi-linear elliptic-parabolic differential equations. *Math. Z.* 1983, *183*, 311-341.

[2] Alt, H.W.; Luckhaus, S.; Visintin, A. On nonstationary flow through porous media. *Ann. Mat. Pura Appl.* 1984, 136, 303-316.

[3] Aronson, D.G. The porous medium equation. In *Some Problems in Nonlinear Diffusion*; Fasano, A; Primicerio, M.; Ed.; *Lecture Notes in Mathematics* **1224**; Springer: Berlin, 1986.

[4] Barbu, V. *Nonlinear Semigroups and Differential Equations in Banach Spaces* ; Editura Academiei & Noordhoff International Publishing: Bucureşti-Leyden, 1976.

[5] Barbu, V.; Marinoschi, G. Existence for a time dependent rainfall infiltration model with a blowing up diffusivity. *Nonlinear Analysis Real World Applications* 2004, 5, 2, 231-245.

[6] Bénilan, Ph.; Crandall, M.G. The continuous dependence on φ of solutions of $u_t - \Delta\varphi(u) = 0$. *Indiana Univ. Math. J.* 1981, 30, 161-177.

[7] Bénilan, Ph.; Krushkov, S.N. Quasilinear first-order equations with continuous nonlinearities. *Russian Acad. Sci. Dokl. Math.* 1995, 50, 3, 391-396.

[8] Borsi, I.; Farina, A.; Fasano, A. On the infiltration of rain water through the soil with runoff of the excess water. *Nonlinear Analysis Real World Applications* 2004, 5, 763-800.

[9] Broadbridge, P.; White, I. Constant rate rainfall infiltration: A versatile nonlinear model. 1. Analytic solution. *Water Resources Research,* 1988, 24, 1, 145-154.

[10] Brezis, H. *Opérateurs Maximaux Monotones et Semi-groupes de Contractions dans les Espaces de Hilbert*, North Holland, 1973.

[11] J. Carillo: Entropy solutions for nonlinear degenerate problems, Arch. Rational Mech. Anal. 1999, *147*, 269-361.

[12] Carillo, J.; Wittbold, P. Uniqueness of renormalized solutions of degenerate elliptic-parabolic problems. *J. Differential Equations* 1999, 156, 93-121.

[13] Chasseigne, E.; Vázquez, J.L. Theory of extended solutions for fast diffusion equations in optimal classes of data. Radiation from singularities. *Archive Rat. Mech. Anal.* 2002, 164, 133-187.

[14] Daskalopoulos, P.; Del Pino, M. On nonlinear parabolic equations of very fast diffusion. *Arch. Rational Mech. Anal*. 1997, 137, 363-380.

[15] Ruiz Goldstein, G. Nonlinear singular diffusion with nonlinear boundary conditions. *Math. Methods Appl. Sci.* 1993, 16, 11, 779-798.

[16] Goldstein, J.A. *Nonlinear semigroups and applications to degenerate parabolic partial differential equations*; Analysis and Geometry; Korea Inst. Tech.: Taejon, Korea, 1989; pp 1-40.

[17] Haraux, A. Nonlinear Evolution Equations-Global Behaviour of Solutions, *Lecture Notes in Mathematics* **841**; Springer Verlag: Berlin, 1981

[18] Krushkov, S.N. Generalized solutions of the Cauchy problem in the large for first-order nonlinear equations. *Soviet Math. Dokl.* 1969, 10, 785-788.

[19] Lions, J.L.; Magenes, E. *Non-homogeneous Boundary Value Problems and Applications*. I. Springer-Verlag: Berlin, 1972.

[20] Marinoschi, G. Nonlinear infiltration with a singular diffusion coefficient. *Differential Integral Equations* 2003, 16, 9, 1093-1110.

[21] Marinoschi, G. A free boundary problem describing the saturated unsaturated flow in a porous medium. *Abstract Applied Analysis.* 2004, 9, 729-755.

[22] Marinoschi, G. On a nonlinear boundary value problem related to infiltration in unsaturated media. In *New Trends in Continuum Mechanics;* Suliciu, M.; Ed.; Theta Publishing: Bucharest 2005, pp. 175-184.

[23] Marinoschi, G. A free boundary problem describing the saturated unsaturated flow in a porous medium. II. Existence of the free boundary in the 3-D case. *Abstract Applied Analysis.* 2005 8, 813-854.

[24] Marinoschi, G. Functional Approach to Nonlinear Models of Water Flow in Soils; *Mathematical Modelling:Theory and Applications* **21**; Springer: Dordrecht, 2006.

[25] Marinoschi, G. Periodic solutions to fast diffusion equations with non Lipschitz convective terms. *Nonlinear Analysis Real World Applications* , in press.

[26] Oleinik, O.A.; Kalashnikov, A.S.; Yui-Lin, C. The Cauchy problem and boundary value problems for equations of the type of nonstationary filtration. *Izv. Akad. Nauk. SSSR Ser. Mat.* 1958, 22, 667-704.

[27] Otto, F. L^1 -contraction and uniqueness for unstationary saturated-unsaturated porous media flow. *Adv. Math. Sci. Appl.* 1997, 7, 537-553.

[28] Temam, R. Infinite Dimensional Dynamical Systems in Mechanics and Physics; *Applied Mathematical Sciences Series* **68**; Springer-Verlag: New-York, 1988.

[29] Vázquez, J.L. Darcy's law and the theory of shrinking solutions of fast diffusion equations. *SIAM J. Math. Anal.* 2003, 35, 4, 1005-1028.

[30] Vázquez, J.L. Symmetrization and mass comparison for degenerate nonlinear parabolic and related elliptic equations. *Advanced Nonlinear Studies* 2005, 5, 87-131.

[31] Vázquez, J.L. Smoothing and Decay Estimates for Nonlinear Diffusion Equations. Equations of Porous Medium Type; *Oxford Lecture Series in Mathematics and its Applications* **33**; Oxford University Press, 2006.

[32] Vázquez, J.L. The Porous Medium Equation. *Mathematical Theory*; Clarendon Press: Oxford, 2007.

Reviewed by Alain Miranville

In: Progress in Nonlinear Analysis Research
Editor: Erik T. Hoffmann, pp. 243-268

ISBN 978-1-60456-359-7

Chapter 10

FUZZY SET BASED MULTICRITERIA DECISION MAKING AND ITS APPLICATIONS

P. Bernardes[1,a], P. Ekel[1,b], J. Kotlarewski[2,c] and R. Parreiras[1,d]

[1]Pontifícia Universidade Católica de Minas Gerais, Av. Dom Jose Gaspar, 500, 30535-610, Belo Horizonte, MG, Brasil

[2]FURNAS Centrais Elétricas, Rua Real Grandeza, 219, Bloco A, 22283-900 Rio de Janeiro, RJ, Brasil

Abstract

This work studies the use of fuzzy sets for handling multicriteria decision making problems. The multicriteria approach is needed to solve:

- problems whose solution consequences cannot be estimated with a single criterion;
- problems that, initially, may require a single criterion, but their unique solutions are unachievable, due to the existence of decision uncertainty regions, which can be contracted using additional criteria.

According to this, two classes of models, $<X, M>$ and $<X, R>$, can be constructed. The analysis of $<X, M>$ models, based on applying the Bellman-Zadeh approach to decision making in a fuzzy environment, is briefly described. The analysis of $<X, R>$ models is based on four techniques for fuzzy preference modeling. These techniques permit the evaluation, comparison, selection, and/or ordering of alternatives with the use of quantitative estimates, as well as qualitative estimates, based on knowledge, experience, and intuition of professionals. With the availability of different techniques, the most appropriate one can be chosen, considering the sources of information and its uncertainty. To extend the results associated with analyzing $<X, R>$ models, two rational consensus schemes are discussed. They permit one to generalize the analysis of $<X, R>$ models to multiperson multicriteria decision making. These schemes can also be used for evaluating priority weights for criteria in analyzing $<X, M>$ models. Finally, the results of the present work are illustrated by using a multiperson

[a] E-mail address: patib@pucminas.br
[b] E-mail address: ekel@pucminas.br
[c] E-mail address: jkotlar@furnas.com.br
[d] E-mail address: roberta.parreiras@terra.com.br

multicriteria decision making framework to solve an enterprise strategy planning problem, generated with the use of the Balanced Scorecard methodology.

1. Introduction

Diverse types of uncertainty are often encountered in a wide range of problems related to the design, planning, and control of complex systems. The consideration of the uncertainty factor in constructing mathematical models serves as a means for increasing their adequacy and, as a result, the credibility and factual efficiency of the decisions based on their analysis. Recent investigations show the benefits of applying fuzzy set theory [1, 2] to deal with diverse types of uncertainty. Its use in problems of optimization character offers advantages of both fundamental nature (the possibility of validly obtaining more effective, less "cautious" solutions) and of computational character. The literature presents examples of successful applications of techniques for decision making in a fuzzy environment in diverse areas. For instance, it is possible to indicate our results related to solving power engineering [3, 4], management [5], and naval engineering [6] problems with the use of these techniques.

The uncertainty of goals is an important kind of uncertainty related to a multicriteria character of many problems of optimization nature. It is possible to classify two major classes of problems, which need the use of a multicriteria approach:

- problems whose solution consequences cannot be estimated with a single criterion: these problems are associated with the analysis of economic, as well as natural indices (when alternatives cannot be reduced to comparable form) and also with the need to consider indices whose cost estimations are hampered;
- problems that, initially, may require the use of a single criterion, but their unique solutions are unachievable, due to the existence of the uncertainty of information, generating the decision uncertainty regions. The convincing means to contract these regions is the introduction of additional criteria, including criteria of qualitative character (based on knowledge, experience, and intuition of involved experts) [7].

In this context, two classes of multicriteria models, so-called $< X, M >$ and $< X, R >$ models, may be constructed. Their analysis is associated with the use of the Bellman-Zadeh approach to decision making in a fuzzy environment and with the use of techniques involving fuzzy preference modeling, respectively.

To extend the results associated with analyzing $<X, R>$ models to multiperson multicriteria decision making, some questions related to the construction of consensus schemes are discussed in the paper. The necessity of developing consensus schemes for multiperson multicriteria decision making is explained by the fact that the expert opinions may be initially discordant. Hence, a consensus scheme is required in order to minimize their divergence and obtain a consistent collective opinion. The growing interest to constructing consensus schemes, as well as aggregation procedures for generating collective decisions is illustrated, for instance by [8-19].

The term "consensus scheme" refers to dynamic and iterative discussion processes among the involved decision makers (DM), conducted by a human or artificial moderator, with the purpose of minimizing the discrepancy between the DM individual opinions. In order to give support to the moderator for guiding the discussion process, these procedures often make use of indexes that reflect the concordance level among the specialists. The term "aggregation procedure", on the other hand, refers to those aggregation models that aims just at obtaining collective consistent solutions, considering the opinions of all involved specialists (without formally involving such kind of rational discussion process).

Two rational consensus schemes are discussed in the paper: one was proposed in [14] and the other was introduced here, being derived from the aggregation model of [15]. They permit one to generalize the analysis of <*X*, *R*> models to multiperson multicriteria decision making. These schemes can also be used for evaluating priority weights for criteria in analyzing <*X*, *M*> models. Finally, the usefulness of extending a fuzzy decision making based on <*X*, *R*> models to handle group decision problems is illustrated by an example, generated with the use of the Balanced Scorecard methodology [20] to solve an enterprise strategy planning problem.

2. Problem Formulation

Multicriteria decision problems involve the following basic elements:

The set of alternatives *X*. When considering the so-called *discrete problems*, *X* corresponds to a finite and discrete list of alternatives. On the other hand, when considering the *continuous problems*, *X* tends to be denser, continuous and limited only by mathematical constraints, being described as a subset of $\Re^n$, named Pareto optimal set [21]. In addition to the execution of a multicriteria decision method, the process of solving continuous problems often requires the execution of an optimization algorithm, in order to obtain a representation of the Pareto optimal set.

The set of criteria $C=\{c_1,...,c_m\}$. Each criterion represents a viewpoint, according to which the alternatives are evaluated and compared. Usually, it is mathematically modeled by a function $c_k(x_j): X \to \Re$ that assigns a grade to each alternative, reflecting the decision maker preferences. In the continuous problems, each criterion is often derived from an objective function. But, it is worth mentioning that additional criteria that do not correspond to any objective function can also be considered.

Table 1. Types and practical examples of multicriteria decision problems [22].

Problem type	Example
Rank alternatives from the best to the worst. This ordering is not necessarily complete.	Ranking candidates for an occupation in an enterprise.
Choose the best alternative or a limited group of satisfactory solutions.	Selecting one from several versions of the same project. The best solutions are not necessarily optimal, but just satisfactory.
Classify alternatives into predefined homogeneous categories.	Medical diagnosis based on the classification of patients into disease groups according to observable symptoms.

Apart from those basic elements, the question raised by the problem must be specified. Table 1 defines and illustrates with practical examples the main types of multicriteria decision problems. This work deals with decision methods suitable for solving ranking and choice problems.

3. Fuzzy Decision with the Bellman-Zadeh Approach

When analyzing $< X, M >$ models, a vector of objective functions $F(X) = \{F_1(X),...,F_m(X)\}$ is considered, and the problem consists in optimizing simultaneously all objective functions, i.e.,

$$F_k(X) \to \underset{X \in L}{\text{extr}}, \quad k = 1,...,m, \tag{1}$$

where L is a feasible region in $\mathfrak{R}^n$.

The first step for solving this problem is associated with determining a set of Pareto solutions $\Omega \subseteq L$ [21]. This step is useful; however it does not permit one to obtain unique solutions. It is necessary to choose a particular Pareto solution, on the basis of the information provided by DM. There are three approaches for using this information: *a priori*, *a posteriori*, and adaptive [23]. The favorite approach is usually the adaptive one. In this approach, the procedure of successive improving the solution quality is performed as a transition from $X_\alpha^0 \in \Omega \subseteq L$ to $X_{\alpha+1}^0 \in \Omega \subseteq L$, taking into consideration the information I_α provided by DM. The solution search may be presented in the following form:

$$X_1^0, F(X_1^0) \xrightarrow{I_1} ... \xrightarrow{I_{\alpha-1}} X_\alpha^0, F(X_\alpha^0) \xrightarrow{I_\alpha} ... \xrightarrow{I_{\omega-1}} X_\omega^0, F(X_\omega^0). \tag{2}$$

The process (2) serves for two types of adaptation: computer to preferences of DM and DM to the problem. The first type of adaptation is based on information received from DM. The second type of adaptation is realized as a result of carrying out several steps $X_\alpha^0, F(X_\alpha^0) \xrightarrow{I_\alpha} X_{\alpha+1}^0, F(X_{\alpha+1}^0)$, which permit DM to understand the correlation between own needs and possibilities of their satisfaction by the model (1).

When analyzing multicriteria problems, it is necessary to solve questions related to normalizing criteria, selecting principles of optimality and considering priorities of criteria. Their solution and, therefore, the development of multicriteria methods is carried out in several directions discussed in [24, 25]. Without discussion of these directions, it is necessary to point out that an important question in multicriteria optimization is the solution quality. It is considered as high if levels of satisfying criteria are equal or close to each other (harmonious solutions), when all objective functions have the same importance [3, 26]. From this point of view, it should be recorded the validity and advisability of the direction related to the principle of guaranteed result, which can be realized [3, 26] on the basis of applying the Bellman-Zadeh approach to decision making under a fuzzy environment [1, 2, 27].

The Bellman-Zadeh approach permits one to realize an effective (from the computational standpoint) as well as rigorous (from the standpoint of obtaining solutions $X^0 \in \Omega \subseteq L$) method of analyzing multicriteria models. Its use also allows one to preserve a natural measure of uncertainty in decision making and to take into account indices, criteria, and constraints of qualitative character.

When using the Bellman-Zadeh approach, each objective function $F_k(X)$ is replaced by a fuzzy objective function or a fuzzy set:

$$A_k = \{X, \mu_{A_k}(X)\}, \quad X \in L, \quad k = 1,\ldots,m, \tag{3}$$

where $\mu_{A_k}(X)$ is the membership function of A_k [1, 2].

A fuzzy solution D with setting up the fuzzy sets (3) is turned out as a result of the intersection $D = \bigcap_{k=1}^{m} A_k$ with a membership function

$$\mu_D(X) = \min_{k=1,\ldots,m} \mu_{A_k}(X), \quad X \in L. \tag{4}$$

Its use permits one to obtain the solution proving the maximum degree

$$\max \mu_D(X) = \max_{X \in L} \min_{k=1,\ldots,m} \mu_{A_k}(X) \tag{5}$$

of belonging to the fuzzy solution D. Therefore, problem (1) is reduced to

$$X^0 = \arg\max_{X \in L} \min_{k=1,\ldots,m} \mu_{A_k}(X). \tag{6}$$

To obtain (6), it is necessary to build membership functions $\mu_{A_k}(X)$, $k = 1,\ldots,m$ reflecting a degree of achieving "own" optima by $F_k(X), X \in L$, $k = 1,\ldots,m$. This condition is satisfied by the use of membership functions

$$\mu_{A_k}(X) = \left[\frac{F_k(X) - \min_{X \in L} F_k(X)}{\max_{X \in L} F_k(X) - \min_{X \in L} F_k(X)} \right]^{w_k} \tag{7}$$

for maximized objective functions or by the use of membership functions

$$\mu_{A_k}(X) = \left[\frac{\max F_k(X) \ - \ F_k(X)}{\max\limits_{X \in L} F_k(X) \ - \min\limits_{X \in L} F_k(X)}\right]^{w_k} \tag{8}$$

for minimized objective functions. In (7) and (8), $w_k, k = 1,...,m$ are importance factors for the corresponding objective functions.

Since the solution X^0 must belong to $\Omega \subseteq L$, it is necessary to build

$$\overline{\mu}_D(X) = \bigwedge_{k=1}^{m} \mu_{A_k}(X) \wedge \mu_\pi(X) = \min\{\min_{k=1,..,m} \mu_{A_k}(X), \mu_\pi(X)\}, \tag{9}$$

where $\mu_\pi(X) = 1$ if $X \in \Omega$ and $\mu_\pi(X) = 0$ if $X \notin \Omega$.

The procedures for solving the problem (5), discussed in [29], provide the line in obtaining $X^0 \in \Omega \subseteq L$ in accordance with (9). Thus, it can be said about equivalence of $\overline{\mu}_D(X)$ and $\mu_D(X)$. It permits one to give up the necessity of implementing a cumbersome procedure for building the set $\Omega \subseteq L$.

Finally, the existence of additional conditions (indices, criteria, and/or constraints) of qualitative character, defined by linguistic variables [1, 2], reduces (6) to

$$X^0 = \arg\max_{X \in L} \min_{k=1,...,m+s} \mu_{A_k}(X), \tag{10}$$

where $\mu_{A_k}(X), X \in L, k = m+1,...,s$ are membership functions of fuzzy values [1, 2] of linguistic variables which reflect these additional conditions.

As it can be noticed, when applying the Bellman-Zadeh approach to analyzing <X, M> models, DM is to specify, as input parameters, only the importance factors for all criteria. Thus, in the group decision context, a consensus scheme is necessary in order to obtain collective values for these importance factors.

3. Fuzzy Decision Based on the Fuzzy Preference Relations

Some criteria, which are to be taken into account in the process of decision making, can be reflected only by the description of qualitative (semantic, contextual) character (for instance, "comfort of operation", "flexibility of development", "investment utility", etc.) based on experience, knowledge, and intuition of the involved experts. This generates the necessity of the evaluation, comparison, choice, and/or ordering of alternatives, which correspond, in full measure, to the preferences of DM. Considering that the human thought and preference appreciation are vague, inexact, and subjective, the theory of fuzzy sets can serve as an important tool in solving many types of problems.

The application of fuzzy set theory to preference modeling provides a flexible structure, which allows one to deal with "fuzziness" of the appreciations and incorporate more human consistency in the applied models. The utility of using fuzzy set theory is also defined by its linguistic aspect applicable for different decision problems and different preference structures. Taking this into account, four approaches for processing fuzzy preference relations are briefly discussed below. The consideration of different techniques allows the choice of the most adequate approach, considering all possible sources of information and its uncertainty.

Assume a set X of alternatives, which are to be examined following m criteria of quantitative and/or qualitative nature. The decision making problem can be represented through <X, R> models, where X is a discrete set of alternatives and $R = \{R_1,...,R_m\}$ is a vector of fuzzy preference relations. In this case, the DM preferences, considering the kth criterion, is described by a fuzzy preference relation $R_k \subset X \times X$, also called in the literature nonstrict fuzzy preference relation, fuzzy binary preference relation, fuzzy weak preference relation and fuzzy binary relation [30]. Its membership function $\mu_{R_k}(X_i, X_j): X \times X \to [0,1]$ indicates the degree to which the alternative X_i weakly dominates (or is at least as good as) X_j.

The entries of the pairwise comparison matrix R_k may be directly or indirectly defined by the specialists. For instance, in [30, 31], it is assumed that DM directly specifies these entries. Reference [28], on the other hand, describes a natural way of constructing these matrices. The availability of fuzzy or linguistic estimates of alternatives $\widetilde{F}_k(X_i)$, k=1,…,m, $X_i \in X$ (constructed on the basis of expert estimation or on the basis of aggregating information arriving from different sources) with the membership functions, $\mu[F_k(X_i)]$ k=1,…,m, $X_i \in X$ permits one to construct elements of R_k, k=1,…,m, as follows:

$$\mu_{R_k}(X_i, X_j) = \sup_{\substack{X_i, X_j \in X \\ F_k(X_i) \geq F_k(X_j)}} \min\{\mu[F_k(X_i)], \mu[F_k(X_j)]\}, \tag{11}$$

$$\mu_{R_k}(X_j, X_i) = \sup_{\substack{X_i, X_j \in X \\ F_k(X_j) \geq F_k(X_i)}} \min\{\mu[F_k(X_i)], \mu[F_k(X_j)]\}, \tag{12}$$

if the kth criterion is associated with maximization. If the kth criterion is associated with minimization, then (11) and (12) are written for $F_k(X_i) \leq F_k(X_j)$ and $F_k(X_j) \leq F_k(X_i)$, respectively.

The nonstrict fuzzy preference relation R_k can be represented by a strict fuzzy preference relation P_k and an indifference relation I_k: $R_k = P_k \cup I_k$. The strict preference relation P_k is constituted by all pairs of alternatives that satisfy the conditions: $(X_i, X_j) \in R_k$ and $(X_j, X_i) \notin R_k$. If $(X_i, X_j) \in P_k$, it can be said that X_i is strictly better than X_j (or X_i dominates X_j) The indifference relation, on the other hand, is constituted by all pairs of alternatives that simultaneously satisfy the conditions: $(X_i, X_j) \in R_k$ and $(X_j, X_i) \in R_k$. If $(X_i, X_j) \in I_k$, it can be said that X_i is indifferent to X_j.

With the use of the inverse relation $(R_k)^{-1}$ ($(X_i, X_j) \in (R_k)^{-1}$ is equivalent to $(X_j, X_i) \in R_k$) the strict preference relation P_k can be defined as a difference of the fuzzy sets: $P_k = R_k \setminus (R_k)^{-1}$, and its membership function can be obtained by [31]:

$$\mu_{P_k}(X_i, X_j) = \max\{\mu_{R_k}(X_i, X_j) - \mu_{R_k}(X_j, X_i), 0\}. \tag{13}$$

The use of (13) permits one to carry out the choice of alternatives. In particular, as $P_k(X_j, X_i)$ describes the set of all alternatives X_i that are strictly dominated by X_j, its compliment $\overline{P}_k(X_j, X_i)$ corresponds to the set of alternatives that are not dominated by other alternatives from X. Therefore, in order to meet the set of alternatives from X that are not dominated by any other alternative, it suffices to obtain the intersection of all $\overline{P}_k(X_j, X_i)$. This intersection is the set of nondominated alternatives with the membership function

$$\mu_{R_k}^{nd}(X_i) = \inf_{X_j \in X}\{1 - \mu_{P_k}(X_j, X_i)\} = 1 - \sup_{X_j \in X} \mu_{P_k}(X_j, X_i), \tag{14}$$

which reflects the level of nondominance of each alternative X_i.

Hence, a good choice for a monocriteria problem based on this model should be the alternatives providing:

$$X_{R_k}^{nd} = \{X_i^{nd} \in X \mid \mu_{R_k}^{nd}(X_i^{nd}) = \sup_{X_i \in X} \mu_{R_k}^{nd}(X_i)\}. \tag{15}$$

It is worth emphasizing that the alternatives satisfying $X_{R_k}^{nd} = \{X_j^{nd} \in X \mid \mu_{R_k}^{nd}(X_j^{nd}) = 1\}$ are actually nonfuzzy nondominated and can be considered as the nonfuzzy solution for the choice problem.

Expressions (13)-(15) may be used to solve choice problems as well as ranking problems with a single criterion. Expressions (13)-(15) may also serve for constructing procedures to solve multicriteria choice and ranking problems.

In the *first procedure* for solving multicriteria problems, when we have nonstrict preference relations for each criterion, a global relation can be obtained through the intersection of these relations:

$$\mu_G(X_i, X_j) = \min(\mu_{R_1}(X_i, X_j), \ldots, \mu_{R_m}(X_i, X_j)). \tag{16}$$

In this case, Equations (13)-(15) can be applied taking into account the global relation (16). The resulting fuzzy set of nondominated alternatives fulfills the role of a Pareto set [31]. This set can be contracted in a subsequent analysis, where the importance of each relation R_k, k=1,...,m is differentiated through their weighted aggregation:

$$\mu_T(X_i,X_j)=\sum_{k=1}^{m} w_k \mu_{R_k}(X_i,X_j). \tag{17}$$

In (17), the weights (or importance factors) of each criterion must satisfy the conditions: $w_k > 0$, k=1,...,m; $\sum_{k=1}^{m} w_k = 1$.

With the construction of $\mu_T(X_i,X_j)$, it is possible to obtain the membership function $\mu_T^{nd}(X_i)$ of the set of nondominated alternatives, following a similar procedure as the one described above for the global relation $\mu_G(X_i,X_j)$, which involves Equations (13)-(15). Finally, the nondominance level can be obtained by performing the intersection between $\mu_R^{nd}(X_i)$ and $\mu_T^{nd}(X_i)$:

$$\mu_W^{nd}(X_i)=\min\{\mu_R^{nd}(X_i),\mu_T^{nd}(X_i)\}, \tag{18}$$

and the following set of fuzzy nondominated solutions can be met:

$$X^{nd}=\{X_i^{nd}\in X \mid \mu_W^{nd}(X_i^{nd})=\sup_{X_i\in X}\mu_W^{nd}(X_i)\}. \tag{19}$$

The *second procedure*, which is of a lexicographic character, consists in a step-by-step introduction of criteria for comparing alternatives. In this case, a sequence $X^1, X^2,\ldots, X^m$, such that $X \supseteq X^1 \supseteq \ldots \supseteq X^m$, is obtained with the following expressions:

$$\mu_{R_k}^{nd}(X_j)=\inf_{X_i\in X^{k-1}}\{1-\mu_{P_k}(X_j,X_i)\}=1-\sup_{x_i\in X^{k-1}}\mu_{P_k}(X_j,X_i),\ k=1,\ldots,m, \tag{20}$$

$$X^k=\{X_i^{nd_k}\in X^{k-1} \mid \mu_{R_k}^{nd}(X_i^{nd_k})=\sup_{X_j\in X^{k-1}}\mu_{R_k}^{nd}(X_i)\}. \tag{21}$$

The described choice and ranking procedures have found practical applications. However, it is possible to propose the *third procedure*. In particular, the use of (14) permits one to construct the membership functions of the set of nondominated alternatives for each fuzzy preference relation. The membership functions $\mu_{R_k}^{nd}(X_i)$, k=1,...,m play a role identical to membership functions replacing objective functions $F_k(X)$, k=1,...,m in analyzing $< X, M >$ models. Therefore, in order to obtain X^{nd}, it is possible to construct:

$$\mu^{nd}(X_i) = \min_{1 \le k \le m} \mu_{R_k}^{nd}(X_i). \tag{22}$$

If necessary to differentiate the importance of each preference relation, it is possible to transform (22) as:

$$\mu^{nd}(X_i) = \min_{1 \le k \le m} [\mu_{R_k}^{nd}(X_i)]^{w_k}. \tag{23}$$

The use of (23) does not require the normalization of w_k, k=1,...,m in the way that they were normalized to apply (17).

The application of the second procedure may lead to solutions different from the results obtained on the basis of the first procedure. However, solutions based on the first and on the third procedures, which have a single generic basis, may at time also be different. At the same time, the third procedure is more preferential from the substantial point of view. In particular, the use of the first procedure can lead to choosing alternatives with the degree of nondominance equal to one, though these alternatives are not the best ones from the point of view of all preference relations. The third procedure can give this result only for alternatives that are the best solutions from the point of view of all fuzzy preference relations. It should be stressed that the possibility to obtain different solutions on the basis of different approaches is to be considered natural, and the choice of the approach is a prerogative of DM.

All procedures directed at analyzing <*X*, *R*> models described above suppose the explicit direct or indirect ordering of all criteria. The first and the third procedures require the specification of an importance factor for each criterion; the second procedure requires the ranking of all criteria based on their respective priority. Taking this into account, it is necessary to distinguish the *fourth procedure* based on the results of [31], which allows one to present the information related to the importance of the criteria in the form of a pairwise comparison matrix:

$$W = [w \times w, \mu_W(w_a, w_b)],\ a,b=1,\ldots,m. \tag{24}$$

With the membership functions of the sets of nondominated alternatives (15) for all preference relations, it is possible to construct the following fuzzy preference relation induced by the preference relations $R_1,\ldots,R_m$:

$$\mu_{R,W}^{nd}(X_i, X_j) = \sup_{w_a, w_b \in W} \min_{X_i, X_j \in X} \{\mu_{R_a}^{nd}(X_i), \mu_{R_b}^{nd}(X_j), \mu_W(w_a, w_b)\}. \tag{25}$$

This induced fuzzy preference relation can be considered as a result of aggregating the vector R_1, ...,R_m, with the use of information reflecting the relative importance of criteria given as described in Equation (24).

4. Consensual Schemes

As already mentioned, in real situations, a set of decision makers $D=\{d_1,...,d_q\}$ may take part in the decision process and it is quite natural that, initially, their opinions disagree. In such cases, a consensual opinion must be achieved, taking into account the information provided by each involved specialist. Classically, the term consensus is defined as a unanimous concordance among all involved individuals. But, in practice, this concept is unsuitable for three main reasons [11]:

- it distinguishes only two situations: the existence or nonexistence of consensus;
- the chances of achieving such level of concordance are very low;
- in practical situations, it is not necessary to achieve such level of concordance.

Taking this into account, the current literature presents several schemes (for instance [10, 11, 14, 16]) that involve measuring consensus levels, based on the concordance among the opinions of DMs. Usually, the use of these schemes involves an iterative discussion, where a (human or artificial) moderator indicates whether the specialists need to discuss their opinions, in an effort to make them more similar. In general, at each step of this iterative process, the least concordant specialist is invited to review his opinion [11].

An important aspect related to consensus schemes lies in the fact that, when defining a collective opinion, sometimes, it is relevant to consider the different levels of influence of each individual. For instance, the opinion of DM who has more experience and knowledge on the problem or who has more authority in the enterprise may be more relevant than the others.

Here, we study the use of two different consensus schemes, based on different aggregation models [14, 15], in order to extend the fuzzy decision techniques described above to the multiperson multicriteria decision context. These schemes have been chosen for the following main reasons: both of them have produced consistent results, have intuitive appeal and are relatively easy to implement and to connect with the decision techniques described here. Both consensus schemes allow differentiating the importance of each expert opinion, which is a very useful resource. Their main difference lies in the kind of information (each one is based on a preference structure [11]) and in the amount of information they require from DM. With the availability of distinct consensus schemes, the most appropriated one may be chosen, taking into account the sources of information and the way the group prefers to express their opinions.

The **consensus scheme 1 (CS1)** is based on the aggregation model from [15] (which is an improved version of the model proposed in [8]) and is briefly described in Section 5. In CS1, the zth DM must provide his own opinion about each alternative, considering the kth criterion, in terms of fuzzy estimates $\mu_z[F_k(X_i)]$. The individual estimates relative to each alternative are aggregated into a collective estimate $\mu[F_k(X_i)]$, which is utilized in the construction of *collective pairwise comparison matrices* R_k, $k=1,\dots,m$, with the use of Equations (11) and (12). These matrices are required by the procedures related to analyzing <*X*, *R*> models. Assuming that the set *X* contains *n* alternatives, it is interesting to observe that each DM must provide at least $m \times n$ fuzzy estimates (if he is not asked to review his opinions), expressing his evaluations of each alternative.

In the case of the fourth procedure, in a similar way, each DM can also provide his opinion about each decision criterion, through fuzzy estimates, and the W matrix (see Expression (24)) can be obtained using Equations (11) and (12). Therefore, in the case of the fourth procedure, at least m additional fuzzy estimates, reflecting the importance of each criterion, are also required.

The **Consensus scheme 2 (CS2)** is based on the aggregation model proposed in [14] and briefly described in Section 6. Initially, each DM performs pairwise comparisons between the alternatives and specifies the preference level of each alternative over another, using the multiplicative preference relations, based on the Saaty scale [32]. Then, a constrained linear fuzzy monoobjective optimization model is constructed from these judgments in order to meet the consensual prioritization weights w_i, i=1,...,n. It is interesting to emphasize that each DM may specify less than $m \times n(n-1)/2$ fuzzy estimates, in this case, as the initial comparison matrix may be incomplete.

In the present work, this consensus scheme is utilized only in order to obtain the importance weights for all criteria. But, it is worth mentioning that it can also be utilized in order to construct the collective pairwise comparison matrices required by the decision methods directed at analyzing <X, R> models. Having met the consensual prioritization weights w_i, i=1,...,n, for all alternatives, a reciprocal pairwise comparison matrix such as the one below can be constructed:

$$M = \begin{bmatrix} \frac{w_1}{w_1} & & \frac{w_1}{w_n} \\ & \ddots & \\ \frac{w_n}{w_1} & & \frac{w_n}{w_n} \end{bmatrix},$$

and the transformation function proposed in [33] can be utilized to convert the multiplicative preference relations $a_{ij} = w_i / w_j$ from M into fuzzy preference relations:

$$\mu_R(X_i, X_j) = f(a_{ij}) = \frac{1}{2}\left(1 + \log_9 a_{ij}\right).$$

5. Aggregation Model Based on Similarity and Distance Measures

The considered aggregation model is based on a measure of concordance that combines both fuzzy distance and fuzzy similarity concepts. According to [15], given the fuzzy estimates $\widetilde{F}_k^y(X)$, $\widetilde{F}_k^z(X)$ informed by d_y and d_z, the weighted similarity between $\widetilde{F}_k^y(X)$ and $\widetilde{F}_k^z(X)$ is given by:

$$S_w(\widetilde{F}_k^y(X),\widetilde{F}_k^z(X)) = \frac{\int_X (\min\{\mu_y[F_k(X)],\mu_z[F_k(X)]\})^2 dX}{\int_X (\max\{\mu_y[F_k(X)]],\mu_z[F_k(X)]]\})^2 dX}, \qquad (26)$$

where $\mu_y[F_k(X)]$ and $\mu_z[F_k(X)]$ are the membership functions relative to the fuzzy estimates $\widetilde{F}_k^y(X)$, $\widetilde{F}_k^z(X)$, respectively.

The distance between $\widetilde{F}_k^y(X)$ and $\widetilde{F}_k^z(X)$ can be calculated as follows [15]:

$$D_h(\widetilde{F}_k^y(X),\widetilde{F}_k^z(X)) = \frac{1}{2}\left[\int_X |\mu_y[F_k(X)] - \mu_z[F_k(X)]|\, dX + d_{inf}(\widetilde{F}_k^y(X),\widetilde{F}_k^z(X))\right]. \qquad (27)$$

In (27) the integral corresponds to the Hamming distance between the fuzzy estimates $\widetilde{F}_k^y(X)$, $\widetilde{F}_k^z(X)$ and the term d_{inf} is given by $d_{inf} = \inf\{d(a,b), a \in \widetilde{F}_k^y(X), b \in \widetilde{F}_k^z(X)\}$, where $d(a,b)$ is the usual distance metric. Thus, supposing the membership functions of $\widetilde{F}^y(X)$ and $\widetilde{F}^z(X)$ correspond to the trapezoidal fuzzy numbers (a_1, a_2, a_3, a_4) and (b_1, b_2, b_3, b_4) [1, 2], respectively, then $d_{inf} = \inf\{d(a,b), a \in [a_1,a_4], b \in [b_1,b_4]\}$. The inclusion of the term d_{inf} in the distance metric from (27) is important to adequately handle the cases where the fuzzy estimates have no intersection [15].

As the next Equation shows, the concordance level $S(\widetilde{F}_k^y(X),\widetilde{F}_k^z(X))$ consists in a linear aggregation of the distance and the weighed similarity metrics. The parameter β, defined in the range $0 \le \beta \le 1$, allows S_w to have more or less influence on the concordance value:

$$S(\widetilde{F}_k^y(X),\widetilde{F}_k^z(X)) = \beta S_w(\widetilde{F}_k^y(X),\widetilde{F}_k^z(X)) + (1-\beta)(1-\overline{D}_h(\widetilde{F}_k^y(X),\widetilde{F}_k^z(X)), \qquad (28)$$

where $\overline{D}_h$ is the normalized distance: $\overline{D}_h = \dfrac{D_h(\widetilde{F}_k^y(X),\widetilde{F}_k^z(X))}{\max\limits_{p,q} D_h(\widetilde{F}_k^p(X),\widetilde{F}_k^q(X))}$.

The weighted consistency degree of the zth DM, given by C_z, depends on the importance e_y of the yth DM and also on the concordance level between the zth DM and the others:

$$C_z = \sum_{y=1}^{q} e_y S(\widetilde{F}_k^z(X),\widetilde{F}_k^y(X)). \qquad (29)$$

The collective estimate is determined by the weighted sum of individual estimates:

$$\widetilde{F}_k^{\,C}(X) = \sum_{z=1}^{q} \overline{C}_z \otimes \widetilde{F}_k^{\,z}(X), \tag{30}$$

where $\otimes$ is the fuzzy multiplication operator [34]. In (30), the weight $\overline{C}_z$ relative to the zth DM is calculated by $\overline{C}_z = \dfrac{C_z}{\sum_{y=1}^{q} C_y}$.

In this paper, a consensus scheme is derived from this aggregation model. Although Reference [15] does not mention it, the concordance metric S can also be used in order to guide the discussion process. Having calculated the concordance between each individual estimate and the collective estimate $S(\widetilde{F}_k^{\,C}(X), \widetilde{F}_k^{\,z}(X))$, the moderator can motivate the least concordant DM to review his opinion, at each step of a discussion process.

In this work, we propose the use of the fuzzy scale described in Table 2, when each specialist evaluates each alternative. After a collective solution is met, the concordance level $S(\widetilde{F}_k^{\,C}(X), \widetilde{F}_k^{\,z}(X))$ can be calculated for each fuzzy estimate using (28), but with the term relative to the normalized distance $\overline{D}_h$, defined as:

$$\overline{D}_h = \frac{D_h(\widetilde{F}_k^{\,y}(X), \widetilde{F}_k^{\,z}(X))}{D_h(EG, EP)}, \tag{31}$$

being $D_h(EG, EP)$ the distance between the extreme fuzzy estimates EG and EP from Table 2. It is worth mentioning that, with this normalization, it became easier to empirically fix a threshold value below which the concordance level was considered unacceptable.

Table 2. Linguistic variables for rating of each alternative according to each criterion.

Linguistic variables	Trapezoid fuzzy numbers
Extreme poor (*EP*)	(0, 0, 1, 2)
Very poor (*VP*)	(1, 2, 3, 4)
Poor (*P*)	(2, 3, 4, 5)
Medium poor (*MP*)	(3, 4, 5, 6)
Fair (*F*)	(4, 5, 6, 7)
Medium good (*MG*)	(5, 6, 7, 8)
Good (*G*)	(6, 7, 8, 9)
Very good (*VG*)	(7, 8, 9, 10)
Extreme good (*EG*)	(8, 9, 10, 10)

6. Aggregation Model Based on Pairwise Comparisons

The CS2 uses a relatively simple and intuitive aggregation method that allows deriving collective priorities from pairwise comparisons [14]. Initially, each DM d_z must provide a set of fuzzy estimates $\tilde{a}_{ij}^z = w_i / w_j$, where the ratio w_i / w_j represents how many times alternative x_i is better than x_j. Then, the prioritization problem is formulated as a constrained linear mono-objective search for the collective priority vector $w=(w_1,\ldots,w_n)$, in conformity with all fuzzy judgments.

The membership function $\mu_{ijz}(w)$ reflects the proximity between each ratio $a_{ij}^z = w_i / w_j$ and its respective estimate $\tilde{a}_{ij}^z$:

$$\mu_{ijz}(w) = \begin{cases} 1 - \dfrac{(w_i - a_{ij}^z w_j)}{\varepsilon_{ijz}}, if (w_i / w_j) \geq a_{ij}^z, \\ 1 + \dfrac{(w_i - a_{ij}^z w_j)}{\varepsilon_{ijz}}, if (w_i / w_j) < a_{ij}^z. \end{cases} \tag{32}$$

In (32), ε_{ijz} delimits the tolerance range admitted by the zth DM. When $\mu_{ijz}(w) = 1$, the judgment is perfectly approximated and the DM is completely satisfied, as $w_i - a_{ij}^z w_j = 0$. If $0 < \mu_{ijz}(w) < 1$, the judgment is partially approximated, as $-\varepsilon_{ijz} \leq (w_i - a_{ij}^z w_j) \leq \varepsilon_{ijz}$. If $\mu_{ijz}(w) < 0$, the DM is completely dissatisfied, as $(w_i - a_{ij}^z w_j) \notin [-\varepsilon_{ijz}, \varepsilon_{ijz}]$.

The existence of a consistent solution for the group prioritization problem requires two assumptions. First, the intersection of all membership functions associated to the zth DM, must be nonempty, otherwise the final results will not be coherent with his judgments. Naturally, for the same reason, this hypothesis must be true for all involved specialists.

Second, there must be an optimal priority vector $w^*=\{w_1,\ldots,w_n\}$, which is associated to the maximum value $\lambda^* = \max \mu_C(w)$, where $\mu_C(w)$ is the group membership function:

$$\mu_C(w) = \min\{\mu_{ijz}(w) \mid i = 1,\ldots,n-1; j = 2,\ldots,n, j > i; z = 1,\ldots,q\}, \tag{33}$$

where $w_i>0$, $i=1,\ldots,n$ and $\sum_{i=1}^{n} w_i = 1$.

This second assumption is true, as (33) defines a convex fuzzy set, which always has a maximum membership degree.

The constrained optimization model is given by:

$$\begin{cases} \text{maximize}\, \lambda, \\ \text{subjected to} \\ \varepsilon_{ijz}\lambda + \mu_{ijz}(w) \le \varepsilon_{ijz}, \\ \varepsilon_{ijz}\lambda - \mu_{ijz}(w) \le \varepsilon_{ijz}, \\ i = 1,2,\ldots,n-1;\ j = 2,3,n;\ j > i;\ z = 1,2,\ldots,q, \\ \sum_{i=1}^{n} w_i = 1, w_i > 0, i = 1,2,\ldots,n. \end{cases} \tag{34}$$

It is interesting to observe that, in this model, λ^* can be used as a measure of the group satisfaction with the obtained optimal priority vector or as a group consistency index:

- $\lambda^* = 1$, only if all group judgments are equal: $a_{ij1}, a_{ij2},\ldots, a_{ijq}$, and the judgments are perfectly consistent, which hardly ever happens in practice;
- $0 \le \lambda^* < 1$, if the group judgments are inconsistent, but the equality constraints are approximately satisfied;
- $\lambda^* < 0$, if the group judgments are very inconsistent and/or the tolerance parameters ε_{ijz} are very small. It is also worth mentioning that the importance of each judgment of each DM may be differentiated by adjusting the parameters ε_{ijz}. Lower values of ε_{ijz} should be associated to the most important judgments.

7. Application Example

The enterprise's board of directors, which includes five members ($d_1, d_2,\ldots,d_5$), is to plan the development of large projects (strategy initiatives) for the following five years. Four possible projects (X_1, X_2, X_3, and X_4) have been marked. It is necessary to compare these projects to select the most important of them, as well as order them from the point of view of their importance, taking into account four criteria (categories) suggested by the Balanced Scorecard methodology [27] (it should be noted that all of them are of the maximization type):

c_1) financial perspective,
c_2) the customer satisfaction,
c_3) internal business process perspective,
c_4) learning and growth perspective.

First, the consensus scheme CS1 was executed in order to construct the collective pairwise comparison matrices $R_1,\ldots,R_4$. So, the specialists were asked to give their opinion relative to each project in terms of fuzzy estimates, using the linguistic variables from Table 2. The parameter β from (28) was fixed at 0.5. The involved professionals were considered of the same importance, except for d_1, whose opinions were judged more important. Therefore, the parameters e_i were set as: e_1=0.3 and e_z=0.175 for z=2,…,5. Table 3 shows all specialists opinions, according to all criteria.

Table 3. Fuzzy estimates of each DM for each alternative, considering the four criteria.

Criteria	Projects	d_1	d_2	d_3	d_4	d_5
c_1	X_1	*F*	*MP*	*F*	*MP*	*MP*
	X_2	*G*	*G*	*MP*	*MG*	*F*
	X_3	*EG*	*G*	*VG*	*G*	*EG*
	X_4	*VP*	*EP*	*VP*	*EP*	*EP*
c_2	X_1	*G*	*MG*	*VG*	*G*	*MG*
	X_2	*MG*	*VG*	*VG*	*VG*	*VG*
	X_3	*MP*	*P*	*P*	*VP*	*VP*
	X_4	*F*	*MP*	*MG*	*F*	*MG*
c_3	X_1	*F*	*G*	*F*	*MG*	*G*
	X_2	*G*	*VG*	*G*	*VG*	*EG*
	X_3	*VP*	*EP*	*P*	*VP*	*MP*
	X_4	*G*	*VG*	*G*	*EG*	*EG*
c_4	X_1	*P*	*VP*	*VP*	*P*	*VP*
	X_2	*F*	*MP*	*MG*	*MP*	*MG*
	X_3	*VP*	*MP*	*MP*	*MP*	*P*
	X_4	*EG*	*VG*	*G*	*VG*	*VG*

Having at hand the specialists' opinion about the alternatives, the concordance level $S(\widetilde{F}_k^{\,C}(X), \widetilde{F}_k^{\,z}(X))$ was calculated for each fuzzy estimate. Table 4 shows the obtained collective estimates and Table 5 contains the concordance levels associated to each judgment performed by each DM.

Table 4. Collective estimates obtained using CS1 (first iteration).

Projects	c_1	c_2	c_3	c_4
X_1	(3.4, 4.4, 5.4, 6.4)	(5.7, 6.7, 7.7, 8.7)	(4.9, 5.9, 6.9, 7.9)	(1.4, 2.4, 3.4, 4.4)
X_2	(5.1, 6.1, 7.1, 8.1)	(6.8, 7.8, 8.8, 9.8)	(6.3, 7.3, 8.3, 9.1)	(4.0, 5.0, 6.0, 7.0)
X_3	(7.1, 8.1, 9.1, 9.7)	(1.8, 2.8, 3.8, 4.8)	(1.3, 2.2, 3.2, 4.2)	(2.2, 3.2, 4.2, 5.2)
X_4	(0.8, 1.4, 2.4, 3.4)	(4.3, 5.3, 6.3, 7.3)	(6.9, 7.9, 8.9, 9.6)	(7.1, 8.1, 9.1, 9.9)

The moderator considered unacceptable the judgments associated to concordance levels lower than 0.4 (see the underlined values from Table 5). This threshold value was empirically chosen by the moderator for this particular decision problem. Therefore, the specialists were asked to review their respective unsatisfactory judgments. It is worth mentioning that, in the specific case of the estimates related to alternative X_3, taking into account the criterion c_3, only d_2 was asked to reevaluate his judgment, as he obtained a lower concordance level than the specialist d_5.The new evaluations are listed below and the new collective estimates are shown in Table 6 and illustrated in Figures 1-4, for each criterion.

d_1: his new evaluation on X_2, taking into account c_2, was *G*;
d_2: his new evaluation on X_3, taking into account c_3, was *P*;
d_3: his new evaluation on X_2, taking into account c_1, was *F*;
d_4: his new evaluation on X_4, taking into account c_1, was *VP*;
d_5: his new evaluation on X_2, taking into account c_3, was *G*.

Table 5. Concordance level for each judgment of each DM.

Criteria	Projects	d_1	d_2	d_3	d_4	d_5
c_1	X_1	0.66	0.77	0.66	0.77	0.77
	X_2	0.56	0.56	0.31	0.91	0.47
	X_3	0.56	0.49	0.92	0.49	0.56
	X_4	0.72	0.40	0.72	0.39	0.40
c_2	X_1	0.84	0.61	0.44	0.84	0.61
	X_2	0.34	0.87	0.87	0.87	0.87
	X_3	0.45	0.86	0.86	0.59	0.59
	X_4	0.83	0.44	0.61	0.83	0.61
c_3	X_1	0.55	0.48	0.55	0.93	0.48
	X_2	0.83	0.61	0.83	0.43	0.37
	X_3	0.84	0.32	0.61	0.84	0.35
	X_4	0.56	0.89	0.56	0.49	0.49
c_4	X_1	0.66	0.77	0.77	0.66	0.77
	X_2	1.00	0.51	0.51	0.51	0.51
	X_3	0.45	0.59	0.59	0.86	0.86
	X_4	0.57	0.96	0.50	0.96	0.96

Table 6. Collective estimates obtained using CS1 (second iteration).

Projects	c_1	c_2	c_3	c_4
X_1	(3.4, 4.4, 5.4, 6.4)	(5.7, 6.7, 7.7, 8.7)	(4.9, 5.9, 6.9, 7.9)	(1.4, 2.4, 3.4, 4.4)
X_2	(5.1, 6.1, 7.1, 8.1)	(6.9, 7.9, 8.9, 9.9)	(6.0, 7.0, 8.0, 9.0)	(4.0, 5.0, 6.0, 7.0)
X_3	(7.1, 8.1, 9.1, 9.7)	(1.8, 2.8, 3.8, 4.8)	(1.7, 2.7, 3.7, 4.7)	(2.2, 3.2, 4.2, 5.2)
X_4	(0.7, 1.5, 2.5, 3.5)	(4.3, 5.3, 6.3, 7.3)	(6.9, 7.9, 8.9, 9.6)	(7.1, 8.1, 9.1, 9.9)

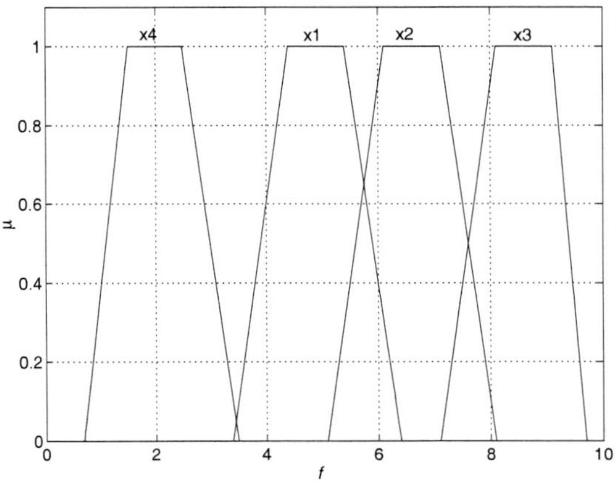

Figure 1. Collective fuzzy estimates related to c_1.

As it can be seen from Table 7, which shows the new concordance levels, computed after the second iteration of CS1, although some of the collective estimates remained exactly the same (compare Tables 4 and 6), the concordance level associated to the least concordant judgments increased. In fact, all judgments from Table 7 have acceptable concordance levels.

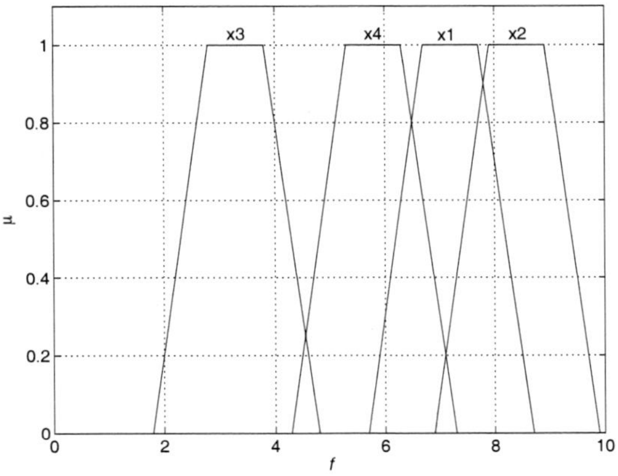

Figure 2. Collective fuzzy estimates related to c_2.

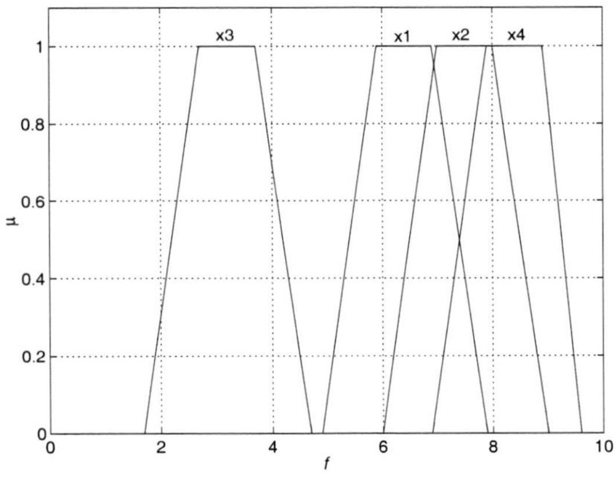

Figure 3. Collective fuzzy estimates related to c_3.

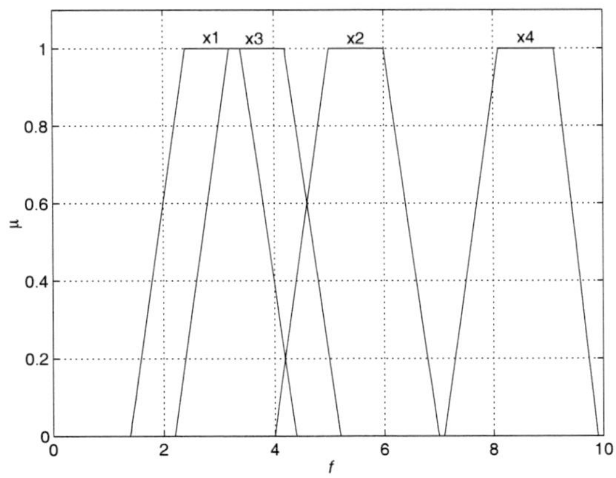

Figure 4. Collective fuzzy estimates related to c_4.

Table 7. Concordance levels calculated after the second iteration of CS1.

Criteria	Projects	d_1	d_2	d_3	d_4	d_5
c_1	X_2	0.55	0.55	0.48	0.92	0.48
c_1	X_4	0.72	0.41	0.72	0.72	0.41
c_2	X_2	0.55	0.92	0.92	0.92	0.92
c_3	x_2	1.00	0.51	1.00	0.51	1.00
c_3	x_3	0.64	0.79	0.79	0.64	0.42

Having calculated the collective estimates, the following comparison matrices, corresponding to the collective nonstrict fuzzy preference relations, were obtained for each criterion, using Equations (11) and (12):

$$R_1 = \begin{bmatrix} 1 & 0.65 & 0 & 1 \\ 1 & 1 & 0.50 & 1 \\ 1 & 1 & 1 & 1 \\ 0.05 & 0 & 0 & 1 \end{bmatrix}, \qquad R_2 = \begin{bmatrix} 1 & 0.90 & 1 & 1 \\ 1 & 1 & 1 & 1 \\ 0 & 0 & 1 & 0.25 \\ 0.80 & 0.20 & 1 & 1 \end{bmatrix},$$

$$R_3 = \begin{bmatrix} 1 & 0.95 & 1 & 0.50 \\ 1 & 1 & 1 & 1 \\ 0 & 0 & 1 & 0 \\ 1 & 1 & 1 & 1 \end{bmatrix}, \qquad R_4 = \begin{bmatrix} 1 & 0.20 & 1 & 0 \\ 1 & 1 & 1 & 0 \\ 1 & 0.60 & 1 & 0 \\ 1 & 1 & 1 & 1 \end{bmatrix}.$$

Let us consider the application of the *first procedure* for analyzing <*X*, *R*> models. In particular, the intersection of all nonstrict fuzzy preference relations ($R_1,\ldots,R_4$), constructed on the basis of Equation (16), can be represented by the following matrix:

$$R = \begin{bmatrix} 1 & 0.20 & 0 & 0 \\ 1 & 1 & 0.50 & 0 \\ 0 & 0 & 1 & 0 \\ 0.05 & 0 & 0 & 1 \end{bmatrix}. \tag{35}$$

The strict fuzzy preference relation derived from (35), using Equation (13), is given by:

$$P = \begin{bmatrix} 0 & 0 & 0 & 0 \\ 0.80 & 0 & 0.50 & 0 \\ 0 & 0 & 0 & 0 \\ 0.05 & 0 & 0 & 0 \end{bmatrix}. \tag{36}$$

The nondominance level of each alternative, obtained on the basis of the strict fuzzy preference relation (36), with the use of Equation (14), is given by: $\mu_R^{nd} = [0.2 \quad 1 \quad 0.5 \quad 1]$. As the alternatives X_2 and X_4 are indistinguishable ($\mu_R^{nd}(X_2) = \mu_R^{nd}(X_4) = 1$), a subsequent analysis was performed, where the importance weight of each criterion was taken into account. Thus, CS2 was executed and each specialist was asked to compare the four categories and specify their respective importance based on the Saaty scale [32]. The pairwise judgments listed below were provided by each specialist. The other input parameters required by CS2 were fixed by the moderator as $\varepsilon_{ij1} = 0.15$ and $\varepsilon_{ijz} = 0.3$, i,j=1,...,4, z=2,...,5 (which means that all information provided by d_1 was considered more important than the information given by the other specialists). The optimal priority vector w^*={0.45, 0.35, 0.13, 0.07} was obtained. As λ^* assumed a positive value ($\lambda^* = 0.37$), this result was considered satisfactory.

d_1: a_{12}= 2, a_{13}=3, a_{23}=4;
d_2: a_{12}=1, a_{13}=2, a_{23}=3;
d_3: a_{13}=5, a_{14}=8;
d_4: a_{23}=4, a_{24}=7;
d_5: a_{13}=3, a_{34}=5.

The weighted aggregation of all nonstrict fuzzy preference relations, constructed on the basis of Equation (17), is given by the following matrix:

$$R_T = \begin{bmatrix} 1.00 & 0.76 & 0.45 & 0.93 \\ 1.00 & 1.00 & 0.72 & 0.99 \\ 0.56 & 0.55 & 1.00 & 0.63 \\ 0.41 & 0.18 & 0.45 & 1.00 \end{bmatrix}. \quad (37)$$

Using (37), on the basis of (13), it is possible to obtain the following strict fuzzy preference relation:

$$P_T = \begin{bmatrix} 0 & 0 & 0.03 & 0.33 \\ 0.25 & 0 & 0.28 & 0.66 \\ 0 & 0 & 0 & 0 \\ 0 & 0 & 0.01 & 0 \end{bmatrix}. \quad (38)$$

The nondominance level of each alternative, obtained on the basis of the matrix (38), with Equation (14), is given by $\mu_T^{nd} = [0.74 \quad 1.00 \quad 0.72 \quad 0.34]$. With the use of (18), the following nondominance level is obtained for each alternative:

$$\mu_W^{nd} = [0.2 \quad 1.00 \quad 0.50 \quad 0.34].$$

Hence, the final ranking, taking into account the information provided by all specialists and according to the four perspectives of the Balanced Scorecard methodology and their respective importance factors is given by: $X_2 \succ X_3 \succ X_4 \succ X_1$.

Next, it will be shown that, when this problem is solved using the same decision procedure, but considering only the information provided by the most important specialist (d_1), a different result is obtained. The following nonstrict fuzzy preference relations were obtained from the d_1 judgments, for each criterion, using Equations (11) and (12):

$$R_1 = \begin{bmatrix} 1 & 0.5 & 0 & 1 \\ 1 & 1 & 0.5 & 1 \\ 1 & 1 & 1 & 1 \\ 0 & 0 & 0 & 1 \end{bmatrix}, \qquad R_2 = \begin{bmatrix} 1 & 1 & 1 & 1 \\ 1 & 1 & 1 & 1 \\ 0 & 0 & 1 & 1 \\ 0.5 & 0.5 & 1 & 1 \end{bmatrix},$$

$$R_3 = \begin{bmatrix} 1 & 0.5 & 1 & 0.5 \\ 1 & 1 & 1 & 1 \\ 0 & 0 & 1 & 0 \\ 1 & 1 & 1 & 1 \end{bmatrix}, \qquad R_4 = \begin{bmatrix} 1 & 0.5 & 1 & 0 \\ 1 & 1 & 1 & 0 \\ 1 & 0 & 1 & 0 \\ 1 & 1 & 1 & 1 \end{bmatrix}.$$

The intersection of these nonstrict fuzzy preference relations ($R_1,\ldots,R_4$) results in:

$$R = \begin{bmatrix} 1 & 0.5 & 0 & 0 \\ 1 & 1 & 0.5 & 0 \\ 0 & 0 & 1 & 0 \\ 0 & 0 & 0 & 1 \end{bmatrix}. \tag{39}$$

The strict fuzzy preference relation obtained from (39) is given by:

$$P = \begin{bmatrix} 0 & 0 & 0 & 0 \\ 0.5 & 0 & 0.5 & 0 \\ 0 & 0 & 0 & 0 \\ 0 & 0 & 0 & 0 \end{bmatrix}. \tag{40}$$

The nondominance level of each alternative, derived from (40), with the use of Equation (14), is given by: $\mu_R^{nd} = [0.5 \quad 1 \quad 0.5 \quad 1]$. As the pairs of alternatives (X_2, X_4) and (X_1, X_3) are indistinguishable, a subsequent analysis, where the importance weight of each criterion is taken into account, is performed in a similar way as in the previous study case. But, now, only the information provided by d_1 was taken into account, when executing CS2 and the parameter ε_{ij1} was fixed as 0.2 for i,j=1,…,n; (which means that all information provided by d_l were considered of the same importance). The optimal priority vector w*={0.46, 0.13, 0.08, 0.33} was obtained and, as $\lambda^* = 1$, this result was considered satisfactory. The weighted aggregation of all nonstrict fuzzy preference relations, constructed on the basis of Equation (17), is given by the following matrix:

$$R_T = \begin{bmatrix} 1 & 0.66 & 0.45 & 0.93 \\ 1 & 1 & 0.71 & 0.99 \\ 0.56 & 0.56 & 1 & 0.88 \\ 0.28 & 0.12 & 0.45 & 1 \end{bmatrix}. \qquad (41)$$

Using Equation (13), it is possible to obtain the following strict fuzzy preference relation from (41):

$$P_T = \begin{bmatrix} 0 & 0 & 0 & 0.65 \\ 0.33 & 0 & 0.15 & 0.87 \\ 0.11 & 0 & 0 & 0.43 \\ 0 & 0 & 0 & 0 \end{bmatrix}. \qquad (42)$$

The nondominance level of each alternative, obtained on the basis of (42), using Equation (14), is given by: $\mu_T^{nd} = [0.56 \quad 1.00 \quad 0.69 \quad 0.82]$. Finally, the following nondominance level is obtained for each alternative:

$$\mu_W^{nd} = [0.50 \quad 1.00 \quad 0.50 \quad 0.13].$$

Hence, the final ranking, according to the opinion of d_1, taking into account the four perspectives of the Balanced Scorecard methodology and their respective importance factors is given by: ($X_2 \succ (X_1 \approx X_3) \succ X_4$). As it can be seen, there is a discrepancy between the ranking obtained according to the information provided by the most important specialist from the group and the ranking obtained according to the information provided by the five professionals.

Conclusions

Since the application of a multicriteria approach is associated with the need to analyze (1) problems in which consequences of obtained solutions cannot be evaluated using a single criterion and (2) problems that may be considered on the basis of a single criterion but their unique solutions are not achieved because the uncertainty of information produces decision uncertainty regions, and the use of additional criteria can serve as a convincing means to contract these regions, two corresponding classes of models (<*X*, *M*> and <*X*, *R*>) have been considered.

The analysis of <*X*, *M*> models is based on the Bellman-Zadeh approach to decision making in a fuzzy environment. The analysis of the <*X*, *R*> models is associated with four distinct techniques based on fuzzy preference modeling.

The results related to constructing and analyzing <*X*, *R*> models have been extended to the group decision context through the use of two flexible consensus schemes: one was proposed in [14] and the other was introduced here, being derived from the aggregation model from [15]. Both consensus schemes allow differentiating the importance of each expert opinion, which is a very useful resource. Their main difference lies in the kind of information (each one is based on a different preference structure) and in the amount of information they require from the DM. This paper presented one way of using these consensus schemes, but suggested other possible ways of combining them with the fuzzy decision techniques considered in the paper. With the availability of distinct consensus schemes, the most appropriated one may be chosen, taking into account the sources of information and the way the group prefer to express their opinions.

Finally, in order to demonstrate the results of the presented research, a decision procedure based on analyzing <*X*, *R*> models has been utilized for solving a hypothetical enterprise strategy planning problem, generated with the use of the Balanced Scorecard methodology.

Acknowledgements

This research is supported by the National Council for Scientific and Technological Development of Brazil (CNPq) - grant 302406/2005-0 and the State of Minas Gerais Research Foundation (FAPEMIG) - grant TEC 00140/07.

References

[1] H. J. Zimmermann, *Fuzzy Set Theory and Its Application*, Kluwer Academic, Boston, 1990.

[2] W. Pedrycz and F. Gomide, *An Introduction to Fuzzy Sets: Analysis and Design*, MIT Press, Cambridge, 1998.

[3] P. Ya. Ekel and E. A. Galperin, Box-triangular multiobjective linear programs for resource allocation with application to load management and energy market problems, *Mathematical and Computer Modelling*, vol. 37, 2003, pp. 1-17.

[4] P. Ya. Ekel, M. Menezes, and F. Schuffner Neto, Decision making in fuzzy environment and its application to power engineering problems, *Nonlinear Analysis: Hybrid Systems*, vol. 1, 2007 2007, pp. 527-536.

[5] R. C. Berredo, P. Ya. Ekel, E. A. Galperin, and A. S. Sant'anna, Fuzzy preference modeling and its management applications, in *Proceedings of the International Conference on Industrial Logistics*, Montevideo, 2005, pp. 41-50.

[6] R. C. Botter and P. Ya. Ekel, Fuzzy preference relations and their naval engineering applications, in *Proceedings of the XIX Congress of the Panamerican Institute of Naval Engineering*. Guayaquil, 2005, Paper 7-8.

[7] P. Ekel, W. Pedrycz, and R. Schinzinger, A general approach to solving a wide class of fuzzy optimization problems, *Fuzzy Sets and Systems*, vol. 97, 1998, pp. 49-66.

[8] H. M. Su and C. T. Chen, Aggregation of fuzzy opinions under group decision making, *Fuzzy Sets and Systems*, vol. 79, 1996, pp. 279-285.

[9] C. T. Chen, Extensions of the TOPSIS for group decision-making under fuzzy environment, *Fuzzy Sets and Systems*, vol. 114, 2000, pp. 1-9.

[10] R. Kwok, J. Ma, D. Zhou, Improving group decision making: a fuzzy GSS approach, *IEEE Transactions on Systems, Man, and Cybernetics - Part C*, vol. 32, 2002, pp. 54-62.

[11] E. Herrera-Viedma, F. Herrera, and F. Chiclana, A consensus model for multiperson decision making with different preference structures, *IEEE Transactions on Systems, Man and Cybernetics – Part A: Systems and Humans*, vol. 32, 2002, pp. 394-402.

[12] P. Prodanovic and S. Simonovic, Fuzzy Compromise Programming for Group Decision Making, *IEEE Transactions on Systems, Man and Cybernetics – Part A: Systems and Humans*, vol. 33, 2003, pp. 358-365.

[13] D. F. Li and J. B. Yang, Fuzzy linear programming technique for multiattribute group decision making in fuzzy environments, *Information Sciences*, vol. 158, 2004, pp. 263-275.

[14] L. Mikhailov, Group prioritization in the AHP by fuzzy preference programming method", *Computers & Operations Research*, vol. 31, 2004, pp. 293-301.

[15] C. Lu, J. Lan, and Z. Wang, Aggregation of fuzzy opinions under group decision-making based on similarity and distance, *Journal of Systems Science and Complexity*, vol. 19, 2006, pp. 63-71.

[16] A.K. Choudhury, Ravi Shankar, M.K. Tiwari, Consensus-based intelligent group decision-making model for the selection of advanced technology, *Decision Support Systems*, vol. 42, 2006, pp. 1776– 1799.

[17] Y. Wang and H. Lee, Generalizing TOPSIS for fuzzy multiple-criteria group decision-making, *Computers and Mathematics with Applications*, vol. 53, 2007, pp. 1762-1772.

[18] K. Jabeur and J. Martel, An ordinal sorting method for group decision-making, *European Journal of Operational Research*, vol. 180, 2007, pp. 1272-1289.

[19] D. F. Li, Compromise ratio method for fuzzy multi-attribute group decision-making, *Applied Soft Computing*, vol. 7, 2007, pp. 807-817.

[20] R. S. Kaplan and D. Norton, *The Balanced Scorecard: Translating Strategy into Action*, Harvard Business School, Boston, 1996.

[21] V. Pareto, *Cours d'Économie Politique*, Lausanne Rouge, Lausanne, 1896.

[22] R. O. Parreiras and J. A. Vasconcelos, Decision making in multiobjective optimization problems, in N. Nedjah and L. M. Mourelle, eds., *ISE Book Series on Real Word Multi-objective Systems Engineering, Nova Science*, New York, 2005, pp. 29-52.

[23] J. Horn, Multicriterion decision making, in: T. Bäck, D. Fogel, and Z. Michalewicz, eds., *Handbook of Evolutionary Computation*, IOP Publishing/ Oxford University Press, New York, 1997,1-15.

[24] P. L. Yu, *Multiple Criteria Decision Making: Concepts, Techniques, and Extensions*, Plenum Press, New York, 1985.

[25] S. Rao, *Engineering Optimization*, Wiley&Sons, New York, 1996.

[26] P. Ya. Ekel, Fuzzy sets and models of decision making, *Computers and Mathematics with Applications*, vol. 44, 2002, pp. 863-875.

[27] R. E. Bellman and L. A. Zadeh, Decision-making in a fuzzy environment, *Management Science*, vol. 17, 1970, pp. 141-164.

[28] P. Ya. Ekel, C. A. P. S. Martins, J.G. Pereira Jr., R. M. Palhares, and L. N. Canha, Fuzzy set based multiobjective allocation of resources and its applications, *Computers and Mathematics with Applications*, vol. 52, 2006, pp. 197-210.

[29] P. Ya. Ekel, M. R. Silva, F. Schuffner Neto, and R. M. Palhares, Fuzzy preference modeling and its application to multiobjective decision making, *Computers and Mathematics with Applications*, vol. 52, 2006, pp. 179-196.

[30] J. Fodor and M. Roubens, *Fuzzy Preference Modelling and Multicriteria Decision Support*, Kluwer Publishers, Boston, 1994.

[31] S. A. Orlovsky, *Problems of Decision Making with Fuzzy Information* (in Russian), Nauka, Moscow, 1981.

[32] T. L. Saaty, Axiomatic foundation of the analytic hierarchy process, *Management Science*, vol. 32, 1986, pp. 841-855.

[33] F. Chiclana, F. Herrera, and E. Herrera-Viedma, Integrating multiplicative preference relations in a multipurpose decision-making model based on fuzzy preference relations, *Fuzzy Sets and Systems*, vol. 122, 2001, pp. 277-291.

[34] A. Kauffman, M. M. Gupta, *Introduction to Fuzzy Arithmetic: Theory and Applications*, Van Nostrand Reinhold, New York, 1985.

In: Progress in Nonlinear Analysis Research
Editor: Erik T. Hoffmann, pp. 269-283
ISBN 978-1-60456-359-7

Chapter 11

On the Secant and Steffensen's Methods for Variational Inclusions

S. Hilout[1]*, ***C. Jean–Alexis***[2]† **and** ***A. Piétrus***[2]‡
[1]Laboratoire d'Applications des Mathématiques,
Université de Poitiers, Bd. Marie et Pierre Curie,
Téléport 2, B.P. 30179, 86962 Futuroscope Chasseneuil Cedex, France
[2]Laboratoire Analyse, Optimisation, Contrôle,
Université des Antilles et de la Guyane,
Département de Mathématiques et Informatique, Campus de Fouillole,
F–97159 Pointe–à–Pitre, France

Abstract

The aim of this paper is in a first time to recall some results existing on Secant–type methods and in a second time to study the Steffensen–type method for solving a variational inclusion in the form $0 \in f(x) + G(x)$ where f is a single function and G is a set–valued map. Under a center–Hölder condition on the first order divided difference and using a well–known fixed point theorem for set–valued maps [13] we prove the existence and the superlinear convergence of a sequence (x_k) satisfying $0 \in f(x_k) + [g_1(x_k), g_2(x_k); f](x_{k+1} - x_k) + G(x_{k+1})$ where g_1 and g_2 are some continuous functions parameter.

Key Words: Variational inclusions, superlinear convergence, Aubin continuity, divided difference, center–Hölder condition, set–valued map, Secant–type method, Steffensen–type method

AMS subject classifications. 49J53, 47H04, 65K10.

*E-mail address: said.hilout@math.univ–poitiers.fr
†E-mail address: celia.jean–alexis@univ–ag.fr
‡E-mail address: apietrus@univ–ag.fr

1. Introduction

This paper is concerned with the problem of approximating a solution x^* of the following variational inclusion

$$0 \in f(x) + G(x) \tag{1}$$

where $f : X \longrightarrow Y$ is a continuous function and $G : X \longrightarrow 2^Y$ is a set–valued map with closed graph, X and Y are Banach spaces.

Variational inclusions are an abstract model of a wide variety including linear and nonlinear complementary problems, systems of nonlinear equations (for example when $G = \{0\}$), systems of inequalities (for example when G is the positive orthant in $\mathbb{R}^m$), variational inequalities (for example first–order necessary conditions for nonlinear programming), etc. In particular, they may characterize optimality or equilibrium and then have several applications. For instance, the Walrasian law of competitive equilibria of exchange economies (see [44]) can be formulated as a variational inclusion and so can the Wardrop principle of user equilibrium in traffic theory (see [45]). For more details and examples, the reader could also be referred to [18, 36, 37] where in particular an extensive documentation of applications of finite–dimensional nonlinear complementary problems in engineering and equilibrium is available.

For solving (1), the authors proposed in [7] and [27] a Steffensen–type method for solving variational inclusions. This method reads:

$$\begin{cases} x_0 \text{ is given as starting point} \\ 0 \in f(x_k) + [g_1(x_k), g_2(x_k); f](x_{k+1} - x_k) + G(x_{k+1}) \end{cases} \tag{2}$$

where g_i $(i = 1, 2)$ are continuous functions from a neighborhood D of x^* into X and $[g_1(x_k), g_2(x_k); f]$ is a first order divided difference of f on the points $g_1(x_k)$ and $g_2(x_k)$ (whose definition is given in the next section).
Let us note that the method (2) recovers a lot of particular methods we will precize bellow.

When $G = \{0\}$ and $g_1(x_k) = x_{k-1}$; $g_2(x_k) = x_k$, algorithm (2) is a Secant method described in [23]. If f is Fréchet differentiable on a neighborhood of x^* and $g_i(x) = x$ for $i = 1; 2$, algorithm (2) is reduced to the classical Newton's method. Let us note that the most popular choice of functions g_1 and g_2 is $g_1(x) = x$ and $g_2(x) = x - f(x)$. In [1], Argyros considers the case $g_1(x) = x$ to solve nonlinear equations under some conditions on divided differences.

When $G \neq \{0\}$ and $g_1(x) = x$, some results of existence and convergence of sequence (2) to a locally unique solution of (1) are given in [7] and [24] using some Hölder and Lipschitz conditions on the first order divided difference. Newton–Steffensen–type method is presented in [25] for solving perturbed generalized equations. Using some ideas given in [3] and [6] for nonlinear equations, an improved local convergence analysis is investigated in [7] under less computational cost.

Some convergence results of algorithm (2) under ω–conditioned divided difference are also presented in [27]. If the function f is Fréchet differentiable and $g_1(x) = g_2(x) = x$ in (2), our method is the satale Newton–type method (see [14, 15]) for solving generalized equations. Superlinear convergence and stability are also obtained in [38, 39] under Hölder continuity condition. Recently, the radius estimates of the convergence balls of Newton's method [14] and the variant of Newton's method [31] are presented in [9] under a weaker Lipschitz conditions. For the special case where $g_1(x_k) = \beta\, x_k + (1-\beta)\, x_{k-1}$ and $g_2(x_k) = x_k$ for a fixed parameter β in $[0, 1)$, algorithm (2) is an uniparametric Secant–type method. This method is developed in the differentiable case in [28] and in [8] under Hölder–type conditions. Some existence–convergence results are developed in [26] under ω–conditioned divided difference which generalizes the usual Lipschitz and Hölder continuous conditions used in [28].

When f is twice Fréchet differentiable, Geoffroy and al. [20, 21] introduced a stable method with a cubic convergence using a second–degree Taylor polynomial expansion of f at the current iterate x_k whenever ∇f and $\nabla^2 f$ are Lipschitz continuous and Hölder. Nevertherless, it is well–known that the computation of the second order derivative of a function in general space is not straightforward and the cost of this computation in application is very expansive. To avoid this drawback, Jean–Alexis [30] inspired by some interpolation formula given by Hummel and Seebeck [29] for one–dimensional functions introduced a method with cubic convergence in which intervene only the first order Fr´echet derivative of f at the points x_k and x_{k+1}.

For a best understanding, let us give some notations. We denote by $\mathbb{B}_r(x)$ the closed ball centered at x with radius r, by $\|.\|$ all the norms, the distance from a point $x \in X$ and a subset $A \subset X$ is defined as $dist(x, A) = \inf_{a \in A} \|x - a\|$ and a set–valued map G from X to Y is indicated by $G : X \to 2^Y$ while its graph is denoted by $\operatorname{gph} G = \{(x, y) \in X \times Y \mid y \in G(x)\}$.

The content of this paper is as follow. In Section 2 we give some preliminaries about fixed points, regularity of set–valued maps and divided differences. In Section 3 we recall results obtained about combination of Newton's method with a Secant–type one for solving variational inclusions, the convergence theorem is ilustrated with a minimization problem. In Section 4, we recall some results about a modification of the Secant–type method. Section 5 is devoted to the convergence analysis of Steffensen–type method (2) in weaker conditions than the methods exposed in the papers [24, 25, 27]. The results obtained in this section are new.

2. Preliminaries

Recall the definition of pseudo–lipschitzness for a set–valued map:

Definition 2.1 *A set–valued map Γ is pseudo–Lipschitz around $(x_0, y_0) \in gph\, \Gamma$ with mod-*

ulus M if there exist constants a and b such that

$$\sup_{z\in\Gamma(y')\cap\mathbb{B}_a(y_0)} dist\,(z,\Gamma(y'')) \le M \parallel y'-y'' \parallel, \quad \text{for all } y' \text{ and } y'' \text{ in } \mathbb{B}_b(x_0). \tag{3}$$

Using the excess, we have an equivalent definition

Definition 2.2 *A set–valued map $\Gamma : X \to 2^Y$ is said to be M–pseudo–Lipschitz around $(x_0, y_0) \in \mathrm{gph}\Gamma$ if there exist constants a and b such that*

$$e(\Gamma(x_1)\cap\mathbb{B}_a(y_0),\Gamma(x_2)) \le M\|x_1-x_2\|, \quad \forall x_1,x_2\in\mathbb{B}_b(x_0),$$

where the excess from the set A to the set C is defined by $e(C,A) = \sup_{x\in C} dist(x,A)$.

The pseudo–lipschitz property has been introduced by Aubin and sometimes this property is also often called “Aubin continuity”, see [10, 11]. This concept is equivalent to the openess with linear rate of Γ^{-1} (the covering property) and to the metric regularity of Γ^{-1} (a basic well–posedness property in optimization). Let us note that when f is a function which is strictly differentiable at some x_0, then the pseudo–lipchitzness of f^{-1} around $(f(x_0), x_0)$ is equivalent to the surjectivity of $\nabla f(x_0)$.
Characterizations of this property are also obtained by Rockafellar [40] using the lipschitz continuity of the distance function $dist(y,\Gamma(x))$ and by Mordukhovich [32, 33, 34] via the concept of coderivative $\mathcal{D}^*\Gamma(x/y)$, i.e.,

$$v\in\mathcal{D}^*\Gamma(x/y)(u) \Longleftrightarrow (v,-u)\in N_{\mathrm{gph}\,\Gamma}(x,y). \tag{4}$$

Then the Mordukhovich criterion says that Γ is pseudo–Lipschitz around (x_0, y_0) if and only if

$$\parallel \mathcal{D}^*\Gamma(x_0/y_0) \parallel^+ = \sup_{u\in\mathbb{B}_1(0)} \quad \sup_{v\in\mathcal{D}^*\Gamma(x_0/y_0)(u)} \parallel v \parallel < \infty. \tag{5}$$

For more details of Aubin’s continuity concept and its applications see for example [16, 17, 41].

Definition 2.3 *An operator $[x, y; f]$ belonging to the space of linear bounded operator from X to Y denoted by $\mathcal{L}(X, Y)$ is called the first order divided difference on the points x and y in X, if we have the following*

$$[x,y;f]\;(x-y) = f(x)-f(y) \;\text{ if }\; x\neq y.$$

Let us note that if f is Fréchet differentiable at x^* then $[x^*, x^*; f] = \nabla f(x^*)$ where ∇f denotes the Fréchet derivative of f.

Definition 2.4 *We say that an operator belonging to the space $\mathcal{L}(X,\mathcal{L}(X,Y))$ denoted by $[x, y, z; g]$ is called the second order divided difference of the operator $g : X \to Y$ on the points $x, y, z \in X$ if the following property hold:*

$[x, y, z; g](z - x) = [y, z; g] - [x, y; g]$ *for the distinct points* x, y *and* z.

Let us note that if g is twice Fréchet differentiable at $x^* \in X$ then $[x^*, x^*, x^*; g] = \frac{1}{2}\nabla^2 g(x^*)$. where $\nabla^2 g$ denotes the second order Fréchet derivative of g.

The main tool of our study is the following fixed point theorem which has been proved in [13].

Lemma 2.1 *Let ϕ be some set–valued map from X into the closed subsets of X, let $\eta_0 \in X$ and let r and λ be such that $0 \le \lambda < 1$ and*
a) dist $(\eta_0, \phi(\eta_0)) \le r(1 - \lambda)$,
b) $e(\phi(x_1) \cap \mathbb{B}_r(\eta_0), \phi(x_2)) \le \lambda \parallel x_1 - x_2 \parallel, \;\; \forall x_1, x_2 \in \mathbb{B}_r(\eta_0)$,
then ϕ has a fixed–point in $\mathbb{B}_r(\eta_0)$. That is, there exists $x \in \mathbb{B}_r(\eta_0)$ such that $x \in \phi(x)$. If ϕ is single–valued, then x is the unique fixed point of ϕ in $\mathbb{B}_r(\eta_0)$.

3. A Combination of Newton's and Secant–Type Methods for Variational Inclusions

Here we consider a variational inclusion of the form:

$$0 \in f(x) + g(x) + G(x), \tag{6}$$

where $f : X \to Y$ is differentiable in a neighborhood of a solution x^* of (6) and $g : X \to Y$ is differentiable at x^* but may be not differentiable in a neighborhood of x^*; whereas G stands for a set–valued map from X into the subsets of Y.

Our purpose is to prove the existence of a sequence which verifies (6) and which is locally super–linearly convergent to x^*.

Here, because of the lack of regularity of g, we can't use the classical methods. Then, to carry out our objective, we propose a combination of Newton's method (applied to f) with the the Secant's one (applied to g). A similar method has been considered by Cătinas [12] for solving nonlinear equations, we extend it here to the set–valued functions framework.

More precisely, we associate to (6) the relation

$$0 \in f(x_k) + g(x_k) + \Big(\nabla f(x_k) + [x_{k-1}, x_k; g]\Big)(x_{k+1} - x_k) + G(x_{k+1}), \tag{7}$$

where $\nabla f(x)$ denotes the Fréchet derivative of f at x and $[x, y; g]$ the *first order divided difference* of g on the points x and y.
One can note that if $x_k \longrightarrow x^*$, then x^* is a solution of (6).

From now on, we make the following assumptions (we recall that x^* denotes a solution of (6)):

($\mathcal{H}00$) *F has closed graph;*

($\mathcal{H}01$) *f is Fréchet differentiable on some neighborhood V of x^*;*

($\mathcal{H}02$) *g is differentiable at x^*;*

($\mathcal{H}03$) *∇f is Lipschitz on V with constant L;*

($\mathcal{H}04$) *There exists $K \in \mathbb{R}$ such that for all $x, y, z \in V$, $\| [x, y, z; g] \| \leq K$;*

($\mathcal{H}05$) *The set–valued map*

$$\Big[f(x^*) + \nabla f(x^*)(x - x^*) + g(x) + G(x) \Big]^{-1}$$

is M–pseudo–Lipschitz around $(0, x^*)$.

The main result of this section is as follow:

Theorem 3.1 *Let x^* be a solution of* (6) *and suppose that the assumptions (*$\mathcal{H}00$*)–(*$\mathcal{H}05$*) are satisfied, then for every $C > M(\frac{L}{2} + K)$, one can find $\delta > 0$ such that for every distinct starting points $x_0, x_1 \in B_\delta(x^*)$, there exists a sequence (x_k), defined by* (7)*, which satisfies*

$$\|x_{k+1} - x^* \| \leq C \| x_k - x^* \| \max\{\|x_k - x^*\|, \|x_{k-1} - x^*\|\}. \tag{8}$$

For the proof of this theorem, the reader could be referred to [22].

As an illustration of our results let us consider the following nonlinear programming problem :

$$\text{minimize } f_0(x) \tag{9}$$

$$\text{subject to} \begin{cases} f_i(x) = 0, & i = 1, \cdots, m \\ f_i(x) \leq 0, & i = m+1, \cdots, p \end{cases}$$

where $f_0 : \mathbb{R}^n \to \mathbb{R}$ is twice continuously differentiable on $\mathbb{R}^n$ while the functions $f_i : \mathbb{R}^n \to \mathbb{R}$, $i = 1, \cdots, p$ are differentiable on $\mathbb{R}^n$ and are twice differentiable at a solution x^* of (9). We consider the sets $\mathcal{I}_1$ and $\mathcal{I}_2$ where $\mathcal{I}_1$ consists of those indices $i \in [1, p]$ such that f_i is twice continuously differentiable on $\mathbb{R}^n$, while $\mathcal{I}_2 = [1, p] \backslash \mathcal{I}_1$.

The lagragian L associated with (9) is defined by

$$L : (x, \lambda) \in \mathbb{R}^n \times \mathbb{R}^p \mapsto f_0(x) + \sum_{i=1}^{p} \lambda_i f_i(x),$$

and we write $L = L_1 + L_2$ where

$$\begin{cases} L_1(x, \lambda) = f_0(x) + \displaystyle\sum_{i \in \mathcal{I}_1} \lambda_i f_i(x) \\ L_2(x, \lambda) = \displaystyle\sum_{i \in \mathcal{I}_2} \lambda_i f_i(x). \end{cases}$$

Then, the Karush–Kuhn–Tucker first order optimality conditions read as follows:

$$\begin{cases} \nabla_x L_1(x,\lambda) + \nabla_x L_2(x,\lambda) = 0 \\ \nabla_\lambda L_1(x,\lambda) + \nabla_\lambda L_2(x,\lambda) \in N_\Lambda(\lambda) \end{cases}$$

where $N_\Lambda(\lambda)$ denotes the normal cone to the set $\Lambda = \mathbb{R}^m \times \mathbb{R}_+^{p-m}$ at the point λ. Then, it is easy to see that Karush–Kuhn–Tucker conditions amount to

$$0 \in (\nabla_x L_1(x,\lambda), -\nabla_\lambda L_1(x,\lambda)) + (\nabla_x L_2(x,\lambda), -\nabla_\lambda L_2(x,\lambda)) + N_C(x,\lambda) \qquad (10)$$

where $C = \mathbb{R}^n \times \Lambda$. Moreover, relation (10) can be reformulated in the following way:

$$0 \in f(x,\lambda) + g(x,\lambda) + G(x,\lambda), \qquad (11)$$

where $f(x,\lambda) = (\nabla_x L_1(x,\lambda), -\nabla_\lambda L_1(x,\lambda))$, $g(x,\lambda) = (\nabla_x L_2(x,\lambda), -\nabla_\lambda L_2(x,\lambda))$ and $F(x,\lambda) = N_C(x,\lambda)$. Hence, Karush–Kuhn–Tucker optimality system is equivalent to (11) which is a variational inclusion of the form of (6) and then can be studied using the method (7).

There are two variants of the previous method.

3.1. A Regula–Falsi–Type Method for Variational Inclusions

The first variant consists in replacing the Secant method by the regula–falsi one. The latter has been considered by Yakoubsohn [43] with only punctual conditions (α–theory of M. Shub and S. Smale [42]) for finding zeroes of single–valued analytic functions. More information about the regula–falsi method for single–valued functions can also be found in [35].

Here, as in Theorem 3.1, we take two starting points x_0 and x_1 in a suitable neighborhood of x^*. Then we fix one of the arguments of the divided difference of g, more precisely, we associate to (6) the relation :

$$0 \in f(x_k) + g(x_k) + \Big(\nabla f(x_k) + [x_0, x_k; g]\Big)(x_{k+1} - x_k) + G(x_{k+1}); \qquad (12)$$

and finally, under the same assumptions ($\mathcal{H}00$)–($\mathcal{H}05$) mentioned at the begining of section 3, we get:

Theorem 3.2 *Let x^* be a solution of* (6) *and suppose that the assumptions ($\mathcal{H}00$)–($\mathcal{H}05$) are satisfied, then for every $C > M(\frac{L}{2} + K)$, one can find $\delta > 0$ such that for every starting point $x_0 \in B_\delta(x^*)$, there exists a sequence (x_k), defined by* (12)*, which satisfies*

$$\|x_{k+1} - x^*\| \le C\,\|x_k - x^*\| \max\{\|x_k - x^*\|, \|x_0 - x^*\|\}. \qquad (13)$$

Both of Theorems 3.1 and 3.2 provide super–linearly convergent sequences but one can note that the convergence in Theorem 3.2 is slower than the one in Theorem 3.1 because in our first result the upper bound of relation (8) involves both x_k and x_{k-1}.

3.2. An Acceleration of the Secant–Type Method

The second method that we would like to consider in this section is based on the following modification of relation (7):

$$0 \in f(x_k) + g(x_k) + \Big(\nabla f(x_k) + [x_{k+1}, x_k; g]\Big)(x_{k+1} - x_k) + G(x_{k+1}). \tag{14}$$

According to definition 2.3 it amounts to :

$$0 \in f(x_k) + \nabla f(x_k)(x_{k+1} - x_k) + g(x_{k+1}) + G(x_{k+1}). \tag{15}$$

Note that the only change made is that we replaced x_{k-1} by x_{k+1} in the expression of the divided difference of g. This can be viewed like an acceleration of the original method in section 3 and actually it amounts to the well–known Newton's method for solving $0 \in f(x) + F(x)$ where $F = g + G$. In this case, we do not need assumption ($\mathcal{H}04$) (about the second order divided difference of g) and we obtain the following improvement of Theorem 3.1 involving a quadratic convergence of (x_k) to x^*:

Theorem 3.3 *Let x^* be a solution of* (6) *and suppose that the assumptions ($\mathcal{H}00$)–($\mathcal{H}03$) and ($\mathcal{H}05$) are satisfied, then for every $C > \dfrac{ML}{2}$, one can find $\delta > 0$ such that for every starting point $x_0 \in B_\delta(x^*)$, there exists a sequence (x_k), defined by* (15)*, which satisfies*

$$\| x_{k+1} - x^* \| \leq C \| x_k - x^* \|^2 . \tag{16}$$

4. A Modification of the Secant–Type Method for Variational Inclusions

This section is concerned with the problem of approximating a solution of the variational inclusion (1) when f is not necesseraly differentiable but posseses a first order divided difference.

For solving (1), we consider in [28], the sequence

$$\begin{cases} x_0 \text{ and } x_1 \text{ are given starting points} \\ y_k = \alpha x_k + (1-\alpha)x_{k-1}; \ \alpha \text{ is fixed in } [0,1[\\ 0 \in f(x_k) + [y_k, x_k; f](x_{k+1} - x_k) + G(x_{k+1}) \end{cases} \tag{17}$$

where $[y_k, x_k; f]$ is a first order divided difference of f on the points y_k and x_k.
In [23], the authors consider a similar iterative method like (17) with $\alpha = 0$ to solve nonlinear equations ($G \equiv 0$), they prove a semilocal convergence result for this method assuming existence of divided differences for f.

Let us recall that Geoffroy in [19] obtained the Q–superlinear convergence of a Secant type method for solving (1) assuming the existence of first and second order divided differences and that the solution satisfies a calmness–type property.

Note that, in this present chapter, the authors don't use the concept of second order divided difference, but only first order divided difference. This means that our method is valid if f possesses a second order divided difference or not.
Here, we have the existence of a sequence defined by (17) which is locally convergent to a solution x^* of (1).

We also make the following assumptions on a neighborhood V of x^*:

($\mathcal{H}11$) There exists $\nu > 0$ such that for all x, y, u and v in V ($x \neq y$ and $u \neq v$)

$$\| [x, y; f] - [u, v; f] \| \leq \nu(\| x - u \|^p + \| y - v \|^p), \; p \in [0, 1],$$

($\mathcal{H}12$) The set–valued map $(f+G)^{-1}$ is pseudo–Lipschitz with modulus M around $(0, x^*)$,

($\mathcal{H}13$) For all $x, y \in V$, we have $\|[x, y; f]\| \leq \kappa$ and $M\kappa < 1$.

Remark 4.1 *The assumption ($\mathcal{H}13$) implies that the function f is κ–Lipschitz on V.*

When a single–valued function f satisfies the assumption ($\mathcal{H}11$), we say that f has a (ν, p)–Hölder continuous divided difference on V. Let us note that in [23], the authors show a semilocal result of convergence of the Secant method to solve a nonlinear equation ($G \equiv 0$) with a new condition relaxing the condition ($\mathcal{H}11$) by replacing (in ($\mathcal{H}11$)) the right term of the inequality by $\omega(\| x - u \|, \| y - v \|)$ where ω from $I\!R_+ \times I\!R_+$ to $I\!R_+$ is a continuous nondecreasing function in both variables.
With the previous assumption, one has :

Theorem 4.1 *Let x^* be a solution of (1). We suppose that assumptions ($\mathcal{H}11$)–($\mathcal{H}13$) are satisfied. For every $C > \dfrac{M\nu[(1-\alpha)^p + \alpha^p]}{1 - M\kappa}$, one can find $\delta > 0$ such that for every distinct starting points x_0 and x_1 in $B_\delta(x^*)$, there exists a sequence (x_k) defined by (2) which satisfies*

$$\| x_{k+1} - x^* \| \leq C \| x_k - x^* \| \max \{\| x_k - x^* \|^p, \| x_{k-1} - x^* \|^p\}. \tag{18}$$

For the proof of the theorem, the reader could be referred to [28].

When $\alpha = 1$, the method is no longer valid, but if we suppose that f is Fréchet differentiable, (17) is equivalent to a Newton–type method to solve (1). In this case, conditions on ∇f give quadratic convergence (see [14]) and superlinear convergence (see [38]) and in both cases the convergence is uniform (see [15] and [39]).

When $\alpha = 0$ the sequence (17) is reduced to the method introduced by M. Geoffroy and A. Piétrus in [22]. Let us note that the problem studied in [22] can be seen as a perturbation of (1) by a Fréchet differentiable function. In both cases, we obtain a superlinear convergence using different assumptions, but in the present paper the existence of second order divided difference is not required.

5. Convergence Analysis of Steffensen–Type Method under Weak Conditions

Let us recall that for Steffensen–type method for variational inclusions, the author in [27] suppose that the first order divided difference of f satisfies a (ν, p)–Hölder condition:

There exists $\nu > 0$ such that for all x, y, u and v belonging to a neighborhood V of a solution x^* of (1) ($x \neq y$ and $u \neq v$),

$$\|[x, y; f] - [u, v; f]\| \leq \nu\ (\|x - u\|^p + \|y - v\|^p), \quad p \in [0, 1]. \tag{19}$$

Here, we suppose that f is Fréchet differentiable at x^*, in this case we have $[x^*, x^*; f] = \nabla f(x^*)$.

Our purpose here is to study the convergence of the method (2) under a weaker condition which is :

there exists $\nu_0 > 0$ such that for all distinct points x and y belonging to a neighborhood V of a solution x^* of (1), the following holds

$$\|[x, y; f] - \nabla f(x^*)\| \leq \nu_0\ (\|x - x^*\|^p + \|y - x^*\|^p), \quad p \in [0, 1], \tag{20}$$

and we say that f satisfies a (ν_0, p)–center–Hölder condition at x^*. Note that in general, $\nu_0 \leq \nu$. So, there are cases where (20) holds but not (19). The inspiration for considering condition (20) comes from Argyros [2, 4, 5].

For studing the convergence of the method (2), we suppose that, for every distinct points x and y in a open convex neighborhood V of x^*, there exists a first order divided difference of f at these points. We also make the following assumptions:

($\mathcal{H}20$) For $i = 1, 2$; g_i is α_i–Lipschitz from V into V; $\alpha_i \in [0, 1)$ and $g_i(x^*) = x^*$.
($\mathcal{H}21$) The first order divided difference of f is (ν_0, p)–center–Hölder continuous at x^*.
($\mathcal{H}22$) The set–valued map $(f(x^*) + G)^{-1}$ is M–pseudo–Lipschitz around $(0, x^*)$.
($\mathcal{H}23$) For all $x, y \in V$, we have $\|[x, y; f]\| \leq \kappa$ and $M\kappa < 1$.

The main result of this chapter is the following:

Theorem 5.1 *We suppose that assumptions ($\mathcal{H}20$)–($\mathcal{H}23$) are satisfied. For every constant* $C > \dfrac{M\,\nu_0\ (1 + \alpha_1^p + \alpha_2^p)}{1 - M\kappa}$, *one can find* $\delta > 0$ *such that for every starting point* x_0 *in* $B_\delta(x^*)$ *(*x_0 *and* x^* *distincts), there exists a sequence* (x_k) *defined by (2) which satisfies*

$$\|\ x_{k+1} - x^*\ \| \leq C\ \|\ x_k - x^*\ \|^{p+1}\ . \tag{21}$$

Before proving Theorem 5.1, we introduce some notations. First, for $k \in \mathbb{N}$ and x_k defined by (2), let us define the set–valued mappings $Q : X \to 2^Y$, $\psi_k : X \to 2^X$ by

$$Q(.) := f(x^*) + G(.); \quad \psi_k(.) := Q^{-1}(Z_k(.)) \tag{22}$$

where Z_k is defined from X to Y by

$$Z_k(x) := f(x^*) - f(x_k) - [g_1(x_k), g_2(x_k); f](x - x_k). \tag{23}$$

The proof of Theorem 5.1 is by induction on k. We first state a result which is the starting point of our algorithm. Let us mention that x_1 is a fixed point of ψ_0 if and only if $0 \in f(x_0) + [g_1(x_0), g_2(x_0); f](x_1 - x_0) + G(x_1)$.

Proposition 5.1 *Under the assumptions of Theorem 5.1, one can find $\delta > 0$ such that for every starting point x_0 in $B_\delta(x^*)$ (x_0 and x^* distincts), the set–valued map ψ_0 has a fixed point x_1 in $B_\delta(x^*)$ satisfying*

$$\| x_1 - x^* \| \leq C \| x_0 - x^* \|^{p+1} . \tag{24}$$

Proof of Proposition 5.1. By ($\mathcal{H}22$) there exist positive numbers M, a and b such that

$$e(Q^{-1}(y') \cap B_a(x^*), Q^{-1}(y'')) \leq M \| y' - y'' \|, \ \forall y', y'' \in B_b(0). \tag{25}$$

Fix $\delta > 0$ such that

$$\delta < \min\left\{ a \ ; \ \sqrt[p+1]{\frac{b}{4\ \nu_0\ (1 + \alpha_1^p + \alpha_2^p)}} \ ; \ \frac{1}{\sqrt[p]{C}} \ ; \ \frac{b}{2\kappa} \right\}. \tag{26}$$

The main idea of the proof is to show that both assertions (a) and (b) of Lemma 2.1 hold where $\eta_0 := x^*$, ϕ is the function ψ_0 defined in (22) and where r and λ are numbers to be set. According to the definition of the excess e, we have

$$\text{dist}\,(x^*, \psi_0(x^*)) \leq e\Big(Q^{-1}(0) \cap B_\delta(x^*), \psi_0(x^*) \Big). \tag{27}$$

Moreover, for all point x_0 in $B_\delta(x^*)$ (x_0 and x^* distinct) we have

$$\| Z_0(x^*) \| = \| f(x^*) - f(x_0) - [g_1(x_0), g_2(x_0); f](x^* - x_0) \| .$$

By assumptions ($\mathcal{H}20$)–($\mathcal{H}21$) we deduce

$$\begin{aligned}
\| Z_0(x^*) \| &= \| \Big([x^*, x_0; f] - [g_1(x_0), g_2(x_0); f] \Big)(x^* - x_0) \| \\
&\leq \| [x^*, x_0; f] - [g_1(x_0), g_2(x_0); f] \| \ \| x^* - x_0 \| \\
&\leq \Big(\| [x^*, x_0; f] - \nabla f(x^*) \| + \| [g_1(x_0), g_2(x_0); f] - \nabla f(x^*) \| \Big) \ \| x^* - x_0 \| \\
&\leq \nu_0 \ (\| x_0 - x^* \|^p + \| g_1(x_0) - x^* \|^p + \| g_2(x_0) - x^* \|^p) \ \| x^* - x_0 \| \\
&\leq \nu_0 \ (1 + \alpha_1^p + \alpha_2^p) \ \| x^* - x_0 \|^{p+1}
\end{aligned} \tag{28}$$

and (26) implies $Z_0(x^*) \in B_b(0)$.

Using (25) we have

$$\begin{aligned}
e\Big(Q^{-1}(0) \cap B_\delta(x^*), \psi_0(x^*) \Big) &= e\Big(Q^{-1}(0) \cap B_\delta(x^*), Q^{-1}[Z_0(x^*)] \Big) \\
&\leq M \ \nu_0 \ (1 + \alpha_1^p + \alpha_2^p) \ \| x^* - x_0 \|^{p+1} .
\end{aligned} \tag{29}$$

With inequality (27), we get

$$\operatorname{dist}(x^*, \psi_0(x^*)) \;\leq\; M\ \nu_0\ (1+\alpha_1^p+\alpha_2^p)\ \parallel x^*-x_0 \parallel^{p+1}. \qquad (30)$$

Since $C(1-M\kappa) > M\ \nu_0\ (1+\alpha_1^p+\alpha_2^p)$, there exists $\lambda \in [M\kappa, 1[$ such that $C(1-\lambda) \geq M\ \nu_0\ (1+\alpha_1^p+\alpha_2^p)$ and

$$\operatorname{dist}(x^*, \psi_0(x^*)) \leq C\ (1-\lambda)\ \parallel x_0 - x^* \parallel^{p+1}. \qquad (31)$$

By setting $r := r_0 = C\ \parallel x_0 - x^* \parallel^{p+1}$ we deduce from the inequality (31) that the assertion (a) in Lemma 2.1 is satisfied.
Now, we show that condition (b) of Lemma 2.1 is satisfied.
By (26), we have $r_0 \leq \delta \leq a$ and moreover for $x \in B_\delta(x^*)$ we have

$$\begin{aligned}
\parallel Z_0(x) \parallel &= \parallel f(x^*) - f(x_0) - [g_1(x_0), g_2(x_0); f](x-x_0) \parallel \\
&= \parallel [x^*, x_0; f](x^* - x + x - x_0) - [g_1(x_0), g_2(x_0); f](x-x_0) \parallel \\
&\leq \parallel [x^*, x_0; f] \parallel \parallel x^* - x \parallel + \parallel [g_1(x_0), g_2(x_0); f] - [x^*, x_0; f] \parallel \parallel x - x_0 \parallel \\
&\leq \parallel [x^*, x_0; f] \parallel \parallel x^* - x \parallel + \Big(\parallel [x^*, x_0; f] - \nabla f(x^*) \parallel + \\
&\qquad \parallel [g_1(x_0), g_2(x_0); f] - \nabla f(x^*) \parallel \Big) \parallel x - x_0 \parallel .
\end{aligned} \qquad (32)$$

Using the assumptions ($\mathcal{H}$20)–($\mathcal{H}$21) and ($\mathcal{H}$23) we obtain

$$\begin{aligned}
\parallel Z_0(x) \parallel &\leq \kappa\ \parallel x^* - x \parallel + \nu_0\ (1+\alpha_1^p+\alpha_2^p)\ \parallel x_0 - x^* \parallel^p \parallel x - x_0 \parallel \\
&\leq \kappa\ \delta + 2\nu_0\ (1+\alpha_1^p+\alpha_2^p)\ \delta^{p+1}.
\end{aligned} \qquad (33)$$

Then by (26) we deduce that for all $x \in B_\delta(x^*)$ we have $Z_0(x) \in B_b(0)$. Then it follows that for all $x', x'' \in B_{r_0}(x^*)$ we have

$$e(\psi_0(x') \cap B_{r_0}(x^*), \psi_0(x'')) \leq e(\psi_0(x') \cap B_\delta(x^*), \psi_0(x'')),$$

which yields using (25)

$$\begin{aligned}
e(\psi_0(x') \cap B_{r_0}(x^*), \psi_0(x'')) &\leq M\ \parallel Z_0(x') - Z_0(x'') \parallel \\
&\leq M\ \parallel [g_1(x_0), g_2(x_0); f] \parallel \parallel x'' - x' \parallel
\end{aligned} \qquad (34)$$

Using ($\mathcal{H}$23) and the fact that $\lambda \geq M\kappa$, we obtain

$$e(\psi_0(x') \cap B_{r_0}(x^*), \psi_0(x'')) \leq M\ \kappa \parallel x'' - x' \parallel \leq \lambda\ \parallel x'' - x' \parallel \qquad (35)$$

and thus condition (b) of Lemma 2.1 is satisfied. Since both conditions of Lemma 2.1 are fulfilled, we can deduce the existence of a fixed point $x_1 \in B_{r_0}(x^*)$ for the map ψ_0. This finishes the proof of Proposition 5.1. □

Proof of Theorem 5.1. Keeping $\eta_0 = x^*$ and setting $r := r_k = C\ \parallel x^* - x_k \parallel^{p+1}$, the application of Proposition 5.1 to the map ψ_k gives the existence of a fixed point x_{k+1} for ψ_k which is an element of $B_{r_k}(x^*)$. This last fact implies the inequality (21), which is the desired conclusion. □

References

[1] I.K. Argyros, A new convergence theorem for Steffensen's method on Banach spaces and applications, Southwest J. of Pure and Appl. Math., 01 (1997), 23–29.

[2] I.K. Argyros, A unifying local–semilocal convergence analysis and applications for two–point Newton–like methods in Banach space, *J. Math. Anal. Appl.*, **298** (2004), 374–397.

[3] I.K. Argyros, Approximate solution of operator equations with applications, *World Scientific Publ. Comp.,* New Jersey, USA, 2005.

[4] I.K. Argyros, Concerning the "terra incognita" between convergence regions of two Newton methods, *Nonlinear Analysis,* **62** (2005), 179–194.

[5] I.K. Argyros, An improved convergence analysis of a superquadratic method for solving generalized equations, *Rev. Colombiana Math.*, **40** (2006), 65–73.

[6] I.K. Argyros, Computational theory of iterative methods, *Studies in Computational Mathematics*, **15**, Elseiver, New York, USA, 2007.

[7] I.K. Argyros and S. Hilout, Steffensen methods for solving generalized equations, *Serdica Math. J.*, to appear.

[8] I.K. Argyros and S. Hilout, On a Secant–like method for solving generalized equations, *Mathematica Bohemica*, to appear.

[9] I.K. Argyros and S. Hilout, Newton's methods for variational inclusions under conditioned Fréchet derivative, *Appl. Math.,* **34** (2007), 349–357.

[10] J–P. Aubin, Lipschitz behavior of solutions to convex minimization problems, *Math. Oper. Res.,* **9** (1984), 87–111 .

[11] J–P. Aubin and H. Frankowska, Set–valued analysis, Birkhäuser, Boston, 1990.

[12] E. Cătinas, On some iterative methods for solving nonlinear equations, *Revue d'analyse numérique et de théorie de l'approximation,* **23** (1994), 17–53.

[13] A.L. Dontchev and W.W. Hager, An inverse mapping theorem for set–valued maps, *Proc. Amer. Math. Soc.*, **121** (1994), 481–489.

[14] A.L. Dontchev, *Local convergence of the Newton method for generalized equation, C.R.A.S.* **322**, Serie I, (1996), 327–331.

[15] A.L. Dontchev, Uniform convergence of the Newton method for Aubin continuous maps, *Serdica Math. J.*, **22**, (1996), 395–398.

[16] A.L. Dontchev and R.T. Rockafellar, Regularity and conditioning of solution mappings in variational analysis, *Set–Valued Anal.*, 1–2 (2004), 79–109.

[17] A.L. Dontchev, M. Quincampoix and N. Zlateva, Aubin criterion for metric regularity, *J. Convex Anal.*, **13** (2006), 281–297.

[18] M.C. Ferris and J.S. Pang, Engineering and economic applications of complementarity problems, *SIAM Rev.,* **39** (1997), 669–713.

[19] M.H. Geoffroy, *A Secant type method for variational inclusions,* Preprint.

[20] M.H. Geoffroy, S. Hilout and A. Piétrus, Acceleration of convergence in Dontchev's iterative method for solving variational inclusions, *Serdica Math. J.,* **29**(1), (2003), 45–54.

[21] M.H. Geoffroy and A. Piétrus, A superquadratic method for solving generalized equations in the Hölder case, *Ricerche Di Matematica* **L II/2**, (2003), 231–240.

[22] M. Geoffroy and A. Piétrus, Local convergence of some iterative methods for generalized equations, *J. of Math. Anal. and Appl.*, **290**(2), (2004), 497–505.

[23] M.A. Hernández and M.J. Rubio, Semilocal convergence of the Secant method under mild convergence conditions of differentiability, *Comp. and Math. with Appl.,* **44** (2002), 277–285.

[24] S. Hilout, Steffensen–type methods on Banach spaces for solving generalized equations, *Adv. Nonlinear Var. Inequa.*, **2**(10), (2007), 105–113.

[25] S. Hilout, An uniparametric Newton–Steffensen–type methods for perturbed generalized equations, *Adv. Nonlinear Var. Inequa.*, **2**(10), (2007), 115–124.

[26] S. Hilout, *An uniparametric Secant–type method for nonsmooth generalized equations*, Positivity, to appear.

[27] S. Hilout, Convergence analysis of a family of Steffensen–type methods for solving generalized equations, *J. Math. Anal. Appl.,* **339** (2008), 753–761.

[28] S. Hilout and A. Piétrus, A semilocal convergence of the Secant-type method for solving a generalized equations, *Positivity,* **10** (2006), 673–700.

[29] P. M. Hummel and C.L. Jr Seebeck, A generalization of Taylor's expansion, *Amer. Math. Monthly*, **56** (1949), 243–247.

[30] C. Jean–Alexis, A cubic method without second order derivative for solving variational inclusions, *C.R. Acad. Bulg. Sci.*, **59** (12), (2006), 1213–1218.

[31] C. Jean–Alexis and A. Piétrus, A variant of Newton's method for generalized equations, *Rev. Colombiana Math.,* **39** (2005), 97–112.

[32] B.S. Mordukhovich, Complete characterization of openness metric regularity and lipschitzian properties of multifunctions, *Trans. Amer. Math. Soc* **340** (1993), 1–36.

[33] B.S. Mordukhovich, Stability theory for parametric generalized equations and variational inequalities via nonsmooth analysis, *Trans. Amer. Math. Soc.,* **343** (1994), 609–657.

[34] B.S. Mordukhovich, *Variational analysis and generalized differentiation* , I. Basic theory, II. Applications (two volumes), Springer, Grundlehren Series, Vol. 330 and 331, 2006.

[35] A. Ostrowski, *Solutions of equations and systems of equations* , Academic Press, third ed, 1973.

[36] M. Patriksson, Cost Approximation Algorithms with nonmonotone line searches for a general class of nonlinear programs, *Optimization,* **44** (1998), 199–217.

[37] M. Patriksson, Sensitivity analysis of traffic equilibria, *Trans. Scien.*, **38** (2004), 258–281.

[38] A. Piétrus, Generalized equations under mild differentiability conditions, *Rev. R. Acad. Cienc. Exact. Fis. Nat.*, **94** (2000), 15–18.

[39] A. Piétrus, Doe's Newton's method converges uniformly in mild differentiability context? *Rev. Colombiana Math.*, **32** (2), (2000), 49–56.

[40] R.T. Rockafellar, Lipschitzian properties of multifunctions, *Nonlinear Analysis*, **9** (1984), 867–885.

[41] R.T. Rockafellar and R. Wets, Variational analysis, *Ser. Com. Stu. Math.*, **317**, Springer, 1998.

[42] M. Shub and S. Smale, Complexity of Bezout's theorem 1 :geometric aspects, *Journal of the Amer. Math. Soc.,* **6** (1993), 459–501.

[43] J.C. Yakoubsohn, Finding zeros of analytic functions : α–theory for Secant type method, *J. of Complexity,* **15**, 2 (1999), 239–281.

[44] L. Walras, Elements of Pure Economics, Alen and Unwin, London, 1954.

[45] L. Wardrop, *Some theoritical aspects of road traffic research, Proceeding of the Institute of Civil Engineers,* Part II (1952), 325–378.

In: Progress in Nonlinear Analysis Research
Editor: Erik T. Hoffmann, pp. 285-308
ISBN 978-1-60456-359-7

Chapter 12

GLOBAL CLASSICAL SOLUTIONS FOR A CLASS OF QUASILINEAR HYPERBOLIC SYSTEMS OF BALANCE LAWS

Zhi-Qiang Shao
Department of Mathematics, Fuzhou University,
Fuzhou 350002, China

Abstract

This paper concerns global classical solutions for a class of quasilinear hyperbolic systems of balance laws in one space dimension. It is shown that the Cauchy problem for a class of quasilinear weakly linearly degenerate hyperbolic systems of balance laws with small and decaying initial data admits a unique global C^1 solution $u = u(t, x)$ on $t \geq 0$. This result is also applied to the flow equations of a model class of fluids with viscosity induced by fading memory.

MSC: 35L45; 35L60

Key Words: cauchy problem, global classical solution, quasilinear hyperbolic systems of balance laws, weakly linearly degenerate characteristics

1. Introduction and Main Result

Consider the following quasilinear hyperbolic system of balance laws in one space dimension:

$$\frac{\partial u}{\partial t} + \frac{\partial f(u)}{\partial x} + Lu = 0, \tag{1.1}$$

where $u = (u_1, \ldots, u_n)^T$ is the unknown vector function of (t, x), $f(u)$ is a given C^3 vector function of u, $L > 0$ is a constant.

It is assumed that system (1.1) is strictly hyperbolic, i.e., for any given u on the domain under consideration, the Jacobian $A(u) = \nabla f(u)$ has n real distinct eigenvalues

$$\lambda_1(u) < \lambda_2(u) < \ldots < \lambda_n(u). \tag{1.2}$$

Let $l_i(u) = (l_{i1}(u), \ldots, l_{in}(u))$ (resp. $r_i(u) = (r_{i1}(u), \ldots, r_{in}(u))^T$) be a left (resp. right) eigenvector corresponding to $\lambda_i(u)(i = 1, \ldots, n)$:

$$l_i(u)A(u) = \lambda_i(u)l_i(u) \qquad (resp.\ A(u)r_i(u) = \lambda_i(u)r_i(u)), \tag{1.3}$$

then we have

$$det|l_{ij}(u)| \neq 0 \qquad (equivalently,\ det|r_{ij}(u)| \neq 0). \tag{1.4}$$

Without loss of generality, we may assume that on the domain under consideration

$$l_i(u)r_j(u) \equiv \delta_{ij} \quad (i, j = 1, \ldots, n) \tag{1.5}$$

and

$$r_i^T(u)r_i(u) \equiv 1 \quad (i = 1, \ldots, n), \tag{1.6}$$

where δ_{ij} stands for the Kronecker's symbol.

Clearly, all $\lambda_i(u)$, $l_{ij}(u)$ and $r_{ij}(u)(i, j = 1, \ldots, n)$ have the same regularity as $A(u)$, i.e., C^2 regularity.

We also assume that system (1.1) is weakly linearly degenerate, i.e., all the characteristics are weakly linearly degenerate. According to Li et al. [25-26], we call that $\lambda_i(u)$ is a weakly linearly degenerate characteristic, if, along the i-th characteristic trajectory $u = u^{(i)}(s)$ passing through $u = 0$, defined by

$$\begin{cases} \frac{du}{ds} = r_i(u), \\ s = 0 : u = 0, \end{cases} \tag{1.7}$$

we have

$$\nabla\lambda_i(u)r_i(u) \equiv 0, \ \ \forall|u| \ small, \tag{1.8}$$

namely,

$$\lambda_i(u^{(i)}(s)) \equiv \lambda_i(0), \quad \forall|s| \ small. \tag{1.9}$$

We are interested in the existence and uniqueness of global C^1 solutions to the Cauchy problem for system (1.1) with the initial data:

$$t = 0 : \quad u = \varphi(x), \quad -\infty < x < \infty, \tag{1.10}$$

where $\phi(x)$ is a C^1 vector function. Also, we assume that there exists a constant $\mu > 0$ such that

$$\theta \stackrel{\triangle}{=} \sup_{x\in R}\{(1 + |x|)^{1+\mu}(|\varphi(x)| + |\varphi'(x)|)\} << 1. \tag{1.11}$$

For the special case where (1.1) is a system of conservation laws, i.e., $L = 0$, such kinds of problems have been extensively studied (for instance, see [1, 9, 13-21, 23-27, 31-35] and references therein). By introducing the concept of weak linear degeneracy, Li et al. [14-21, 23, 25-26] have obtained a series of important results on the existence and life-span of global C^1 solutions for quasilinear hyperbolic systems of conservation laws. On the other hand, for quasilinear hyperbolic systems of balance laws, many results on the global existence of weak solutions have also been obtained by T.-P. Liu, Dafermos, G.-Q. Chen, L. Hsiao and others (for instance, see [2-6, 10-12, 22, 28-30, 34-40] and references therein), and some methods have been established. So the following question arises naturally: when can we obtain the existence and uniqueness of global C^1 solutions for quasilinear hyperbolic systems of balance laws? According to the general theory of hyperbolic systems of balance laws [4], if the flux function $f(u)$ and the source term are smooth enough, it is well known that problem (1.1) and (1.10) has a unique local smooth solution, at least for some time interval $[0, T]$ with $T > 0$, if the initial data are also sufficiently smooth. In the general case, and even for very good initial data, smooth solutions may break down in finite time, due to the appearance of singularities, either discontinuities or blow-up. In some cases, however, dissipative mechanisms due to the source term can prevent the formation of singularities, the global existence in time of classical solutions can be obtained, at least for some restricted classes of initial data, as observed for many models which arise to describe physical phenomena. Therefore any sufficient condition guaranteeing the global existence in time of classical solutions is interesting. In this paper, we prove that the presence of strong linear damping, plus "weak linear degeneracy" of the characteristics, with small smooth initial data, produces the global existence in time of C^1 solutions. Our main results in this paper can be stated as:

Theorem 1.1. Suppose that system (1.1) is strictly hyperbolic, i.e., (1.2) holds on the domain under consideration. Suppose furthermore that system (1.1) is weakly linearly degenerate. Suppose finally that $A(u) \in C^2$ in a neighborhood of $u = 0$ and $\phi(x)$ is a C^1 vector function satisfying that there exists a constant $\mu > 0$ such that

$$\theta \triangleq \sup_{x \in R} (1 + |x|)^{1+\mu} (|\phi(x)| + |\phi'(x)|) < +\infty. \tag{1.12}$$

Then there exists a sufficiently small $\theta_0 > 0$ such that for any given $\theta \in [0, \theta_0]$, Cauchy problem (1.1) and (1.10) admits a unique global C^1 solution $u = u(t, x)$ on $t \geq 0$.

Remark 1.2. Suppose that system (1.1) is non-strictly hyperbolic but each characteristic has a constant multiplicity. Suppose furthermore that system (1.1) is weakly linearly degenerate. Then, if there exist normalized coordinates, similar conclusion holds as in Theorem 1.1 (some related results can be found in [20]).

According to the local existence and uniqueness of the C^1 solutions to the Cauchy problem for nonhomogeneous quasilinear hyperbolic systems (see [8, Theorem VI], see also [7, 24]), the key point to prove Theorem 1.1 is to obtain a uniform a priori estimate for the C^0 norm of u and u_x on any given domain of existence of the C^1 solution $u = u(t, x)$. The biggest difficulty we are faced with here is how to estimate the lower-order terms, caused by the nonhomogeneous term. Using the weakly linearly degenerate condition and combining the normalized coordinates, we overcome successfully this difficulty. Moreover, the uniform a priori estimates is much more complicated than those for homogeneous quasilinear hyperbolic systems. The rest of this paper is organized as follows. In Section 2, we will construct an example to show that if system (1.1) is genuinely nonlinear, then the C^1 solution $u = u(t, x)$ may blow up in a finite time even for arbitrary small C^1 initial data satisfying the decay property as shown in (1.11). In Section 3, we give the main tools of the proof, that is several formulas on the decomposition of waves for system (1.1). Then, the main result will be proved in Section 4. Finally, an application is given in Section 5.

2. An Example

Consider the following Burger equation

$$\frac{\partial u}{\partial t} + \frac{\partial(\frac{u^2}{2})}{\partial x} = -u \tag{2.1}$$

with

$$u(0,x) = \varepsilon\psi(x), \tag{2.2}$$

where $\varepsilon > 0$ is a arbitrary small constant, $\psi(x)$ is a nontrivial C^1 function with compact support and

$$\psi(x) = e^{-x}, \quad for \quad x \in [-L, L], \tag{2.3}$$

where $L > 0$ is a constant. It is clear that (2.1) is genuinely nonlinear. Now we show that for any given, arbitrary small $\varepsilon > 0$, there exists a suitable large constant $L > 0$ such that the first-order derivative of the C^1 solution $u = u(t,x)$ blows up in a finite time.

For this purpose, on the domain of existence of the C^1 solution, let $\beta \in (-\infty, ln\varepsilon)$ and $x = x(t,\beta)$ be the characteristic passing through $(0,\beta)$. Hence,

$$\varepsilon - e^{\beta} > 0.$$

We choose L such that $L \geq |\beta| + 1$. Let $w = \frac{\partial u}{\partial x}$, then it follows from (2.1)-(2.3) that

$$\frac{\partial w}{\partial t} + u\frac{\partial w}{\partial x} + w = -w^2, \tag{2.4}$$

$$w(0,x) = -\varepsilon e^{-x}, \quad for \quad x \in (-L, L). \tag{2.5}$$

Thus, along the characteristic $x = x(t,\beta)$ we have

$$\begin{cases} \frac{d}{dt}(e^t w) = -e^t w^2, \\ t = 0 : w = -\varepsilon e^{-\beta}. \end{cases} \tag{2.6}$$

Hence, we get

$$w(t, x(t,\beta)) = \frac{\varepsilon}{(\varepsilon - e^{\beta})e^t - \varepsilon}. \tag{2.7}$$

Therefore $w(t, x(t,\beta))$ and then $\frac{\partial u}{\partial x}$ blow up at time $t = \ln\frac{\varepsilon}{\varepsilon - e^{\beta}}$.

3. Decomposition of Waves

Suppose that on the domain under consideration system (1.1) is strictly hyperbolic and (1.2)-(1.6) hold.

Suppose that $A(u) \in C^k$, where k is an integer ≥ 1. By Lemma 2.5 in [25], there exists an invertible C^{k+1} transformation $u = u(\tilde{u})(u(0) = 0)$ such that in $\tilde{u}$-space, for each $i = 1, \ldots, n$, the ith characteristic trajectory passing through $\tilde{u} = 0$ coincides with the $\tilde{u}_i$-axis at least for $|\tilde{u}_i|$ small, namely,

$$\tilde{r}_i(\tilde{u}_i e_i) \equiv e_i, \quad \forall\, |\tilde{u}_i| \; small \; (i = 1, \ldots, n), \tag{3.1}$$

where

$$e_i = (0, \ldots, 0, \overset{(i)}{1}, 0, \ldots, 0)^T.$$

This transformation is called the normalized transformation, and the corresponding unknown variables $\widetilde{u} = (\widetilde{u}_1, \ldots, \widetilde{u}_n)^T$ are called the normalized variables or normalized coordinates(see [26]).

Let

$$v_i = l_i(u)u \quad (i = 1, \ldots, n) \tag{3.2}$$

and

$$w_i = l_i(u)u_x \quad (i = 1, \ldots, n), \tag{3.3}$$

where

$$l_i(u) = (l_{i1}(u), \ldots, l_{in}(u))$$

denotes the ith left eigenvector.

By (1.5), it is easy to see that

$$u = \sum_{k=1}^{n} v_k r_k(u) \tag{3.4}$$

and

$$u_x = \sum_{k=1}^{n} w_k r_k(u). \tag{3.5}$$

Let

$$\frac{d}{d_i t} = \frac{\partial}{\partial t} + \lambda_i(u)\frac{\partial}{\partial x} \tag{3.6}$$

be the directional derivative along the ith characteristic. Our aim in this section is to prove several formulas on the decomposition of waves for system (1.1), which will play an important role in our discussion.

Lemma 3.1(Generalized John's formula).

$$\frac{d(e^{Lt}w_i)}{d_i t} = \sum_{j,k=1}^{n} e^{Lt}\gamma_{ijk}(u)w_j w_k + \sum_{j,k=1}^{n} e^{Lt}\widetilde{\gamma}_{ijk}(u)v_j w_k \quad (i = 1, \ldots, n), \tag{3.7}$$

where

$$\gamma_{ijk}(u) = (\lambda_k(u) - \lambda_j(u))r_j^T(u)\nabla l_i(u)r_k(u) - \nabla\lambda_i(u)r_j(u)\delta_{ik}, \tag{3.8}$$

$$\widetilde{\gamma}_{ijk}(u) = -Lr_j^T(u)\nabla l_i(u)r_k(u). \tag{3.9}$$

Hence, we have

$$\gamma_{ijj}(u) \equiv 0, \quad \forall j \neq i (i, j = 1, \ldots, n), \tag{3.10}$$

$$\gamma_{iii}(u) = -\nabla\lambda_i(u)r_i(u) \;\; (i = 1, \ldots, n). \tag{3.11}$$

Moreover, in the normalized coordinates,

$$\widetilde{\gamma}_{ijj}(u_j e_j) \equiv 0, \quad \forall |u_j| \; small, \; \forall i, j; \tag{3.12}$$

while, when the ith characteristic $\lambda_i(u)$ is weakly linearly degenerate, in the normalized coordinates,

$$\gamma_{iii}(u_ie_i) \equiv 0, \quad \forall\, |u_i| \ small, \ \forall\, i. \tag{3.13}$$

Proof. By (3.3), (1.1) and (1.3), it is easy to see that

$$\begin{aligned}\frac{dw_i}{d_it} &= u_t^T\nabla l_i(u)u_x + l_i(u)u_{xt} + \lambda_i(u)[u_x^T\nabla l_i(u)u_x + l_i(u)u_{xx}]\\ &= (-A(u)u_x - Lu)^T\nabla l_i(u)u_x + l_i(u)(-A(u)u_x - Lu)_x + \lambda_i(u)[u_x^T\nabla l_i(u)u_x + l_i(u)u_{xx}]\\ &= -u_x^TA^T(u)\nabla l_i(u)u_x - Lu^T\nabla l_i(u)u_x - l_i(u)\frac{\partial}{\partial x}[A(u)]u_x - Lw_i + \lambda_i(u)u_x^T\nabla l_i(u)u_x. \end{aligned}\tag{3.14}$$

Moreover, by (3.5), we have

$$u_x^T = \sum_{j=1}^n w_jr_j^T(u). \tag{3.15}$$

Thus, noting (3.5), (3.15) and the second equality in (1.3), it follows from (3.14) that

$$\begin{aligned}\frac{dw_i}{d_it} &= \sum_{j,k=1}^n(\lambda_i(u)-\lambda_j(u))r_j^T(u)\nabla l_i(u)r_k(u)w_jw_k - \sum_{k=1}^n l_i(u)\frac{\partial}{\partial x}[A(u)]r_k(u)w_k\\ &\quad -L\sum_{j,k=1}^n r_j^T(u)\nabla l_i(u)r_k(u)v_jw_k - Lw_i.\end{aligned}\tag{3.16}$$

Differentiating the first equality in (1.3) with respect to x yields

$$u_x^T\nabla l_i(u)A(u) + l_i(u)\frac{\partial}{\partial x}[A(u)] = \nabla\lambda_i(u)u_xl_i(u) + \lambda_i(u)u_x^T\nabla l_i(u). \tag{3.17}$$

Multiplying (3.17) by $r_k(u)$ and noting (1.5) and the second equality in (1.3), we get

$$l_i(u)\frac{\partial}{\partial x}[A(u)]r_k(u) = [\lambda_i(u)-\lambda_k(u)]u_x^T\nabla l_i(u)r_k(u) + \nabla\lambda_i(u)u_x\delta_{ik}. \tag{3.18}$$

Therefore, noting (3.5) and (3.15), it follows from (3.16) and (3.18) that

$$\begin{aligned}\frac{dw_i}{d_it} &= \sum_{j,k=1}^n[(\lambda_k(u)-\lambda_j(u))r_j^T(u)\nabla l_i(u)r_k(u) - \nabla\lambda_i(u)r_j(u)\delta_{ik}]w_jw_k\\ &\quad -L\sum_{j,k=1}^n r_j^T(u)\nabla l_i(u)r_k(u)v_jw_k - Lw_i.\end{aligned}\tag{3.19}$$

Hence, from (3.19) we immediately get (3.7).

Moreover, in the normalized coordinates, by differentiating the relation $l_i(u_je_j)r_j(u_je_j) \equiv \delta_{ij}$ with respect to u_j and using (3.1), we have

$$r_j^T(u_je_j)\nabla l_i(u_je_j)r_j(u_je_j) = -l_i(u_je_j)\nabla r_j(u_je_j)r_j(u_je_j). \tag{3.20}$$

By (3.1), it is easy to see that (see also [26] or [23])

$$\widetilde{\gamma}_{ijj}(u_je_j) = Ll_i(u_je_j)\nabla r_j(u_je_j)r_j(u_je_j) \equiv 0, \qquad \forall |u_j| \ small \ \ \forall i,j. \tag{3.21}$$

This proves (3.12). The proof of Lemma 3.1 is finished.

On the other hand, we have

Lemma 3.2. For $i = 1, \ldots, n$, it follows that

$$\frac{d(e^{Lt}v_i)}{d_i t} = \sum_{j,k=1}^{n} e^{Lt}\beta_{ijk}(u)v_j w_k + \sum_{j,k=1}^{n} e^{Lt}\widetilde{\beta}_{ijk}(u)v_j v_k, \tag{3.22}$$

where

$$\beta_{ijk}(u) = (\lambda_i(u) - \lambda_k(u))r_j^T(u)\nabla l_i(u)r_k(u), \tag{3.23}$$

$$\widetilde{\beta}_{ijk}(u) = -Lr_j^T(u)\nabla l_i(u)r_k(u). \tag{3.24}$$

Thus, we have

$$\beta_{iji}(u) \equiv 0, \qquad \forall i, j(i, j = 1, \ldots, n). \tag{3.25}$$

Moreover, by (3.1), in the normalized coordinates we have

$$\beta_{ijj}(u_j e_j) \equiv 0, \qquad \forall |u_j| \ \ small, \quad \forall i, j \tag{3.26}$$

and

$$\widetilde{\beta}_{ijj}(u_j e_j) \equiv 0, \qquad \forall |u_j| \ \ small, \quad \forall i, j. \tag{3.27}$$

Proof. In a way completely similar to the proof of Lemma 3.1, we can prove (3.22)-(3.27) without any essential difficulty. Here we omit the details.

For any given $y \geq 0$, on the domain of existence of C^1 solution, let $x = \widetilde{x}_i(t, y)$ be the ith characteristic passing through point $(\frac{y}{a}, y)(a > 0$, constant):

$$\begin{cases} \frac{d\widetilde{x}_i(t,y)}{dt} = \lambda_i(u(t, \widetilde{x}_i(t, y))), \\ \widetilde{x}_i(\frac{y}{a}, y) = y. \end{cases} \tag{3.28}$$

Lemma 3.3. Let $q_i(t, x)$ be defined by $q_i(t, \widetilde{x}_i(t, y)) = w_i(t, \widetilde{x}_i(t, y))\frac{\partial \widetilde{x}_i(t,y)}{\partial y}$. Then along the ith characteristic $x = \widetilde{x}_i(t, y)$ we have

$$\frac{d(e^{Lt}q_i)}{d_i t} = \sum_{j,k=1}^{n} e^{Lt}\Gamma_{ijk}(u)\frac{\partial \widetilde{x}_i(t, y)}{\partial y} w_j w_k + \sum_{j,k=1}^{n} e^{Lt}\widetilde{\gamma}_{ijk}(u)\frac{\partial \widetilde{x}_i(t, y)}{\partial y} v_j w_k, \tag{3.29}$$

where $\widetilde{\gamma}_{ijk}(u)$ is given by (3.9) and

$$\Gamma_{ijk}(u) = (\lambda_k(u) - \lambda_j(u))r_j^T(u)\nabla l_i(u)r_k(u). \tag{3.30}$$

Hence,

$$\Gamma_{ijj}(u) \equiv 0, \qquad \forall i, j. \tag{3.31}$$

Proof. Differentiating the first equation of (3.28) with respect to y gives

$$\frac{d}{dt}\left(\frac{\partial \widetilde{x}_i(t, y)}{\partial y}\right) = \nabla\lambda_i(u(t, \widetilde{x}_i(t, y)))\frac{\partial u}{\partial x}(t, \widetilde{x}_i(t, y))\frac{\partial \widetilde{x}_i(t, y)}{\partial y}. \tag{3.32}$$

Then, noting (3.7), it follows from (3.32) that

$$\begin{aligned}\frac{d(e^{Lt}q_i)}{d_it} &= \frac{d(e^{Lt}w_i)}{d_it}\frac{\partial\widetilde{x}_i(t,y)}{\partial y}+e^{Lt}w_i\frac{d}{d_it}\left(\frac{\partial\widetilde{x}_i(t,y)}{\partial y}\right)\\ &= \left(\sum_{j,k=1}^{n}e^{Lt}[\gamma_{ijk}(u)w_jw_k+\widetilde{\gamma}_{ijk}(u)v_jw_k]+e^{Lt}w_i\nabla\lambda_i(u)u_x\right)\frac{\partial\widetilde{x}_i(t,y)}{\partial y}. \qquad (3.33)\end{aligned}$$

Thus, from (3.5), (3.8) and (3.33), we immediately get (3.29)-(3.30). This completes the proof.

Similarly, noting (3.5), by (3.22) and (3.32), we have

Lemma 3.4. Let $p_i(t,x)$ be defined by $p_i(t,\widetilde{x}_i(t,y))=v_i(t,\widetilde{x}_i(t,y))\frac{\partial\widetilde{x}_i(t,y)}{\partial y}$. Then along the ith characteristic $x=\widetilde{x}_i(t,y)$ we have

$$\frac{d(e^{Lt}p_i)}{d_it}=\sum_{j,k=1}^{n}e^{Lt}B_{ijk}(u)\frac{\partial\widetilde{x}_i(t,y)}{\partial y}v_jw_k+\sum_{j,k=1}^{n}e^{Lt}\widetilde{\beta}_{ijk}(u)\frac{\partial\widetilde{x}_i(t,y)}{\partial y}v_jv_k, \qquad (3.34)$$

where $\widetilde{\beta}_{ijk}(u)$ is given by (3.24) and

$$B_{ijk}(u)=\beta_{ijk}(u)+\nabla\lambda_i(u)r_k(u)\delta_{ij}. \qquad (3.35)$$

By (3.25), it is easy to see that

$$B_{iji}(u)\equiv 0, \qquad \forall i\neq j(i,j=1,\ldots,n), \qquad (3.36)$$

$$B_{iii}(u)=\nabla\lambda_i(u)r_i(u), \qquad \forall i(i=1,\ldots,n). \qquad (3.37)$$

Moreover, by (3.26), in the normalized coordinates we have

$$B_{ijj}(u_je_j)\equiv 0, \qquad \forall|u_j|\ small, \quad \forall j\neq i; \qquad (3.38)$$

while, when the ith characteristic $\lambda_i(u)$ is weakly linearly degenerate, in the normalized coordinates,

$$B_{iii}(u_ie_i)\equiv 0, \qquad \forall|u_i|\ small, \quad \forall i. \qquad (3.39)$$

4. Proof of Theorem 1.1

For the sake of simplicity and without loss of generality, we may assume that

$$0<\lambda_1(0)<\lambda_2(0)<\ldots<\lambda_n(0). \qquad (4.1)$$

By the existence and uniqueness of a local C^1 solution for nonhomogeneous quasilinear hyperbolic systems (see [8, Theorem VI] or [7, 24]), in order to prove Theorem 1.1 it suffices to establish a uniform a priori estimate for the C^0 norm of u and u_x on the domain of existence of the C^1 solution $u=u(t,x)$.

By (4.1), there exist positive sufficiently small constants δ and δ_0 such that

$$\lambda_{i+1}(u)-\lambda_i(v)\geq 4\delta_0, \qquad \forall|u|,|v|\leq\delta(i=1,\ldots,n-1) \qquad (4.2)$$

and

$$|\lambda_i(u) - \lambda_i(v)| \leq \frac{\delta_0}{2}, \qquad \forall |u|, |v| \leq \delta (i = 1, \ldots, n). \tag{4.3}$$

For the time being we assume that on the domain of existence of the C^1 solution $u = u(t, x)$ we have

$$|u(t, x)| \leq \delta. \tag{4.4}$$

At the end of the proof of Lemma 4.3, we will explain that this hypothesis is reasonable. Thus, in order to prove Theorem 1.1, we only need to establish a uniform a priori estimate for the C^0 norm of v and w defined by (3.2)-(3.3) on the domain of existence of the C^1 solution $u = u(t, x)$.

By (4.1) and (4.4), on the domain of existence of the C^1 solution we have

$$0 < \lambda_1(u) < \lambda_2(u) < \ldots < \lambda_n(u), \tag{4.5}$$

provided that δ is suitably small.

For any fixed $T > 0$, let

$$D_+^T = \{(t, x) | 0 \leq t \leq T, x \geq (\lambda_n(0) + \delta_0)t\}, \tag{4.6}$$

$$D_-^T = \{(t, x) | 0 \leq t \leq T, x \leq (\lambda_1(0) - \delta_0)t\}, \tag{4.7}$$

$$D^T = \{(t, x) | 0 \leq t \leq T, (\lambda_1(0) - \delta_0)t \leq x \leq (\lambda_n(0) + \delta_0)t\} \tag{4.8}$$

and for $i = 1, \ldots, n$, let

$$\begin{aligned} D_i^T = \{(t, x) | 0 \leq t \leq T, -[\delta_0 + \eta(\lambda_i(0) - \lambda_1(0))]t \\ \leq x - \lambda_i(0)t \leq [\delta_0 + \eta(\lambda_n(0) - \lambda_1(0))]t\}, \end{aligned} \tag{4.9}$$

where $\eta > 0$ is suitably small, see Figure 1.

Noting that $\eta > 0$ is small, by (4.2), it is easy to see that

$$D_i^T \cap D_j^T = \emptyset, \quad \forall i \neq j \tag{4.10}$$

and

$$\bigcup_{i=1}^n D_i^T \subset D^T. \tag{4.11}$$

Let

$$V(D_\pm^T) = \max_{i=1,\ldots,n} \|(1 + |x|)^{1+\mu} v_i(t, x)\|_{L^\infty(D_\pm^T)}, \tag{4.12}$$

$$W(D_\pm^T) = \max_{i=1,\ldots,n} \|(1 + |x|)^{1+\mu} w_i(t, x)\|_{L^\infty(D_\pm^T)}, \tag{4.13}$$

$$V_\infty^c(T) = \max_{i=1,\ldots,n} \sup_{(t,x)\in D^T \setminus D_i(T)} \{(1 + |x - \lambda_i(0))t|)^{1+\mu} |v_i(t, x)|\}, \tag{4.14}$$

$$W_\infty^c(T) = \max_{i=1,\ldots,n} \sup_{(t,x)\in D^T \setminus D_i(T)} \{(1 + |x - \lambda_i(0))t|)^{1+\mu} |w_i(t, x)|\}, \tag{4.15}$$

$$U_\infty^c(T) = \max_{i=1,\ldots,n} \sup_{(t,x)\in D^T \setminus D_i(T)} \{(1 + |x - \lambda_i(0))t|)^{1+\mu} |u_i(t, x)|\}, \tag{4.16}$$

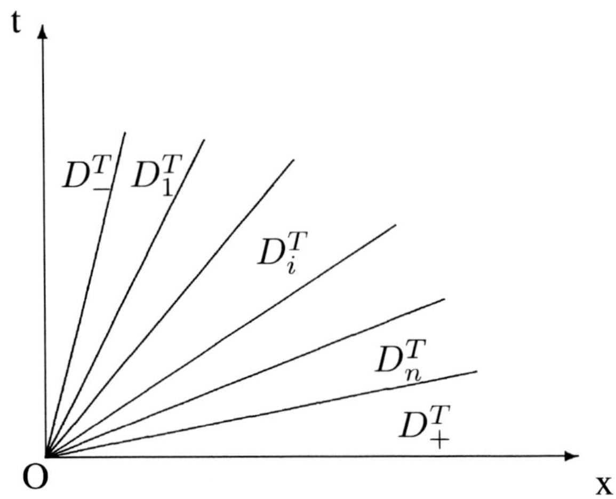

Figure 1. Domains $D_i^T (i = 1, \cdots, n)$ in (t, x)-plane.

$$\widetilde{V}_1(T) = \max_{i=1,\dots,n} \max_{j\neq i} \sup_{\widetilde{C}_j} \int_{\widetilde{C}_j} |v_i(t,x)|dt, \tag{4.17}$$

$$\widetilde{W}_1(T) = \max_{i=1,\dots,n} \max_{j\neq i} \sup_{\widetilde{C}_j} \int_{\widetilde{C}_j} |w_i(t,x)|dt, \tag{4.18}$$

where $\widetilde{C}_j (j \neq i)$ denotes any given jth characteristic in D_i^T,

$$V_1(T) = \max_{i=1,\dots,n} \sup_{0\leq t\leq T} \int_{D_i^T(t)} |v_i(t,x)|dx, \tag{4.19}$$

$$W_1(T) = \max_{i=1,\dots,n} \sup_{0\leq t\leq T} \int_{D_i^T(t)} |w_i(t,x)|dx, \tag{4.20}$$

where $D_i^T(t) (t \geq 0)$ denotes the t-section of D_i^T :

$$D_i^T(t) = \{(t,x) | \tau = t, \quad (\tau, x) \in D_i^T\}, \tag{4.21}$$

$$V_\infty(T) = \max_{i=1,\dots,n} \sup_{\substack{0\leq t\leq T\\ x\in R}} |v_i(t,x)| \tag{4.22}$$

and

$$W_\infty(T) = \max_{i=1,\dots,n} \sup_{\substack{0\leq t\leq T\\ x\in R}} |w_i(t,x)|. \tag{4.23}$$

Clearly, $V_\infty(T)$ is equivalent to

$$U_\infty(T) = \max_{i=1,\dots,n} \sup_{\substack{0\leq t\leq T\\ x\in R}} |u_i(t,x)|. \tag{4.24}$$

By the definitions of D_i^T and D^T, it is easy to get the following lemma.

Lemma 4.1. For each $i = 1, \ldots, n$, on the domain $D^T \setminus D_i^T$ we have

$$ct \leq |x - \lambda_i(0)t| \leq Ct, \quad cx \leq |x - \lambda_i(0)t| \leq Cx, \tag{4.25}$$

where c and C are positive constants independent of T.

Lemma 4.2. Suppose that in a neighborhood of $u = 0$, $A(u) \in C^2$ and (1.2) holds. Suppose furthermore that $\phi(x)$ satisfies (1.11). Then there exists a sufficiently small $\theta_0 > 0$ such that for any fixed $\theta \in [0, \theta_0]$, on any given domain of existence $0 \leq t \leq T$ of the C^1 solution $u = u(t, x)$ to the Cauchy problem (1.1) and (1.10) we have the following uniform a priori estimates:

$$V(D_{\pm}^T), \quad W(D_{\pm}^T) \leq k_1\theta, \tag{4.26}$$

where here and henceforth, $k_i (i = 1, 2, \ldots)$ are positive constants independent of θ and T.

Proof. We now prove (4.26) for D_+^T. The proof of (4.26) for D_-^T is similar.

For each $i = 1, \ldots, n$, let $\xi = x_i(s, y)$ be the ith characteristic passing through any fixed point $(t, x) \in D_+^T$ and intersecting the x-axis at a point $(0, y)$. Noting (4.4), by (4.2)-(4.3), it is easy to see that the whole characteristic $\xi = x_i(s, y)(0 \leq s \leq t)$ is included in D_+^T.

Noting (4.5), by (4.3) we get

$$y \leq x_i(s, y) \leq y + (\lambda_i(0) + \delta_0/2)s, \qquad \forall s \in [0, t]. \tag{4.27}$$

By (4.6), it is easy to see that

$$s \leq t \leq t_0, \tag{4.28}$$

where t_0 denotes the t-coordinate of the intersection point of the straight line $x = (\lambda_n(0) + \delta_0)t$ with the straight line $x = y + (\lambda_i(0) + \delta_0/2)t$ passing through the point $(0, y)$. Clearly,

$$t_0 = \frac{y}{\lambda_n(0) - \lambda_i(0) + \delta_0/2}. \tag{4.29}$$

Thus, it follows from (4.27) and (4.28) that

$$y \leq x_i(s, y) \leq \frac{\lambda_n(0) + \delta_0}{\lambda_n(0) - \lambda_i(0) + \delta_0/2} y, \quad \forall s \in [0, t]. \tag{4.30}$$

By integrating (3.7) along this ith characteristic, we have

$$\begin{aligned} w_i(t, x) \quad &= \quad e^{-Lt} w_i(0, y) + \int_0^t e^{-L(t-s)} \Big[\sum_{j,k=1}^n \gamma_{ijk}(u) w_j w_k \\ &\quad + \sum_{j,k=1}^n \tilde{\gamma}_{ijk}(u) v_j w_k \Big] (s, x_i(s, y)) ds. \end{aligned} \tag{4.31}$$

Thus, using(4.28)-(4.30), noting (4.4) and the fact that $L > 0$, it follows from (4.31) that

$$\begin{aligned} (1+x)^{1+\mu} |w_i(t, x)| &\leq C_1 \{(1+y)^{1+\mu} |w_i(0, y)| + (1+y)^{-(1+\mu)} t [(W(D_+^T))^2 + W(D_+^T) V(D_+^T)]\} \\ &\leq C_2 \{\theta + (1+y)^{-\mu} W(D_+^T) [W(D_+^T) + V(D_+^T)]\} \\ &\leq C_2 \{\theta + W(D_+^T) [W(D_+^T) + V(D_+^T)]\}, \end{aligned} \tag{4.32}$$

where here and henceforth, $C_i (i = 1, 2, \ldots)$ will denote positive constants independent of θ and T. Hence, we get

$$W(D_+^T) \le C_2\{\theta + W(D_+^T)[W(D_+^T) + V(D_+^T)]\}. \tag{4.33}$$

Similarly, by integrating (3.22) along this ith characteristic, we have

$$\begin{aligned} v_i(t,x) &= e^{-Lt} v_i(0,y) + \int_0^t e^{-L(t-s)} \Big[\sum_{j,k=1}^n \beta_{ijk}(u) v_j w_k \\ &\quad + \sum_{j,k=1}^n \widetilde{\beta}_{ijk}(u) v_j v_k \Big] (s, x_i(s,y)) ds. \end{aligned} \tag{4.34}$$

Similar to (4.33), we have

$$V(D_+^T) \le C_3\{\theta + V(D_+^T)[W(D_+^T) + V(D_+^T)]\}. \tag{4.35}$$

By (4.33) and (4.35), it is easy to prove that for $\theta > 0$ suitably small, there exists a positive constant k_1 independent of θ and T, such that for any fixed T_0 $(0 < T_0 \le T)$, if

$$W(D_+^{T_0}), \ \ V(D_+^{T_0}) \le 2k_1\theta, \tag{4.36}$$

then

$$W(D_+^{T_0}), \ \ V(D_+^{T_0}) \le k_1\theta. \tag{4.37}$$

Hence, noting (1.11), by continuity we immediately get (4.26) for D_+^T. This completes the proof of Lemma 4.2.

Lemma 4.3. Under the assumptions of Lemma 4.2, suppose furthermore that system (1.1) is weakly linearly degenerate. In the normalized coordinates there exists a sufficiently small $\theta_0 > 0$ such that for any fixed $\theta \in [0, \theta_0]$, on any given existence domain $0 \le t \le T$ of the C^1 solution $u = u(t,x)$ to the Cauchy problem (1.1) and (1.10), we have the following uniform a priori estimates:

$$\widetilde{W}_1(T), W_1(T), \widetilde{V}_1(T), V_1(T) \le k_2\theta, \tag{4.38}$$

$$W_\infty^c(T), V_\infty^c(T) \le k_3\theta, \tag{4.39}$$

$$W_\infty(T), V_\infty(T) \le k_4\theta, \tag{4.40}$$

$$U_\infty^c(T) \le k_5\theta \tag{4.41}$$

and

$$U_\infty(T) \le k_4\theta. \tag{4.42}$$

Proof. We first estimate $\widetilde{W}_1(T)$.

Let $\widetilde{C}_j : x = x_j(t)$ $(t_1 \le t \le t_2,\ j \ne i)$ be any given jth characteristic in D_i^T. By (4.3), the whole ith characteristic $x = x_i(t)$ passing through $O(0,0)$ is included in D_i^T. Let $(t_0, x_j(t_0))$ be the intersection point of this characteristic with $\widetilde{C}_j$. Passing through any given point $(t, x_j(t))$ on $\widetilde{C}_j$, we draw the ith characteristic $\xi = \widetilde{x}_i(s,y)$ which intersects one of the boundaries of D^T, say, $x = (\lambda_n(0) + \delta_0)t$ (resp. $x = (\lambda_1(0) - \delta_0)t$) at a point $A_y(y/(\lambda_n(0)+\delta_0), y)$ (resp. $B_y(y/(\lambda_1(0)-\delta_0), y)$ if $t_0 \le t \le t_2$ (resp. $t_1 \le t \le t_0$).

Clearly, we have

$$\widetilde{x}_i(t,y) = x_j(t) \tag{4.43}$$

which gives a one-to-one correspondence $t = t(y)$ between the segment $\overline{OA_{y2}}$ (resp.$\overline{B_{y_1}O}$) and $\widetilde{C}_j(t_0 \le t \le t_2)$(resp. $\widetilde{C}_j(t_1 \le t \le t_0)$). Thus, the integral on $\widetilde{C}_j$ with respect to t can be reduced to the integral with respect to y. Differentiating (4.43) with respect to t gives

$$dt = \frac{1}{\lambda_j(u(t,\widetilde{x}_i(t,y))) - \lambda_i(u(t,\widetilde{x}_i(t,y)))}\frac{\partial \widetilde{x}_i(t,y)}{\partial y}dy, \tag{4.44}$$

in which $t = t(y)$. Then, noting (4.2) and (4.4), it is easy to see that in order to estimate

$$\int_{\widetilde{C}_j}|w_i(t,x)|dt = \int_{t_1}^{t_0}|w_i(t,x_j(t))|dt + \int_{t_0}^{t_2}|w_i(t,x_j(t))|dt$$

$$= \int_{t_1}^{t_0}|w_i(t,\widetilde{x}_i(t,y))|dt + \int_{t_0}^{t_2}|w_i(t,\widetilde{x}_i(t,y))|dt, \tag{4.45}$$

it suffices to estimate

$$\int_0^{y_1}|q_i(t,\widetilde{x}_i(t,y))|_{t=t(y)}dy \ \ and \ \ \int_0^{y_2}|q_i(t,\widetilde{x}_i(t,y))|_{t=t(y)}dy. \tag{4.46}$$

We now estimate $\int_0^{y_2}|q_i(t,\widetilde{x}_i(t,y))|_{t=t(y)}dy$.

By integrating (3.29) along $\xi = \widetilde{\ }x_i(s,y)$, and noting (3.31) and the fact that $\widetilde{x}_i(y/(\lambda_n(0)+\delta_0),y) = y$, we have

$$q_i(t,\widetilde{x}_i(t,y))|_{t=t(y)} = e^{-L(t(y)-\frac{y}{\lambda_n(0)+\delta_0})}w_i\Big(\frac{y}{\lambda_n(0)+\delta_0},y\Big)\Big\{1-\frac{\lambda_i(u(y/(\lambda_n(0)+\delta_0),y)}{\lambda_n(0)+\delta_0}\Big\}$$
$$+\int_{y/(\lambda_n(0)+\delta_0)}^{t(y)} e^{-L(t(y)-s)}\frac{\partial\widetilde{x}_i(s,y)}{\partial y}\Big[\sum_{j=k\ /}\Gamma_{ijk}(u)w_jw_k + \widetilde{\gamma}_{ijk}(u)v_jw_k\Big](s,\widetilde{x}_i(s,y))ds$$
$$+\int_{y/(\lambda_n(0)+\delta_0)}^{t(y)} e^{-L(t(y)-s)}\frac{\partial\widetilde{x}_i(s,y)}{\partial y}\Big[\sum_{j=1}^n\widetilde{\gamma}_{ijj}(u)v_jw_j\Big](s,\widetilde{x}_i(s,y))ds. \tag{4.47}$$

By Hadamard's formula and (3.12), we have

$$\widetilde{\gamma}_{ijj}(u) = \widetilde{\gamma}_{ijj}(u) - \widetilde{\gamma}_{ijj}(u_je_j) = \int_0^1\sum_{l\neq j}\frac{\partial\widetilde{\gamma}_{ijj}}{\partial u_l}(\tau u_1,\dots,\tau u_{j-1},u_j,\tau u_{j+1},\dots,\tau u_n)u_ld\tau. \tag{4.48}$$

Noting (4.4)-(4.5), (4.10), (4.25) and the fact that $L > 0$ and $\frac{\partial\widetilde{x}_i(s,y)}{\partial y} > 0$, we obtain from (3.47) and (4.48) that

$$q_i(t,\widetilde{x}_i(t,y))|_{t=t(y)} \le \Big|w_i\Big(\frac{y}{\lambda_n(0)+\delta_0},y\Big)\Big| + C_4\Big\{[W_\infty^c(T)^2 + W_\infty^c(T)V_\infty^c(T) + W_\infty^c(T)V_\infty^c(T)V_\infty(T)]$$
$$\times\int_{y/(\lambda_n(0)+\delta_0)}^{t(y)}(1+s)^{-(1+\mu)}(1+|\widetilde{x}_i(s,y)|)^{-(1+\mu)}\frac{\partial\widetilde{x}_i(s,y)}{\partial y}ds$$
$$+[W_\infty^c(T) + V_\infty^c(T) + U_\infty^c(T)V_\infty(T)]$$
$$\times\sum_{k=1}^n\int_{(s,\ \widetilde{x}_i(s,y))\in D_k^T}(1+s)^{-(1+\mu)}|w_k(s,\widetilde{x}_i(s,y))|\frac{\partial\widetilde{x}_i(s,y)}{\partial y}ds$$
$$+W_\infty^c(T)\sum_{j=1}^n\int_{(s,\ \widetilde{x}_i(s,y))\in D_j^T}(1+s)^{-(1+\mu)}|v_j(s,\widetilde{x}_i(s,y))|\frac{\partial\widetilde{x}_i(s,y)}{\partial y}ds\Big\}. \tag{4.49}$$

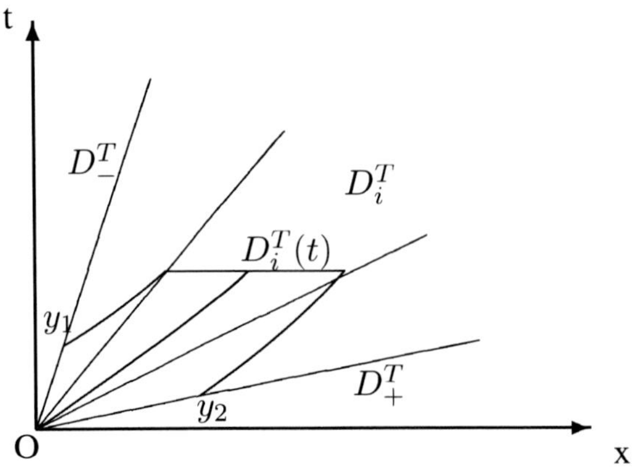

Figure 2. Points $y_i (i = 1, 2)$ in (t, x)-plane.

Noting that the transformation $\begin{cases} x = \widetilde{x}_i(s, y) \\ t = s \end{cases}$ gives the area element

$$dtdx = \frac{\partial \widetilde{x}_i(s, y))}{\partial y} dsdy, \tag{4.50}$$

by Lemma 4.2, it easily follows from (4.50) that

$$\begin{aligned} \int_0^{y_2} |q_i(t, \widetilde{x}_i(t, y))|_{t=t(y)} dy \quad \leq \quad & C_5\{\theta + W_\infty^c(T)[W_\infty^c(T) + V_\infty^c(T) + W_1(T) + V_1(T) \\ & + V_\infty^c(T)V_\infty(T)] + W_1(T)[V_\infty^c(T) + U_\infty^c(T)V_\infty(T)]\}. \end{aligned} \tag{4.51}$$

In a similar way we can estimate $\int_0^{y_2} |q_i(t, \widetilde{x}_i(t, y))|_{t=t(y)} dy$. Thus, we obtain

$$\begin{aligned} \widetilde{W}_1(T) \quad \leq \quad & C_6\{\theta + W_\infty^c(T)[W_\infty^c(T) + V_\infty^c(T) + W_1(T) + V_1(T) + V_\infty^c(T)V_\infty(T)] \\ & + W_1(T)[V_\infty^c(T) + U_\infty^c(T)V_\infty(T)]\}. \end{aligned} \tag{4.52}$$

We next estimate $W_1(T)$.

Passing through any given point $(t, x) \in D_i^T(t)$ we draw the ith characteristic $\xi = \widetilde{x}_i(s, y)$ which intersects one of the boundaries of D^T, say, $x = (\lambda_n(0) + \delta_0)t$ (resp. $x = (\lambda_1(0) - \delta_0)t$ at a point $A_y(y/(\lambda_n(0) + \delta_0), y)$ (resp. $B_y(y/(\lambda_1(0) - \delta_0), y)$). Clearly, we have $x = \widetilde{x}_i(t, y)$. Thus, we get

$$\int_{D_i^T(t)} |w_i(t, x)| dx = \int_0^{y_1} |q_i(t, \widetilde{x}_i(t, y)) dy + \int_0^{y_2} |q_i(t, \widetilde{x}_i(t, y))| dy, \tag{4.53}$$

where y_1, y_2 are shown in Figure 2.

Similar to (4.52), it follows from (4.53) that

$$\begin{aligned} W_1(T) \quad \leq \quad & C_7\{\theta + W_\infty^c(T)[W_\infty^c(T) + V_\infty^c(T) + W_1(T) + V_1(T) \\ & + V_\infty^c(T)V_\infty(T)] + W_1(T)[V_\infty^c(T) + U_\infty^c(T)V_\infty(T)]\}. \end{aligned} \tag{4.54}$$

We next estimate $W^c_\infty(T)$.

For any given point $(t,x) \in D^T$ but $(t,x) \notin D^T_i$, by the definition of D^T_i, for fixing the idea we may suppose that

$$x - \lambda_i(0)t > [\delta_0 + \eta(\lambda_n(0) - \lambda_i(0))]t, \tag{4.55}$$

which implies $i < n$. Let $\xi = x_i(s;t,x)$ be the ith characteristic passing through (t,x) which intersects the boundary $x = (\lambda_n(0) + \delta_0)t$ of D^T at a point (t_0, y), see Figure 3.

Noting (4.3), it is easy to see that

$$x - (\lambda_i(0) + \frac{\delta_0}{2})t \leq y - (\lambda_i(0) + \frac{\delta_0}{2})t_0. \tag{4.56}$$

Since

$$y = (\lambda_n(0) + \delta_0)t_0, \tag{4.57}$$

noting (4.55) and the fact that $t \geq t_0$, it follows from (4.56) that

$$t \geq t_0 \geq \eta t. \tag{4.58}$$

By integrating (3.7) along $\xi = x_i(s;t,x)$ and noting (3.11), we have

$$\begin{aligned} w_i(t,x) &= e^{-Lt}w_i(t_0,y) \\ &+ \int_{t_0}^t e^{-L(t-s)}[\sum_{j,k=1}^n (\gamma_{ijk}(u)w_jw_k](s, x_i(s;t,x)ds \\ &+ \int_{t_0}^t e^{-L(t-s)}[\sum_{j\neq k} \widetilde{\gamma}_{ijk}(u)v_jw_k](s, x_i(s;t,x)ds \\ &+ \int_{t_0}^t e^{-L(t-s)}[\sum_{j=1}^n(\widetilde{\gamma}_{ijj}(u) - \widetilde{\gamma}_{ijj}(u_je_j))v_jw_j](s, x_i(s;t,x))ds. \end{aligned} \tag{4.59}$$

By Lemma 4.2 and noting (4.57)-(4.58), it is easy to see that

$$|w_i(t_0,y)| \leq k_1\theta(1+y)^{-(1+\mu)} \leq C_8\theta(1+t_0)^{-(1+\mu)} \leq C_9\theta(1+t)^{-(1+\mu)}. \tag{4.60}$$

Thus, noting (3.10) and the fact that $L > 0$, and using (4.25), (4.58) and Hadamard's formula, it follows from (4.59) that

$$\begin{aligned} W^c_\infty(T) \leq\ & C_{10}\{\theta + W^c_\infty(T)[W^c_\infty(T) + V^c_\infty(T) + \widetilde{W}_1(T) + \widetilde{V}_1(T) \\ & + V^c_\infty(T)V_\infty(T)] + \widetilde{W}_1(T)[V^c_\infty(T) + U^c_\infty(T)V_\infty(T)]\}. \end{aligned} \tag{4.61}$$

We next estimate $\widetilde{V}_1(T)$ and $V_1(T)$.

Similar to (4.45)-(4.46), in order to estimate $\widetilde{V}_1(T)$ it suffices to estimate

$$\int_0^{y_1} |p_i(t, \widetilde{x}_i(t,y))|_{t=t(y)}dy \quad and \int_0^{y_2} |p_i(t, \widetilde{x}_i(t,y))|_{t=t(y)}dy. \tag{4.62}$$

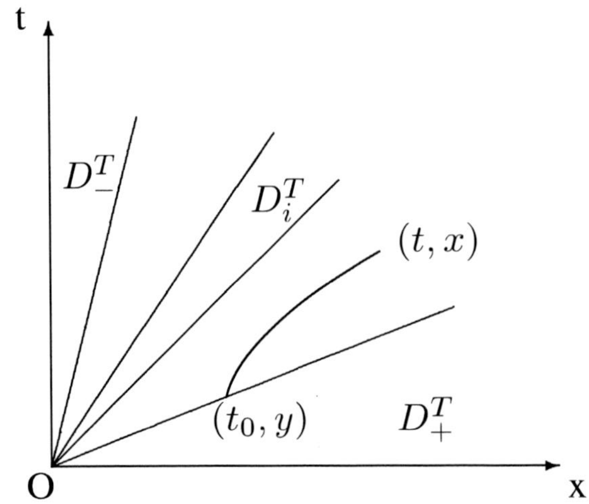

Figure 3. The ith characteristic issuing from (t, x).

By integrating (3.34) along $\xi = \widetilde{x}_i(s, y)$, and noting (3.36) and the fact that $\widetilde{x}_i(\frac{y}{\lambda_n(0)+\delta_0}, y) = y$, we have

$$\begin{aligned} p_i(t, \widetilde{x}_i(t, y))|_{t=t(y)} &= e^{-L(t(y)-\frac{y}{\lambda_n(0)+\delta_0})} v_i\Big(\frac{y}{\lambda_n(0)+\delta_0}, y\Big)\Big\{1 - \frac{\lambda_i(u(y/(\lambda_n(0)+\delta_0), y)}{\lambda_n(0)+\delta_0}\Big\} \\ &+ \int_{y/(\lambda_n(0)+\delta_0)}^{t(y)} e^{-L(t(y)-s)} \frac{\partial \widetilde{x}_i(s, y)}{\partial y} [\sum_{k \neq i,j} B_{ijk}(u) v_j w_k](s, \widetilde{x}_i(s, y)) ds \\ &+ \int_{y/(\lambda_n(0)+\delta_0)}^{t(y)} e^{-L(t(y)-s)} \frac{\partial \widetilde{x}_i(s, y)}{\partial y} [\sum_{j=1}^{n} B_{ijj}(u) v_j w_j](s, \widetilde{x}_i(s, y)) ds \\ &+ \int_{y/(\lambda_n(0)+\delta_0)}^{t(y)} e^{-L(t(y)-s)} \frac{\partial \widetilde{x}_i(s, y)}{\partial y} [\sum_{j \neq k}^{n} \widetilde{\beta}_{ijk}(u) v_j v_k](s, \widetilde{x}_i(s, y)) ds \\ &+ \int_{y/(\lambda_n(0)+\delta_0)}^{t(y)} e^{-L(t(y)-s)} \frac{\partial \widetilde{x}_i(s, y)}{\partial y} [\sum_{j=1}^{n} \widetilde{\beta}_{ijj}(u) v_j^2](s, \widetilde{x}_i(s, y)) ds. \end{aligned} \quad (4.63)$$

By Hadamard's formula, and noting (3.27) and (3.38)-(3.39), we have

$$\begin{aligned} \widetilde{\beta}_{ijj}(u) &= \widetilde{\beta}_{ijj}(u) - \widetilde{\beta}_{ijj}(u_j e_j) \\ &= \int_0^1 \sum_{l=j} \frac{\partial \widetilde{\beta}_{ijj}}{\partial u_l} (\tau u_1, \ldots, \tau u_{j-1}, u_j, \tau u_{j+1}, \ldots, \tau u_n) u_l d\tau \end{aligned} \quad (4.64)$$

and

$$\begin{aligned} B_{ijj}(u) &= B_{ijj}(u) - B_{ijj}(u_j e_j) \\ &= \int_0^1 \sum_{l \neq j} \frac{\partial B_{ijj}}{\partial u_l} (\tau u_1, \ldots, \tau u_{j-1}, u_j, \tau u_{j+1}, \ldots, \tau u_n) u_l d\tau. \end{aligned} \quad (4.65)$$

In a way similar to the proof of (4.52), we get from (4.63) that

$$\begin{aligned}\widetilde{V}_1(T) \ \le\ & C_{11}\{\theta + V^c_\infty(T)[W^c_\infty(T) + V^c_\infty(T) + W_1(T) + V_1(T)\\ & + V^c_\infty(T)V_\infty(T) + W^c_\infty(T)V_\infty(T)] + U^c_\infty(T)[V_\infty(T)W_1(T)\\ & + V_1(T)V_\infty(T)] + W^c_\infty(T)V_1(T)\}. \end{aligned} \tag{4.66}$$

Similar to (4.54), we have

$$\begin{aligned}V_1(T) \ \le\ & C_{12}\{\theta + V^c_\infty(T)[W^c_\infty(T) + V^c_\infty(T) + W_1(T) + V_1(T)\\ & + V^c_\infty(T)V_\infty(T) + W^c_\infty(T)V_\infty(T)] + U^c_\infty(T)[V_\infty(T)W_1(T)\\ & + V_1(T)V_\infty(T)] + W^c_\infty(T)V_1(T)\}. \end{aligned} \tag{4.67}$$

We next estimate $V^c_\infty(T)$.
Similar to (4.59), we have

$$\begin{aligned} v_i(t,x) \ =\ & e^{-L(t-t_0)}v_i(t_0,y)\\ & + \int_{t_0}^t e^{-L(t-s)}\left[\sum_{k\neq i,j}^n \beta_{ijk}(u)v_jw_k\right](s,x_i(s;t,x))ds\\ & + \int_{t_0}^t e^{-L(t-s)}\left[\sum_{j\neq k}^n \widetilde{\beta}_{ijk}(u)v_jv_k\right](s,x_i(s;t,x))ds\\ & + \int_{t_0}^t e^{-L(t-s)}\left[\sum_{j=1}^n (\beta_{ijj}(u)-\beta_{ijj}(u_je_j))v_jw_j\right](s,x_i(s;t,x))ds\\ & + \int_{t_0}^t e^{-L(t-s)}\left[\sum_{j=1}^n (\widetilde{\beta}_{ijj}(u)-\widetilde{\beta}_{ijj}(u_je_j))v_j^2\right](s,x_i(s;t,x))ds. \end{aligned} \tag{4.68}$$

In a way similar to the proof of (4.61), we get from (4.68) that

$$\begin{aligned}V^c_\infty(T) \ \le\ & C_{13}\{\theta + V^c_\infty(T)[W^c_\infty(T) + V^c_\infty(T) + \widetilde{W}_1(T) + \widetilde{V}_1(T)\\ & + V^c_\infty(T)V_\infty(T) + W^c_\infty(T)V_\infty(T)] + U^c_\infty(T)[V_\infty(T)\widetilde{W}_1(T)\\ & + \widetilde{V}_1(T)V_\infty(T)] + W^c_\infty(T)\widetilde{V}_1(T)\}. \end{aligned} \tag{4.69}$$

We next estimate $U^c_\infty(T)$.
For any given point $(t,x) \in D^T \setminus D^T_i$, by using (3.4), we have

$$u_i(t,x) = u^T(t,x)e_i = \sum_{k=1}^n v_k r_k^T(u)e_i. \tag{4.70}$$

Thus, if $(t,x) \notin D^T_k (k = 1,\dots,n)$, noting Lemma 4.1 and (4.4), it is easy to see that

$$(1 + |x - \lambda_i(0)t|)^{1+\mu}|u_i(t,x)| \le C_{14}V^c_\infty(T). \tag{4.71}$$

On the other hand, if there exists some $j(\neq i)$ such that $(t,x) \in D^T_j$, then $(t,x) \notin D^T_k (k \neq j)$, and noting (3.1), (4.70) can be rewritten as

$$u_i(t,x) = \sum_{k\neq j} v_k r_k^T(u)e_i + v_j(r_j^T(u) - r_j^T(u_je_j))e_i. \tag{4.72}$$

By Hadamard's formula, we have

$$r_j^T(u) - r_j^T(u_je_j) = \int_0^1 \sum_{l\neq j} \frac{\partial r_j^T}{\partial u_l}(su_1,\ldots,su_{j-1},u_j,su_{j+1},\ldots,su_n)u_l ds. \qquad (4.73)$$

Thus, it is easy get

$$U_\infty^c(T) \leq C_{15}\{V_\infty^c(T) + U_\infty^c(T)V_\infty(T)\}. \qquad (4.74)$$

We now estimate $V_\infty(T)$.

Passing through any given point $(t,x) \in D_i^T$, we draw the ith characteristic $\xi = x_i(s;t,x)$ which intersects one of the boundaries of D^T at one point. For fixing the idea, suppose that this characteristic intersects $x = (\lambda_n(0)+\delta_0)t$ at a point $(y/(\lambda_n(0)+\delta_0), y)$. By integrating (3.22) along this characteristic and noting (3.25)-(3.27), we have

$$\begin{aligned} v_i(t,x) &= e^{-L(t-\frac{y}{\lambda_n(0)+\delta_0})}v_i\Big(\frac{y}{\lambda_n(0)+\delta_0},y\Big) \\ &+ \int_{y/(\lambda_n(0)+\delta_0)}^t e^{-L(t-s)}\Big[\sum_{k\neq j,i}^n \beta_{ijk}(u)v_jw_k\Big](s,x_i(s;t,x))ds \\ &+ \int_{y/(\lambda_n(0)+\delta_0)}^t e^{-L(t-s)}\Big[\sum_{j=1}^n \Big(\beta_{ijj}(u)-\beta_{ijj}(u_je_j)\Big)v_jw_j\Big](s,x_i(s;t,x))ds \\ &+ \int_{y/(\lambda_n(0)+\delta_0)}^t e^{-L(t-s)}\Big[\sum_{j\neq k}^n \widetilde{\beta}_{ijk}(u)v_jv_k\Big](s,x_i(s;t,x))ds \\ &+ \int_{y/(\lambda_n(0)+\delta_0)}^t e^{-L(t-s)}\Big[\sum_{j=1}^n \Big(\widetilde{\beta}_{ijj}(u)-\widetilde{\beta}_{ijj}(u_je_j)\Big)v_j^2\Big](s,x_i(s;t,x))ds. \qquad (4.75) \end{aligned}$$

Noting (4.25)-(4.26) and using Hadamard's formula, it follows from (4.75) that

$$\begin{aligned} |v_i(t,x)| \leq\ & C_{16}\{\theta + V_\infty^c(T)[W_\infty^c(T) + V_\infty^c(T) \\ & + V_\infty(T) + W_\infty(T) + V_\infty^c(T)V_\infty(T) + W_\infty^c(T)V_\infty(T)] \\ & + U_\infty^c(T)V_\infty(T)[W_\infty(T) + V_\infty(T)] + W_\infty^c(T)V_\infty(T)\}. \qquad (4.76) \end{aligned}$$

On the other hand, for any given point $(t,x) \notin D_i^T (i = 1,\ldots,n)$, $|v_i(t,x)|$ can be controlled by $V_\infty^c(T)$ or $V(D_\pm^T)$. Thus, by using Lemma 4.2 again, we have

$$\begin{aligned} V_\infty(T) \leq\ & C_{17}\{\theta + V_\infty^c(T) + V_\infty^c(T)[W_\infty^c(T) + V_\infty^c(T) \\ & + W_\infty(T) + V_\infty(T) + W_\infty^c(T)V_\infty(T) + V_\infty^c(T)V_\infty(T)] \\ & + U_\infty^c(T)V_\infty(T)[W_\infty(T) + V_\infty(T)] + W_\infty^c(T)V_\infty(T)\}. \qquad (4.77) \end{aligned}$$

We finally estimate $W_\infty(T)$.

Similar to (4.75), by integrating (3.7) along $\xi = x_i(s;t,x)$, and noting the fact that $\lambda_i(u)$ is weakly linearly degenerate and (3.10), we have

$$w_i(t,x) = e^{-L(t-y/(\lambda_n(0)+\delta_0))}w_i\Big(\frac{y}{\lambda_n(0)+\delta_0},y\Big)$$

$$+\int_{y/(\lambda_n(0)+\delta_0)}^{t} e^{-L(t-s)}\Big[\sum_{j\neq k}^{n}\gamma_{ijk}(u)w_jw_k\Big](s,x_i(s;t,x))ds$$
$$+\int_{y/(\lambda_n(0)+\delta_0)}^{t} e^{-L(t-s)}[(\gamma_{iii}(u)-\gamma_{iii}(u_ie_i))w_i^2](s,x_i(s;t,x))ds$$
$$+\int_{y/(\lambda_n(0)+\delta_0)}^{t} e^{-L(t-s)}\Big[\sum_{j\neq k}^{n}\widetilde{\gamma}_{ijk}(u)v_jw_k\Big](s,x_i(s;t,x))ds$$
$$+\int_{y/(\lambda_n(0)+\delta_0)}^{t} e^{-L(t-s)}\Big[\sum_{j=1}^{n}(\widetilde{\gamma}_{ijj}(u)-\widetilde{\gamma}_{ijj}(u_je_j))v_jw_j\Big](s,x_i(s;t,x))ds.\quad(4.78)$$

Similar to (4.77), we have

$$\begin{aligned}W_\infty(T) \ \le\ & C_{18}\{\theta+W_\infty^c(T)+W_\infty^c(T)[W_\infty^c(T)+V_\infty^c(T)\\ &+W_\infty(T)+V_\infty(T)+W_\infty^c(T)V_\infty(T)+V_\infty^c(T)V_\infty(T)]\\ &+U_\infty^c(T)W_\infty(T)[W_\infty(T)+V_\infty(T)]+V_\infty^c(T)W_\infty(T)\}.\end{aligned}\quad(4.79)$$

We now prove (4.38)-(4.42).

Noting (1.11), evidently we have

$$W_\infty^c(0),V_\infty^c(0),U_\infty^c(0)\le C_{19}\theta,\quad(4.80)$$
$$W_1(0)=V_1(0)=\widetilde{W}_1(0)=\widetilde{V}_1(0)=0\quad(4.81)$$

and

$$W_\infty(0),V_\infty(0)\le C_{20}\theta.\quad(4.82)$$

Thus, by continuity there exist positive constants $k_2,\ k_3,\ k_4$ and k_5 independent of θ, such that (4.38)-(4.41) hold at least for $0\le T\le\tau_0$, where τ_0 is a small positive number. Hence, in order to prove (4.38)-(4.41) it suffices to show that we can choose $k_2,\ k_3,\ k_4$ and k_5 in such a way that for any fixed T_0 $(0<T_0\le T)$ with

$$\widetilde{W}_1(T_0),W_1(T_0),\widetilde{V}_1(T_0),V_1(T_0)\le 2k_2\theta,\quad(4.83)$$
$$W_\infty^c(T_0),V_\infty^c(T_0)\le 2k_3\theta,\quad(4.84)$$
$$W_\infty(T_0),V_\infty(T_0)\le 2k_4\theta,\quad(4.85)$$
$$U_\infty^c(T_0)\le 2k_5\theta,\quad(4.86)$$

we have

$$\widetilde{W}_1(T_0),W_1(T_0),\widetilde{V}_1(T_0),V_1(T_0)\le k_2\theta,\quad(4.87)$$
$$W_\infty^c(T_0),V_\infty^c(T_0)\le k_3\theta,\quad(4.88)$$
$$W_\infty(T_0),V_\infty(T_0)\le k_4\theta,\quad(4.89)$$
$$U_\infty^c(T_0)\le k_5\theta,\quad(4.90)$$

To this end, substituting (4.83)-(4.86) into the right-hand sides of (4.52), (4.54), (4.61), (4.66)-(4.67), (4.69), (4.74), (4.77) and (4.79)(in which we take $T=T_0$), it is easy to see that, when $\theta_0>0$ is suitably small, we have

$$\widetilde{W}_1(T_0)\le 2C_6\theta,\quad(4.91)$$

$$W_1(T_0) \le 2C_7\theta, \tag{4.92}$$

$$W_\infty^c(T_0) \le 2C_{10}\theta, \tag{4.93}$$

$$\widetilde{V}_1(T_0) \le 2C_{11}\theta, \tag{4.94}$$

$$V_1(T_0) \le 2C_{12}\theta, \tag{4.95}$$

$$V_\infty^c(T_0) \le 2C_{13}\theta, \tag{4.96}$$

$$U_\infty^c(T_0) \le 3C_{15}k_3\theta, \tag{4.97}$$

$$V_\infty(T_0) \le 2C_{17}(1+k_3)\theta, \tag{4.98}$$

$$W_\infty(T_0) \le 2C_{18}(1+k_3)\theta. \tag{4.99}$$

Hence, if $k_2 \ge 2\max\{C_6, C_7, C_{11}, C_{12}\}$, $k_3 \ge 2\max\{C_{10}, C_{13}\}$, $k_4 \ge 2\max\{C_{17}, C_{18}\}(1+k_3)$ and $k_5 \ge 3C_{15}k_3$, then we get (4.87)-(4.90). This proves (4.38)-(4.42).

Finally, we observe that when $\theta_0 > 0$ is suitably small, by (4.42) we have

$$U_\infty(T) \le k_4\theta \le k_4\theta_0 \le \frac{\delta}{2}. \tag{4.100}$$

This implies the validity of hypothesis (4.4). The proof of Lemma 4.3 is finished.

Proof of Theorem 1.1. It suffices to prove Theorem 1.1 in the normalized coordinates. Under the assumptions of Theorem 1.1, by (4.40) and (4.42), we know that there is a sufficiently small $\theta_0 > 0$ such that for any fixed $\theta \in (0, \theta_0]$, on any given domain of existence $0 \le t \le T$ of the C^1 solution $u = u(t,x)$ to the Cauchy problem (1.1) and (1.10), we have the following uniform a priori estimate for the C^1 norm of the solution:

$$\|u(t,.)\|_{C^1} \triangleq \|u(t,.)\|_{C^0} + \|u_x(t,.)\|_{C^0} \le k_6\theta. \tag{4.101}$$

Thus we immediately get the conclusion of Theorem 1.1. The proof of Theorem 1.1 is finished.

5. Application

Consider the following Cauchy problem for the system of the flow equations of a model class of fluids with viscosity induced by fading memory (cf. [6, 22, 34, 36-38]):

$$\begin{cases} w_t - v_x + w = 0, \\ v_t - (\sigma(w))_x + v = 0, \end{cases} \tag{5.1}$$

$$t = 0 : (w, v) = (w_0(x), \widetilde{v}_0 + v_0(x)). \tag{5.2}$$

Here, w is the displacement gradient and v is the velocity of a model class of fluids, the stress-strain function $\sigma(w)$ is a suitably smooth function of w such that

$$\sigma'(0) > 0, \tag{5.3}$$

$\widetilde{v}_0$ is a constant. Also, we assume that $w_0(x), v_0(x)$ are C^1 functions satisfying the decaying property as shown in (1.11).

Let

$$u = \begin{pmatrix} w \\ v \end{pmatrix}. \tag{5.4}$$

By (5.3), it is easy to see that in a neighborhood of $u_0 = \begin{pmatrix} 0 \\ \widetilde{v}_0 \end{pmatrix}$, system (5.1) is strictly hyperbolic and has the following two distinct real eigenvalues:

$$\lambda_1(u) \triangleq -\sqrt{\sigma'(w)} < \lambda_2(u) \triangleq \sqrt{\sigma'(w)}. \tag{5.5}$$

The corresponding right and left eigenvectors are

$$r_1(u)//(1, \sqrt{\sigma'(w)})^T, \quad r_2(u)//(1, -\sqrt{\sigma'(w)})^T \tag{5.6}$$

and

$$l_1(u)//(\sqrt{\sigma'(w)}, 1), \quad l_2(u)//(-\sqrt{\sigma'(w)}, 1), \tag{5.7}$$

respectively. It is easy to see that in a neighborhood of $u_0 = \begin{pmatrix} 0 \\ \widetilde{v}_0 \end{pmatrix}$, all characteristics are linearly degenerate, then weakly linearly degenerate, provided that

$$\sigma''(w) \equiv 0, \quad \forall \ |w| \ small. \tag{5.8}$$

By Theorem 1.1 we get

Theorem 5.1. Suppose that (5.8) holds. Then there is a sufficiently small $\theta_0 > 0$ such that for any given $\theta \in [0, \theta_0]$, Cauchy problem (5.1) and (5.2) admits a unique global C^1 solution $u = u(t, x)$ on $t \geq 0$.

Acknowledgment

The author would like to thank Prof. Ta-tsien Li very much for his valuable lectures and guidance in Fuzhou in December, 2003. The author also want to thank Prof. Jiaxing Hong very much for his guidance and constant encouragement.

References

[1] A. Bressan, Hyperbolic Systems of Conservation Laws: *The one-dimensional Cauchy Problem*, Oxford University Press, Oxford, 2000.

[2] G.-Q. Chen, C. D. Levermore, T.-P. Liu, Hyperbolic conservation laws with stiff relaxation terms and entropy, *Comm. Pure Appl. Math.* **47**, 787-830(1994).

[3] G.-Q. Chen, D. H. Wagner, Global entropy solutions to exothermically reacting compressible Euler equations, *J. Differential Equations* **191**, 277-322(2003).

[4] C. M. Dafermos, *Hyperbolic Conservation Laws in Continuum Physics*, Springer, Berlin, 2000.

[5] C. M. Dafermos, Hyperbolic systems of balance laws with weak dissipation, *J. Hyperbolic Differential Equations* **3**, 505-527(2006).

[6] C. M. Dafermos, L. Hsiao, Hyperbolic systems of balance laws with inhomogeneity and dissipation, *Indiana Univ. Math. J.* **31**, 471-491(1982).

[7] A. Douglis, Some existence theorems for hyperbolic systems of partial differential equations in two independent variables, *Comm. Pure Appl. Math.* **2**, 119-154(1952).

[8] P. Hartman, A. Wintner, Hyperbolic partial differential equations, *Amer. J. Math.* **74**, 834-864(1952).

[9] L. Hörmander, The Lifespan of Classical Solutions of Nonlinear Hyperbolic Equations, *Lecture Notes in Mathematics,* vol. 1256, Springer, Berlin, 1987.

[10] L. Hsiao, P. Marcati, Nonhomogeneous quasilinear hyperbolic system arising in chemical engineering, *Ann. Scuola Norm. Sup. Pisa Cl. Sci.* **15**(4), 65-97(1988).

[11] L. Hsiao, T.-P. Liu, Convergence to nonlinear diffusion waves for solutions of a system of hyperbolic conservation laws with damping, *Comm. Math. Phys.* **143**, 599-605(1992).

[12] L. Hsiao, S.-Q. Tang, Construction and qualitative behavior of the solution of the perturbated Riemann problem for the system of one-dimensional isentropic flow with damping, *J. Differential Equations* **123**, 480-503(1995).

[13] F. John, Formation of singularities in one-dimensional nonlinear wave propagation, *Comm. Pure Appl. Math.* **27**, 377-405(1974).

[14] D.-X. Kong, Cauchy Problem for Quasilinear Hyperbolic Systems, *MSJ Memoirs*, vol. 6, The Mathematical society of Japan, Tokyo, 2000.

[15] D.-X. Kong, Life-span of classical solutions to quasilinear hyperbolic systems with slow decay initial data, *Chinese Ann. Math. Ser. B* **21**, 413-440 (2000).

[16] D.-X. Kong, Global structure instability of Riemann solutions of quasilinear hyperbolic systems of conservation laws: Rarefaction waves, *J. Differential Equations* **219**, 421-450(2005).

[17] T.-T. Li, Global Classical Solutions for Quasilinear Hyperbolic Systems, *Research in Applied Mathematics*, vol. 32, Wiley/Masson, New York, 1994.

[18] T.-T. Li, Une remarque sur les coordonnées et ses applications aux systèmes hyperboliques quasi linéaires, *C. R. Acad. Sci. Paris, Série* **331**(1), 447-452(2000).

[19] T.-T. Li, D.-X. Kong, Breakdown of classical solutions to quasilinear hyperbolic systems, *Nonlinear Analysis TMA* **40**, 407-437(2000).

[20] T.-T. Li, D.-X. Kong, Y. Zhou, Global classical solutions for general quasilinear non-strictly hyperbolic systems with decay initial data, *Nonlinear Studies* **3**, 203-229(1996).

[21] T.-T. Li, Y.-J. Peng, Cauchy problem for weakly linearly degenerate hyperbolic systems in diagonal form. *Nonlinear Analysis* **55**, 937-949 (2003).

[22] T.-T. Li, Z.-Q. Shao, D.-X. Kong, Global existence of classical solutions to the Cauchy problem on a semi-bounded initial axis for a nonhomogeneous quasilinear hyperbolic system, *J. Math. Anal. Appl.* **325** (1), 205-225(2007).

[23] T.-T. Li, L.-B. Wang, Global existence of classical solutions to the Cauchy problem on a semi-bounded initial axis for quasilinear hyperbolic systems. *Nonlinear Analysis* **56** (2004), 961-974.

[24] T.-T. Li, W.-C. Yu, Boundary Value Problems for Quasilinear Hyperbolic Systems, *Duke University Mathematics Series* **V**, 1985.

[25] T.-T. Li, Y. Zhou, D.-X. Kong, Weak linear degeneracy and global classical solutions for general quasilinear hyperbolic systems, *Comm. Partial Differential Equations* **19**, 1263-1317(1994).

[26] T.-T. Li, Y. Zhou, D.-X. Kong, Global classical solutions for general quasilinear hyperbolic systems with decay initial data, *Nonlinear Analysis TMA* **28**, 1299-1332(1997).

[27] T.-P. Liu, Development of singularties in the nonlinear waves for quasilinear hyperbolic partial differential equations, *J. Differential Equations* **33**, 92-111(1979).

[28] T.-P. Liu, Quasilinear hyperbolic systems, *Comm. Math. Phys*. **68**, 141-172(1979).

[29] M. Luskin, On the existence of global smooth solutions for a model equation for fluid flow in a pipe, *J. Math. Anal. Appl.* **84**, 614-630(1981).

[30] M. Luskin, J. B. Temple, The existence of a global weak solution to the nonlinear waterhammer problem, *Comm. Pure Appl. Math.* **35**, 697-735(1982).

[31] Z.-Q. Shao, Global structure instability of Riemann solutions for general quasilinear hyperbolic systems of conservation laws in the presence of a boundary, *J. Math. Anal. Appl.* **330**, 511-540(2007).

[32] Z.-Q. Shao, Blow-up of solutions to the initial-boundary value problem for quasilinear hyperbolic systems of conservation laws, *Nonlinear Anal. Theory Methods Appl.* **68**, 716-740(2008).

[33] Z.-Q. Shao, Global solution to the generalized Riemann problem in the presence of a boundary and contact discontinuities, *Journal of Elasticity* **87**, 277-310(2007).

[34] Z.-Q. Shao, Shock reflection for a system of hyperbolic balance laws, *J. Math. Anal. Appl.* **343**, 1131-1153 (2008).

[35] Z.-Q. Shao, Global weakly discontinuous solutions for hyperbolic conservation laws in the presence of a boundary, *J. Math. Anal. Appl.* (2008), doi: 10.1016/j.jmaa.2008.03.054.

[36] Z.-Q. Shao, Global solutions with shock waves to the generalized Riemann problem for a class of quasilinear hyperbolic systems of balance laws II, *Math. Nachr.* **281**, 879-902 (2008).

[37] Z.-Q. Shao, D.-X. Kong, Y.-C. Li, Global solutions with shock waves to the generalized Riemann problem for a class of quasilinear hyperbolic systems of balance laws, *Nonlinear Anal. Theory Methods Appl.* **64**, 2575-2603(2006).

[38] Z.-Q. Shao, Y.-C. Li, D.-X. Kong, Global weakly discontinuous solutions to quasilinear hyperbolic systems of conservation laws with damping with a kind of non-smooth initial data, *Z. Angew. Math. Phys.* (2007), doi: 10.1007/s00033-007-5096-0.

[39] L.-A. Ying, C.-H. Wang, Global solutions of the Cauchy problem for a nonhomogeneous quasilinear hyperbolic system, *Comm. Pure Appl. Math.* **33**, 579-597(1980).

[40] L.-A. Ying, C.-H. Wang, Solution in the large for nonhomogeneous quasilinear hyperbolic systems of equations, *J. Math. Anal. Appl.* **78**, 440-454(1980).

In: Progress in Nonlinear Analysis Research
Editor: Inès N. Roux, pp. 309-329
ISBN 978-1-60456-359-7

Chapter 13

IMMERSED BOUNDARY METHOD: THE EXISTENCE OF APPROXIMATE SOLUTION IN TWO-DIMENSIONAL CASE

Ling Rao[1,2] * ***and Hongquan Chen***[1]
[1]College of Aerospace Engineering
Nanjing University of Aeronautics and Astronautics
Nanjing 210016, China
[2]Department of Applied Mathematics
Nanjing University of Science and Technology
Nanjing 210094, China

Abstract

This paper deals with the two-dimensional Navier-Stokes equations in which the source term involves a Dirac delta function and describes the elastic reaction of the immersed boundary. We analyze the existence of the approximate solution with Dirac delta function approximated by differentiable function. We obtain the result via the Banach Fixed Point Theorem and the properties of the solutions to the Navier-Stokes equations of viscous incompressible fluids with periodic boundary conditions.

Key Words: Navier-Stokes Equations, immersed boundary method, nonlinear ordinary differential equations. AMS Subject Classification: 35K10, 46N20, 34A12

1. Introduction

Problems involving the interaction between a fluid flow and elastic interfaces may appear in several branches of science such as engineering, physics, biology, and medicine. Regardless the field, as a rule, they share in common a high degree of complexity, often displaying intricate geometry or time-dependent elastic properties, turning the problem into a real challenge for applied scientists, from both the mathematical modeling and the numerical simulation points of view.

*E-mail address: lingrao@sina.com

In the early 70s, Peskin [2, 8] introduced a mathematical model and a computational method to study the flow patterns of the blood around the heart valves. Through years, Peskin's immersed boundary (IB) method was developed for the computer simulation of general problems [9–12] involving a transient incompressible viscous fluid containing an immersed elastic interface, which may have time-dependent geometry or elastic properties, or both.

The IB method is at the same time a mathematical formulation and a numerical scheme. The mathematical formulation is based on the use of Eulerian variables to describe the dynamic of fluid and of Lagrangean variables along the moving structure. The force exerted by the structure on the fluid is taken into account by means of a Dirac delta function constructed according to certain principles. The main idea is to use a regular Eulerian mesh for the fluid dynamics simulation, coupled with a Lagrangian representation of the immersed boundary. The advantage of this method is that the fluid domain can have a simple shape, so that structured grids can be used. The Lagrangian mesh is independent of the Eulerian mesh. The interaction between the fluid and the immersed boundary is modeled using a well-chosen discrete approximation to the Dirac delta function.

Although the immersed boundary method is a particularly effective approach in scientific computation, but very little theoretical analysis has been performed on either the underlying model equations or numerical methods. Daniele Boffi [3] gave a variational formulation of the problem and provided a suitable modification of the IB method which made use of finite elements method. Because the source term in the Navier-Stokes equations involves a Dirac delta function, the problem is highly nonlinear and presents several difficulties related with the lacking of regularity of the solution of the Navier-Stokes equations due to such source term. Daniele Boffi analyzed the existence of the solution in a very simple one-dimensional heat equation. In [13] we dealt with the two-dimensional heat equation. We analyzed the existence of the approximate solution with Dirac delta function approximated by differentiable function. In this paper will deal with two-dimensional Navier-Stokes equations. We analyze the existence of the approximate solution still with Dirac delta function approximated by differentiable function. We obtain the results via the Banach Fixed Point Theorem and a theorem in nonlinear ordinary differential equations in abstract space and the properties of the solutions to the Navier-Stokes equations of viscous incompressible fluids with periodic boundary conditions.

In section 2 we will present the mathematical model of the problem. In order to prove the existence of the approximate solution, some properties of the Navier-Stokes equations with the space-periodic boundary conditions are reviewed in section 3. In section 4 we will introduce the corresponding variational formulation of the problem. In section 5 we will prove the existence of the approximate solution of the problem via a fixed point argument.

2. Problem

The authors (see [3]) considered the model problem of a viscous incompressible fluid in a two-dimensional square domain Ω containing an immersed massless elastic boundary in the form of a curve. To be more precise, for all $t \in [0, T]$, let Γ_t be a simple curve, the configuration of which is given in a parametric form, $\mathbf{X}(s,t), 0 \leq s \leq L, \mathbf{X}(0,t) =$

$\mathbf{X}(L,t)$. The equations of motion of the system are

$$\frac{\partial \mathbf{u}}{\partial t} - \mu\Delta\mathbf{u} + \mathbf{u}\cdot\Delta\mathbf{u} + \Delta p = \mathbf{F} \qquad \text{in } \Omega\times(0,T), \tag{1}$$

$$\Delta\cdot\mathbf{u} = 0 \qquad \text{in } \Omega\times(0,T), \tag{2}$$

$$\mathbf{F}(\mathbf{x},t) = \int_0^L \mathbf{f}(s,t)\delta(x-\mathbf{X}(s,t))ds \qquad \forall\mathbf{x}\in\Omega,\ t\in(0,T), \tag{3}$$

$$\frac{\partial\mathbf{X}}{\partial t}(s,t) = \mathbf{u}(\mathbf{X}(s,t),t) \qquad \forall s\in[0,L],\ t\in(0,T). \tag{4}$$

Here $\mathbf{u}$ is the fluid velocity and p is the fluid pressure. The coefficient μ is the fluid viscosity constant. Eqs. (1) and (2) are the usual incompressible Navier-Stokes equations. $\mathbf{F}$ is the force density generated by the boundary on the fluid. The force exerted by the element of boundary on the fluid is $\mathbf{f}$. The function δ in the integrals is the two-dimensional Dirac delta function concentrate at the origin. Eq. (4) is equivalent to the no-slip condition that the fluid sticks to the boundary.

In [3] and [13] the problem was supplemented with the following boundary and initial conditions:

$$\mathbf{u}(\mathbf{x},0) = \mathbf{u}_0(\mathbf{x}) \qquad \forall\mathbf{x}\in\Omega, \tag{5}$$

$$\mathbf{X}(s,0) = \mathbf{X}_0(s) \qquad \forall s\in[0,L], \tag{6}$$

$$\mathbf{u}(\mathbf{x},t) = 0 \qquad \forall(\mathbf{x},t)\in\partial\Omega\times(0,T). \tag{7}$$

When we perform the numerical simulation with the IB method the fluid domain Ω often can be chosen to have a simple big enough shape such as square or rectangle, then we can make space-periodic extension of the problem and suppose that the fluid satisfies space-periodic boundary condition. Such assumption is also useful for idealizations and needed in the following proofs. In this paper, we use the space-periodic boundary condition instead of (7). We use Ω to denote the space-period:

$$\Omega = (-\frac{L_1}{2},\frac{L_1}{2})\times(-\frac{L_2}{2},\frac{L_2}{2}).$$

We assume that the fluid fills the entire space R^2 but with condition that

$\mathbf{u}$, $\mathbf{F}$ and p are periodic in each direction ox_i, $i=1,2$, with corresponding periods $L_i>0$.

It is sometimes useful and simpler to assume that average flow is zero, that is

$$\frac{1}{|\Omega|}\int_\Omega \mathbf{u}(\mathbf{x})d\mathbf{x} = 0.$$

Moreover, the general case-where the volume forces and the initial condition do not average to zero - can be reduced to this case. We may refer to [4, 6] for the method. We show it as follows.

First, averaging each term in (1). Using integration by parts and periodicity condition, several terms vanish, we get

$$\frac{d}{dt}\int_\Omega \mathbf{u}(\mathbf{x},t)d\mathbf{x} = \int_\Omega \mathbf{F}(\mathbf{x},t)d\mathbf{x}.$$

Therefore if we set

$$M_u(t) = \frac{1}{|\Omega|}\int_\Omega \mathbf{u}(\mathbf{x},t)d\mathbf{x} \tag{8}$$

and

$$M_F(t) = \frac{1}{|\Omega|}\int_\Omega \mathbf{F}(\mathbf{x},t)d\mathbf{x},$$

we obtain

$$\frac{dM_u(t)}{dt} = M_F(t),$$

thus

$$M_u(t) = \frac{1}{|\Omega|}\int_\Omega \mathbf{u}_0(\mathbf{x})d\mathbf{x} + \int_0^t M_F(t)dt.$$

Let

$$\begin{aligned}
&\hat{\mathbf{u}} = \mathbf{u} - M_u,\\
&\hat{\mathbf{F}} = \mathbf{F} - M_F,\\
&I_u(t) = \textstyle\int_0^t M_u(s)ds,\\
&\hat{\mathbf{u}}(\mathbf{x},t) = \hat{\mathbf{u}}(\mathbf{x}+I_u(t),t),\\
&\hat{\hat{p}}(\mathbf{x},t) = p(\mathbf{x}+I_u(t),t),\\
&\hat{\hat{\mathbf{F}}}(\mathbf{x},t) = \hat{\mathbf{F}}(\mathbf{x}+I_u(t),t).
\end{aligned} \tag{9}$$

Notice that $\hat{\mathbf{u}}$, $\hat{\mathbf{p}}$, and $\hat{\hat{\mathbf{F}}}$ are also periodic with period Ω and $\hat{\mathbf{u}}$, $\hat{\hat{\mathbf{F}}}$ have zero space average. With $\mathbf{u}$, p and $\mathbf{F}$ in (1)-(6) replaced by $\hat{\mathbf{u}}$, $\hat{\mathbf{p}}$ and $\hat{\hat{\mathbf{F}}}$, we need to solve the following equations instead of (1)-(6) for $\hat{\mathbf{u}}$, $\hat{\mathbf{p}}$ and $\mathbf{X}$:

$$\frac{\partial \mathbf{u}}{\partial t} - \mu\Delta\mathbf{u} + \mathbf{u}\cdot\Delta\mathbf{u} + \Delta p = \hat{\hat{\mathbf{F}}} \qquad \text{in } \Omega\times(0,T), \tag{10}$$

$$\Delta\cdot\mathbf{u} = 0 \qquad \text{in } \Omega\times(0,T), \tag{11}$$

$$\frac{\partial \mathbf{X}}{\partial t}(s,t) = \mathbf{u}(\mathbf{X}(s,t) - I_u(t),t) + M_u(t) \qquad \forall s\in[0,L],\quad t\in(0,T), \tag{12}$$

$$\mathbf{u}(\mathbf{x},0) = \hat{\mathbf{u}}_0(x), \qquad \forall x\in\Omega, \tag{13}$$

$$\mathbf{X}(s,0) = \mathbf{X}_0(s) \qquad \forall s\in[0,L], \tag{14}$$

$$\mathbf{X}(0,t) = \mathbf{X}(L,t) \qquad \forall t\in(0,T), \tag{15}$$

where for given $\mathbf{f}$ and $\mathbf{u}_0$, $\hat{\mathbf{u}}_0$ and $\hat{\hat{\mathbf{F}}}$ may be calculated by (3) and (9),

$$\begin{aligned}
\hat{\mathbf{u}}_0(x) &= \mathbf{u}_0(x) - M_u(0) = \mathbf{u}_0(x) - \frac{1}{|\Omega|}\int_\Omega \mathbf{u}_0(x)d\mathbf{x},\\
\hat{\hat{\mathbf{F}}}(\mathbf{x},t) &= \mathbf{F}(\mathbf{x}+I_u(t),t) - M_F(t)\\
&= \int_0^L \mathbf{f}(s,t)\delta(\mathbf{x}+I_u(t)-\mathbf{X}(s,t))ds\\
&\quad -\frac{1}{|\Omega|}\int_0^L \mathbf{f}(s,t)(\int_\Omega \delta(x-\mathbf{X}(s,t))dx)ds.
\end{aligned} \tag{16}$$

And with the solutions $\mathbf{u}(=\hat{\mathbf{u}})$, $p(=\hat{\mathbf{p}})$ of (10)-(16) we can recover the solutions $\mathbf{u}$, p of (1)-(6) by (9).

In order to obtain the existence of the approximate solution to equations (10)-(16), we review some properties of the Navier-Stokes equations with the space-periodic boundary conditions as follows.

3. Review of the mathematical theory of the Navier-Stokes equations with the space-periodic boundary conditions

We introduce some basic mathematical properties about Navier-Stoks equations with the space-periodic boundary conditions. For more details of them, the readers may refer to [4,6]. We shall be concerned with the spaces of two-dimensional vector functions. We use the notations

$$\mathbf{H}^m(\Omega) = \{H^m(\Omega)\}^2, \quad \mathbf{L}^2(\Omega) = \{L^2(\Omega)\}^2,$$

and we suppose that these product spaces are equipped with the usual product norm. The norm on $\mathbf{L}^2(\Omega)$ is denoted by $|\cdot|$ (also denoted by $\|\cdot\|_0$). The norm on $\mathbf{H}^m(\Omega)$ is denoted by $\|\cdot\|_m$. $(\cdot,\cdot)$ stands for the scalar product on $\mathbf{L}^2(\Omega)$. We denote by $\mathbf{H}^m_{per}(\Omega)$, $m \in \mathbb{N}$, the space of functions which are in $\mathbf{H}^m_{Loc}(R^n)$ (i.e. $\mathbf{u}|_O \in \mathbf{H}^m(O)$ for every open bound set O) and which are periodic with period Ω. For $m = 0$, $\mathbf{H}^0_{per}(\Omega)$ is also denoted by $\mathbf{L}^2_{per}(\Omega)$ and coincides simply with $\mathbf{L}^2(\Omega)$ (the restrictions of the functions in $\mathbf{H}^0_{per}(\Omega)$ to Ω are the whole space $\mathbf{L}^2(\Omega)$). For an arbitrary $m \in \mathbb{N}$, $\mathbf{H}^m_{per}(\Omega)$ is a Hilbert space for the scalar product and the norm

$$(\mathbf{u},\mathbf{v})_m = \sum_{[\alpha]\le m}\int_\Omega D^\alpha\mathbf{u}(\mathbf{x})D^\alpha\mathbf{v}(\mathbf{x})d\mathbf{x}, \qquad \|\mathbf{u}\|_m = \{(\mathbf{u},\mathbf{u})_m\}^{\frac{1}{2}},$$

where $\alpha = (\alpha_1,\ldots,\alpha_n)$, $\alpha_i \in \mathbb{N}$, $[\alpha] = \alpha_1 + \ldots + \alpha_n$ and $D^\alpha = \dfrac{\partial^{[\alpha]}}{\partial x^{\alpha_1}\ldots\partial^{\alpha_n}}$.

We work with complex representation, for which we take $i = \sqrt{-1}$. Then a square integrable vector field $\mathbf{u} = \mathbf{u}(x)$ on Ω can be represented by the Fourier series expansion

$$\mathbf{u}(\mathbf{x}) = \sum_{\mathbf{k}\in\mathbb{Z}^n} c_{\mathbf{k}}e^{2\pi i\frac{\mathbf{k}}{\mathbf{L}}\cdot\mathbf{x}},$$

where

$$\frac{\mathbf{k}}{\mathbf{L}} = (\frac{k_1}{L_1},\frac{k_2}{L_2}).$$

The functions in $\mathbf{H}^m_{per}(\Omega)$ are easily characterized by their Fourier series expansion

$$\mathbf{H}^m_{per}(\Omega) = \{\mathbf{u},\ \mathbf{u} = \sum_{\mathbf{k}\in\mathbb{Z}^n} c_{\mathbf{k}} e^{2\pi i \frac{\mathbf{k}}{\mathbf{L}}\cdot\mathbf{x}} \quad \bar{c}_{\mathbf{k}} = c_{-\mathbf{k}},\ \sum_{\mathbf{k}\in\mathbb{Z}^n} |\mathbf{k}|^{2m}|c_{\mathbf{k}}|^2 < \infty\}, \tag{17}$$

and the norm $\|\mathbf{u}\|_m$ is equivalent to the norm $\{\sum_{\mathbf{k}\in\mathbb{Z}^n}(1+|\mathbf{k}|^{2m})|c_{\mathbf{k}}|^2\}^{\frac{1}{2}}$.

We denote by V the space of smooth($\mathcal{C}^\infty$) divergence-free vector fields on R^2 that are periodic with period Ω. Let H be the closure of V in $\mathbf{L}^2(\Omega)$ and let V be the closure of V in $\mathbf{H}^1(\Omega)$. The space H is equipped with the scalar product $(\cdot,\cdot)$ induced by $\mathbf{L}^2(\Omega)$; the space V with the scalar product

$$((\mathbf{u},\mathbf{v})) = \int_\Omega \sum_{i=1}^{2} \frac{\partial \mathbf{u}}{\partial x_i}\cdot\frac{\partial \mathbf{v}}{\partial x_i} d\mathbf{x}$$

and the associated norms, denoted by

$$|\mathbf{u}| = (\mathbf{u},\mathbf{u})^{\frac{1}{2}} \text{ for } \mathbf{u}\in H, \qquad \|\mathbf{u}\| = ((\mathbf{v},\mathbf{v}))^{\frac{1}{2}} \text{ for } \mathbf{v}\in V.$$

For the sake of simplicity, we restrict ourselves to the space-periodic case with vanishing space average. For the vanishing space average case, we have the additional condition

$$c_0 = \int_\Omega \mathbf{u}(\mathbf{x})d\mathbf{x} = 0.$$

Then

$$H = \{\mathbf{u}\in \mathbf{L}^2_{per}(\Omega);\ \nabla\cdot\mathbf{u} = 0,\ \int_\Omega \mathbf{u}(\mathbf{x}) = 0\}(\triangleq \dot{H});$$

$$V = \{\mathbf{u}\in \mathbf{H}^1_{per}(\Omega);\ \nabla\cdot\mathbf{u} = 0,\ \int_\Omega \mathbf{u}(\mathbf{x}) = 0\}(\triangleq \dot{V});$$

and

$$\mathbf{H}^m_{per}(\Omega) = \{\mathbf{u}\in \mathbf{H}^m_{per}(\Omega) \text{ of type } (17),\ c_0 = 0\}(\triangleq \dot{\mathbf{H}}^m_{per}(\Omega)).$$

The functions in $\dot{H}$ and $\dot{V}$ also can be characterized by there Fourier series expansion

$$\dot{H} = \{\mathbf{u},\ \mathbf{u} = \sum_{\mathbf{k}\in\mathbb{Z}^2\backslash\{0\}} c_{\mathbf{k}} e^{2\pi i \frac{\mathbf{k}}{\mathbf{L}}\cdot\mathbf{x}},\quad \bar{c}_{\mathbf{k}} = c_{-\mathbf{k}},\ \frac{\mathbf{k}}{\mathbf{L}}\cdot c_{\mathbf{k}} = 0,\ \sum_{\mathbf{k}\in\mathbb{Z}^2\backslash\{0\}} |c_{\mathbf{k}}|^2 < \infty\};$$

$$\dot{V} = \{\mathbf{u},\ \mathbf{u} = \sum_{\mathbf{k}\in\mathbb{Z}^2\backslash\{0\}} c_{\mathbf{k}} e^{2\pi i \frac{\mathbf{k}}{\mathbf{L}}\cdot\mathbf{x}},\quad \bar{c}_{\mathbf{k}} = c_{-\mathbf{k}},\ \frac{\mathbf{k}}{\mathbf{L}}\cdot c_{\mathbf{k}} = 0,\ \sum_{\mathbf{k}\in\mathbb{Z}^2\backslash\{0\}} |\frac{\mathbf{k}}{\mathbf{L}}|^2 |c_{\mathbf{k}}|^2 < \infty\}.$$

For all $s\in R$ we may consider the space

$$V_s = \{\mathbf{u}\in \dot{\mathbf{H}}^s_{per}(\Omega),\ div\mathbf{u} = 0\} =$$
$$\{\mathbf{u},\ \mathbf{u} = \sum_{\mathbf{k}\in\mathbb{Z}^2\backslash\{0\}} c_{\mathbf{k}} e^{2\pi i \frac{\mathbf{k}}{\mathbf{L}}\cdot\mathbf{x}},\quad \bar{c}_{\mathbf{k}} = c_{-\mathbf{k}},\ \frac{\mathbf{k}}{\mathbf{L}}\cdot c_{\mathbf{k}} = 0,\ \sum_{\mathbf{k}\in\mathbb{Z}^2\backslash\{0\}} |\frac{\mathbf{k}}{\mathbf{L}}|^{2s} |c_{\mathbf{k}}|^2 < \infty\}.$$

Note that, V_s is a closed subspace of $\dot{\mathbf{H}}^s_{per}(\Omega)$. $V_{s_1} \subset V_{s_2}$ for $s_1 \geq s_2$, $V_1 = \dot{V}$ and $V_0 = \dot{H}$. Hence, $\dot{V} \subset V_s \subset \dot{H}$ for $0 \leq s \leq 1$, $V_s \subset \dot{V}$ for $s \geq 1$, and $V_s \supset \dot{H}$ for $s < 0$. It can be shown that V_s is a Hilbert space for the norm

$$\|\mathbf{u}\|_{V_s} = \Big(\sum_{\mathbf{k}\in\mathbb{Z}^2\setminus\{0\}} |\frac{\mathbf{k}}{\mathbf{L}}|^{2s} |c_{\mathbf{k}}|^2 \Big)^{\frac{1}{2}}.$$

Of particular interest are the spaces V_2 and V_{-1}. The space V_{-1} is the dual space of $\dot{V}$, usually denoted V'; this is the space of linear continuous forms on $\dot{V}$. More generally, for all $s \geq 0$, V_{-s} is the dual of V_s.

We have

$$H^{\perp} = \{\mathbf{u} \in \mathbf{L}^2(\Omega); \mathbf{u} = \nabla p,\ p \in \mathbf{H}^1_{per}(\Omega)\}.$$

We definite the so-called Leray projector by

$$P_L : \mathbf{L}^2(\Omega) \to H,$$

which is the orthogonal projector onto H in $\mathbf{L}^2(\Omega)$.

We definite the Stokes operator A by

$$A\mathbf{u} = -P_L \triangle \mathbf{u} \quad \forall \mathbf{u} \in D(A) = \dot{V} \cap \dot{\mathbf{H}}^2_{per}(\Omega),$$

where $\triangle$ is the Laplacian. And we can see that

$$-P_L \triangle \mathbf{u} = -\triangle \mathbf{u} \quad \forall \mathbf{u} \in D(A) = \dot{V} \cap \dot{\mathbf{H}}^2_{per}(\Omega).$$

The Stokes operator is a positive self-adjoint operator, so we can work with fractional powers of A. A is just the mapping

$$\mathbf{u} = \sum_{\mathbf{k}\in\mathbb{Z}^2\setminus\{0\}} c_{\mathbf{k}} e^{2\pi i \frac{\mathbf{k}}{\mathbf{L}}\cdot\mathbf{x}} \to A\mathbf{u} = \sum_{\mathbf{k}\in\mathbb{Z}^2\setminus\{0\}} |\frac{\mathbf{k}}{\mathbf{L}}|^2 c_{\mathbf{k}} e^{2\pi i \frac{\mathbf{k}}{\mathbf{L}}\cdot\mathbf{x}}.$$

$A^r\mathbf{u}$ is the operator

$$\mathbf{u} = \sum_{\mathbf{k}\in\mathbb{Z}^2\setminus\{0\}} c_{\mathbf{k}} e^{2\pi i \frac{\mathbf{k}}{\mathbf{L}}\cdot\mathbf{x}} \to A^r\mathbf{u} = \sum_{\mathbf{k}\in\mathbb{Z}^2\setminus\{0\}} |\frac{\mathbf{k}}{\mathbf{L}}|^{2r} c_{\mathbf{k}} e^{2\pi i \frac{\mathbf{k}}{\mathbf{L}}\cdot\mathbf{x}}.$$

It is straightforward to see that A^r maps V_{2s} continuously onto V_{2s-2r} $(s, r \in \mathbb{R})$. In particular, for $s \geq 0$, we have $A^s V_{2s} = \dot{H}$ and so $V_{2s} = D(A^s)$ is the domain of the (unbounded) operator A^s in $\dot{H}$. The norm $|A^s\mathbf{u}| (= \|\mathbf{u}\|_{V_{2s}})$ is equivalent to the norm induced by $\dot{\mathbf{H}}^{2s}_{per}(\Omega)$,

$$c\|\mathbf{u}\|_{2s} \leq |A^s\mathbf{u}| \leq c'\|\mathbf{u}\|_{2s} \quad \forall \mathbf{u} \in D(A^s), \tag{18}$$

with positive constants c, c' depending on L_1, L_2 and s.

We use some notations as follows. Let a, b be two extended real numbers, $-\infty \leq a < b \leq +\infty$, and let X be a Banach space. For given α, $1 \leq \alpha < +\infty$, $L^{\alpha}(a, b; X)$ denotes the space of L^α-integrable functions from [a, b] into X, which is equipped with the Banach norm

$$|\int_a^b \|f(t)\|_X^\alpha dt\, |^{\frac{1}{\alpha}} .$$

The space $C([a,b];X)$ is the space of continuous functions from $[a,b]\ (-\infty < a < b < \infty)$ into X and is equipped with the Banach norm

$$\sup_{t\in[a,b]} \|f(t)\|_X.$$

For $\mathbf{u},\ \mathbf{v},\ \mathbf{w} \in \mathbf{L}^1(\Omega)$, we set

$$b(\mathbf{u},\mathbf{v},\mathbf{w}) = \sum_{i,j=1}^{2} \int_\Omega \mathbf{u}_i(D_i\mathbf{v}_j)\mathbf{w}_j dx,$$

whenever the integrals make sense.

The trilinear operator $b = b(\mathbf{u},\mathbf{v},\mathbf{w})$ can be extended to a continuous trilinear operator defined on V. Moreover,

$$\begin{aligned} &b(\mathbf{u},\mathbf{v},\mathbf{v}) = 0 && \mathbf{u},\mathbf{v} \in V. \\ &b(\mathbf{u},\mathbf{v},\mathbf{w}) = -b(\mathbf{u},\mathbf{w},\mathbf{v}) && \mathbf{u},\mathbf{v},\mathbf{w} \in V. \end{aligned} \tag{19}$$

In the periodic case,

$$b(\mathbf{v},\mathbf{v},\mathbf{A}v) = 0 \quad \mathbf{v} \in D(A) = \dot{\mathbf{H}}^2_{per}(\Omega) \cap \dot{V}. \tag{20}$$

Now we consider the Navier-Stokes equations

$$\begin{cases} \dfrac{\partial \mathbf{u}}{\partial t} - \mu\Delta\mathbf{u} + \mathbf{u}\cdot\Delta\mathbf{u} + \Delta p = \mathbf{f}, \\ \Delta\cdot\mathbf{u} = 0, \end{cases} \tag{21}$$

in

$$\Omega = (-\frac{L_1}{2},\frac{L_1}{2}) \times (-\frac{L_2}{2},\frac{L_2}{2}), \quad L_1, L_2 > 0,$$

with $\mu > 0$ and require that

$$\mathbf{u} = \mathbf{u}(\mathbf{x},t),\ p = p(\mathbf{x},t), \text{ and } \mathbf{f} = \mathbf{f}(\mathbf{x},t) \text{ with } \mathbf{x} = (x_1,x_2), \\ \text{are } L_i\text{-periodic in each variable } x_i$$

and

$$\int_\Omega \mathbf{f}(\mathbf{x},t)d\mathbf{x} = \int_\Omega \mathbf{u}(\mathbf{x},t)d\mathbf{x} = 0.$$

Let $T > 0$ be given. Assume that $\mathbf{u} \in C^2(R^2 \times [0,T]),\ p \in C^1(R^2 \times [0,T])$ are the classical solutions of (21). For each $v \in$ V, by multiplying momentum equation in (21) by v, integrating over Ω and using the Stokes formula we find that (cf. [5] Chap.III for the details)

$$\frac{d}{dt}(\mathbf{u},\mathbf{v}) + \mu((\mathbf{u},\mathbf{v})) + b(\mathbf{u},\mathbf{u},\mathbf{v}) = \langle \mathbf{f},\mathbf{v}\rangle \qquad \forall\, \mathbf{v} \in \mathrm{V}. \tag{22}$$

By continuity, (22) holds also for each $v \in V$.

This suggests the following variational formulation of (21) (strong solutions).

For $\mathbf{f}$ and $\mathbf{u}_0$ given,

$$\mathbf{f} \in L^2(0, T; \dot{H}),$$

$$\mathbf{u}_0 \in \dot{V},$$

find $\mathbf{u}$ satisfying

$$\mathbf{u} \in L^2(0, T; D(A) \cap L^\infty(0, T; \dot{V})$$

such that

$$\begin{cases} \dfrac{d}{dt}(\mathbf{u}, \mathbf{v}) + \mu((\mathbf{u}, \mathbf{v})) + b(\mathbf{u}, \mathbf{u}, \mathbf{v}) = \langle \mathbf{f}, \mathbf{v} \rangle & \forall\, \mathbf{v} \in V,\ t \in (0, T), \\ \mathbf{u}(0) = \mathbf{u}_0. \end{cases} \tag{23}$$

If f is square integrable but not in H then we can replace it by its Leray projection on H, so that f is always assumed to be in H. We may refer to [5], p.307 for the relation of equations (21) and its variational formulation (23). One can show that if $\mathbf{u}$ is a solution of (23) then there exists p such that (21) is satisfied in a weak sense.

Lemma 1. *(Existence and Uniqueness of Strong Solutions in Two Dimensions) [4, 6]*
(i) *Assume that $\mathbf{u}_0, \mathbf{f}, T > 0$ are given and satisfy*

$$\mathbf{u}_0 \in \dot{V}, \quad \mathbf{f} \in L^2(0, T; \dot{H}).$$

Then there exists a unique solution $\mathbf{u}$ of (23), satisfying

$$u_i, \frac{\partial u_i}{\partial t}, \frac{\partial u_i}{\partial x_j}, \frac{\partial^2 u_i}{\partial x_j \partial x_k} \in L^2(\Omega \times (0, T)), \quad i, j, k = 1, 2,$$

and $\mathbf{u}$ is a continuously function from [0,T] into $\dot{V}$. We also can see that

$$\mathbf{u} \in L^2(0, T; D(A)) \cap \mathrm{C}([0, T], \dot{V}).$$

And $\mathbf{u}$ satisfies the following priori estimates

$$|\mathbf{u}(s)|^2 \le |\mathbf{u}(0)|^2 + \frac{1}{\mu} \int_0^s \|\mathbf{f}(t)\|_{V'}^2 dt \qquad \forall s \in [0, T],$$

$$\sup_{t \in [0,T]} \|\mathbf{u}(t)\|^2 \le K_1, \qquad \int_0^T |A\mathbf{u}(t)|^2 dt \le K_2,$$

where constants K_1, K_2 are dependent on $\|\mathbf{u}_0\|$, $\int_0^T |\mathbf{f}(t)|^2 dt$, μ, L_1, L_2.
(ii) *For $r \ge 1$, if the initial data $\mathbf{u}_0 \in V_r$ and $\mathbf{f} \in L^\infty(0, T; V_{r-1})$, then the solution $\mathbf{u} = \mathbf{u}(t)$ belongs to $\mathrm{C}([0, T]; V_r)$. And there exists constant K dependent on μ, $\|\mathbf{u}_0\|_r$, $|\mathbf{f}|_{L^\infty(0,T;V_{r-1})}$, such that*

$$\sup_{t \in [0,T]} \|\mathbf{u}(t)\|_r \le K. \tag{24}$$

Gronwall's Lemma:

(A) *[7] Let $a \in L_1(\tau, T)$, b be absolutely continuous on $[\tau, T]$. If $x \in L^\infty(\tau, T)$ satisfies*

$$x(t) \leq b(t) + \int_\tau^t a(s)x(s)ds,$$

then for $t \in (0, T)$

$$x(t) \leq b(\tau)exp\int_\tau^t a(s)ds + \int_\tau^t b'(s)exp\left(\int_s^t a(\rho)d\rho\right) ds.$$

(B) *[4] Let $a, \theta \in L_1(0, T)$. If y satisfies*

$$y' \leq a + \theta y,$$

then for $t \in (0, T)$

$$y(t) \leq y(0)exp(\int_0^t \theta(\tau)d\tau) + \int_0^t a(s)exp(\int_s^t \theta(\tau)d\tau)ds. \quad \square$$

Lemma 2. *Assume that $0 < T \leq \widetilde{T}$, c is a positive constant and $\mathbf{u}_0, \mathbf{f}_1, \mathbf{f}_2$ satisfy*

$$\mathbf{u}_0 \in \dot{V}, \quad \mathbf{f}_i, \in L^2(0, T; \dot{H}), \quad \int_0^T |\mathbf{f_i}(t)|^2 dt \leq c \quad for\ i = 1, 2.$$

Suppose that $\mathbf{u}_1$, $\mathbf{u}_2$ are the two solutions to (23) corresponding forcing terms $\mathbf{f}_1$, $\mathbf{f}_2$ respectively. Then we have the following estimate

$$\int_0^T \|\mathbf{u}_1(t) - \mathbf{u}_2(t)\|_2^2 dt \leq c_1 \int_0^T |\mathbf{f}_1(t) - \mathbf{f}_2(t)|^2 dt$$

where constant c_1 is dependent on μ, $\|\mathbf{u}_0\|$, $\widetilde{T}$, c.

Proof According to the assumptions and **Lemma 1**, we have

$$\mathbf{u}_1, \mathbf{u}_2 \in L^2(0, T; D(A))$$

and

$$\sup_{t\in[0,T]} \|\mathbf{u}_1(t)\|^2 \leq K_1, \qquad \int_0^T |A\mathbf{u}_1(t)|^2 dt \leq K_2, \tag{25}$$

where constants K_1, K_2 are dependent on μ, $\|\mathbf{u}_0\|$, c.

By the assumption, we have

$$\begin{cases} \frac{d}{dt}((\mathbf{u}_1 - \mathbf{u}_2)(t), \mathbf{v}) + \mu((\mathbf{u}_1 - \mathbf{u}_2, \mathbf{v})) + b(\mathbf{u}_1, \mathbf{u}_1, \mathbf{v}) \\ \qquad -b(\mathbf{u}_2, \mathbf{u}_2, \mathbf{v}) = \langle \mathbf{f}_1 - \mathbf{f}_2, \mathbf{v}\rangle \qquad \forall\, \mathbf{v} \in V,\ t \in (0, T), \\ (\mathbf{u}_1 - \mathbf{u}_2)(0) = 0. \end{cases} \tag{26}$$

Let $\mathbf{u} = \mathbf{u}_1 - \mathbf{u}_2,\ \mathbf{f} = \mathbf{f}_1 - \mathbf{f}_2$. Using (19), we have

$$\begin{aligned} &b(\mathbf{u}_1, \mathbf{u}_1, \mathbf{u}_1 - \mathbf{u}_2) - b(\mathbf{u}_2, \mathbf{u}_2, \mathbf{u}_1 - \mathbf{u}_2) = -b(\mathbf{u}_1, \mathbf{u}_1, \mathbf{u}_2) + b(\mathbf{u}_2, \mathbf{u}_1, \mathbf{u}_2) \\ &= -b(\mathbf{u}, \mathbf{u}_1, \mathbf{u}_2) = b(\mathbf{u}, \mathbf{u}_1, \mathbf{u}_1) - b(\mathbf{u}, \mathbf{u}_1, \mathbf{u}_2) = -b(\mathbf{u}, \mathbf{u}, \mathbf{u}_1) \end{aligned}$$

Replacing $\mathbf{v}$ by $\mathbf{u}$ in (26), we obtain

$$\frac{1}{2}\frac{d}{dt}|\mathbf{u}(t)|^2 + \mu\|\mathbf{u}\|^2 - b(\mathbf{u}, \mathbf{u}, \mathbf{u}_1) = \langle \mathbf{f}, \mathbf{u} \rangle \quad \forall t \in (0, T). \tag{27}$$

Using Poicare inequality (see [6]), we obtain

$$\langle \mathbf{f}, \mathbf{u} \rangle \le |\mathbf{f}||\mathbf{u}| \le \frac{1}{\lambda_1^{\frac{1}{2}}}|\mathbf{f}|\|\mathbf{u}\| \le \frac{1}{\mu\lambda_1}|\mathbf{f}|^2 + \frac{\mu}{4}\|\mathbf{u}\|^2. \tag{28}$$

And by the following inequality (see [4], page 13)

$$|b(\mathbf{u}, \mathbf{u}, \mathbf{u}_1)| \le k|\mathbf{u}|\|\mathbf{u}\|\|\mathbf{u}_1\|^{\frac{1}{2}}|A\mathbf{u}_1|^{\frac{1}{2}},$$

we obtain

$$|b(\mathbf{u}, \mathbf{u}, \mathbf{u}_1)| \le \frac{\mu}{4}\|\mathbf{u}\|^2 + \frac{k^2}{\mu}\|\mathbf{u}_1\||A\mathbf{u}_1||\mathbf{u}|^2. \tag{29}$$

By (27),(28),(29) we obtain, for $t \in [0, T]$,

$$\frac{d}{dt}|\mathbf{u}(t)|^2 + \mu\|\mathbf{u}\|^2 \le \frac{2}{\mu\lambda_1}|\mathbf{f}|^2 + \frac{2k^2}{\mu}\|\mathbf{u}_1\||A\mathbf{u}_1||\mathbf{u}|^2. \tag{30}$$

By **Gronwall's Lemma (B)** and (25), we deduce that, for $t \in [0, T]$,

$$|\mathbf{u}(t)|^2 \le \int_0^t \left(\frac{2}{\mu\lambda_1}|\mathbf{f}(\mathbf{s})|^2\right)\left(exp\int_s^t \frac{2k^2}{\mu}\|\mathbf{u}_1(\tau)\||A\mathbf{u}_1(\tau)|d\tau\right)ds,$$

then

$$\sup_{0\in[0,T]} |\mathbf{u}(t)|^2 \le K_3 \int_0^T |\mathbf{f}(\mathbf{s})|^2 ds, \tag{31}$$

where constant K_3 is dependent on $\mu,\ \|\mathbf{u}_0\|,\ c$.

Replacing $\mathbf{v}$ by $A\mathbf{u}$ in (26), we have, for $t \in (0, T)$,

$$\frac{1}{2}\frac{d}{dt}\|\mathbf{u}(t)\|^2 + \mu|A\mathbf{u}|^2 + b(\mathbf{u}_1, \mathbf{u}_1, A\mathbf{u}) - b(\mathbf{u}_2, \mathbf{u}_2, A\mathbf{u}) = \langle \mathbf{f}(t), A\mathbf{u} \rangle.$$

Since

$$\begin{aligned} &b(\mathbf{u}_1, \mathbf{u}_1, A\mathbf{u}) - b(\mathbf{u}_2, \mathbf{u}_2, A\mathbf{u}) \\ &= b(\mathbf{u}_1, \mathbf{u}_1, A\mathbf{u}) - b(\mathbf{u}_2, \mathbf{u}_1, A\mathbf{u}) + b(\mathbf{u}_2, \mathbf{u}_1, A\mathbf{u}) - b(\mathbf{u}_2, \mathbf{u}_2, A\mathbf{u}) \\ &= b(\mathbf{u}, \mathbf{u}_1, A\mathbf{u}) + b(\mathbf{u}_2, \mathbf{u}, A\mathbf{u}), \end{aligned}$$

then

$$\frac{1}{2}\frac{d}{dt}\|\mathbf{u}(t)\|^2 + \mu|A\mathbf{u}|^2 + b(\mathbf{u}, \mathbf{u}_1, A\mathbf{u}) + b(\mathbf{u}_2, \mathbf{u}, A\mathbf{u}) = \langle \mathbf{f}(t), A\mathbf{u} \rangle. \tag{32}$$

And using (20), we obtain the following equality (see [6], page 135)

$$\frac{1}{2}\frac{d}{dt}\|\mathbf{u}(t)\|^2 + \mu|A\mathbf{u}|^2 = b(\mathbf{u},\mathbf{u},A\mathbf{u}_1) + \langle \mathbf{f}(t), A\mathbf{u}\rangle. \tag{33}$$

Notice that

$$\langle \mathbf{f}, A\mathbf{u}\rangle \leq |\mathbf{f}||A\mathbf{u}| \leq \frac{1}{\mu}|\mathbf{f}|^2 + \frac{\mu}{4}|A\mathbf{u}|^2. \tag{34}$$

And by the following inequality (see [4], page 13)

$$|b(\mathbf{u},\mathbf{u},A\mathbf{u}_1)| \leq k|\mathbf{u}|^{\frac{1}{2}}|A\mathbf{u}|^{\frac{1}{2}}\|\mathbf{u}\||A\mathbf{u}_1|,$$

we obtain

$$|b(\mathbf{u},\mathbf{u},A\mathbf{u}_1)| \leq \frac{\mu}{4}|A\mathbf{u}|^2 + \frac{1}{2}|A\mathbf{u}_1|^2\|\mathbf{u}\|^2 + \frac{k^4}{4\mu}|\mathbf{u}|^2. \tag{35}$$

By (33),(34),(35), we deduce that

$$\frac{d}{dt}\|\mathbf{u}(t)\|^2 + \mu|A\mathbf{u}|^2 \leq |A\mathbf{u}_1|^2\|\mathbf{u}\|^2 + \frac{k^4}{2\mu}|\mathbf{u}|^2 + \frac{2}{\mu}|\mathbf{f}|^2. \tag{36}$$

By **Gronwall's Lemma (B)**, we obtain, for $t \in [0,T]$,

$$\|\mathbf{u}(t)\|^2 \leq (\int_0^t (\frac{k^4}{2\mu}|\mathbf{u}(s)| + \frac{2}{\mu}|\mathbf{f}(s)|^2)exp(\int_s^t |A\mathbf{u}_1(\tau)|^2 d\tau)ds.$$

Then

$$\sup_{t\in[0,T]} \|\mathbf{u}(t)\|^2 \leq exp(\int_0^T |A\mathbf{u}_1(\tau)|^2 d\tau)(\int_0^T (\frac{k^4}{2\mu}|\mathbf{u}(s)| + \frac{2}{\mu}|\mathbf{f}(s)|^2)ds).$$

By (25), (31), there exists constant K_4 dependent on $\mu,\ \|\mathbf{u}_0\|,\ c,\ \widetilde{T}$, such that

$$\sup_{t\in[0,T]} \|\mathbf{u}(t)\|^2 \leq K_4 \int_0^T |\mathbf{f}(\mathbf{s})|^2 ds. \tag{37}$$

Then by integration in t of (36) from 0 to T, after dropping unnecessary terms we obtain

$$\int_0^T |A\mathbf{u}(t)|^2 dt \leq K_5 \int_0^T |\mathbf{f}(\mathbf{t})|^2 dt,$$

that is

$$\int_0^T |A\mathbf{u}_1(t) - A\mathbf{u}_2(t)|^2 dt \leq K_5 \int_0^T |\mathbf{f}_1(t) - \mathbf{f}_2(t)|^2 dt, \tag{38}$$

where constant K_5 is dependent on $\mu,\ \|\mathbf{u}_0\|,\ c,\ \widetilde{T}$. Then using (18) with s=1, we obtain

$$\int_0^T \|\mathbf{u}_1(t) - \mathbf{u}_2(t)\|_2^2 dt \leq c_1 \int_0^T |\mathbf{f}_1(t) - \mathbf{f}_2(t)|^2 dt$$

where constant c_1 is dependent on $\mu,\ \|\mathbf{u}_0\|,\ c,\ \widetilde{T}$.□

4. Variational Formulation of the Problem

$\mathbf{C}[0, L]$ denotes the space of continuous functions from $[0, L]$ into R^2 and is equipped with the norm

$$\|\mathbf{x}\|_c = \max_{0\le s\le L} |\mathbf{x}(s)|.$$

In what follows, we always suppose that $\widetilde{T} > 0$ is a given constant, and constant $T \in (0, \widetilde{T}]$. Let

$$G = \mathrm{L}([0,T]; \mathbf{C}[0,L]); \quad E = \{\mathbf{x} \in \mathbf{C}[0,L] : \mathbf{x}(0) = \mathbf{x}(L)\};$$
$$\mathrm{F} = \{\mathbf{x} \in E | \mathbf{x}(s) \in \Omega,\ 0 \le s \le L\},$$

where

$$\Omega = (-\frac{L_1}{2}, \frac{L_1}{2}) \times (-\frac{L_2}{2}, \frac{L_2}{2}).$$

Notice that

$$\overline{\mathrm{F}} = \{\mathbf{x} \in E | \mathbf{x}(s) \in \overline{\Omega},\ 0 \le s \le L\}.$$

For $\mathbf{x} \in G, \mathbf{x}(\cdot, t) \in \mathbf{C}[0, L],\ \forall\, t \in [0, T], \mathbf{x}(\cdot, t)$ is simply denoted by $\mathbf{x}(t)$.

Given $\mathbf{X}_0 \in \mathrm{F}$, let

$$\mathbb{X} = \{\mathbf{X} \in \mathrm{L}([0,T]; E) : \mathbf{X}(0) = \mathbf{X}_0\}.$$

$\mathbb{X}$ is a closed subset of Banach space G.

Given $\delta_h : R^2 \to R^2,\ \mathbf{f} : [0, L] \times [0, \widetilde{T}] \to R^2$, we assume that they satisfy *Condition* ***A***:

$$\delta_h \in \mathbf{H}_0^2(R^2), \quad \mathbf{f} \in \mathbf{C}([0, L] \times [0, \widetilde{T}]).$$

In this paper, we let Dirac delta function δ be approximated by differentiable function δ_h. In fact, some authors do so when performing the numerical simulation in order to get the approximate solution of the problem (1)-(7). For example, δ_h in [9] is chosen as follows

$$\delta_h(\mathbf{x}) = d_h(x) d_h(y), \qquad \forall \mathbf{x} \in \mathbf{R}^2$$

where

$$d_h(z) = \begin{cases} \dfrac{0.25}{h}\left[1 + \cos(\dfrac{\pi z}{2h})\right] & |z| \le 2h, \\ 0, & |z| > 2h. \end{cases}$$

It is clear that $\delta_h \in \mathbf{H}_0^2(R^2)$.

According to the theory and notations in **Section 2, Section3**, we introduce the corresponding variational formulation of the equations (10)-(16):

Problem 1. Given $\mathbf{u}_0 \in V,\ \mathbf{X}_0 \in \mathrm{F}$ and $\delta_h : R^2 \to R^2,\ \mathbf{f} : [0, L] \times [0, \widetilde{T}] \to R^2$ satisfying *Condition* ***A***, find $\mathbf{u} \in \mathbf{L}^2(0, T; D(A)) \cap \mathbf{L}^\infty(0, T; \dot{V})$, and $\mathbf{X} : [0, L] \times [0, T] \to \overline{\Omega}$, such that

$$\begin{aligned}
&\frac{d}{dt}(\mathbf{u}(t), \mathbf{v}) + \mu((\mathbf{u}, \mathbf{v})) + b(\mathbf{u}, \mathbf{u}, \mathbf{v}) = \langle \hat{\hat{\mathbf{F}}}(t), \mathbf{v}\rangle && \forall\, \mathbf{v} \in V,\ t \in (0, T),\\
&\frac{\partial \mathbf{X}}{\partial t}(s, t) = \mathbf{u}(\mathbf{X}(s, t) - I_u(t), t) + M_u(t) && \forall\, s \in [0, L], t \in (0, T),\\
&\mathbf{u}(\mathbf{x}, 0) = \hat{\mathbf{u}}_0(\mathbf{x}) && \forall\, \mathbf{x} \in \Omega,\\
&\mathbf{X}(s, 0) = \mathbf{X}_0(s) && \forall s \in [0, L],\\
&\mathbf{X}(0, t) = \mathbf{X}(L, t) && \forall t \in (0, T),
\end{aligned}$$

where

$$\hat{\mathbf{u}}_0(\mathbf{x}) = \mathbf{u}_0(\mathbf{x}) - \frac{1}{|\Omega|}\int_\Omega \mathbf{u}_0(\mathbf{x})d\mathbf{x},$$
$$\hat{\hat{\mathbf{F}}}(\mathbf{x},t)$$
$$= \int_0^L \mathbf{f}(s,t)\delta_h(\mathbf{x}+I_u(t)-\mathbf{X}(s,t))ds - \frac{1}{|\Omega|}\int_0^L \mathbf{f}(s,t)(\int_\Omega \delta_h(\mathbf{x}-\mathbf{X}(s,t))dx)ds.$$

5. Conclusion

In this section we shall prove the existence of the solution of **Problem 1** via a fixed point argument. We define an operator $\mathbf{T}$ on $\mathbb{X}$ as follows.

Given $\mathbf{u}_0 \in V$, $\overline{\mathbf{X}} \in \mathbb{X}$. $\forall x \in \Omega, t \in (0,T)$, let

$$\mathbf{F}_{\overline{X}}(\mathbf{x},t) = \int_0^L \mathbf{f}(s,t)\delta_h(\mathbf{x}-\overline{\mathbf{X}}(s,t))ds.$$

Hence

$$\hat{\hat{\mathbf{F}}}_{\overline{X}}(\mathbf{x},t) = \int_0^L \mathbf{f}(s,t)\delta_h(\mathbf{x}+I_{\overline{X}}(t)-\overline{\mathbf{X}}(s,t))ds - \frac{1}{|\Omega|}\int_0^L \mathbf{f}(s,t)(\int_\Omega \delta_h(x-\overline{\mathbf{X}}(s,t))dx)ds$$

where, according to (9),

$$I_{\overline{X}}(t) = \frac{t}{|\Omega|}\int_\Omega \mathbf{u}_0(\mathbf{x})d\mathbf{x} + \int_0^t\int_0^s M_{\mathbf{F}_{\overline{X}}}(\tau)d\tau ds, \tag{39}$$

$$M_{\mathbf{F}_{\overline{X}}}(t) = \frac{1}{|\Omega|}\int_\Omega \mathbf{F}_{\overline{X}}(\mathbf{x},t)d\mathbf{x} = \frac{1}{|\Omega|}\int_0^L \mathbf{f}(s,t)(\int_\Omega \delta_h(x-\overline{\mathbf{X}}(s,t))dx)ds. \tag{40}$$

If *Condition* **A** is satisfied, we can verify that $\hat{\hat{\mathbf{F}}}_{\overline{X}} \in \mathbf{L}^\infty(0,T;\dot{\mathbf{H}}^2_{per}(\Omega))$.

Then let $\overline{\mathbf{u}}$ (also denoted by $\mathbf{u}_{\overline{\mathbf{X}}}$) be the solution $\mathbf{u} \in \mathbf{L}^\infty(0,T;\dot{V})$ of

$$\begin{cases} \frac{d}{dt}(\mathbf{u}(t),\mathbf{v}) + \mu(\nabla\mathbf{u},\nabla\mathbf{v}) = \langle\hat{\hat{\mathbf{F}}}_{\overline{X}}(t),\mathbf{v}\rangle \quad \forall\mathbf{v}\in V,\ t\in(0,T), \\ \mathbf{u}(0) = \hat{\mathbf{u}}_0. \end{cases} \tag{41}$$

Finally, let $\overline{\overline{\mathbf{X}}}$ be the solution $\mathbf{X} \in \mathbb{X}$ of

$$\begin{cases} \mathbf{X}'(t) = \overline{\mathbf{u}}(\mathbf{X}(s,t) - I_{\overline{X}}(t),t) + M_{u\overline{X}}(t)) \quad \forall t\in(0,T), \\ \mathbf{X}(0) = \mathbf{X}_0, \end{cases} \tag{42}$$

where

$$M_{u\overline{X}}(t)) = \frac{1}{|\Omega|}\int_\Omega \mathbf{u}_0(\mathbf{x})d\mathbf{x} + \int_0^t M_{\mathbf{F}_{\overline{X}}}(\tau)d\tau, \tag{43}$$

and let $\mathbf{T}(\overline{\mathbf{X}}) = \overline{\overline{\mathbf{X}}}$.

We recall that the definition of $\mathbb{X}$ is

$$\mathbb{X} = \{\mathbf{X} \in \mathrm{L}([0,T];E) : \mathbf{X}(0) = \mathbf{X}_0\}.$$

For $t \in [0,T]$, $\mathbf{X}(t) = \mathbf{X}(\cdot,t) \in E$. (42) is an ordinary equation in the Banach space E and equivalent to

$$\begin{cases} \dfrac{\partial \mathbf{X}}{\partial t}(s,t) = \overline{\mathbf{u}}(\mathbf{X}(s,t) - I_{\overline{X}}(t),t) + M_{u\overline{X}}(t)) & \forall s \in [0,L], \forall t \in (0,T), \\ \mathbf{X}(s,0) = \mathbf{X}_0(s) & \forall s \in [0,L], \\ \mathbf{X}(0,t) = \mathbf{X}(L,t) & \forall t \in (0,T). \end{cases}$$

We observe that, if $\mathbf{T}$ has a fixed point $\widetilde{\mathbf{X}}$ in $\mathbb{X}$, then this fixed point and the solution $\widetilde{\mathbf{u}}$ to (41) corresponding to $\overline{\mathbf{X}} = \widetilde{\mathbf{X}}$, give the solutions to **Problem 1.**

The following lemma guarantees the existence of the solution to (42):

Lemma 3. *(see [1]) Let $x : [t_0, t_0 + a] \to Y$ be a mapping into the Banach-space Y. Consider the initial value problem*

$$\frac{dx}{dt} = f(x,t), \quad x(t_0) = x_0. \tag{44}$$

Let $Q = [t_0, t_0 + a] \times Y$. Suppose $f : Q \to Y$ is continuous and satisfies

$$\|f(t,x) - f(t,y)\| \leq L\|x - y\|, \qquad for\ all \quad (t,x),(t,y) \in Q,\ and\ fixed\ L \geq 0.$$

Then initial problem (44) has exactly one continuously differential solution on $[t_0, t_0 + a]$ for each initial value $x_0 \in Y$. □

Theorem 1. *Assume that $T \in (0, \widetilde{T}]$, $\mathbf{u}_0 \in \mathbf{H}^3_{per}(\Omega) \cap H$, $\mathbf{X}_0 \in$ F and Condition **A** holds.*
***(a)** For given $\overline{\mathbf{X}} \in \mathbb{X}$, (41) has a unique solution $\overline{\mathbf{u}} \in \mathrm{L}([0,T], \mathbf{C}^1(\overline{\Omega}))$. Furthermore there exist constants c_1, c_2, c_3 dependent on μ, $\widetilde{T}$, δ_h, $|\mathbf{f}\|_{\mathbf{C}([0,L]\times[0,\widetilde{T}])}$, $\|\mathbf{u}_0\|_{V_3}$, such that*

$$\sup_{\substack{z\in\overline{\Omega}\\ 0\leq t\leq T}} |\overline{\mathbf{u}}(z,t)| \leq c_1 < \infty, \tag{45}$$

$$\sup_{\substack{z\in\overline{\Omega}\\ 0\leq t\leq T}} |D\overline{\mathbf{u}}(z,t)| \leq c_2 < \infty, \tag{46}$$

$$\sup_{\substack{z\in\overline{\Omega}\\ 0\leq t\leq T}} |\overline{\mathbf{u}}(z,t) + M_{u\overline{X}}(t)| \leq c_3 < \infty.$$

***(b)** Suppose that $\overline{\mathbf{u}}$ is the solution of (41). Then (42) has a unique solution $\overline{\overline{\mathbf{X}}} \in \mathbb{X}$. Furthermore, if $0 < T < \min\{\frac{L_m - \|\mathbf{X}_0\|_c}{c_3}, \widetilde{T}\}$ holds, where $L_m = min\{L_1, L_2\}$, then $\overline{\overline{\mathbf{X}}}(s,t) \in \Omega$, for all $(s,t) \in [0,L] \times [0,\widetilde{T}]$.*

Proof:

(a) Since $\mathbf{u}_0 \in \mathbf{H}^3_{per}(\Omega) \cap H$, then $\hat{\mathbf{u}}_0 \in V_3$. Since *Condition **A*** is satisfied, then $\hat{\hat{\mathbf{F}}}_{\overline{X}} \in \mathbf{L}^\infty(0,T;\dot{\mathbf{H}}^2_{per}(\Omega))$ and $P_L\hat{\hat{\mathbf{F}}}_{\overline{X}} \in \mathbf{L}^\infty(0,T;V_2)$. We can verify that there exists constant C dependent on δ_h, such that

$$\|P_L\hat{\hat{\mathbf{F}}}_{\overline{X}}\|_{\mathbf{L}^\infty(0,T;V_2)} \leq \|\hat{\hat{\mathbf{F}}}_{\overline{X}}\|_{\mathbf{L}^\infty(0,T;\dot{\mathbf{H}}^2_{per}(\Omega))} \leq C\|\mathbf{f}\|_{\mathbf{C}([0,L]\times[0,\widetilde{T}])}. \tag{47}$$

By **Lemma 1 (ii)** with $r = 3$, there exists $\overline{\mathbf{u}}$ belonging to $\mathrm{C}([0,T];V_3)$ satisfying

$$\begin{cases} \dfrac{d}{dt}(\mathbf{u}(t),\mathbf{v}) + \mu(\nabla\mathbf{u},\nabla\mathbf{v}) = \langle P_L\hat{\hat{\mathbf{F}}}_{\overline{X}}(t),\mathbf{v}\rangle \quad \forall\mathbf{v}\in V,\ t\in(0,T), \\ \mathbf{u}(0) = \hat{\hat{\mathbf{u}}}_0. \end{cases}$$

Since P_L is the orthogonal projector onto H in $\mathbf{L}^2(\Omega)$, we have

$$\langle P_L\hat{\hat{\mathbf{F}}}_{\overline{X}}(t),\mathbf{v}\rangle = \langle\hat{\hat{\mathbf{F}}}_{\overline{X}}(t),\mathbf{v}\rangle \quad \forall\mathbf{v}\in V.$$

Then $\overline{\mathbf{u}}$ is the solution of (41). Since $\overline{\mathbf{u}}$ belonging to $\mathrm{C}([0,T];V_3)$ and V_3 is a closed subspace of $\dot{\mathbf{H}}^3_{per}(\Omega)$, we obtain that $\overline{\mathbf{u}}\in \mathrm{L}([0,T],\mathbf{C}^1(\overline{\Omega}))$ due to (18) and Sobolev embedding theorem (see [7]). Furthermore by (24) and (47) there exist constants c_1, c_2, c_3 dependent on μ, $\widetilde{T}$, δ_h, $|\mathbf{f}\|_{\mathbf{C}([0,L]\times[0,\widetilde{T}])}$, $\|\mathbf{u}_0\|_3$, such that

$$\sup_{\substack{z\in\overline{\Omega}\\0\le t\le T}} |\overline{\mathbf{u}}(z,t)| \le c_1 < \infty,$$

$$\sup_{\substack{z\in\overline{\Omega}\\0\le t\le T}} |D\overline{\mathbf{u}}(z,t)| \le c_2 < \infty,$$

$$\sup_{\substack{z\in\overline{\Omega}\\0\le t\le T}} |\overline{\mathbf{u}}(z,t) + M_{u\overline{X}}(t)| \le c_3 < \infty.$$

(b) We shall prove the conclusion by **Lemma 3** with $Y = E, f = \overline{\mathbf{u}}, x_0 = \mathbf{X}_0, t_0 = 0$. Below we verify that the conditions in **Lemma 3** hold.

First we prove that $\overline{\mathbf{u}} : E\times[0,T]\to E$ is continuous. Given $(x_0,t_0)\in E\times[0,T]$, $(x,t)\in E\times[0,T]$. It is easy to see $\overline{\mathbf{u}}(x,t)\in E$, $\overline{\mathbf{u}}(x_0,t_0)\in E$. We have

$$\begin{aligned}\|\overline{\mathbf{u}}(x,t) - \overline{\mathbf{u}}(x_0,t_0)\|_c &= \sup_{0\le s\le L} |\overline{\mathbf{u}}(x(s),t) - \overline{\mathbf{u}}(x_0(s),t_0)| \\ \le \sup_{z\in\overline{\Omega}} |\overline{\mathbf{u}}(z,t) - \overline{\mathbf{u}}(z,t_0)| &+ \sup_{z\in\overline{\Omega}} |\overline{\mathbf{u}}(z,t_0) - \overline{\mathbf{u}}(z_0,t_0)|.\end{aligned}$$

Then $\overline{\mathbf{u}}(x,t)$ is continuous at (x_0,t_0) due to $\overline{\mathbf{u}}\in \mathrm{L}([0,T],\mathbf{C}^1(\overline{\Omega}))$. For $t\in[0,T], x,y\in E$, by (46) we have

$$\begin{aligned}\|\overline{\mathbf{u}}(x,t) - \overline{\mathbf{u}}(y,t)\|_c &= \sup_{0\le s\le L} |\overline{\mathbf{u}}(x(s),t) - \overline{\mathbf{u}}(y(s),t)| \\ \le \sup_{\substack{z\in\overline{\Omega}\\0\le t\le T}} |D\overline{\mathbf{u}}(z,t)| \|x-y\|_c &\le c_2\|x-y\|_c.\end{aligned}$$

Hence the conditions in **Lemma 3** hold. (42) has a unique solution $\overline{\overline{\mathbf{X}}}\in\mathbb{X}$.

By (42), we have

$$\overline{\overline{\mathbf{X}}}(t) = \mathbf{X}_0 + \int_0^t (\overline{\mathbf{u}}(\overline{\overline{\mathbf{X}}}(s,\tau) - I_{\overline{X}}(\tau),\tau) + M_{u\overline{X}}(\tau))d\tau.$$

Notice that $\mathbf{X}_0 \in \mathrm{F}$, then $L_m - \|\mathbf{X}_0\|_c > 0$. If $0 < T < \frac{L_m - \|\mathbf{X}_0\|_c}{c_3}$ holds, for $t \in [0, T]$, we have

$$\|\overline{\overline{\mathbf{X}}}(t)\|_c \leq \|\mathbf{X}_0\|_c + t \sup_{\substack{z\in\overline{\Omega} \\ 0\leq\tau\leq T}} |\overline{\mathbf{u}}(z,\tau) + M_{u\overline{X}}(\tau)|$$
$$< \|\mathbf{X}_0\|_c + c_3 \frac{L_m - \|\mathbf{X}_0\|_c}{c_3} = L_m.$$

Then $\overline{\overline{\mathbf{X}}}(t) \in \mathrm{F}$. That is, if $0 < T < \frac{L_m - \|\mathbf{X}_0\|_c}{c_3}$ holds, for all $(s,t) \in [0, L] \times [0, T]$, $\overline{\overline{\mathbf{X}}}(s,t) \in \Omega$. $\square$

Follow the above argument, we can prove the main conclusion:

Theorem 2. *Assume that* $\mathbf{u}_0 \in \mathbf{H}^3_{per}(\Omega) \cap H$, $\mathbf{X}_0 \in \mathrm{F}$ *and Condition* **A** *holds. Then there exists* $T'' \in (0, \widetilde{T}]$ *dependent on* $\mu,\ \widetilde{T}, \delta_h, \mathbf{f}, \mathbf{u}_0, L_m, \|\mathbf{X}_0\|_c$, *such that if* $T \in (0, T'')$, ***Problem 1*** *has a unique pair of solutions* $\widetilde{\mathbf{u}}$ *and* $\widetilde{\mathbf{X}}$. *And* $\widetilde{\mathbf{X}}$ *may be solved by successive approximations*

$$\mathbf{Y}_{n+1} = \mathbf{T}\mathbf{Y}_n \quad \forall \mathbf{Y_0} \in B \quad n = 0, 1, 2, \dots,$$

where

$$B = \{\mathbf{X} \in \mathbb{X} \mid \mathbf{X}(t) \in \overline{\mathrm{F}} \quad \forall t \in [0, T]\}.$$

Proof: Below we take T satisfying $0 < T < \min\{\frac{L_m - \|\mathbf{X}_0\|_c}{c_3},\ \widetilde{T}\}$. By **Theorem 1 (b)**, we can observe that $\mathbf{T}$ maps B into itself. The set B is a bounded closed convex subset of $\mathbb{X}$. We will prove that $\mathbf{T} : B \to B$ is a contractive map, and then by the Banach Fixed-Point Theorem (see [1]), $\mathbf{T}$ has a unique fixed point.

Arbitrarily given two elements $\mathbf{X}, \mathbf{Y} \in \mathbf{B}$, we denote the corresponding solutions of (41) by $\mathbf{u_X}$ and $\mathbf{u_Y}$ respectively. Then $\mathbf{u}_X - \mathbf{u}_Y$ solves the following equation

$$\begin{cases} \dfrac{d}{dt}((\mathbf{u_X} - \mathbf{u_Y})(t), \mathbf{v}) + \mu(\nabla(\mathbf{u_X} - \mathbf{u_Y})(t), \nabla\mathbf{v}) = \langle \hat{\hat{\mathbf{F}}}_\mathbf{X}(t) - \hat{\hat{\mathbf{F}}}_\mathbf{Y}(t), \mathbf{v}\rangle, \\ \hfill \forall v \in V,\ t \in (0, T), \\ (\mathbf{u_X} - \mathbf{u_Y})(0) = 0. \end{cases}$$

By the definitions of $\hat{\hat{\mathbf{F}}}_\mathbf{X},\ \hat{\hat{\mathbf{F}}}_\mathbf{Y}$ and using *Condition* **A**, we can deduce that there exists a constant k dependent on $\mathbf{f},\ \delta_h$, such that

$$\int_0^T \|\hat{\hat{\mathbf{F}}}_X(t)\|_0^2 dt \leq k, \quad \int_0^T \|\hat{\hat{\mathbf{F}}}_Y(t)\|_0^2 dt \leq k.$$

Then according to **Lemma 2**, we have

$$\int_0^T \|\mathbf{u}_X(t) - \mathbf{u}_Y(t)\|_2^2 dt \leq c_0 \int_0^T \|\hat{\hat{\mathbf{F}}}_X(t) - \hat{\hat{\mathbf{F}}}_Y(t)\|_0^2 dt \tag{48}$$

where constant c_0 is dependent on $\mu,\ \|\mathbf{u}_0\|,\ \widetilde{T},\ \mathbf{f},\ \delta_h$.

By the definitions of $\hat{\hat{\mathbf{F}}}_\mathbf{X},\ \hat{\hat{\mathbf{F}}}_\mathbf{Y}$ and using *Condition* **A**, we can deduce that there exists a constant l_1 dependent on $\mathbf{f},\ \delta_h,\ \widetilde{T}$, such that, for each $t \in (0, T)$,

$$\|\hat{\hat{\mathbf{F}}}_\mathbf{X}(t) - \hat{\hat{\mathbf{F}}}_\mathbf{Y}(t)\|_0^2 \leq l_1 \sup_{0\leq s\leq L} |\mathbf{Y}(s,t) - \mathbf{X}(s,t)|^2 = l_1 \|\mathbf{Y}(t) - \mathbf{X}(t)\|_c^2. \tag{49}$$

By (48) and (49),

$$\int_0^T \|\mathbf{u}_X(t) - \mathbf{u}_Y(t)\|_2^2 dt \leq c_0 l_1 \int_0^T \|\mathbf{X}(t) - \mathbf{Y}(t)\|_c^2 dt. \tag{50}$$

Let $\widetilde{X} = \mathbf{TX}$ and $\widetilde{Y} = \mathbf{TY}$. According to (42), we have

$$\begin{cases} \widetilde{\mathbf{X}}'(t) = \mathbf{u}_X(\widetilde{\mathbf{X}}(t) - I_X(t), t) + M_{uX}(t) \quad \forall t \in (0, T), \\ \widetilde{\mathbf{X}}(0) = \mathbf{X}_0, \end{cases} \tag{51}$$

$$\begin{cases} \widetilde{\mathbf{Y}}'(t) = \mathbf{u}_Y(\widetilde{\mathbf{Y}}(t) - I_Y(t), t) + M_{uY}(t) \quad \forall t \in (0, T), \\ \widetilde{\mathbf{Y}}(0) = \mathbf{X}_0, \end{cases} \tag{52}$$

Subtracting (51) from (52) we obtain

$$\begin{aligned} &\widetilde{\mathbf{X}}'(t) - \widetilde{\mathbf{Y}}'(t) \\ &= \mathbf{u}_\mathbf{X}(\widetilde{\mathbf{X}}(t) - I_X(t), t) - \mathbf{u}_\mathbf{Y}(\widetilde{\mathbf{Y}}(t) - I_Y(t), t) + M_{uY}(t) - M_{uX}(t) \\ &= \mathbf{u}_\mathbf{X}(\widetilde{\mathbf{X}}(t) - I_X(t), t) - \mathbf{u}_\mathbf{Y}(\widetilde{\mathbf{X}}(t) - I_X(t), t) \\ &\quad + \mathbf{u}_\mathbf{Y}(\widetilde{\mathbf{X}}(t) - I_X(t), t) - \mathbf{u}_\mathbf{Y}(\widetilde{\mathbf{Y}}(t) - I_Y(t), t) \\ &\quad + M_{uY}(t) - M_{uX}(t) \end{aligned}$$

Integrating from 0 to t in the two sides of above equation, we have

$$\begin{aligned} \widetilde{\mathbf{X}}(t) - \widetilde{\mathbf{Y}}(t) &= \int_0^t [\mathbf{u}_\mathbf{X}(\widetilde{\mathbf{X}}(\tau) - I_X(\tau), \tau) - \mathbf{u}_\mathbf{Y}(\widetilde{\mathbf{X}}(\tau) - I_X(\tau), \tau)] d\tau \\ &+ \int_0^t [\mathbf{u}_\mathbf{Y}(\widetilde{\mathbf{X}}(\tau) - I_X(\tau), \tau) - \mathbf{u}_\mathbf{Y}(\widetilde{\mathbf{Y}}(\tau) - I_Y(\tau), \tau)] d\tau + \int_0^t [M_{uY}(\tau) - M_{uX}(\tau)] d\tau. \end{aligned}$$

Hence

$$\begin{aligned} \|\widetilde{\mathbf{X}}(t) - \widetilde{\mathbf{Y}}(t)\|_c &\leq \int_0^t \|\mathbf{u}_\mathbf{X}(\widetilde{\mathbf{X}}(\tau) - I_X(\tau), \tau) - \mathbf{u}_\mathbf{Y}(\widetilde{\mathbf{X}}(\tau) - I_X(\tau), \tau)\|_c d\tau \\ &+ \int_0^t \|\mathbf{u}_\mathbf{Y}(\widetilde{\mathbf{X}}(\tau) - I_X(\tau), \tau) - \mathbf{u}_\mathbf{Y}(\widetilde{\mathbf{Y}}(\tau) - I_Y(\tau), \tau)\|_c d\tau \\ &+ \int_0^t |M_{uY}(\tau) - M_{uX}(\tau)| d\tau. \end{aligned} \tag{53}$$

By Sobolev embedding theorem, there exists a constant l_2 such that

$$\begin{aligned} &\|\mathbf{u}_\mathbf{X}(\widetilde{\mathbf{X}}(\tau) - I_X(\tau), \tau) - \mathbf{u}_\mathbf{Y}(\widetilde{\mathbf{X}}(\tau) - I_X(\tau), \tau)\|_c \\ &= \sup_{0 \leq s \leq L} |\mathbf{u}_\mathbf{X}(\widetilde{\mathbf{X}}(s, \tau) - I_X(\tau), \tau) - \mathbf{u}_\mathbf{Y}(\widetilde{\mathbf{X}}(s, \tau) - I_X(\tau), \tau)| \\ &\leq \sup_{z \in \overline{\Omega}} |\mathbf{u}_\mathbf{X}(z, \tau) - \mathbf{u}_\mathbf{Y}(z, \tau)| \\ &\leq l_2 \|\mathbf{u}_\mathbf{X}(\tau) - \mathbf{u}_\mathbf{Y}(\tau)\|_2. \end{aligned} \tag{54}$$

By (50) and (54), there exists a constant l_3, dependent on μ, $\mathbf{f}, \delta_h, \widetilde{T}$, $\|\mathbf{u}_0\|$, such that for each $t \in [0,T]$,

$$\begin{aligned}
&\textstyle\int_0^t \|\mathbf{u}_{\mathbf{X}}(\widetilde{\mathbf{X}}(\tau) - I_X(\tau), \tau) - \mathbf{u}_{\mathbf{Y}}(\widetilde{\mathbf{X}}(\tau) - I_X(\tau), \tau)\|_c d\tau \\
&\le l_2 \textstyle\int_0^t \|\mathbf{u}_{\mathbf{X}}(\tau) - \mathbf{u}_{\mathbf{Y}}(\tau)\|_2 d\tau \\
&\le l_2 t^{1/2} \left(\textstyle\int_0^T \|\mathbf{u}_{\mathbf{X}}(\tau) - \mathbf{u}_{\mathbf{Y}}(\tau)\|_2^2 d\tau\right)^{1/2} \\
&\le l_3 t^{1/2} \left(\textstyle\int_0^T \|\mathbf{X}(t) - \mathbf{Y}(t)\|_c^2 dt\right)^{\frac{1}{2}} \\
&\le l_3 T \sup_{0\le t\le T} \|\mathbf{X}(t) - \mathbf{Y}(t)\|_c.
\end{aligned} \tag{55}$$

By (46), there exists c_2 dependent on μ, $\widetilde{T}, \delta_h, \mathbf{f}, \mathbf{u}_0$, such that

$$\sup_{\substack{z\in\overline{\Omega} \\ 0\le t\le T}} |D\mathbf{u}_{\mathbf{Y}}(z,t)| \le c_2.$$

Hence for $\tau \in [0,T]$,

$$\begin{aligned}
&\|\mathbf{u}_{\mathbf{Y}}(\widetilde{\mathbf{X}}(\tau) - I_X(\tau), \tau) - \mathbf{u}_{\mathbf{Y}}(\widetilde{\mathbf{Y}}(\tau) - I_Y(\tau), \tau)\|_c \\
&= \sup_{0\le s\le L} |\mathbf{u}_{\mathbf{Y}}(\widetilde{\mathbf{X}}(s,\tau) - I_X(\tau), \tau) - \mathbf{u}_{\mathbf{Y}}(\widetilde{\mathbf{Y}}(s,\tau) - I_Y(\tau), \tau)| \\
&\le \sup_{\substack{z\in\overline{\Omega} \\ 0\le \tau\le T}} |D\mathbf{u}_{\mathbf{Y}}((z,\tau)|(\|\widetilde{\mathbf{X}}(\tau) - \widetilde{\mathbf{Y}}(\tau)\|_c + |I_Y(\tau) - I_X(\tau)|) \\
&\le c_2\|\widetilde{\mathbf{X}}(\tau) - \widetilde{\mathbf{Y}}(\tau)\|_c + c_2|I_Y(\tau) - I_X(\tau)|.
\end{aligned} \tag{56}$$

By (53), (55) and (56), for $t \in [0,T]$, we have

$$\begin{aligned}
\|\widetilde{\mathbf{X}}(t) - \widetilde{\mathbf{Y}}(t)\| \le l_3 T \sup_{0\le t\le T} \|\mathbf{X}(t) - \mathbf{Y}(t)\|_c + c_2 \int_0^t \|\widetilde{\mathbf{X}}(\tau) - \widetilde{\mathbf{Y}}(\tau)\| d\tau \\
+ c_2 \int_0^t |I_Y(\tau) - I_X(\tau)| d\tau + \int_0^t |M_{uY}(\tau) - M_{uX}(\tau)| d\tau.
\end{aligned}$$

By (39, (40),(43) and using *Condition* **A**, we can deduce that there exists constant l_4 dependent on $\mathbf{f}$, $\delta_h, \widetilde{T}$, such that, for $\tau \in [0,T]$,

$$c_2|I_Y(\tau) - I_X(\tau)| + |M_{uY}(\tau) - M_{uX}(\tau)| \le l_4 \sup_{0\le t\le T} \|\mathbf{X}(t) - \mathbf{Y}(t)\|_c.$$

Then for $t \in [0,T]$,

$$c_2 \int_0^t |I_Y(\tau) - I_X(\tau)| d\tau + \int_0^t |M_{uY}(\tau) - M_{uX}(\tau)| d\tau \le l_4 T \|\mathbf{X}(t) - \mathbf{Y}(t)\|_c.$$

Hence there exists constant l_5 dependent on μ, $\mathbf{f}$, $\delta_h, \widetilde{T}$, $\mathbf{u}_0$, such that

$$\|\widetilde{\mathbf{X}}(t) - \widetilde{\mathbf{Y}}(t)\| \le l_5 T \sup_{0\le t\le T} \|\mathbf{X}(t) - \mathbf{Y}(t)\|_c + c_2 \int_0^t \|\widetilde{\mathbf{X}}(\tau) - \widetilde{\mathbf{Y}}(\tau)\| d\tau.$$

By **Gronwall's Lemma (A)**,

$$\|\widetilde{\mathbf{X}}(t) - \widetilde{\mathbf{Y}}(t)\| \leq l_5 T e^{c_2 T} \sup_{0\leq t\leq T} \|\mathbf{X}(t) - \mathbf{Y}(t)\|_c. \tag{57}$$

The inequality (57) implies for each constant $l_6 \in (0,1)$, there exists T' dependent on μ, $\widetilde{T}, \delta_h, \mathbf{f}, \mathbf{u}_0$, such that when $0 < T \leq T'$,

$$l_5 T e^{c_2 T} \leq l_5 T e^{c_2 \widetilde{T}} \leq l_6,$$

$$\|\widetilde{\mathbf{X}}(t) - \widetilde{\mathbf{Y}}(t)\| \leq l_6 \sup_{0\leq t\leq T} \|\mathbf{X}(t) - \mathbf{Y}(t)\|_c \qquad 0 \leq t \leq T.$$

Let $\|\mathbf{x}\|_C$ denote the norm in $\mathbb{X}$,

$$\|\mathbf{x}\|_C = \sup_{0\leq t\leq T} \max_{0\leq s\leq L} |\mathbf{x}(s,t)|.$$

Then

$$\sup_{0\leq t\leq T} \|\widetilde{\mathbf{X}}(t) - \widetilde{\mathbf{Y}}(t)\| \leq l_6 \|\mathbf{X} - \mathbf{Y}\|_C,$$

that is

$$\|\mathbf{TX} - \mathbf{TY}\|_C \leq l_6 \|\mathbf{X} - \mathbf{Y}\|_C. \tag{58}$$

Let $T'' = min\{T', \frac{L_m - \|\mathbf{X}_0\|_c}{c_3}, \widetilde{T}\}$. Notice that T'' is dependent on μ, $\widetilde{T}, \delta_h, \mathbf{f}, \mathbf{u}_0, L_m, \|\mathbf{X}_0\|_c$. The inequality (58) implies that if $T \in (0, T'')$, $\mathbf{T} : B \to B$ is a contractive map. By the Banach Fixed-Point Theorem (see [1]), $\mathbf{T}$ has a unique fixed point $\widetilde{\mathbf{X}} \in B$, and $\widetilde{\mathbf{X}}$ may be solved by successive approximations

$$\mathbf{Y}_{n+1} = \mathbf{TY}_n, \quad \forall \mathbf{Y_0} \in B, \quad n = 0, 1, 2, \ldots.$$

Let $\widetilde{\mathbf{u}}$ be the unique solution to (41) corresponding $\widetilde{\mathbf{X}}$. Then $\widetilde{\mathbf{u}}$ and $\widetilde{\mathbf{X}}$ is the unique pair of solutions we need. □

Remark: In our conclusion we only obtain the existence of the approximate solution with Dirac delta function approximated by differentiable function. How the solution of the original problem is approximated by this approximate solution is still an open problem.

Acknowledgement

This work is partially supported by National Natural Science Foundation of China(No: 10671092).

References

[1] Ebhard Zeidler. *Nonlinear functional analysis and its applications, I: Fixed-point theorems*. Berlin: Springer-Verlag, 1986.

[2] Peskin C.S. *The immersed boundary method, Acta Numerica*. Cambridge University Press, 2002.

[3] Daniele Boffi, Lucia Gastaldi. A finite element approach of the immersed boundary method. *Computers and Structures*, 2003, Vol.81, 491-501.

[4] Roger Temam. *Navier-Stokes equations and nonlinear functional analysis* (CBM-SNSF Regional Conf. Ser. in Aplied Mathematics 66), SIAM, Philadelphia, 1983.

[5] Roger Temam. *Navier-Stokes equations, theory and numerical analysis, Studies in Mathematics and its applications*. volume 2, North-Holland Publishing Company 1977.

[6] C.Foias, O.Manley, R.Rosa, Roger Temam. *Navier-Stokes equations and turbulence*. Cambridge University Press 2001.

[7] Ruth F.Curtain, A.J.Pritchard. *Functional Analysis in Modern Applied Mathematics*. Academic Press INC. (London) LTD. 1977.

[8] C.S. Peskin. Numerical analysis of blood flow in the heart, *J. Comp. Phys.* 1977, Vol.25, 220-252.

[9] Santos Alberto Enriquez-Remigio and Alexandre Megiorin Roma. Incompressible flows in elastic domains: an immersed boundary method approach. *Applied Mathematical Modeling*, January 2005, Volume 29, Issue 1, 35-54.

[10] Lima, A.L.F., Silva, E., Silveira-Neto A.,& Damasceno J.J.R. Numerical simulation of two-dimensional flows over a circular cylinder using the immersed boundary method. *Journal of Computational Physics*, 2003, Vol.189, 351-370.

[11] Arthurs, K.M., Moore, L.C., Peskin, C.S., Pitman, E.B., &Layton, H.E. Modeling arteriolar flow and mass transport using the immersed boundary method. *J. Comp. Physiol.* 1998, Vol.147, 402-440.

[12] Roma, A.M., Peskin, C.S., & Berger, M.J. An adaptive version of the immersed boundary method. *J. Comp. Physiol.*, 1999, Vol.153, 509-534

[13] Ling Rao and Hongquan Chen. Immersed boundary method: The existence of approximate solution of the two-dimensional heat equation. *Nonlinear Analysis: Real World Applications*, April 2008, Volume 9, Issue 2, 384-393.

Reviewed by Professor Yuanguo Zhu, Department of Applied Mathematics, Nanjing University of Science and Technology.

In: Progress in Nonlinear Analysis Research
Editor: Erik T. Hoffmann, pp. 331-345
ISBN 978-1-60456-359-7

Chapter 14

EXACT PENALTY FUNCTIONS FOR CONSTRAINED OPTIMIZATION PROBLEMS

Alexander J. Zaslavski
Department of Mathematics,
The Technion-Israel Institute of Technology, Israel

Abstract

In this paper we use the penalty approach for constrained minimization problems in infinite dimensional spaces. A penalty function is said to have the exact penalty property if there is a penalty coefficient for which a solution of an unconstrained penalized problem is a solution of the corresponding constrained problem. We discuss the exact penalty property for several large classes of constrained minimization problems.

1. Introduction

Penalty methods are an important and useful tool in constrained optimization. See, for example, [1, 2, 3, 5, 8] and the references mentioned there. In this paper we use the penalty approach in order to study constrained minimization problems in infinite dimensional spaces. A penalty function is said to have the exact penalty property if there is a penalty coefficient for which a solution of an unconstrained penalized problem is a solution of the corresponding constrained problem. The notion of exact penalization was introduced by Eremin [7] and Zangwill [10] for use in the development of algorithms for nonlinear constrained optimization. For a detailed historical review of the literature on exact penalization see [1, 2, 5]. In this paper we discuss the results of [12, 13, 14] which establish the existence of exact penalty for several large classes of constrained minimization problems on a Banach space X.

In [12, 13, 14] we establish the existence of a penalty coefficient for which approximate solutions of the unconstrained penalized problem are close enough to approximate solutions of the corresponding constrained problem. This is a novel approach in the penalty type methods.

Consider a minimization problem $h(z) \to \min$, $z \in X$ where $h : X \to R^1$ is a lower semicontinuous bounded from below function. If the space X is infinite-dimensional, then the existence of solutions of the problem is not guaranteed and in this situation we consider δ-approximate solutions. Namely, $x \in X$ is a δ-approximate solution of the problem $h(z) \to \min$, $z \in X$, where $\delta > 0$, if $h(x) \le \inf\{h(z) : z \in X\} + \delta$.

Since in this paper we consider minimization problems on a general Banach space the existence of their solutions is not guaranteed. Therefore we are interested in approximate solutions of the unconstrained penalized problem and in approximate solutions of the corresponding constrained problem. Under certain mild assumptions we show the existence of a constant $\Lambda_0 > 0$ such that the following property holds:

For each $\epsilon > 0$ there exists $\delta(\epsilon) > 0$ which depends only on ϵ such that if x is a $\delta(\epsilon)$-approximate solution of the unconstrained penalized problem whose penalty coefficient is larger than Λ_0, then there exists an ϵ-approximate solution y of the corresponding constrained problem such that $||y - x|| \le \epsilon$.

This property implies that any exact solution of the unconstrained penalized problem whose penalty coefficient is larger than Λ_0, is an exact solution of the corresponding constrained problem. Indeed, let x be a solution of the unconstrained penalized problem whose penalty coefficient is larger than Λ_0. Then for any $\epsilon > 0$ the point x is also a $\delta(\epsilon)$-approximate solution of the same unconstrained penalized problem and in view of the property above there is an ϵ-approximate solution y_ϵ of the corresponding constrained problem such that $||x - y_\epsilon|| \le \epsilon$. Since ϵ is an arbitrary positive number we can easily deduce that x is an exact solution of the corresponding constrained problem. Therefore our results also include the classical penalty result as a special case.

Comparing the results of [12-14] with exact penalty results known in the literature, we note the following significant improvements.

Our results are established for constrained optimization problems on a general Banach space X. Since we do not assume that this space is finite-dimensional, no compactness property is used in our study.

Our results are established for very large classes of constrained minimization problems.

In the classical approach it is established the existence of a penalty coefficient for which solutions of the unconstrained penalized problem are solutions of the corresponding constrained problem. In [12-14] we establish the existence of a penalty coefficient for which approximate solutions of the unconstrained penalized problem are close enough to approximate solutions of the corresponding constrained problem. In particular, we show the existence of a penalty coefficient such that for any minimizing sequence $\{x_i\}_{i=1}^\infty$ of the unconstrained penalized problem there exists a minimizing sequence $\{y_i\}_{i=1}^\infty$ of the corresponding constrained problem such that $\lim_{i\to\infty} ||y_i - x_i|| = 0$. This property is important and useful even when optimization problems have exact solutions because if one uses methods in order to solve optimization problems these methods usually provide only approximate solutions of the problems.

We begin with a large class of inequality-constrained minimization problems

$$f(x) \to \min \text{ subject to } x \in A$$

where

$$A = \{x \in X : g_i(x) \le c_i \text{ for } i = 1, \dots, n\}$$

studied in [14]. Here X is a Banach space, c_i, $i = 1, \dots, n$ are real numbers, the constraint functions g_i, $i = 1, \dots, n$ are convex and lower semicontinuous and the objective function f belongs to a space of functions $\mathcal{M}$. This space of objective functions $\mathcal{M}$ is a convex cone in the vector space of all functions on X. It includes the set of all convex bounded from below semicontinuous functions $f : X \to R^1$ which satisfy the growth condition $\lim_{||x|| \to \infty} f(x) = \infty$ and the set of all functions f on X which satisfy the growth condition above and which are Lipschitzian on all bounded subsets of X. This class of problems is considered in Section 2. We discuss the existence of the exact penalty property established in [14]. In Section 3 we use the penalty approach in order to study two constrained nonconvex minimization problems with Lipschitzian (on bounded sets) objective functions. The first problem is an equality-constrained problem in a Banach space with a locally Lipschitzian constraint function and the second problem is an inequality-constrained problem in a Banach space with a locally Lipschitzian constraint function. In Section 3 we discuss the results of [12] which establish a very simple sufficient condition for the exact penalty property. In Section 4 we use the penalty approach in order to study two constrained nonconvex minimization problems with smooth cost functions. The first problem is an equality-constrained problem in a Hilbert space with a smooth constraint function and the second problem is an inequality-constrained problem in a Hilbert space with a smooth constraint function.

2. Inequality-Constrained Minimization Problems with Convex Constraint Functions

In this section we discuss the exact penalty property for a large class of inequality-constrained minimization problems

$$f(x) \to \min \text{ subject to } x \in A$$

where

$$A = \{x \in X : \ g_i(x) \leq c_i \text{ for } i = 1, \dots, n\}$$

established in [14]. Here X is a Banach space, c_i, $i = 1, \dots, n$ are real numbers, the constraint functions g_i, $i = 1, \dots, n$ are convex and lower semicontinuous and the objective function f belongs to a space of functions $\mathcal{M}$ decsribed below. This space of objective functions $\mathcal{M}$ is a convex cone in the vector space of all functions on X. It includes the set of all convex bounded from below lower semicontinuous functions $f : X \to R^1$ which satisfy the growth condition $\lim_{||x|| \to \infty} f(x) = \infty$ and the set of all functions f on X which satisfy the growth condition above and which are Lipschitzian on all bounded subsets of X. It should be mentioned that if f belongs to this class of functions and $g : R^1 \to R^1$ is an increasing Lipschitzian function, then $g \circ f$ also belongs to $\mathcal{M}$. Moreover, if $f \in \mathcal{M}$ and if a function $g : X \to R^1$ is Lipschitzian on all bounded subsets of X and the infimum of g is positive, then $f \cdot g \in \mathcal{M}$.

We associate with the inequality-constrained minimization problem above the corresponding family of unconstrained minimization problems

$$f(z) + \gamma \sum_{i=1}^{n} \max\{g_i(z) - c_i, 0\} \to \min, \ z \in X$$

where $\gamma > 0$ is a penalty.

We use the convention that $\lambda \cdot \infty = \infty$ for all $\lambda \in (0, \infty)$, $\lambda + \infty = \infty$ and $\max\{\lambda, \infty\} = \infty$ for any real number λ and that supremum over empty set is $-\infty$. We use the following notation and definitions.

Let $(X, ||\cdot||)$ be a Banach space. For each $x \in X$ and each $r > 0$ set

$$B(x, r) = \{y \in X : \ ||x - y|| \le r\}.$$

For each function $f : X \to R^1 \cup \{\infty\}$ and each nonempty set $A \subset X$ put

$$\mathrm{dom}(f) = \{x \in X : \ f(x) < \infty\},$$

$$\inf(f) = \inf\{f(z) : \ z \in X\}$$

and

$$\inf(f; A) = \inf\{f(z) : \ z \in A\}.$$

For each $x \in X$ and each $B \subset X$ set

$$d(x, B) = \inf\{||x - y|| : \ y \in B\}.$$

Let n be a natural number. For each $\kappa \in (0, 1)$ denote by Ω_κ the set of all $\gamma = (\gamma_1, \ldots, \gamma_n) \in R^n$ such that

$$\kappa \le \min\{\gamma_i : \ i = 1, \ldots, n\} \text{ and } \max\{\gamma_i : \ i = 1, \ldots, n\} = 1.$$

Let $g_i : X \to R^1 \cup \{\infty\}$, $i = 1, \ldots, n$ be convex lower semicontinuous functions and $c = (c_1, \ldots, c_n) \in R^n$. Set

$$A = \{x \in X : \ g_i(x) \le c_i \text{ for } i = 1, \ldots, n\}.$$

Let $f : X \to R^1 \cup \{\infty\}$ be a bounded from below lower semicontinuous function which satisfies the following growth condition

$$\lim_{||x|| \to \infty} f(x) = \infty. \tag{2.1}$$

We suppose that there is $\tilde{x} \in X$ such that

$$g_j(\tilde{x}) < c_j \text{ for } j = 1, \ldots, n \text{ and } f(\tilde{x}) < \infty. \tag{2.2}$$

In this paper we consider the following constrained minimization problem

$$f(x) \to \min \text{ subject to } x \in A. \tag{P}$$

Clearly, $A \neq \emptyset$ and $\inf(f; A) < \infty$. For each vector $\gamma = (\gamma_1, \ldots, \gamma_n) \in (0, \infty)^n$ define

$$\psi_\gamma(z) = f(z) + \sum_{i=1}^{n} \gamma_i \max\{g_i(z) - c_i, 0\}, \ z \in X.$$

Clearly for each $\gamma \in (0, \infty)^n$ the function $\psi_\gamma : X \to R^1 \cup \{\infty\}$ is bounded from below and lower semicontinuous and satisfies $\inf(\psi_\gamma) < \infty$. We associate with problem (P) the corresponding family of unconstrained minimization problems

$$\psi_\gamma(z) \to \min, \ z \in X \tag{P_γ}$$

where $\gamma \in (0, \infty)^n$.

In this section we assume that there exists a function $h : X \times \mathrm{dom}(f) \to R^1 \cup \{\infty\}$ such that the following assumptions hold:

(A1) $h(z, y)$ is finite for each $y, z \in \mathrm{dom}(f)$ and $h(y, y) = 0$ for each $y \in \mathrm{dom}(f)$.

(A2) For each $y \in \mathrm{dom}(f)$ the function $h(\cdot, y) \to R^1 \cup \{\infty\}$ is convex.

(A3) For each $z \in \mathrm{dom}(f)$ and each $r > 0$

$$\sup\{h(z, y) : \ y \in \mathrm{dom}(f) \cap B(0, r)\} < \infty.$$

(A4) For each $M > 0$ there exists $M_1 > 0$ such that for each $y \in X$ satisfying $f(y) \le M$ there exists a neighborhood V of y in X such that if $z \in V$, then

$$f(z) - f(y) \le M_1 h(z, y).$$

Remark 2.1. Note that if f is convex, then assumptions (A1)-(A4) hold with $h(z, y) = f(z) - f(y)$, $z \in X$, $y \in \mathrm{dom}(f)$. In this case $M_1 = 1$ for all $M > 0$. If the function f is finite-valued and Lipschitzian on all bounded subsets of X, then assumptions (A1)-(A4) hold with $h(z, y) = ||z - y||$ for all $z, y \in X$.

Let $\kappa \in (0, 1)$. The main result of this section (Theorem 2.1) stated below imply that if λ is sufficiently large, then any solution of problem $(P_{\lambda\gamma})$ with $\gamma \in \Omega_\kappa$ is a solution of problem (P). Note that if the space X is infinite-dimensional, then the existence of solutions of problems $(P_{\lambda\gamma})$ and (P) is not guaranteed. In this case Theorem 2.1 implies that for each $\epsilon > 0$ there exists $\delta(\epsilon) > 0$ which depends only on ϵ such that the following property holds:

If $\lambda \ge \Lambda_0$, $\gamma \in \Omega_\kappa$ and if x is a δ-approximate solution of $(P_{\lambda\gamma})$, then there exists an ϵ-approximate solution y of (P) such that $||y - x|| \le \epsilon$.

Here Λ_0 is a positive constant which does not depend on ϵ.

It should be mentioned that in our results we deal with penalty functions whose penalty parameters for constraints $g_1, \ldots, g_n$ are $\lambda\gamma_1, \ldots, \lambda\gamma_n$ respectively, where $\lambda > 0$ and $(\gamma_1, \ldots, \gamma_n) \in \Omega_\kappa$ for a given $\kappa \in (0, 1)$. Note that the vector $(1, 1, \ldots, 1) \in \Omega_\kappa$ for any $\kappa \in (0, 1)$. Therefore our result also includes the case $\gamma_1 = \ldots = \gamma_n = 1$ where one single parameter λ is used for all constraints. Note that sometimes it is an advantage from numerical consideration to use penalty coefficients $\lambda\gamma_1, \ldots, \lambda\gamma_n$ with different parameters γ_i, $i = 1, \ldots, n$. For example, in the case when some of the constrained functions are very "small" and some of the constraint functions are very "large".

The next theorem is the main result of [14] and of this section.

Theorem 2.1 Let $\kappa \in (0, 1)$. Then there exists a positive number Λ_0 such that for each $\epsilon > 0$ there exists $\delta \in (0, \epsilon)$ such that the following assertion holds:

If $\gamma \in \Omega_\kappa$, $\lambda \ge \Lambda_0$ and if $x \in X$ satisfies

$$\psi_{\lambda\gamma}(x) \le \inf(\psi_{\lambda\gamma}) + \delta,$$

then there exists $y \in A$ such that

$$||y - x|| \leq \epsilon \text{ and } f(y) \leq \inf(f; A) + \epsilon.$$

Theorem 2.1 implies the following result.

Corollary 2.1 Let $\kappa \in (0,1)$. Then there exists $\Lambda_0 > 0$ such that for each $\gamma \in \Omega_\kappa$, each $\lambda \geq \Lambda_0$ and each sequence $\{x_i\}_{i=1}^\infty \subset X$ which satisfies

$$\lim_{i\to\infty} \psi_{\lambda\gamma}(x_i) = \inf(\psi_{\lambda\gamma})$$

there exists a sequence $\{y_i\}_{i=1}^\infty \subset A$ such that

$$\lim_{i\to\infty} f(y_i) = \inf(f; A) \text{ and } \lim_{i\to\infty} ||y_i - x_i|| = 0.$$

The next result follows from Corollary 2.1.

Corollary 2.2 Let $\kappa \in (0,1)$. Assume that there exists $\bar{x} \in A$ for which the following conditions hold:

$$f(\bar{x}) = \inf(f; A);$$

any sequence $\{x_n\}_{n=1}^\infty \subset A$ which satisfies $\lim_{n\to\infty} f(x_n) = \inf(f; A)$ converges to $\bar{x}$ in the norm topology.

Then there exists $\Lambda_0 > 0$ such that for each $\gamma \in \Omega_\kappa$ and each $\lambda \geq \Lambda_0$ the point $\bar{x}$ is a unique solution of the minimization problem

$$\psi_{\lambda\gamma}(z) \to \min, \; z \in X.$$

Corollary 2.1 implies the following result.

Corollary 2.3 Let $\kappa \in (0,1)$. Then there exists $\Lambda_0 > 0$ such that if $\gamma \in \Omega_\kappa$, $\lambda \geq \Lambda_0$ and if x is a solution of the minimization problem

$$\psi_{\lambda\gamma}(z) \to \min, \; z \in X,$$

then $x \in A$ and $f(x) = \inf(f; A)$.

It is clear that Corollary 2.3 is the classical exact penalty result.

Denote by $\mathcal{M}$ the set of all bounded from below lower semicontinuous functions $f : X \to R^1 \cup \{\infty\}$ satisfying (2.1) which are not identically infinity and for which there exists a function $h_f : X \times \text{dom}(f) \to R^1 \cup \{\infty\}$ such that assumptions (A1)-(A4) hold with $h = h_f$.

We have already mentioned (see Remark 2.1) that $f \in \mathcal{M}$ if at least one of the following conditions holds:

$f : X \to R^1 \cup \{\infty\}$ is a convex bounded from below lower semicontinuous function satisfying (2.1) which is not identically infinity;

$f : X \to R^1$ is a function which satisfies (2.1) and which is Lipschitzian on all bounded subsets of X.

Obviously, if $f \in \mathcal{M}$ and $\lambda > 0$, then $\lambda f \in \mathcal{M}$. It is not difficult to see that the following properties hold:

if $f_1 \in \mathcal{M}$ and $f_2 : X \to R^1$ is a convex lower semicontinuous bounded from below function, then $f_1 + f_2 \in \mathcal{M}$;

if $f_1 \in \mathcal{M}$ and $f_2 : X \to R^1$ is a bounded from below function which is Lipschitzian on all bounded subsets of X, then $f_1 + f_2 \in \mathcal{M}$.

Note that Theorem 2.1 holds for any $f \in \mathcal{M}$ such that (2.2) is valid with some $\tilde{x} \in X$.

In [14] we prove the following three results which show that the class $\mathcal{M}$ is essentially larger than the union of the set of all convex bounded from below lower semicontinuous functions satisfying the growth condition (2.1) and the set of all functions satisfying (2.1) which are Lipschitzian on all bounded subsets of X.

Proposition 2.1 Let $f_1, f_2 \in \mathcal{M}$, $\mathrm{dom}(f_1) \subset \mathrm{dom}(f_2)$, $h_i : X \times \mathrm{dom}(f_i) \to R^1 \cup \{\infty\}$, $i = 1, 2$ and let for $i = 1, 2$ assumptions (A1)-(A4) hold with $f = f_i$, $h = h_i$. Then $f_1 + f_2 \in \mathcal{M}$ and (A1)-(A4) hold with $f = f_1 + f_2$ and with a function $h : X \times \mathrm{dom}(f_1) \to R^1 \cup \{\infty\}$ defined by

$$h(z, y) = \max\{h_1(z, y), 0\} + \max\{h_2(z, y), 0\},\ x \in X,\ y \in \mathrm{dom}(f_1).$$

Corollary 2.4 Let $f_1 \in \mathcal{M}$ be a convex function and $f_2 \in \mathcal{M}$ be a finite-valued function which is Lipschitzian on all bounded subsets of X. Then $f_1 + f_2 \in \mathcal{M}$ and (A1)-(A4) hold with $f = f_1 + f_2$ and with a function $h : X \times \mathrm{dom}(f_1) \to R^1 \cup \{\infty\}$ defined by

$$h(z, y) = \max\{f_1(z) - f_1(y), 0\} + ||z - y||,\ x \in X,\ y \in \mathrm{dom}(f_1).$$

Proposition 2.2 Let $f \in \mathcal{M}$ and $g : R^1 \to R^1$ be an increasing Lipschitzian function such that $\lim_{t \to \infty} g(t) = \infty$. Then $g \circ f \in \mathcal{M}$.

(Here we assume that $g(\infty) = \infty$).

Proposition 2.3 Let $f \in \mathcal{M}$ and let a function $g : X \to R^1$ be Lipschitzian on all bounded subsets of X and satisfy

$$\inf\{g(x) : \ x \in X\} > 0.$$

Then $f \cdot g \in \mathcal{M}$.

3. Constrained Nonconvex Minimization Problems with One Constraint

In this section we use the penalty approach in order to study two constrained nonconvex minimization problems with Lipschitzian (on bounded sets) objective functions. The first problem is an equality-constrained problem in a Banach space with a locally Lipschitzian constraint function and the second problem is an inequality-constrained problem in a Banach space with a locally Lipschitzian constraint function. In this section we discuss the results of [12] which establish a very simple sufficient condition for the exact penalty property.

Let $(X, ||\cdot||)$ be a Banach space, $(X^*, ||\cdot||_*)$ its dual space and let $f : X \to R^1$ be a locally Lipschitzian function. For each $x \in X$ let

$$f^0(x,h) = \limsup_{t\to 0^+, y\to x} [f(y+th) - f(y)]/t, \ h \in X$$

be the Clarke generalized directional derivative of f at the point x [3], let

$$\partial f(x) = \{l \in X^* : \ f^0(x,h) \geq l(h) \text{ for all } h \in X\}$$

be Clarke's generalized gradient of f at x [3] and set

$$\Xi_f(x) = \inf\{f^0(x,h) : \ h \in X \text{ and } ||h|| = 1\}$$

(see [11]).

A point $x \in X$ is called a critical point of f if $0 \in \partial f(x)$. It is not difficult to see that $x \in X$ is a critical point of f if and only if $\Xi_f(x) \geq 0$.

A real number $c \in R^1$ is called a critical value of f if there is a critical point x of f such that $f(x) = c$.

It is known [3, Chapter 2, Section 2.3] that $\partial(-f)(x) = -\partial f(x)$ for any $x \in X$. This equality implies that $x \in X$ is a critical point of f if and only if x is a critical point of $-f$ and $c \in R^1$ is a critical value of f if and only if $-c$ is a critical value of $-f$.

For each function $f : X \to R^1$ set $\inf(f) = \inf\{f(z) : \ z \in X\}$. For each $x \in X$ and each $B \subset X$ put

$$d(x,B) = \inf\{||x-y|| : \ y \in B\}.$$

Let $f : X \to R^1$ be a function which is Lipschitzian on all bounded subsets of X and which satisfies the following growth condition

$$\lim_{||x||\to\infty} f(x) = \infty.$$

Clearly, f is bounded from below. Let $g : X \to R^1$ be a locally Lipschitzian function which satisfies the following Palais-Smale (P-S) condition [9, 11]:

If $\{x_i\}_{i=1}^\infty \subset X$, the sequence $\{g(x_i)\}_{i=1}^\infty$ is bounded and if

$$\liminf_{i\to\infty} \Xi_g(x_i) \geq 0,$$

then there is a norm convergent subsequence of $\{x_i\}_{i=1}^\infty$.

Let $c \in R^1$ be such that $g^{-1}(c)$ is nonempty.

We consider the constrained problems

$$f(x) \to \min \ \text{subject to } x \in g^{-1}(c) \tag{P_e}$$

and

$$f(x) \to \min \ \text{subject to } x \in g^{-1}((-\infty, c]). \tag{P_i}$$

We associate with these two problems the corresponding families of unconstrained minimization problems

$$f(x) + \lambda|g(x) - c| \to \min, \ x \in X \tag{$P_{\lambda e}$}$$

and

$$f(x) + \lambda \max\{g(x) - c,\ 0\} \to \min,\ x \in X \tag{$P_{\lambda i}$}$$

where $\lambda > 0$.

Set

$$\inf(f; c) = \inf\{f(z) :\ z \in g^{-1}(c)\},$$

$$\inf(f; (-\infty, c]) = \inf\{f(z) :\ z \in X \text{ and } g(z) \le c\}.$$

The next theorem is the main result of [12].

Theorem 3.1 Assume that the number c is not a critical value of the function g. Then there exist positive numbers λ_0 and λ_1 such that for each $\epsilon > 0$ there exists $\delta \in (0, \epsilon)$ such that the following assertions hold:

1. If $\lambda > \lambda_0$ and if $x \in X$ satisfies

$$f(x) + \lambda|g(x) - c| \le \inf\{f(z) + \lambda|g(z) - c| :\ z \in X\} + \delta,$$

then there exists $y \in g^{-1}(c)$ such that

$$||y - x|| \le \epsilon \text{ and } f(y) \le \inf(f; c) + \lambda_1\epsilon.$$

2. If $\lambda > \lambda_0$ and if $x \in X$ satisfies

$$f(x) + \lambda \max\{g(x) - c,\ 0\} \le \inf\{f(z) + \lambda \max\{g(z) - c,\ 0\} :\ z \in X\} + \delta,$$

then there exists $y \in g^{-1}((-\infty, c])$ such that

$$||y - x|| \le \epsilon \text{ and } f(y) \le \inf(f; (-\infty, c]) + \lambda_1\epsilon.$$

Theorem 3.1 implies the following result.

Theorem 3.2 Assume that the number c is not a critical value of the function g. Then there is $\lambda_0 > 0$ such that the following assertions hold:

1. For each $\lambda > \lambda_0$ and each sequence $\{x_i\}_{i=1}^{\infty} \subset X$ which satisfies

$$\lim_{i \to \infty} [f(x_i) + \lambda|g(x_i) - c|] = \inf\{f(z) + \lambda|g(z) - c| :\ z \in X\}$$

there exists a sequence $\{y_i\}_{i=1}^{\infty} \subset g^{-1}(c)$ such that

$$\lim_{i \to \infty} f(y_i) = \inf(f; c) \text{ and } \lim_{i \to \infty} ||y_i - x_i|| = 0.$$

2. For each $\lambda > \lambda_0$ and each sequence $\{x_i\}_{i=1}^{\infty} \subset X$ which satisfies

$$\lim_{i \to \infty} [f(x_i) + \lambda \max\{g(x_i) - c, 0\}] = \inf\{f(z) + \lambda \max\{g(z) - c, 0\} :\ z \in X\}$$

there exists a sequence $\{y_i\}_{i=1}^{\infty} \subset g^{-1}((-\infty, c])$ such that

$$\lim_{i \to \infty} f(y_i) = \inf(f; (-\infty, c]) \text{ and } \lim_{i \to \infty} ||y_i - x_i|| = 0.$$

Assertion 1 of Theorem 3.2 implies the following result.

Theorem 3.3 Assume that the number c is not a critical value of the function g and that there exists $\bar{x} \in g^{-1}(c)$ for which the following conditions hold:

$f(\bar{x}) = \inf(f; c)$;

any sequence $\{x_n\}_{n=1}^{\infty} \subset g^{-1}(c)$ which satisfies $\lim_{n\to\infty} f(x_n) = \inf(f; c)$ converges to $\bar{x}$ in the norm topology.

Then there exists $\lambda_0 > 0$ such that for each $\lambda > \lambda_0$ the point $\bar{x}$ is a unique solution of the minimization problem

$$f(z) + \lambda|g(z) - c| \to \min, \;\; z \in X.$$

Assertion 2 of Theorem 3.2 implies the following result.

Theorem 3.4 Assume that the number c is not a critical value of the function g and that there exists $\bar{x} \in g^{-1}((-\infty, c])$ for which the following conditions hold:

$f(\bar{x}) = \inf(f; (-\infty, c])$;

any sequence $\{x_n\}_{n=1}^{\infty} \subset g^{-1}((-\infty, c])$ which satisfies

$$\lim_{n\to\infty} f(x_n) = \inf(f; (-\infty, c])$$

converges to $\bar{x}$ in the norm topology.

Then there exists $\lambda_0 > 0$ such that for each $\lambda > \lambda_0$ the point $\bar{x}$ is a unique solution of the minimization problem

$$f(z) + \lambda \max\{g(z) - c, 0\} \to \min, \;\; z \in X.$$

The next result follows from Theorem 3.2.

Theorem 3.5 Assume that $X = R^n$ and the number c is not a critical value of the function g. Then there is $\lambda_0 > 0$ such that the following assertions hold:

1. If $\lambda > \lambda_0$ and if x is a solution of the minimization problem

$$f(z) + \lambda|g(z) - c| \to \min, \;\; z \in X,$$

then $x \in g^{-1}(c)$ and $f(x) = \inf(f; c)$.

2. If $\lambda > \lambda_0$ and if x is a solution of the minimization problem

$$f(z) + \lambda \max\{g(z) - c, 0\} \to \min, \;\; z \in X,$$

then $g(x) \leq c$ and $f(x) = \inf(f; (-\infty, c])$.

Example 3.1. Assume that $X = R^n$, $g \in C^1(R^n)$ and that the gradient of g is not zero at any point $x \in R^n$. Then Theorems 3.1, 3.2 and 3.5 hold.

Example 3.2. Assume that $X = R^n$, g is convex and bounded from below and $c > \inf(g)$. Then Theorems 3.1, 3.2 and 3.5 hold.

Now we give an example which shows that exactness fails when c is a critical value of g.

Example 3.3.

Let $X = R^1$ and consider the minimization problem

$$f(x) \to \min, \ x \in R^1, \ g(x) = 0$$

where $f(x) = (x - 10)^2, x \in R^1$ and

$$g(x) = (x-1)^2, \ x \in [1, \infty), \ g(x) = (x+1)^2, \ x \in (-\infty, -1], \ g(x) = 0, \ x \in (-1, 1).$$

This problem is equivalent to the problem

$$f(x) \to \min, \ x \in R^1, \ g(x) \le 0.$$

Clearly zero is a critical value of g and $\bar{x} = 1$ is a unique solution of the problem.

We show that for each $\lambda > 0$, $\inf(f + \lambda g) < f(1) = 81$. Fix $\lambda > 0$. For each $x \in [1, \min\{2, 1 + 4/\lambda\}]$

$$(f + \lambda g)'(x) = 2(x - 10) + 2\lambda(x - 1) \le -16 + 2\lambda(x - 1) \le -16 + 2\lambda(4/\lambda) = -8.$$

This relation implies that $\inf(f + \lambda g) < f(1)$.

4. Existence of Exact Penalty for Constrained Smooth Optimization Problems in Hilbert Spaces

In this section we use the penalty approach in order to study two constrained nonconvex minimization problems with smooth cost functions. The first problem is an equality-constrained problem in a Hilbert space with a smooth constraint function and the second problem is an inequality-constrained problem in a Hilbert space with a smooth constraint function.

Let $(X, < \cdot, \cdot >)$ be a Hilbert space with the inner product $< \cdot, \cdot >$ and the norm $||x|| = < x, x >^{1/2}, \ x \in X$. Denote by $C^1(X; R^1)$ the set of all Frechet differentiable functions $f : X \to R^1$ such that the mapping $x \to f'(x), x \in X$ is continuous. Here $f'(x) \in X$ is a Frechet derivative of f at $x \in X$.

Denote by $C^2(X; R^1)$ the set of all Frechet differentiable functions $f \in C^1(X; R^1)$ such that the mapping $x \to f'(x), x \in X$ is also Frechet differentiable and that the mapping $x \to f''(x), x \in X$ is continuous. Here $f''(x)$ is a Frechet second order derivative of f at $x \in X$. It is a linear continuous self-mapping of X.

For each $x \in X$ and each $r > 0$ set

$$B(x, r) = \{y \in X : \ ||x - y|| \le r\}.$$

For each function $f : X \to R^1$ and each $A \subset X$ put

$$\inf(f) = \inf\{f(z) : \ z \in X\}$$

and

$$\inf(f; A) = \inf\{f(x) : \ x \in A\}.$$

For each $x \in X$ and each $B \subset X$ set

$$d(x, B) = \inf\{||x - y|| : y \in B\}.$$

Let $g \in C^2(X; R^1)$. A point $x \in X$ is called a critical point of g if $g'(x) = 0$. Denote by $Cr(g)$ the set of all critical points of g. A real number c is a critical value of g if there exists $x \in Cr(g)$ such that $g(x) = c$. Denote by $Cr(g, +)$ the set of all $x \in Cr(g)$ such that

$$< g''(x)u, u > \geq 0 \text{ for all } u \in X$$

and by $Cr(g, -)$ the set of all $x \in Cr(g)$ such that

$$< g''(x)u, u > \leq 0 \text{ for all } u \in X.$$

Let $c \in R^1$, $\gamma > 0$ and let $g^{-1}(c) \neq \emptyset$. We assume that g satisfies the following Palais-Smale (P-S) condition [9] on the set $g^{-1}([c - \gamma, c + \gamma])$:

(PS) If $\{x_i\}_{i=1}^{\infty} \subset g^{-1}([c - \gamma, c + \gamma])$ is a bounded (with respect to the norm of X) sequence and if $\lim_{i \to \infty} ||g'(x_i)|| = 0$, then there is a norm convergent subsequence of $\{x_i\}_{i=1}^{\infty}$.

Let a function $f \in C^1(X; R^1)$ be Lipschitzian on all bounded subsets of X and satisfy the growth condition

$$\lim_{||x|| \to \infty} f(x) = \infty$$

and let the mapping $f' : X \to X$ be locally Lipschitzian. Clearly f is bounded from below.

We consider the constrained problems

$$f(x) \to \min \text{ subject to } x \in g^{-1}(c) \qquad (P_e)$$

and

$$f(x) \to \min \text{ subject to } x \in g^{-1}((-\infty, c]). \qquad (P_i)$$

We associate with these two problems the corresponding families of unconstrained minimization problems

$$f(x) + \lambda|g(x) - c| \to \min, \ x \in X \qquad (P_{\lambda e})$$

and

$$f(x) + \lambda \max\{g(x) - c, \ 0\} \to \min, \ x \in X \qquad (P_{\lambda i})$$

where $\lambda > 0$.

In [12] we showed the existence of exact penalty for problems (P_i) and (P_e) if c is not a critical value of g. In this section we discuss the results of [13] which establish the existence of exact penalty under essentially weaker condition. These results imply that exact penalty exists for the problem (P_e) if $g^{-1}(c) \cap (Cr(g, +) \cup Cr(g, -)) = \emptyset$ and that exact penalty exists for the problem (P_i) if $g^{-1}(c) \cap Cr(g, +) = \emptyset$.

Remark 4.1. Note that the results of [12] were established for a Banach space X and locally Lipschitzian functions f and g.

The next two theorems are the main results of [13].

Theorem 4.1 Assume that

$$\{x \in g^{-1}(c) : f(x) = \inf(f; g^{-1}(c))\} \cap (Cr(g, +) \cup Cr(g, -)) = \emptyset.$$

Then there exists a positive number Λ_0 such that for each $\epsilon > 0$ there exists $\delta \in (0, \epsilon)$ such that the following assertion holds:

If $\lambda \geq \Lambda_0$ and if $x \in X$ satisfies

$$f(x) + \lambda|g(x) - c| \leq \inf\{f(z) + \lambda|g(z) - c| : z \in X\} + \delta,$$

then there exists $y \in g^{-1}(c)$ such that

$$||y - x|| \leq \epsilon \text{ and } f(y) \leq \inf(f; g^{-1}(c)) + \epsilon.$$

Theorem 4.2 Assume that

$$\{x \in g^{-1}(c) : f(x) = \inf(f; g^{-1}((-\infty, c]))\} \cap Cr(g, +) = \emptyset.$$

Then there is $\Lambda_0 > 0$ such that for each $\epsilon > 0$ there exists $\delta \in (0, \epsilon)$ such that the following assertion holds:

If $\lambda \geq \Lambda_0$ and if $x \in X$ satisfies

$$f(x) + \lambda \max\{g(x) - c, 0\} \leq \inf\{f(z) + \lambda \max\{g(z) - c, 0\} : z \in X\} + \delta,$$

then there exists $y \in g^{-1}((-\infty, c])$ such that

$$||y - x|| \leq \epsilon \text{ and } f(y) \leq \inf(f; g^{-1}((-\infty, c])) + \epsilon.$$

Theorems 4.1 and 4.2 imply the following result.

Theorem 4.3 1. Assume that

$$\{x \in g^{-1}(c) : f(x) = \inf(f; g^{-1}(c))\} \cap (Cr(g, +) \cup Cr(g, -)) = \emptyset.$$

Then there exists $\Lambda_0 > 0$ such that for each $\lambda \geq \Lambda_0$ and each sequence $\{x_i\}_{i=1}^{\infty} \subset X$ which satisfies

$$\lim_{i \to \infty} [f(x_i) + \lambda|g(x_i) - c|] = \inf\{f(z) + \lambda|g(z) - c| : z \in X\}$$

there exists a sequence $\{y_i\}_{i=1}^{\infty} \subset g^{-1}(c)$ such that

$$\lim_{i \to \infty} f(y_i) = \inf(f; g^{-1}(c)) \text{ and } \lim_{i \to \infty} ||y_i - x_i|| = 0.$$

2. Assume that

$$\{x \in g^{-1}(c) : f(x) = \inf(f; g^{-1}((-\infty, c]))\} \cap Cr(g, +) = \emptyset.$$

Then there exists $\Lambda_0 > 0$ such that for each $\lambda \geq \Lambda_0$ and each sequence $\{x_i\}_{i=1}^{\infty} \subset X$ which satisfies

$$\lim_{i \to \infty} [f(x_i) + \lambda \max\{g(x_i) - c, 0\}] = \inf\{f(z) + \lambda \max\{g(z) - c, 0\} : z \in X\}$$

there exists a sequence $\{y_i\}_{i=1}^{\infty} \subset g^{-1}((-\infty, c])$ such that

$$\lim_{i\to\infty} f(y_i) = \inf(f; g^{-1}((-\infty, c])) \text{ and } \lim_{i\to\infty} ||y_i - x_i|| = 0.$$

The next result easily follows from Theorem 4.3.

Theorem 4.4 1. Assume that

$$\{x \in g^{-1}(c) : \ f(x) = \inf(f; g^{-1}(c))\} \cap (Cr(g,+) \cup Cr(g,-)) = \emptyset.$$

Then there exists $\Lambda_0 > 0$ such that if $\lambda \geq \Lambda_0$ and if $x \in X$ satisfies

$$f(x) + \lambda|g(x) - c| = \inf\{f(z) + \lambda|g(z) - c| : \ z \in X\},$$

then $g(x) = c$ and $f(x) = \inf(f; g^{-1}(c))$.

2. Assume that

$$\{x \in g^{-1}(c) : \ f(x) = \inf(f; g^{-1}((-\infty, c])\} \cap Cr(g,+) = \emptyset.$$

Then there exists $\Lambda_0 > 0$ such that if $\lambda \geq \Lambda_0$ and if $x \in X$ satisfies

$$f(x) + \lambda \max\{g(x) - c, 0\} = \inf\{f(z) + \lambda \max\{g(z) - c, 0\} : \ z \in X\},$$

then $g(x) \leq c$ and $f(x) = \inf(f; g^{-1}((-\infty, c]))$.

Theorem 4.3 implies the following result.

Theorem 4.5 1. Assume that

$$\{x \in g^{-1}(c) : \ f(x) = \inf(f; g^{-1}(c))\} \cap (Cr(g,+) \cup Cr(g,-)) = \emptyset$$

and that $\bar{x} \in g^{-1}(c)$ satisfies the following conditions:

$f(\bar{x}) = \inf(f; g^{-1}(c))$;

any sequence $\{x_n\}_{n=1}^{\infty} \subset g^{-1}(c)$ which satisfies $\lim_{n\to\infty} f(x_n) = \inf(f; g^{-1}(c))$ converges to $\bar{x}$ in the norm topology.

Then there exists $\Lambda_0 > 0$ such that for each $\lambda \geq \Lambda_0$ the point $\bar{x}$ is a unique solution of the minimization problem

$$f(z) + \lambda|g(z) - c| \to \min, \ z \in X.$$

2. Assume that

$$\{x \in g^{-1}(c) : \ f(x) = \inf(f; g^{-1}((-\infty, c])\} \cap Cr(g,+) = \emptyset$$

and that $\bar{x} \in g^{-1}((-\infty, c])$ satisfies the following conditions:

$f(\bar{x}) = \inf(f; g^{-1}((-\infty, c]))$;

any sequence $\{x_n\}_{n=1}^{\infty} \subset g^{-1}((-\infty, c])$ which satisfies

$$\lim_{n\to\infty} f(x_n) = \inf(f; g^{-1}((-\infty, c]))$$

converges to $\bar{x}$ in the norm topology.

Then there exists $\Lambda_0 > 0$ such that for each $\lambda \geq \Lambda_0$ the point $\bar{x}$ is a unique solution of the minimization problem

$$f(z) + \lambda \max\{g(z) - c, 0\} \to \min, \ z \in X.$$

References

[1] D. Boukari and A.V. Fiacco, Survey of penalty, exact-penalty and multiplier methods from 1968 to 1993, *Optimization* **32**, 301-334 (1995).

[2] J. V.Burke, An exact penalization viewpoint of constrained optimization, *SIAM J. Control Optim.* **29**, 968-998 (1991).

[3] F. H. Clarke, *Optimization and Nonsmooth Analysis* , Willey Interscience (1983).

[4] R. Deville, R. Godefroy and V. Zizler, *Smoothness and Renorming in Banach Spaces* , Longman (1993).

[5] G. Di Pillo and L. Grippo, Exact penalty functions in constrained optimization, *SIAM J. Control Optim.* **27**, 1333-1360 (1989).

[6] I. Ekeland, On the variational principle, *J. Math. Anal. Appl.* **47**, 324-353 (1974).

[7] I. I. Eremin, The penalty method in convex programming, *Soviet Math. Dokl.* **8**, 459-462 (1966).

[8] B. S. Mordukhovich, *Variational Analysis and Generalized Differentiation, I: Basic Theory, II: Applications*, Springer, Berlin (2006).

[9] R. Palais, Lusternik-Schnirelman theory of Banach manifolds, *Topology* **5**, 115-132 (1966).

[10] W. I. Zangwill, Nonlinear programming via penalty functions, *Management Sci.* **13**, 344-358 (1967).

[11] A. J. Zaslavski, On critical points of Lipschitz functions on smooth manifolds, *Siberian Math. J.* **22**, 63-68 (1981).

[12] A. J. Zaslavski, A sufficient condition for exact penalty in constrained optimization, *SIAM Journal on Optimization* **16**, 250-262 (2005).

[13] A. J. Zaslavski, Existence of exact penalty for constrained optimization problems in Hilbert spaces, *Nonlinear Analysis* **67**, 238-248 (2007).

[14] A. J. Zaslavski, Existence of approximate exact penalty in constrained optimization, *Mathematics of Operations Research* **32**, 484-495 (2007).

In: Progress in Nonlinear Analysis Research
Editor: Erik T. Hoffmann, pp. 347-362
ISBN 978-1-60456-359-7

Chapter 15

ESTIMATION OF VALUE AT RISK FOR HETEROSCEDASTIC AND HEAVY-TAILED ASSET TIME SERIES: EVIDENCE FROM EMERGING ASIAN STOCK MARKETS

***Tzu-Chuan Kao*[1] *and Chu-Hsiung Lin*[2]**
[1]Department of Finance and Banking, Kun Shan University, Taiwan
[2]Department of Risk Management and Insurance, National Kaohsiung First University of Science and Technology, Taiwan

Abstract

We propose a two-stage approach for estimating Value-at-Risk (VaR) that can simultaneously reflect two stylized facts displayed by most asset return series: volatility clustering and the heavy-tailedness of conditional return distributions over short horizons. The proposed method combines the bias-corrected exponentially weighted moving average (EWMA) model for estimating the conditional volatility and the extreme value theory (EVT) for estimating the tail of the innovation distribution. In particular, for minimizing bias in the estimation procedure, the proposed method makes minimal assumptions about the underlying innovation distribution and concentrates on modeling its tail using the non-parametric Hill estimator and uses the moment-ratio Hill estimator for the shape parameter of the extreme value distribution. To validate the model, we conducted an empirical investigation on the daily stock market returns of eight emerging Asian markets: China, India, Indonesia, Malaysia, Philippines, South Korea, Taiwan, and Thailand. In addition, the proposed method was compared with J.P. Morgan's RiskMetrics approach. The empirical results show that the proposed method provides a more accurate forecast of VaR for lower probabilities of VaR violation from 0.1% to 1%. Furthermore, we demonstrate that applying the Hill estimator to estimate the tail of the innovation distribution can better capture additional downside risk faced during times of greater fluctuation than the second-order moment-ratio Hill estimator.

Key Words: value-at-risk, bias-corrected EWMA estimator, extreme value theory, Hill estimator, moment-ratio Hill estimator

1. Introduction

To ensure a stable financial system, it is important to identify and measure market risk accurately, and to control it. According to the Capital Adequacy Directive produced by the Bank of International Settlement in Basle (Basle Committee, 1996), the risk capital of a bank must be sufficient to cover losses on the bank's trading portfolio over a 10-day holding period on 99% of occasions. The value of risk capital is referred to as Value at Risk (VaR). From a statistical standpoint, VaR is defined as the possible maximum loss over a given holding period with a fixed confidence level. Hence, the VaR at the $100(1-\alpha)$ percent confidence level is the lower 100α percentile of the return distribution. Thus, the VaR is estimated on the basis of the distribution of the expected returns. This entails that one needs to make assumptions concerning the form of the expected return distribution. However, ad-hoc assumptions about the form of return distribution will generate a biased estimate of the VaR. To provide the investment community and risk managers with a more accurate VaR model, this study provides a two-stage approach for reducing the estimation bias that may result from a misspecified model and applies it to the stock market returns of eight emerging Asian markets.

In the literature, we find three approaches to estimating the return distribution (McNeil and Frey, 2000). First, there is the *non-parametric historical simulation method*, in which the actual empirical distribution is considered using observations of past returns as a basis. The method is easy to implement and avoids misspecifications of the form of the return distribution. However, when using it, it is difficult to estimate the extreme VaR quantiles (Beder, 1995; Pritisker, 1997; Danielsson and de Vries, 1997). Second, there are *fully parametric methods* that are based on an econometric model of time varying volatility and that assume conditional normality for the returns. Examples of these methods are J.P. Morgan's RiskMetrics[1], and GARCH-type models. GARCH-type models capture the effects of volatility clustering in the returns, but have the weakness that the assumption of conditional normality does not seem to hold for actual return distributions (Baillie and DeGennaro, 1990; Poon and Taylor, 1992; Danielsson and de Vries, 1997). Third, there are *methods based on extreme value theory* (EVT). EVT-based models model the tails of the return distribution directly and thus have the potential to yield better estimates and forecasts of extreme VaR. However, given the conditional heteroscedasticity of asset return data, EVT-based models cannot reflect the conditional volatility (McNeil and Frey, 2000; Cotter, 2001). A number of authors have attempted to alleviate the weaknesses of each of the above approaches. Barone-Adesi et al. (1998) combine the GARCH model with an historical simulation method to estimate the VaR. They fit a GARCH model to a return series and then apply an historical simulation to infer the distribution of the innovations. McNeil and Frey (2000) estimate the VaR by filtering return series with a GARCH model and then apply threshold methods from EVT to fit the innovations distribution. McNeil and Frey's (2000) study indicates that the method proposed by Barone-Adesi et al. (1998) may work well for large data sets, but for smaller data sets, threshold methods from EVT can give better estimates of the tails of the residuals. Recently, Gencay et al. (2003), Byström (2004), and

[1] The RiskMetrics approach uses a EWMA model to forecast conditional volatility. Notice that the EWMA model can be viewed as a special case of the simple GARCH model. See the JP Morgan Bank RiskMetrics Technical (1996).

Kuester et al. (2006) have demonstrated that McNeil and Frey's (2000) method can yield accurate estimates of the VaR.

In practice, the most widely used approach estimating the VaR is J.P. Morgan's RiskMetrics (Deloitte Touche Tohmatsu, 2002). The RiskMetrics approach to calculating the VaR assumes that returns are conditional normal and uses an exponentially weighted moving average (EWMA) model to forecast conditional volatility. The main advantage of the EWMA model is the simplicity of the estimation procedure with a small number of available observations. However, Harris and Shen (2004) indicate that the EWMA model implicitly assumes that the conditional distribution of real returns is normal, if as is suggested by empirical evidence, the conditional distribution is heavy-tailed, the EWMA estimator will be inefficient.

Taking into account the foregoing, the study presented herein extends the work of McNeil and Frey (2000) and proposes a two-stage approach for estimating the VaR. The proposed method combines the bias-corrected EWMA model proposed by Harris and Shen (2004) for estimating the conditional volatility and EVT for estimating the tail of the innovation distribution. Unlike the methods of J.P. Morgan's RiskMetrics and McNeil and Frey (2000), the two-stage approach presented here possesses two advantages: (a) it can simultaneously reflect two stylized facts that are displayed by most asset return series, namely, volatility clustering and the heavy-tailedness of conditional return distributions over short horizons; and (b) it can minimize the bias in the estimation procedure. These advantages are gained because the proposed method makes minimal assumptions about the underlying innovation distribution and concentrates on modeling its tail using a non-parametric Hill estimator (Hill, 1975) and uses a moment-ratio Hill estimator (Danielsson et al., 1996) for the shape parameter of the extreme value distribution.

In an empirical investigation, the proposed two-stage approach was used to measure the downside risk for daily stock market returns of eight emerging Asian markets: China, India, Indonesia, Malaysia, Philippines, South Korea, Taiwan, and Thailand. In addition, the proposed method was compared with the RiskMetrics approach from J.P. Morgan. To assess the predictive performance of the different methods, the study employed a backtesting procedure, which included the test of predictive versus theoretical violation probability, Kupiec's (1995) unconditional coverage testing, and Christofferson's (1998) conditional coverage testing.

The remainder of this paper is organized as follows. Section 2 presents the theoretical foundations used to estimate the downside risk. Section 3 describes the stylized fact of the empirical data. Section 4 presents the empirical results. Section 5 concludes.

2. Methodology

The following subsections (i) describe the bias-corrected EWMA model proposed by Harris and Shen (2004), which was used to measure the conditional volatility; (ii) describe the non-parametric Hill estimator (Hill, 1975) and moment-ratio Hill estimator (Danielsson et al., 1996) based on EVT for estimating the tail index of the innovations distribution; and (iii) present the backtesting procedure.

2.1. An Alternative VaR Model

Let ($X_t, t \in Z$) be a stationary time series that presents daily observations of the log return on a financial asset price. This study assumes that the dynamics of X are given by:

$$X_t = \mu_t + \sigma_t Z_t \tag{1}$$

where the innovations Z_t are a white noise process with zero mean and unit variance and are independent and identical according to a distribution function $F_Z(z)$ (i.e. iid). This study assumes that the conditional mean μ_t and the conditional standard deviation σ_t are measurable with respect to I_{t-1}, where I_{t-1} is the set of all information through time *t-1*. From Equation (1), the time *t* quantile, x_q^t, can be derived for a given probability level *q*. Therefore, x_q^t is defined such that $q = P(X_t \le x_q | I_{t-1}) = F_{X_t|I_{t-1}}(x_q)$, where $F_X(x)$ denotes the cumulative probability distribution of X_t. Using Equation (1) the probability level *q* can be rewritten as

$$\begin{aligned} q &= P\{\mu_t + \sigma_t Z_t \le x_q | I_{t-1}\} \\ &= P\{Z_t \le (x_q - \mu_t)/\sigma_t | I_{t-1}\} \\ &= F_Z\{(x_q - \mu_t)/\sigma_t\} \ (2) \end{aligned} \tag{2}$$

If the quantile associated with the distribution $F_Z(z)$ is denoted as z_q , namely $z_q = (x_q - \mu_t)/\sigma_t$, the time *t* quantile (or the time *t* VaR at the $100(1-q)$ % confidence level), x_q^t, can be calculated as

$$x_q^t = \mu_t + \sigma_t z_q \tag{3}$$

Via Equation (3), x_q^t is obtained and must estimate μ_t, σ_t, and z_q. To measure x_q^t this study proposes a two-stage approach. The first stage uses the equal-weight moving-average approach to estimate μ_t and applies the bias-corrected EWMA estimator to estimate σ_t. The second stage uses the Hill estimator (Hill, 1975) and moment-ratio Hill estimator based on EVT to estimate z_q. The following subsections detail the proposed two-stage approach.

2.1.1. First Stage: Estimating μ_t and σ_t

For predictive purposes, a rolling window of 500 observations was used to dynamically estimate conditional mean μ_t and conditional standard deviation σ_t . First, this study considers using the equal-weight moving-average approach to estimate the conditional mean μ_t .

$$\mu_t = \sum_{n=1}^{500} x_{t-n} \Big/ 500 \tag{4}$$

Sequentially, this study considers that the conditional variance σ_t^2 follows the bias-corrected EWMA model proposed by Harris and Shen (2004). The EWMA model, known as an integrated GARCH or IGARCH model, is used to estimate the conditional variance σ_t^2 because it relies on one parameter only and thus facilitates estimation. The EWMA estimator is given by:

$$\sigma_t^2 = \lambda \sigma_{t-1}^2 + (1-\lambda) X_{t-1}^2 \tag{5}$$

where the parameter λ is called the decay factor and must be less than unity. The empirical study set $\lambda = 0.94$, which is the value used by RiskMetrics to estimate daily volatility. The EWMA estimator is based on the maximum likelihood estimator of the variance of the normal distribution; hence, if the EWMA model is misspecified, it will generate biased conditional variance forecasts. To generate bias-corrected conditional variance forecasts, Harris and Shen (2004) first estimate the conditional bias in the EWMA model using a realized volatility regression with squared returns, and then apply the estimated regression parameters to obtain new bias-corrected conditional variance forecasts. The theoretical framework for the estimation is detailed in Theil (1966), Mincer and Zarnovitz (1969), and Harris and Shen (2004). The estimation process is summarized as follows:

Step 1: Use Equation (5) to obtain the forecasts of conditional variance $\hat{\sigma}_t^2$.

Step 2: Use the realized volatility measured by the squared returns (X_t^2) and the conditional variance forecasts ($\hat{\sigma}_t^2$) to construct a regression model, as follows:

$$X_t^2 = a + b\hat{\sigma}_t^2 + v_t \tag{6}$$

where v_t denotes the error term. Equation (6) is widely used in the literature on volatility modeling to evaluate the explanatory power of a particular conditional variance model and to correct conditional variance forecasts (Theil, 1966; Mincer and Zarnovitz, 1969). In order to correct the bias of conditional variance forecasts that are generated by the EWMA model, we need to estimate the parameters a and b.

Step 3: Once we have estimated parameter $\hat{a}$ and $\hat{b}$, we can define a new bias-corrected conditional variance forecasts, $\hat{\hat{\sigma}}_t^2$ given by

$$\hat{\hat{\sigma}}_t^2 = \hat{a} + \hat{b}\hat{\sigma}_t^2 \tag{7}$$

2.1.2. Second Stage: Estimating z_q

Following McNeil and Frey (2000), this study assumes that the residuals distribution is not normal, and that perhaps it has heavy tails or is leptokurtic. Therefore this study used EVT to model the tail of the residuals distribution and applied this EVT model to estimate z_q. EVT investigates the distribution of tail observations in large samples. In the limit, the shape of the tail follows a Pareto law for a general class of heavy-tailed distributions. The tail fatness of the distribution is characterized by the tail index. As indicated by Dewachter and Gielens (1999) and Cotter (2001), for minimizing model risk, the Hill non-parametric method for estimating the tail index is superior to parametric procedures that require assumptions about the exact distribution type of extreme values. Consequently, this study applies a non-parametric Hill estimator (Hill, 1975) and moment-ratio Hill estimator (Danielsson et al., 1996) to estimate the tail index of residuals distribution and z_q. The process used to estimate z_q is presented below.

This study assumes that the standardized residuals series $(z_{t-n}, z_{t-n+1}, \dots, z_{t-1}) = \left(\frac{x_{t-n} - \hat{\mu}_{t-n}}{\hat{\sigma}_{t-n}}, \dots, \frac{x_{t-1} - \hat{\mu}_{t-1}}{\hat{\sigma}_{t-1}} \right)$ follow an extreme value distribution with a tail-index parameter γ and applies the Hill estimator to estimate γ, the desired quantile z_q can be estimated by:

$$\hat{z}_q = z_{(m+1)} \left(\frac{1-q}{m/n} \right)^{-\gamma}. \tag{8}$$

Let $z_{(i)}$ denote the i th decreasing order statistic of the absolute standardized residuals such that $z_{(i)} \geq z_{(i+1)}$ for $i = 2,\dots,n$. Specifying m as the number of tail observations, the tail estimator γ proposed by Hill (1975) is obtained as follows:

$$\gamma = \frac{1}{m} \sum_{i=1}^{m} \ln z_{(i)} - \ln z_{(m)} \tag{9}$$

For simplicity, we follow Quintos et al. (2001) and set the threshold m to be 10% of the sample size.

In addition, Danielsson et al. (1996) propose a generalization of the Hill estimator, the so-called k-order moment-ratio Hill estimator, to estimate the tail index of the distribution. They point out that the moment-ratio Hill estimator has a lower asymptotic bias than the Hill estimator. Moreover, Wagner and Marsh (2004) demonstrate that a second-order moment-ratio Hill estimator performs better. Therefore this study also uses a second-order moment-ratio Hill estimator, ω_2, to estimate the tail-index parameter γ.The k-order moment-ratio Hill estimator is given by

$$\omega_k(z_{(m)}) = \frac{\phi_k(z_{(m)})}{k\phi_{k-1}(z_{(m)})} \qquad \mathrm{k} = 1,2,... \tag{10}$$

where $\phi_0(z_{(m)}) = 1$, $\phi_k(z_{(m)}) = \frac{1}{m}\sum_{i=1}^{k}(\ln z_{(i)} - \ln z_{(m)})^k$. When k=1, the first-order moment-ratio Hill estimator ω_1 is equal to the Hill estimator (Hill, 1975).

Finally, we substitute $\hat{\mu}_t$, $\hat{\sigma}_t$ and $\hat{z}_q$ into Equation (3). Then, the forecasted time t VaR at $100(1-q)\%$ condition level, $\hat{x}_q^t$, is given by

$$\hat{x}_q^t = \hat{\mu}_t + \hat{\hat{\sigma}}_t \hat{z}_q \tag{11}$$

2.2. Backtesting VaRs

To assess the predictive performance of the different risk models and obtain a robust evaluation result, this study employed three backtesting methods, as follows:

First, this study used the test of predictive versus theoretical probability that a VaR violation will occur that is recommended by the Basle Committee of Banking Supervision. Using each of risk models for estimating VaR, given different probabilities that a VaR violation[2] q^* will occur, this study forecasts the daily VaR. These forecasts of the VaR are then compared with the actual returns on the days in question and the number of days on which the actual returns exceed the forecasted VaR is counted. The number of such days, N, is called the number of exceedences. The predictive probability that a VaR violation will occur can then be obtained using the number of exceedences divided by the total number of observed returns, T. If the predictive probability that a VaR violation will occur that is estimated by a particular risk model is close to the theoretical probability, the risk model in question may be said to perform well for estimating VaRs.

Second, this study used the unconditional coverage test developed by Kupiec (1995). The unconditional coverage hypothesis states that the probability that a VaR violation will occur that is obtained for a particular risk model, call it q, is significantly different from the given

[2] A VaR violation occurs when the actual returns exceed the forecasted VaR.

probability that a VaR violation q^* will occur. Namely, $H_0 : q = q^*$. The likelihood ratio (LR) statistic is

$$LR_{uc} = -2\ln\left[(1-q^*)^{T-N}(q^*)^N\right] + 2\ln\left\{[1-(N/T)]^{T-N}(N/T)^N\right\} \sim \chi^2(1) \qquad (12)$$

The LR_{uc} test statistic has a chi-squared distribution with one degree of freedom.

Third, this study used the conditional coverage test developed by Christofferson (1998). The conditional coverage hypothesis is to simultaneously test whether the VaR violations are independent and the average number of violations is correct. Namely, $H_0 : \pi_{01} = \pi_{11} = q^*$. The LR statistic is

$$LR_{cc} = -2\ln(L_A - L_I) \sim \chi^2(2) \qquad (13)$$

$$L_A = (1-q^*)^{T-N}(q^*)^N$$

$$L_I = (1-\pi_{01})^{T_{00}}(\pi_{01})^{T_{01}}(1-\pi_{11})^{T_{10}}(\pi_{11})^{T_{11}}$$

$$\pi_{ij} = T_{ij}/(T_{i0} + T_{i1})$$

where T_{ij}, $i, j = 0, 1$ is the number of times state j follows state i, state 0 denotes an actual return less than the forecasted VaR, and state 1 denotes an actual return that exceeds the forecasted VaR. The LR_{cc} test statistic has a chi-squared distribution with two degrees of freedom.

3. Data Description

To verify the performance of the proposed approach, we conducted an empirical investigation that examined the daily stock market indices of eight emerging Asian markets (China, India, Indonesia, Malaysia, Philippines, South Korea, Taiwan, and Thailand[3]) for the period August 23, 1991 to August 23, 2007. The daily returns were measured by the first difference of the natural logarithm of the stock market index. The dataset was downloaded from Datastream.

Table 1 lists the preliminary statistics of the returns for the eight stock markets. The mean returns for the entire period are almost zero. According to the standard deviation for the returns,

[3] The names of the eight emerging Asian stock market indices are as follows: (1) China: Shanghai SE Composite Price Index (2) India: India BSE National Price Index (3) Indonesia: Jakarta SE Composite Price Index (4) Malaysia: KLCI Composite Price Index (5) Philippines: Philippine SE Price Index (6) South Korea: Korea SE Composite Price Index (7) Taiwan: TSEC Weighted Index (8) Thailand: Bangkok S.E.T. Price Index.

Table 1. Summary statistics of the daily returns for eight emerging Asian stock markets

Statistics	China	India	Indonesia	Malaysia	Philippines	South Korea	Taiwan	Thailand
Mean	0.0004	0.0005	0.0005	0.0002	0.0003	0.0002	0.0002	0.0000
Maximum	0.1522	0.1664	0.1313	0.1492	0.1618	0.0898	0.0617	0.1135
Minimum	-0.1401	-0.1194	-0.1273	-0.1424	-0.0974	-0.1280	-0.0698	-0.1606
Std. Dev.	0.0221	0.0162	0.0152	0.0138	0.0147	0.0182	0.0151	0.0167
Skewness	0.0378	0.0473	0.0114	0.3799	0.5894	-0.1578	-0.1017	0.1949
Kurtosis	10.5611	10.7223	12.5916	19.1659	13.2976	6.5508	5.0820	9.8330
Jarque-Bera [a]	9550.72*	10375.36*	15494.22*	44317.97*	18267.50*	2230.79*	778.79*	7918.29*
ADF [b]	-34.14	-58.48	-53.14	-53.75	-53.87	-60.68	-62.06	-57.45
Ljung-Box Q(6) [c]	43.94*	62.73*	131.59*	66.96*	120.76*	131.59*	30.02*	52.42*
Ljung-Box $Q^2(6)$ [d]	1078*	497.24*	630.63*	1913.80*	2044.6*	871.09*	558.60*	592.84*
Observations	4009	4175	4042	4061	4081	4213	4271	4057

Note: * significant at 1% level

[a] Jarque-Bera is a test statistic for testing whether the series is normally distributed.

[b] ADF indicates augmented Dickey and Fuller (1979,1981) unit root tests for whether the series is stationary.

[c] Ljung-Box $Q(6)$ indicates the Ljung-Box Q-statistics at lag 6 by log return series, it is a test statistic for the null hypothesis that there is no autocorrelation up to order 6.

[d] Ljung-Box $Q^2(6)$ indicates the Ljung-Box Q-statistic at lag 6 by squared log return series, it is a test statistic for the null hypothesis that there is no autocorrelation up to order 6.

investing in firms that are listed on the Chinese stock market carries with it a higher risk than for the stock markets of other countries. The returns distributions are heavy-tailed or leptokurtic, as demonstrated by high kurtosis and highly significant Jargue-Bera statistics. The returns distributions for the South Korean and Taiwanese stock markets display a negative skewness, and for the stock markets of other countries they display a positive skewness. This implies that negative extreme returns are more likely to occur than the normal distribution forecasts in the stock markets of South Korea and Taiwan, while in other countries positive extreme returns are more likely to occur than the normal distribution forecasts. These findings indicate that the returns distributions for the eight emerging Asian stock markets can be characterized by fat-tailed distributions and that the left (or negative) and right (or positive) tails should be treated, respectively[4]. In addition, this investigation also reports the ADF statistics, which indicate that the eight emerging Asian stock market return series are stationary. Ljung-Box statistics for the returns themselves and for the squared returns are also presented. These statistics confirm that the empirical return series contains autocorrelation and volatility clustering, which suggests that the conditional modeling of short run returns is beneficial.

4. Empirical Results

To implement Equation (11), we used a moving window of 500 observations to measure dynamically the forecasts of the VaR given different probabilities of VaR violation, which range from 0.1% to 5%. The proposed two-stage approaches, the B-EWMA-Hill method and B-EWMA-moment-Hill method, were compared with RiskMetrics approach (named EWMA). The B-EWMA-Hill method combines the bias-corrected EWMA model with the Hill estimator, while the B-EWMA-moment-Hill method combines the bias-corrected EWMA model with the second-order moment-ratio Hill estimator. To assess the predictive performance of different VaR models, we used a backtesting procedure that included the test of predictive versus theoretical violation probability, Kupiec's (1995) unconditional coverage test and Christofferson's (1998) conditional coverage test.

Table 2 lists the empirical results for predictive versus theoretical probability that a VaR violation will occur. For all the emerging Asian stock markets, the B-EWMA-Hill method outperformed the other methods when the violation probabilities were below 1%. Furthermore, for violation probabilities of from 0.1% to 1%, the predictive probabilities that a VaR violation will occur that computed by RiskMetrics and the B-EWMA-moment-Hill method exceeded the theoretical probabilities and the predictive probabilities that a VaR violation will occur that were calculated using the B-EWMA-Hill method conditional were equal to or slight higher than the theoretical probabilities. As a result, the forecasts of VaR estimated by RiskMetrics and B-EWMA-moment-Hill methods have an underestimation bias and those estimated by the B-EWMA-Hill method have a slight underestimation bias. Consequently, using the B-EWMA-Hill method to measure VaR can capture the additional downside risk that is faced during times of greater fluctuation.

[4] Although both the left and right tails of the return distributions are interesting from the perspective of risk management, we chose to focus solely on estimating the left tails of the return distributions to measure the downside risk for the eight emerging Asian stock markets

Table 2. Backtesting results: predictive vs. theoretical VaR violation probability

Country	Method	Theoretical VaR violation probability							
		0.05	0.025	0.01	0.005	0.004	0.003	0.002	0.001
China	EWMA	0.0568#	0.0369#	0.0202#	0.0127	0.0099	0.0091	0.0075	0.0067
	B-EWMA-Hill	0.0742	0.0472	0.0218	0.0115#	0.0075#	0.0060#	0.0044#	0.0024#
	B-EWMA-moment-Hill	0.0798	0.0580	0.0306	0.0187	0.0171	0.0147	0.0099	0.0060
India	EWMA	0.0557#	0.0347#	0.0232	0.0183	0.0157	0.0146	0.0101	0.0090
	B-EWMA-Hill	0.0706	0.0388	0.0198#	0.0086#	0.0078#	0.0064#	0.0045#	0.0026#
	B-EWMA-moment-Hill	0.0740	0.0489	0.0258	0.0157	0.0127	0.0112	0.0075	0.0034
Indonesia	EWMA	0.0535#	0.0366#	0.0228	0.0185	0.0161	0.0149	0.0138	0.0106
	B-EWMA-Hill	0.0704	0.0444	0.0200#	0.0106#	0.0079#	0.0059#	0.0031#	0.0016#
	B-EWMA-moment-Hill	0.0755	0.0507	0.0283	0.0145	0.0122	0.0102	0.0079	0.0028
Malaysia	EWMA	0.0503#	0.0355	0.0226	0.0191	0.0179	0.0152	0.0117	0.0097
	B-EWMA-Hill	0.0659	0.0339#	0.0160#	0.0074#	0.0051#	0.0039#	0.0023#	0.0012#
	B-EWMA-moment-Hill	0.0748	0.0452	0.0269	0.0144	0.0117	0.0105	0.0066	0.0035
Philippines	EWMA	0.0538#	0.0348#	0.0232	0.0151	0.0139	0.0124	0.0108	0.0085
	B-EWMA-Hill	0.0712	0.0418	0.0186#	0.0077#	0.0062#	0.0039#	0.0027#	0.0008#
	B-EWMA-moment-Hill	0.0763	0.0534	0.0294	0.0186	0.0143	0.0108	0.0077	0.0035
South Korea	EWMA	0.0622#	0.0350#	0.0206#	0.0125#	0.0099#	0.0092	0.0070	0.0044
	B-EWMA-Hill	0.0685	0.0468	0.0265	0.0133	0.0099#	0.0081#	0.0055#	0.0022#
	B-EWMA-moment-Hill	0.0700	0.0534	0.0343	0.0210	0.0177	0.0147	0.0118	0.0070
Taiwan	EWMA	0.0552#	0.0386#	0.0216	0.0155	0.0148	0.0133	0.0105	0.0076
	B-EWMA-Hill	0.0700	0.0433	0.0180#	0.0087#	0.0065#	0.0032#	0.0022#	0.0007#
	B-EWMA-moment-Hill	0.0765	0.0523	0.0296	0.0180	0.0141	0.0115	0.0069	0.0040
Thailand	EWMA	0.0449#	0.0297#	0.0156#	0.0106	0.0086	0.0078	0.0066	0.0051
	B-EWMA-Hill	0.0598	0.0344	0.0164	0.0074#	0.0047#	0.0039#	0.0031#	0.0012#
	B-EWMA-moment-Hill	0.0641	0.0426	0.0215	0.0117	0.0109	0.0078	0.0051	0.0035

Note: " # " indicates the predictive VaR violation probability is close to theoretical VaR violation probability.

Table 3. Backtesting results: unconditional coverage test statistics (LR_{uc})

Country	Method	Theoretical VaR violation probability							
		0.05	0.025	0.01	0.005	0.004	0.003	0.002	0.001
China	EWMA	2.3322#	12.8328	20.5967	21.0165	15.6769	20.4081	22.5960	36.0394
	B-EWMA-Hill	27.2845	40.6976	26.6352	15.6694	6.2865#	5.7034#	5.2696#	3.4576#
	B-EWMA-moment-Hill	40.1815	82.3225	69.5371	55.4481	59.3757	59.0054	40.3283	28.6257
India	EWMA	1.7439#	9.3104	34.1521	56.4215	52.5857	61.6824	44.2555	62.8056
	B-EWMA-Hill	21.3250	18.0694	20.1995	5.7073*	7.7441	7.5889	6.0941#	4.8180#
	B-EWMA-moment-Hill	28.3847	49.4017	46.8783	39.1355	32.1852	35.3166	23.5040	9.1945
Indonesia	EWMA	0.6273#	12.2273	30.9007	54.7617	52.9988	61.6270	75.5222	78.8775
	B-EWMA-Hill	19.8201	32.0895	20.0816	12.1644	7.4181	5.5564#	1.4202#	0.7093#
	B-EWMA-moment-Hill	30.2593	53.3971	57.5555	30.6866	27.5888	27.1362	25.0177	5.2662*
Malaysia	EWMA	0.0040#	10.2228	30.3326	59.4982	67.0299	64.3435	56.4484	69.1553
	B-EWMA-Hill	12.4116	7.5244	7.8412	2.5957#	0.6749#	0.6306#	0.1395#	0.0697#
	B-EWMA-moment-Hill	29.0789	34.8110	50.5695	30.2653	25.0341	29.3053	17.0415	9.7359
Philippines	EWMA	0.7724#	9.1644	33.2556	34.2991	38.7966	42.4901	49.1808	99.7529
	B-EWMA-Hill	21.7930	24.9928	15.3405	3.3433#	2.6715#	0.6004#	0.5867#	4.1445#
	B-EWMA-moment-Hill	32.5519	64.9135	64.6887	56.3411	41.3411	31.5975	24.5629	27.6823
South Korea	EWMA	7.9808	9.9086	23.6959	21.7377	16.9971	22.4688	20.5231	17.1286
	B-EWMA-Hill	17.6400	42.2164	51.4942	25.5575	16.9971	16.0811	11.3772	2.9496#
	B-EWMA-moment-Hill	20.4086	68.1483	98.9340	77.4215	68.8995	63.9844	60.6444	41.4626
Taiwan	EWMA	1.5148#	18.0584	28.4604	39.3763	47.7116	53.3724	49.2360	48.6991
	B-EWMA-Hill	20.8025	31.2905	14.5920	6.1042#	3.6306#	0.0545#	0.0363#	0.2391#
	B-EWMA-moment-Hill	35.4062	64.7873	70.3495	56.4709	42.5393	39.0618	19.9493	13.8857
Thailand	EWMA	1.4262#	2.1887	6.9963	11.9956	10.1922	13.7138	17.1067	21.4194
	B-EWMA-Hill	4.8713	8.2987	8.9058	2.6297#	0.2891#	0.6432#	1.3861#	0.0721#
	B-EWMA-moment-Hill	9.8557	26.9299	25.6872	16.8351	20.9485	13.7138	8.4972	9.7712

Note: The critical value of the LR_{uc} statistics at 1% significant level is 6.6349. "#" indicates test results not reject null hypothesis at 1% significant level.

Table 4. Backtesting results: conditional coverage test statistics (LR_{cc})

Country	Method	Theoretical VaR violation probability							
		0.05	0.025	0.01	0.005	0.004	0.003	0.002	0.001
China	EWMA	10.9224	22.0800	21.3435	21.6519	17.0067	22.0044	24.8468	38.6934
	B-EWMA-Hill	42.5895	42.6171	26.6726	16.3452	6.5754#	5.8831#	5.3662#	3.4862#
	B-EWMA-moment-Hill	52.0718	88.4909	69.7111	55.4652	59.4654	59.3231	41.6580	28.8054
India	EWMA	30.5209	25.0596	42.8937	62.5868	57.2685	67.1173	53.7739	73.7257
	B-EWMA-Hill	51.2713	38.0375	36.2947	23.5123	27.1014	22.8058	17.3364	4.8547#
	B-EWMA-moment-Hill	55.7928	73.6935	67.0903	51.6681	43.6462	48.7470	36.6623	14.4683
Indonesia	EWMA	20.8270	15.4334	37.3363	61.1549	61.2960	67.0755	81.8390	79.9883
	B-EWMA-Hill	47.9754	38.0748	22.7437	13.2753	7.7352#	5.7344#	1.4707#	0.7219#
	B-EWMA-moment-Hill	58.3914	66.8305	65.6353	36.4131	31.1858	28.3601	25.3347	5.3049*
Malaysia	EWMA	26.9185	39.8781	51.8354	77.1518	86.4028	82.9127	75.2672	85.2367
	B-EWMA-Hill	36.5472	28.5272	25.2184	4.8787#	4.3849#	5.3897#	7.0446#	0.0767#
	B-EWMA-moment-Hill	60.0595	51.1958	74.0748	40.1318	28.8857	33.9049	19.7286	14.9271
Philippines	EWMA	28.1307	28.7716	45.7548	43.4304	49.1232	54.6324	58.0770	101.5379
	B-EWMA-Hill	45.6153	51.5482	16.4065	10.2407	5.5964#	0.6782#	0.6247#	4.1476#
	B-EWMA-moment-Hill	58.7291	89.9826	77.5188	59.5942	41.6828	32.6242	26.6757	27.7453
South Korea	EWMA	7.9808#	9.9086	23.6959	21.7377	16.9971	22.4688	20.5231	17.1286
	B-EWMA-Hill	19.1412	46.0865	57.5430	31.9085	26.5940	22.4210	20.7926	2.9762#
	B-EWMA-moment-Hill	21.5149	72.8063	106.7779	84.6469	75.4326	73.0936	73.1556	48.9540
Taiwan	EWMA	19.2337	26.8263	30.1819	39.5255	47.9325	53.7853	49.8492	49.0197
	B-EWMA-Hill	36.7875	50.5594	15.6314	6.5234#	3.8659#	0.1131#	0.0623#	0.2420#
	B-EWMA-moment-Hill	50.0684	83.5100	76.5823	57.5103	42.8478	39.8093	20.2115	13.9733
Thailand	EWMA	5.4015#	4.8638#	11.9489	21.2636	21.9125	26.6115	25.2672	25.1242
	B-EWMA-Hill	12.6733	12.4725	13.3721	9.8975	4.3096#	5.3971#	7.0615#	0.0791#
	B-EWMA-moment-Hill	15.0062	33.9947	30.1047	24.8886	29.7933	26.6115	12.2021	14.9570

Note: The critical value of the LR_{cc} statistics at 1% significant level is 9.2103. "#" indicates test results not reject null hypothesis at 1% significance level.

To demonstrate that the proposed two-stage approach is reasonably accurate for estimating VaR, we also checked its validity systematically by backtesting, according to the procedures proposed by Kupiec (1995) and Christoffersen (1998). Kupiec's (1995) methodology is an unconditional coverage testing and Christoffersen's (1998) methodology is a conditional coverage testing. Their null hypothesis is that the risk model is correct. Thus, if the test statistics reject the null hypothesis, the proposed risk model is not adequate. The empirical results are presented in Tables 3 and 4. For all emerging Asian stock markets (the Indian and South Korean stock markets excepted), the B-EWMA-Hill method exhibit excellent results given probabilities of VaR violation from 0.1% to 1% because the unconditional and conditional coverage test statistics (i.e. LR_{uc} and LR_{cc}) fail to reject the null hypothesis in almost all cases. However, for the Indian and South Korean stock markets, the use of RiskMetrics, the B-EWMA-Hill method and the B-EWMA-moment-Hill method could not provide more accurate VaR forecasts. Given that the problem of estimating VaR is still related to the estimation of conditional volatility, the results suggest that the decay factor λ in the EWMA and bias-corrected EWMA models may vary across returns series and that the choice of decay factor may lead to variations in performance[5].

Summarizing the above findings, given lower probabilities of VaR violation, the B-EWMA-Hill method for measuring VaR is more accurate than the RiskMetrics and B-EWMA-moment-Hill methods. Furthermore, the results show that using the Hill estimator to estimate the tail of the innovation distribution can better capture the nature of downside risk than can the second-order moment-ratio Hill estimator.

5. Conclusions

Previous empirical research has found that combining the GARCH model and EVT to determine VaR can fully protect against default that results from volatility clustering and extreme price movements. The main contribution of the study reported herein is to provide a two-stage approach that incorporates the non-parametric Hill estimator and the moment-ratio Hill estimator based on EVT into a bias-corrected EWMA model for estimating VaR. This approach is especially appealing because it makes minimal assumptions about the underlying innovation distribution and concentrates on modeling its tail using the non-parametric Hill estimator and use the moment-ratio Hill estimator for the shape parameter of the extreme value distribution. Moreover, rather than using the RiskMetrics approach developed by J.P. Morgan, applying the bias-corrected EWMA model to estimate conditional volatility, as we have done, can reduce the misspecified model risk. The empirical study that involved the eight emerging Asian stock markets demonstrated the accuracy of the proposed approach.

For lower probabilities of VaR violation from 0.1% to 1%, the backtesting showed that the proposed B-EWMA-Hill method produces more accurate forecasts of VaR than the RiskMetrics and B-EWMA-moment-Hill methods. Thus, the proposed B-EWMA-Hill method can ensure adequate prudence for the setting of risk capital. Furthermore, we demonstrate that using the Hill estimator to estimate the tail of the innovation distribution can better capture the additional downside risk faced during times of greater fluctuation than the second-order moment-ratio Hill estimator.

---

[5] We set $\lambda = 0.94$.

References

Baillie, R. & DeGennaro, R. (1990). Stock Returns and Volatility. *Journal of financial and Quantitative Analysis*, **25**, 203-214.

Barone-Adesi, G., Boutgoin, F., & Giannopoulos, K. (1998). Don't Look Back. *Risk*, 11(8).

Base Committee on Banking Supervision (1996). Ammendment to the Capital Accord to Incorporate Market Risks. *Basle Report* **(24),** BIS.

Beder, T, S. (1995). Seductive but Dangerous. *Financial Analysis Journal*, September-October, 12-24.

Byström, H.N.E. (2004). Managing Extreme Risks in Tranquil and Volatile Markets Using Conditional Extreme Value Theory. *International Review of Financial Analysis*, **13**, 133-152.

Christoffersen, P. (1998). Evaluating Interval Forecasts. *International Economic Review*, **39**, 841-862.

Cotter, J. (2001). Margin Exceedences for European Stock Index Futures Using Extreme Value Theory. *Journal of Banking & Finance*, **25**, 1475-1502.

Danielsson, J., Jansen, D.W., & de Vries, C.G. (1996). The Method of Moments Ratio Estimator for the Tail Shape Parameter. *Communication in Statistics Theory and Methods*, **25**(4), 711-720.

Danielsson, J. & De Vires, C. G. (1997). Tail Index and Quantile Estimation with Very High Frequency Data. *Journal of Empirical Finance*, **4**, 241-257.

Dewachter , H. & Gielens, G. (1999). Setting Futures Margins: the Extremes Approach. *Applied Financial Economics*, **9**, 173-181.

Deloitte Touche Tohmatsu. (2002). *Global Risk Management Survey*.

Gencay, R., Selcuk, F., & Ulugülyağci, A. (2003). High Volatility, Thick Tails and Extreme Value Theory in Value-at-Risk Estimation. *Insurance: Mathematics and Economics*, **33**, 337-356.

Harris, R.D.F. & Shen, J. (2004). Estimation of VaR with Bias-Corrected Forecasts of Conditional Volatility. *Journal of Derivatives*, **11**(4), 10-20.

Hill, B.M. (1975). A Simple Approach to Inference About The Tail of A Distribution. *The Annals of Mathematical Statistics*, **3**, 1163-1174.

JP Morgan. (1996). Riskmetrics^{TM} Technical Document. (4th ed), New York.

Kuester, K., Mittnik, S., & Paolella, M. S. (2006). Value-at Risk Prediction: A Comparison of Alternative Strategies. *Journal of Financial Econometrics*, **4**, 53-89.

Kupiec, P. (1995). Technique for Verifying the Accuracy of Risk Management Models. *Journal of Derivative*, 3, 73-84.

McNeil, A. & Frey, R. (2000). Estimation of Tail-Related Risk Measure for Heteroscedastic Financial Time Series: An Extreme Value Approach. *Journal of Empirical finance*, **7,** 271-300.

Mincer, J., & Zarnovitz, V. (1969). The Valuation of Economic Forecasts. In J. Mincer, ed., Economic Forecasts and Expectations. Cambridge: National Bureau of Economic Research.

Poon, S. H. & Taylor, S. J. (1992). Stock Returns and Volatility: An Empirical study of the UK Stock Market. *Journal of Banking and Finance*, **16**, 37-59.

Pritsker, M. (1997). Evaluating Value at Risk Methodologies: Accury versus Computational Time. *Journal of Financial Services Research*, **12**(3), 201-243.

Theil, H. (1966). Applied Economic Forecasting. Amsterdam: North Holland.

Quintos, C., Fan, Z., & Phillips, P. C. B. (2001). Structural Change Tests in Tail Behavior and the Asian Crisis. *Review of Economic Studies*, **68**, 633-663.

Wanger, N., & Marsh, T. A. (2004). Tail Index Estimation in Small Samples Simulation Results for Independent and ARCH-type Financial Return Models. *Statistical Paper*, **45**, 545-562.

In: Progress in Nonlinear Analysis Research
Editor: Erik T. Hoffmann, pp. 363-389
ISBN 978-1-60456-359-7

Chapter 16

STABILITY OF SOLUTIONS OF SYSTEMS WITH IMPULSE EFFECT

***Alexander O. Ignatyev*[1]* and *Oleksiy A. Ignatyev*[2]†**
[1]Institute of Applied Mathematics and Mechanics,
R.Luxemburg Street,74, Donetsk-83114, Ukraine
[2]Department of Statistics and Probability,
Michigan State University, East Lansing, MI 48824, USA

Abstract

In this chapter, a system of ordinary differential equations with impulse effect at fixed instants is considered. The system is assumed to have the zero solution. It is shown that the existence of a corresponding Lyapunov function is a necessary and sufficient condition for the uniform asymptotic stability of the zero solution. Restrictions on perturbations of the right-hand sides of differential equations and impulse effect are obtained under which the uniform asymptotic stability of the zero solution of the "unperturbed" system implies the uniform asymptotic stability of the zero solution of the "perturbed" system. In the case of a periodic system with impulse effect, it is shown that if the trivial solution of the system is stable or asymptotically stable, then it is uniformly stable or uniformly asymptotically stable, respectively. By using the method of Lyapunov functions, the criteria of asymptotical stability and instability are obtained.

1. Introduction and Main Definitions

Many evolution processes are characterized by the fact that at certain moments of time they experience a change of state abruptly. These processes are subject to short-term perturbations which duration is negligible in comparison with the duration of the process. Consequently, it is natural to assume that these perturbations act instantaneously, that is, in the form of impulses. It is known, for example, that many biological phenomena involving

*E-mail address: aoignat@mail.ru; ignat@iamm.ac.donetsk.ua
†E-mail address: ignatyev@stt.msu.edu; lehaignat@yahoo.com

thresholds, bursting rhythm models in medicine and biology, optimal control models in economics, pharmacokinetics and frequency modulated systems, do exhibit impulsive effects. Thus impulsive differential equations, that is, differential equations involving impulse effects, appear as a natural description of observed evolution phenomena of several real world problems [6, 7, 19, 20, 21, 44, 46, 48, 49, 51, 50, 56, 54, 55]. The investiagation of systems with impulse effect was first carried out in monograph [36] where an example of the pendulum with impulse disturbance was considered. Dynamical systems with discontinuous trajectories were also studied in [42, 43]. The early work on differential equations with impulse effect were summarized in monograph [46] in which the foundations of this theory were described. In recent years, the study of impulsive systems has received an increasing interest [1, 2, 3, 4, 5, 8, 12, 16, 17, 23, 25, 24, 26, 33, 34, 35, 37, 38, 39, 47, 52].

The theory of impulsive differential equations is much richer than the corresponding theory of differential equations without impulse effects. Consequently, the theory of impulsive differential equations is interesting in itself, and it is easy to see that it will assume greater importance in the near future since the application of the theory to various fields is also increasing. Thus there is every reason for studying the theory of impulsive differential equations as a well deserved discipline.

Consider the system of differential equations with impulse effect at fixed moments of time:

$$\frac{dx}{dt} = X(t, x), \ \ t \neq \tau_i, \ \ t \in \mathbb{R}_+ \tag{1.1}$$

$$x(\tau_i^+) - x(\tau_i) = J_i(x), \ \ i \in \mathbb{N} \tag{1.2}$$

where $t \in \mathbb{R}_+ := [0, \infty)$ is time, $x = (x^1, \ldots, x^n) \in \mathbb{R}^n$, $0 = \tau_0 < \tau_1 < \tau_2 < \ldots, \ \tau_i \to \infty$, $\mathbb{N}$ is the set of positive integers, $x(\tau_i^+)$ means the right-hand limit of x at τ_i.

Let $F(t) = F(t, t_0, x_0)$ be any solution of system (1.1) and (1.2) starting at (t_0, x_0) where $t_0 \in (\tau_0, \tau_1)$. The evolution process behaves as follows: the point $(t, F(t))$ begins its motion from the initial point (t_0, x_0) and moves along the curve $\{(t, x) : t \geq t_0, x = F(t)\}$ where $F(t)$ is the solution of system (1.1) with initial data (t_0, x_0) until the time $\tau_1 > t_0$ at which the position $x_1 = F(\tau_1)$ is transfered to the position $x_1^+ = x_1 + J_1(x_1)$. Then the point $(t, F(t))$ continues to move along the curve describing by differential equations (1.1) with initial conditions (τ_1, x_1^+) until the time τ_2, etc.

The union of relations (1.1) and (1.2) characterizing the evolutionary process is called a system of differential equations with impulse effect (or an impulsive system). The curve described by the point $(t, F(t))$ in the extended phase space is called an integral curve, and the function defining this curve is called a solution of the system of differential equations with impulse effect. The instants τ_k are called the instants of impulse effect. It is assumed that a solution $F(t)$ of a system of differential equations with impulse effect is a left continuous function at the instants of impulse effect, i.e. $F(\tau_i^-) = F(\tau_k - 0) = F(\tau_k)$.

One of the main problems in the investigation of dynamical systems (in particular, systems with impulse effect) is the stability problem. If one needs to study the stability of some solution $F(t)$ of system (1.1) and (1.2), this problem can be reduced to the problem of the stability investigation of the zero solution of the system of perturbed motion. So throughout this chapter we assume that $X(t, 0) \equiv 0, \ J_i(0) = 0$, and system (1.1) and (1.2) has the trivial solution

$$x = 0. \tag{1.3}$$

Consider system (1.1) and (1.2). Let us denote by $F(t) = F(t, t_0, x_0)$ the solution of system (1.1) and (1.2) satisfying the identities $F(t_0, t_0, x_0) = x_0$ where $x_0 \in B_H$ if $t_0 \in \bigcup_{i=1}^{\infty}(\tau_{i-1}, \tau_i)$. If, however, $t_0 = \tau_i$, $i \in \mathbb{N}$, then we denote by $F(t, t_0, x_0)$ for $t > t_0$ the solution of system (1.1) and (1.2) with initial conditions

$$F(t_0^+, t_0^+, x_0 + J_i(x_0)) = x_0 + J_i(x_0).$$

According to the existing tradition [3, 8, 25, 37], the solution $F(t)$ is assumed to be continuous from the left at the points τ_i : $F(\tau_i) = F(\tau_i^-)$, $i \in \mathbb{N}$.

Denote $\|x\| = \max_{1 \le s \le n} |x^s|$,

$$B_H = \{x \in \mathbb{R}^n : \|x\| \le H,\}, \quad G = \bigcup_{i=1}^{\infty}(\tau_{i-1}, \tau_i) \times B_H.$$

Let us introduce next properties.

- (P1) A function X is continuous in each domain $(\tau_{i-1}, \tau_i) \times B_H \quad (i \in \mathbb{N})$, satisfies the Lipschitz condition

$$\|X(t, x_1) - X(t, x_2)\| \le L\|x_1 - x_2\|,$$

  and for any $k \in \mathbb{N}$ there exist the finite limits

$$\lim_{t \to \tau_k^-} X(t, x) = X(\tau_k, x), \qquad \lim_{t \to \tau_k^+} X(t, x) = X(\tau_k^+, x).$$

- (P2) Functions $J_i(x)$ $(i \in \mathbb{N})$ are continuous and

$$\|J_i(x_1) - J_i(x_2)\| \le L\|x_1 - x_2\|, \quad x_1 \in B_H,\ x_2 \in B_H,\ i \in \mathbb{N}.$$

- (P3) Constants $\tau_i \ (i \in \mathbb{N})$ satisfy conditions $0 = \tau_0 < \tau_1 < \tau_2 < \ldots$, $\lim_{i \to \infty} \tau_i = +\infty$, and in each segment $[t, \tau] \subset \mathbb{R}_+$, there are not more then p points τ_i where p depends on the length of the segment $[t, \tau]$.

Let us introduce some necessary definitions.

Definition 1.1. *The trivial (zero) solution of system (1.1) and (1.2) is said to be stable if for any $\varepsilon > 0$ and $t_0 \in \mathbb{R}_+$ there exists a $\delta = \delta(\varepsilon, t_0) > 0$ such that $\|x_0\| < \delta$ implies $\|F(t, t_0, x_0)\| < \varepsilon$ for $t \ge t_0$.*

Definition 1.2. *If in last definition δ can be chosen independent of t_0 (i.e. $\delta = \delta(\varepsilon)$), then the zero solution of (1.1) and (1.2) is called uniformly stable.*

Definition 1.3. *The solution $x = 0$ of system (1.1) and (1.2) is said to be attractive if for any $t_0 \in \mathbb{R}_+$ there exists an $\eta = \eta(t_0) > 0$ such that for any $\varepsilon > 0$ and $x_0 \in B_\eta$ there exists a $\sigma = \sigma(\varepsilon, t_0, x_0) > 0$ such that $\|F(t, t_0, x_0)\| < \varepsilon$ for all $t \ge t_0 + \sigma$. We say that B_η is contained in the domain of attraction of the zero solution of system (1.1) and (1.2) at the moment t_0.*

In other words, solution (1.3) of system (1.1) and (1.2) is attractive if

$$\lim_{t \to \infty} \|F(t, t_0, x_0)\| = 0. \tag{1.4}$$

Definition 1.4. *The zero solution of system (1.1) and (1.2) is said to be uniformly attractive if for some* $\eta > 0$ *and any* $\varepsilon > 0$ *there exists a* $\sigma = \sigma(\varepsilon) > 0$ *such that* $\|F(t, t_0, x_0)\| < \varepsilon$ *for all* $t_0 \in \mathbb{R}_+$, $x_0 \in B_\eta$ *and* $t \geq t_0 + \sigma$.

In other words, the zero solution of system (1.1) and (1.2) is uniformly attractive if (1.4) holds uniformly with respect to $t_0 \geq 0, x_0 \in B_\eta$.

Definition 1.5. *The zero solution of system (1.1) and (1.2) is said to be:*

- *asymptotically stable if it is stable and attractive;*
- *uniformly asymptotically stable if it is uniformly stable and uniformly attractive.*

Let $\mathcal{K}$ denote the class of Hahn functions [45], that is $g \in \mathcal{K}$ if $g : \mathbb{R}_+ \to \mathbb{R}_+$ is a continuous increasing function such that $g(0) = 0$. Note that in [13, 14, 15, 27, 28] these functions are called *wedges*.

Definition 1.6. *We say that a function* $V : \mathbb{R}_+ \times B_H \to \mathbb{R}$ *belongs to the class* $\mathcal{V}_0$ *if it is continuous on G, satisfies the condition* $|V(t, x_1) - V(t, x_2)| \leq L\|x_1 - x_2\|$ *uniformly with respect to* $t \in \mathbb{R}_+$*;* $V(t, 0) \equiv 0$ *for* $t \in \mathbb{R}_+$*, and for any* $k \in \mathbb{N}$ *there exist the finite limits*

$$\lim_{t \to \tau_k^-} V(t, x) = V(\tau_k, x), \qquad \lim_{t \to \tau_k^+} V(t, x) = V(\tau_k^+, x).$$

We say that a function $V \in \mathcal{V}_0$ *belongs to the class* $\mathcal{V}_1$ *if it is a* C^1 *function on G. For* $(t, x) \in G$ *we define the derivative of the function* $V \in \mathcal{V}_1$ *as*

$$\frac{dV}{dt} = \frac{\partial V}{\partial t} + \frac{\partial V}{\partial x} \cdot X(t, x).$$

Definition 1.7. *System (1.1) and (1.2) is said to be periodic with respect to* t *with the period* ω *if there exists a* $q \in \mathbb{N}$*, such that*

$$J_{i+q}(x) \equiv J_i(x), \ \tau_{i+q} = \tau_i + \omega \ \ (i = 1, 2, ...); \quad X(t + \omega, x) \equiv X(t, x). \tag{1.5}$$

Samoilenko and Perestyuk [46, p.17] proved the following lemma.

Lemma 1.1. *Assume that there are* p *points of impulse effect in* $(t_0, t_1]$*, system (1.1) and (1.2) satisfies properties (P1) and (P2) in the domain* $\mathbb{R}_+ \times B_H$*, and solutions* $F(t, t_0, x_1)$ *and* $F(t, t_0, x_2)$ *lie in* B_H *for* $t \in (t_0, t_1]$*. Then*

$$\|F(t, t_0, x_1) - F(t, t_0, x_2)\| \leq (1 + L)^p \|x_1 - x_2\| e^{L(t_1 - t_0)}. \tag{1.6}$$

In [22] the following theorem was proved.

Theorem 1.1. *Let system (1.1) and (1.2) be such that properties (P1), (P2), and (P3) hold and there exist functions* $V \in \mathcal{V}_1, g \in \mathcal{K}, b \in \mathcal{K}, c \in \mathcal{K}$*, such that*

$$V(t, x) \geq g(\|x\|), \quad t \in \mathbb{R}_+, \ x \in B_H, \tag{1.7}$$

$$V(t, x) \leq b(\|x\|), \quad t \in \mathbb{R}_+, \ x \in B_H, \tag{1.8}$$

$$\left.\frac{dV}{dt}\right|_{(1.1)} \leq -c(\|x\|), \quad t \in \mathbb{R}_+,\ x \in B_H,\ t \neq \tau_i\ (i \in \mathbb{N}), K \tag{1.9}$$

$$V(\tau_i^+, x + J_i(x)) \leq V(\tau_i, x), \quad i \in \mathbb{N}, x \in B_H. \tag{1.10}$$

Then solution (1.3) of system (1.1) and (1.2) is asymptotically stable.

This chapter is organized as follows. In section 2 we prove Theorem 2.1 which is analogous to Theorem 1.1, but we assume that $V \in \mathcal{V}_0$. Moreover, we prove that the asymptotic stability is uniform, and this theorem is conversible. Section 3 is devoted to stability conditions in the case system (1.1) and (1.2) to be periodic. In section 4 we obtain stability and instability criteria by means of Lyapunov functions with derivatives of high orders. Section 5 completes the chapter; we consider the stability problem of the zero solution of perturbed systems with impulse effect there.

2. The Theorem on the Uniform Asymptotic Stability and the Converse Theorem

Theorem 2.1. *Let system (1.1) and (1.2) be such that properties (P1), (P2), and (P3) hold and there exist functions $V \in \mathcal{V}_0, g \in \mathcal{K}, b \in \mathcal{K}, c \in \mathcal{K}$, such that inequalities (1.7), (1.8), (1.10) and*

$$D^+V(t,x) \leq -c(\|x\|), \quad t \in \mathbb{R}_+,\ x \in B_H,\ t \neq \tau_i\ (i \in \mathbb{N}) \tag{2.1}$$

hold where $D^+V(t,x)$ is the right upper Dini derivative of the function V along the solution $x(t)$.

Then solution (1.3) of system (1.1) and (1.2) is uniformly asymptotically stable and there exists $H_0 > 0$ $(H_0 < H)$ such that the domain of attraction of (1.3) contains the set B_{H_0}.

Proof. First let us prove the uniform stability of the zero solution. Pick any $\varepsilon_1 > 0$ $(\varepsilon_1 < H)$, and choose $\delta = b^{-1}(g(\varepsilon_1))$. If $\|x_0\| < \delta$, then from properties (1.7), (1.10), (2.1), and (1.8) we have

$$g(\|F(t,t_0,x_0)\|) \leq V(t, F(t,t_0,x_0)) \leq V(t_0,x_0) \leq b(\|x_0\|) < b(b^{-1}(g(\varepsilon_1))) = g(\varepsilon_1),$$

whence it follows $\|F(t,t_0,x_0)\| < \varepsilon_1$ for $t > t_0$. This proves the uniform stability of the zero solution.

We shall show that solution (1.3) is uniformly attractive. The uniform stability of the zero solution implies that for every positive ε_2 $(\varepsilon_2 < H)$ there exists $H_0 = H_0(\varepsilon_2) > 0$ such that for any $t_0 \in R_+$, $x_0 \in B_{H_0}$ the inequality $\|F(t,t_0,x_0)\| < \varepsilon_2$ holds for $t > t_0$. Consider the solution $F(t,t_0,x_0)$ of equations (1.1) and (1.2) with $x_0 \in B_{H_0}$. Since $x_0 \in B_{H_0}$ then

$$V(t_0,x_0) \leq b(\|x_0\|) \leq b(H_0). \tag{2.2}$$

Let ε be any sufficiently small positive number. Denote $T(\varepsilon) := b(H_0)/c(b^{-1}(g(\varepsilon)))$. Let us show that there exists $\sigma \in [0,T]$ such that

$$V(t_0+\sigma, F(t_0+\sigma,t_0,x_0)) < g(\varepsilon). \tag{2.3}$$

Suppose the opposite: for any $\sigma \in [0, T]$ the inequality $V(t_0 + \sigma, F(t_0 + \sigma, t_0, x_0)) \geq g(\varepsilon)$ holds, whence we have

$$\|F(t, t_0, x_0)\| \geq b^{-1}(V(t, x(t, t_0, x_0)) \geq b^{-1}(g(\varepsilon)) \tag{2.4}$$

for $t_0 \leq t \leq t_0 + T$. From inequalities (2.4) we derive that

$$\frac{dV(t, F(t, t_0, x_0))}{dt} \leq -c(b^{-1}(g(\varepsilon))), \quad t \in \mathbb{R}_+,\ x \in B_{\varepsilon_2},\ t \neq \tau_i\ (i \in \mathbb{N}),\ c \in \mathcal{K}.$$

This inequality together with (1.7) and (1.10) imply

$$0 < g(\|F(t, t_0, x_0)\|) \leq V(t, F(t, t_0, x_0)) \leq V(t_0, x_0) - c(b^{-1}(g(\varepsilon)))(t - t_0).$$

For $t - t_0 = T$ we have $V(t_0, x_0) - b(H_0) > 0$, but this contradicts to (2.2). This proves the existence of $\sigma \in [0, T]$ such that inequality (2.3) is valid. Since V does not increase along the solution $F(t, t_0, x_0)$, then $V(t, F(t, t_0, x_0)) < g(\varepsilon)$ for $t \geq t_0 + \sigma$. This implies $\|F(t, t_0, x_0)\| < \varepsilon$ for $t \geq t_0 + \sigma$. Hence solution (1.3) is uniformly attractive, and its domain of attraction contains the set B_{H_0}. This completes the proof. □

Let us show that Theorem 2.1 is conversible. For this we shall prove the following auxiliary assertions.

Lemma 2.1. *Suppose that $\psi(\tau) : \mathbb{R}_+ \to \mathbb{R}_+$ is a non-negative bounded piecewise continuous function approaching zero as $\tau \to \infty$, with discontinuity points of the first kind $\tau_1, \ldots, \tau_n, \ldots$, such that $0 < \tau_1 < \tau_2 < \ldots$ and $\lim_{i\to\infty} \tau_i = +\infty$. Suppose that $\psi(\tau_i) = \psi(\tau_i^-)$, $i \in \mathbb{N}$, and on the set $\bigcup_{i=1}^{\infty} (\tau_{i-1}, \tau_i)$ the function $\psi(\tau)$ has a derivative $\psi'(\tau)$, satisfying the inequality $|\psi'(\tau)| \leq P$. Then the function $f(t) = \sup_{t\leq\tau<\infty} \psi(\tau)$ at any value of $t \in \mathbb{R}_+$ has one-sided derivatives such that*

$$-P \leq f'(t^-) \leq 0, \qquad -P \leq f'(t^+) \leq 0. \tag{2.5}$$

Proof. Note that the curve $y = f(t)$ for $t \in \mathbb{R}_+$ consists of alternating parts of the curve $y = \psi(t)$, where $\psi(t)$ is non-increasing and segments where the function $f(t)$ is constant; that is $f(t)$ is a piecewise continuous monotonically non-increasing function approaching zero as $t \to \infty$. The discontinuity points can occur only at the points $t = \tau_i$ $(i \in \mathbb{N})$. For $t \in \mathbb{R}_+$ this function has the one-sided derivatives $f'(t^{\pm}) = f'(t \pm 0)$ satisfying conditions (2.5), as required. □

Lemma 2.2. *Suppose that $f_1 : \mathbb{R}_+ \to \mathbb{R}_+$, and $f_2 : \mathbb{R}_+ \to \mathbb{R}_+$ are two bounded nonnegative piecewise continuous functions having one-sided limits at the discontinuity points and such that*

$$\lim_{t\to\infty} f_1(t) = 0, \qquad \lim_{t\to\infty} f_2(t) = 0.$$

Then

$$|\sup_{t\in\mathbb{R}_+} f_1(t) - \sup_{t\in\mathbb{R}_+} f_2(t)| \leq \sup_{t\in\mathbb{R}_+} |f_1(t) - f_2(t)|.$$

Proof. Lemma 2.2 is obviously true in the case $\sup\limits_{t\in\mathbb{R}_+} f_1(t) = \sup\limits_{t\in\mathbb{R}_+} f_2(t)$. Suppose that $\sup\limits_{t\in\mathbb{R}_+} f_1(t) \neq \sup\limits_{t\in\mathbb{R}_+} f_2(t)$. Without loss of generality we assume that $\sup\limits_{t\in\mathbb{R}_+} f_1(t) > \sup\limits_{t\in\mathbb{R}_+} f_2(t)$. Since the functions $f_1(t), f_2(t)$, being non-negative and bounded, approach zero as $t \to \infty$, there exist finite values t_1, t_2, t_3 such that $\sup\limits_{t\in\mathbb{R}_+} f_1(t) = f_1(t_1^{\pm})$, $\sup\limits_{t\in\mathbb{R}_+} f_2(t) = f_2(t_2^{\pm})$, and $\sup\limits_{t\in\mathbb{R}_+} |f_1(t) - f_2(t)| = |f_1(t_3^{\pm}) - f_2(t_3^{\pm})|$. Here $f_k(t_i^{\pm})$ $(k = 1, 2;\ i = 1, 2, 3)$ denotes either the value or the one-sided limit of the function f_k at the point t_i. Consequently, $f_1(t_1^{\pm}) - f_2(t_2^{\pm}) \le f_1(t_1^{\pm}) - f_2(t_1^{\pm}) \le |f_1(t_3^{\pm}) - f_2(t_3^{\pm})|$. This proves Lemma 2.2. $\square$

Theorem 2.2. *Suppose that the properties (P1) – (P3) hold, solution (1.3) of system (1.1) and (1.2) is uniformly asymptotically stable, and its domain of attraction contains the set B_{H_*} $(0 < H_* < H)$. Then there exist constants $P > 0$, $L_1 > 0$ and functions $g \in \mathcal{K}$, $b \in \mathcal{K}$, $c \in \mathcal{K}$, $V : \mathbb{R}_+ \times B_{H_*} \to \mathbb{R}_+$ such that $V \in \mathcal{V}_0$,*

$$|V(t, x_1) - V(t, x_2)| \le L_1 \|x_1 - x_2\| \tag{2.6}$$

for $t \in \mathbb{R}_+$, $x_1 \in B_{H_*}$, $x_2 \in B_{H_*}$, $|V(t_1, x) - V(t_2, x)| \le L_1|t_1 - t_2|$ for $x \in B_{H_*}$, $t_1 \in (\tau_{i-1}, \tau_i)$, $t_2 \in (\tau_{i-1}, \tau_i)$, $(i \in \mathbb{N})$, conditions (1.7), (1.8), (1.10), (2.1) are satisfied, and $D^+V(t, x) \ge -P$.

If system (1.1) and (1.2) is periodic with the period ω, then the function V can also be chosen to be periodic in t with the period ω.

Proof. To prove this theorem, let us use the method of [9]. Denote $\varphi(t)$ a scalar monotonically decreasing function satisfying the inequality

$$\|F(t, t_0, x_0)\| \le \varphi(t - t_0) \quad \text{for } t \ge t_0 \tag{2.7}$$

for any $x_0 \in B_{H_*}$, and such that $\lim\limits_{t\to\infty} \varphi(t) = 0$. The existence of such function $\varphi(t)$ follows from the property of uniform asymptotic stability of solution (1.3) of system (1.1) and (1.2) in the sense of Definition 1.5. (It suffices to choose for $\varphi(t)$ any continuous positive function that is monotonically decreasing to zero and satisfies the inequality $\varphi(t) > \varepsilon$ for $t \in [\sigma(\varepsilon), \sigma(\frac{\varepsilon}{2})]$).

Let $Q(t) : \mathbb{R}_+ \to \mathbb{R}_+$ be a monotonically increasing continuous function such that $\lim\limits_{t\to\infty} Q(t) = +\infty$. In [40, p.452-458] it is shown that there exists a continuously differentiable function $g = g(\varphi) : \mathbb{R}_+ \to \mathbb{R}_+$, such that

$$g \in \mathcal{K}, \qquad g' \in \mathcal{K}, \tag{2.8}$$

$$\int_0^\infty g(\varphi(\tau))d\tau = N_1 < +\infty, \tag{2.9}$$

$$\int_0^\infty g'(\varphi(\tau))M(\tau)d\tau = N_2 < +\infty, \tag{2.10}$$

$$g'(\varphi(\tau))M(\tau) < N_3 \quad \text{for} \quad \text{all} \quad \tau \geq 0, \tag{2.11}$$

where N_1, N_2, N_3 are positive constants.

Let us show that the function

$$V(t,x) = \int_t^\infty g(\|F(\tau,t,x)\|)d\tau + \sup_{t\leq\tau<\infty} g(\|F(\tau,t,x)\|) \tag{2.12}$$

satisfies all the conditions of the theorem.

Integral (2.9) converges; hence by estimate (2.7) the integral in the right-hand side of (2.12) converges. Consequently, the function V is defined in the domain

$$\mathbb{R}_+ \times B_{H_*}. \tag{2.13}$$

Note that $\sup\limits_{t\leq\tau<\infty} \|F(\tau,t,x)\| \geq \|x\|$. By (2.8) we obtain

$$\int_t^\infty g(\|F(\tau,t,x)\|)d\tau \geq 0, \qquad \sup_{t\leq\tau<\infty} g(\|F(\tau,t,x)\|) \geq g(\|x\|),$$

that is, the function V satisfies inequality (1.7).

From estimate (2.7) we obtain that $\|F(t,t_0,x_0)\| \leq \varphi(0)$ for all t_0, x_0 from domain (2.13); hence

$$V(t,x) \leq \int_0^\infty g(\varphi(\tau))d\tau + g(\varphi(0)) = N_4 = \text{const}\,.$$

Consequently, the function V is uniformly bounded in domain (2.13). Let us show that the function V satisfies inequality (2.6). Using Lemma 2.2 we obtain

$$|V(t,x_1) - V(t,x_2)| = \left| \int_t^\infty [g(\|F(\tau,t,x_1)\|) - g(\|F(\tau,t,x_2)\|)]d\tau \right.$$

$$\left. +[\sup_{t\leq\tau<\infty} g(\|F(\tau,t,x_1)\|) - \sup_{t\leq\tau<\infty} g(\|F(\tau,t,x_2)\|)] \right|$$

$$\leq \int_t^\infty g'_\varphi(\sup(\|F(\tau,t,x_1)\|, \|F(\tau,t,x_2)\|)\|F(\tau,t,x_1) - F(\tau,t,x_2)\|d\tau$$

$$+ \sup_{t\leq\tau<\infty} \big|g(\|F(\tau,t,x_1)\|) - g(\|F(\tau,t,x_2)\|)\big|. \tag{2.14}$$

According to Lemma 1.1, we have

$$\|F(\tau,t,x_1) - F(\tau,t,x_2)\| < Q(\tau - t)\|x_1 - x_2\|, \tag{2.15}$$

where $Q : \mathbb{R}_+ \to \mathbb{R}_+$ is monotonically increasing positive continuous function satisfying the inequality

$$Q(\tau - t) > (1+L)^p e^{L(\tau - t)};$$

here p is the number of points τ_i in the segment $[t,\tau]$. By property (P3) such a function does exist. Taking into account inequality (2.14), applying to the second summand in the right-hand side of (2.14) the Mean Value Theorem, and using estimates (2.10) and (2.11) we obtain

$$|V(t,x_1)-V(t,x_2)| \leq \|x_1-x_2\|\Bigg[\int_t^\infty g'_\varphi(\sup(\|F(\tau,t,x_1)\|,\ \|F(\tau,t,x_2)\|)Q(\tau-t)d\tau$$

$$+\sup_{t\leq\tau<\infty}(g'_\varphi(\varphi(\tau-t)))Q(\tau-t)\Bigg] \leq (N_2+N_3)\|x_1-x_2\|, \tag{2.16}$$

which proves that V satisfies condition (2.6). This implies that there exists a function $b\in\mathcal{K}$ such that inequality (1.8) holds. One can choose $b(\|x\|)=(N_2+N_3)(\|x\|)$.

We now verify that $|V(t_1,x)-V(t_2,x)| \leq L_1|t_1-t_2|$ for $x\in B_{H_*}$, $t_1\in(\tau_{i-1},\tau_i)$, $t_2\in(\tau_{i-1},\tau_i)$, $i\in\mathbb{N}$, where L_1 is a constant that does not depend on i. Indeed, this follows from the fact that the first summand in the right-hand side of (2.12) is a continuous function with respect to t for $t\in\mathbb{R}_+$ and a differentiable function with respect to t for $t\neq\tau_i$ with the absolute value of the derivative uniformly bounded, while the second summand is continuous with respect to t for $t\neq\tau_i$, has bounded non-positive left and right derivatives with respect to t for $t\in\mathbb{R}_+$ by Lemma 2.1, and the absolute values of these derivatives are also uniformly bounded.

We consider $D^+V(t,x)$ along solutions of system (1.1) and (1.2). We have $D^+V = D^+\overline{V}$, where $\overline{V}$ is the result of substituting an arbitrary solution $F(t,t_0,x_0)$ of system (1.1) and (1.2) into the function V. But

$$\overline{V}=\int_t^\infty g(\|F(\tau,t,F(t,t_0,x_0))\|)d\tau+\sup_{t\leq\tau<\infty} g(\|F(\tau,t,F(t,t_0,x_0))\|)$$

$$=\int_t^\infty g(\|F(\tau,t_0,x_0))\|)d\tau+\sup_{t\leq\tau<\infty} g(\|F(\tau,t,F(t,t_0,x_0))\|),$$

since $F(\tau,t,F(t,t_0,x_0))\equiv F(\tau,t_0,x_0)$. Hence for $t=t_0$ we obtain

$$D^+V(t,F(t,t_0,x_0))\Bigg|_{t=t_0}=\frac{d}{dt}\int_t^\infty g(\|F(\tau,t_0,x_0)\|)d\tau\Bigg|_{t=t_0}+$$

$$+\lim_{\Delta t\to 0+}\sup\frac{1}{\Delta t}\Bigg(\sup_{t_0+\Delta t\leq\tau<\infty} g(\|F(\tau,t_0,x_0)\|)-\sup_{t_0\leq\tau<\infty} g(\|F(\tau,t_0,x_0)\|)\Bigg).$$

The second summand in the right-hand side of the last equality is non-positive; hence

$$D^+V(t,F(t,t_0,x_0))\Bigg|_{t=t_0}\leq -g(\|F(t_0,t_0,x_0)\|)=-g(\|x_0\|),$$

that is, D^+V satisfies (2.1).

For $x = F(\tau_k, t_0, x_0)$ we have $x + J_k(x) = F(\tau_k^+, t_0, x_0)$, whence, bearing in mind that the second summand in (2.12) is a non-increasing function, we get that inequality (1.10) holds along the solution $F(t, t_0, x_0)$ of system (1.1) and (1.2).

Now suppose that system (1.1) and (2.2) is periodic with respect to t with the period ω. This means that $X(t + \omega, x) \equiv X(t, x)$, and there exists a $q \in \mathbb{N}$, such that $J_k(x) \equiv J_{k+q}(x)$ and $\tau_{k+q} = \tau_k + \omega$ for any positive integer k. We shall prove that in this case the function $V(t, x)$ defined by equality (2.12) has the property $V(t+\omega, x) \equiv V(t, x)$. Indeed,

$$V(t+\omega, x) = \int_{t+\omega}^{\infty} g(\|F(\tau, t+\omega, x)\|)d\tau + \sup_{t+\omega \le \tau < \infty} g(\|F(\tau, t+\omega, x)\|).$$

Introducing a new variable s by the formula $\tau = s + \omega$ we obtain

$$V(t+\omega, x) = \int_{t}^{\infty} g(\|F(s+\omega, t+\omega, x)\|)ds + \sup_{t \le s < \infty} g(\|F(s+\omega, t+\omega, x)\|). \quad (2.17)$$

Using the obvious property of solutions of periodic systems

$$F(t+\omega, t_0+\omega, x_0) = F(t, t_0, x_0), \quad (2.18)$$

by equalities (2.17) and (2.18) we obtain $V(t + \omega, x) \equiv V(t, x)$, as required. The theorem is proved. □

3. Stability of the Zero Solution of Periodic Systems

In this section, we assume that system (1.1) and (1.2) is periodic in t with the period ω.

Theorem 3.1. *If the zero solution of system (1.1) and (1.2) is stable, then it is uniformly stable.*

Proof. Conditions (1.5) imply that

$$F(t+\omega, t_0+\omega, x_0) \equiv F(t, t_0, x_0), \quad (3.1)$$

hence it suffices to prove that for any $\varepsilon > 0$ there exists $\delta = \delta(\varepsilon) > 0$ such that the inequality $\|F(t, t_0, x_0)\| \le \varepsilon$ holds for $t \ge t_0$ and for all $t_0 \in [0, \omega)$, $x_0 \in B_\delta$. By assumption, for any $\varepsilon > 0$ there exists $\delta_1 > 0$ such that if $x_\omega = F(\omega)$ satisfies the condition $x_\omega \in B_{\delta_1}$, then $F(t, \omega, x_\omega) \in B_\varepsilon$ for $t \ge \omega$. If the condition $x_0 = F(t_0, t_0, x_0) \in B_\delta$ holds at any time $t_0 \in [0, \omega)$, then from (1.6) it follows that if $\delta = (1 + L)^{-q} e^{-L\omega} \delta_1$, then $F(t, t_0, x_0) \in B_\varepsilon$. This completes the proof of the theorem. □

Theorem 3.2. *If the zero solution of system (1.1) and (1.2) is asymptotically stable, then it is uniformly asymptotically stable.*

Proof. By the assumptions, solution $x = 0$ is asymptotically stable; hence (1.4) holds in the domain

$$t_0 \in \mathbb{R}_+, \quad x_0 \in B_\lambda, \quad (3.2)$$

where λ is a sufficiently small positive number. Now, let us prove that (1.4) holds uniformly in t_0, x_0, i.e. for every $\varepsilon > 0$ there exists $\sigma = \sigma(\varepsilon) > 0$ such that the inequality

$\|F(t, t_0, x_0)\| \leq \varepsilon$ holds for all $t \geq t_0 + \sigma$. Since the system is periodic, we assume that t_0 belongs to the segment $[0, \omega]$. First, let us define the number $\eta = \eta(\varepsilon) > 0$ from the condition

$$\|F(t, t_0, x_0)\| \leq \varepsilon \text{ for } x_0 \in B_\eta, \ t > t_0. \tag{3.3}$$

This is possible because of the uniform stability of the zero solution. Arguing by contradiction, assume that the number $\sigma = \sigma(\varepsilon)$ does not exist. Then, for an arbitrary large integer m, there exists a $t_m > m\omega$, initial values $t_{0m} \in [0, \omega]$ and $x_{0m} \in B_\lambda$ such that

$$\|F(t_m, t_{0m}, x_{0m})\| > \varepsilon. \tag{3.4}$$

Since the sequence of points $\{t_{0m} \times x_{0m}\}$ belongs to a compact set, one can choose a subsequence of this sequence which converges to some point $t_* \times x_*$ where $t_* \in [0, \omega]$, $\|x_*\| \leq \lambda$. Without loss of generality, we assume that the sequence $\{t_{0m}\}$ itself converges to the point $t_* \in [0, \omega]$, and the sequence $\{x_{0m}\}$ converges to the point x_*, $\|x_*\| \leq \lambda$. Hence (1.4) holds for the initial values $t_0 = t_*$, $x_0 = x_*$. Then there exists large enough $k = k(\varepsilon)$ such that

$$\|F(t_* + k\omega, t_*, x_*)\| < \frac{1}{2}\eta(\varepsilon). \tag{3.5}$$

Denote $x^{(k)} = x(t_{0m}+k\omega, t_{0m}, x_{0m})$. Since $t_{0m} \to t_*$, $x_{0m} \to x_*$, there exist arbitrary large values of m for which we have the inequality

$$\|x^{(k)}\| < \eta(\varepsilon), \tag{3.6}$$

where $x^{(k)} = F(t_{0m} + k\omega, t_{0m}, x_{0m})$. Estimates (3.6) and (3.3) imply that for all $t > t_{0m}$ we have $\|F(t, t_{0m}, x^{(k)})\| \leq \varepsilon$, hence identity (3.1) and uniqueness property of the solution imply the estimate

$$\varepsilon \geq \|F(t, t_{0m}, x^{(k)})\| \equiv \|F(t + k\omega, t_{0m} + k\omega, x^{(k)})\| \equiv \|F(t + k\omega, t_{0m}, x_{0m})\|.$$

The obtained inequality contradicts assumption (3.4), because there exists an instant t_m, such that $t_m > k\omega$. This contradiction proves that (1.4) is uniform in t_0 and x_0. This completes the proof of the theorem. □

Next, we apply the ideas of [18, 29, 30] to the stability analysis of the trivial solution of a system of impulsive differential equations by using Lyapunov's second method.

Following [31], we introduce the following definitions.

Definition 3.1. *We say that the function $g : \mathbb{R}_+ \to \mathbb{R}^s$, $s \in \mathbb{N}$ is not eventually vanishing if for any $M > 0$ there exists a $t > M$ such that $g(t) \neq 0$.*
We say that the sequence of numbers $\{u_k\}_{k=1}^{\infty}$ is not eventually vanishing if for any positive integer M there exists a $k > M$ such that $u_k \neq 0$.

Theorem 3.3. *Suppose that every solution of system (1.1) and (1.2) with initial data $(t_0, x_0) \in \mathbb{R}_+ \times B_H$ is bounded, there exists a function $V(t, x) \in \mathcal{V}_1$ which is periodic in t with the period ω and satisfies the conditions*

$$a(\|x\|) \leq V(t, x) \leq b(\|x\|), \ a \in \mathcal{K}, \ b \in \mathcal{K}, \tag{3.7}$$

and

$$\frac{dV}{dt} \le 0 \text{ for } (t, x) \in G,$$

$$\Delta V_i(x) = V(\tau_i^+, x + J_i(x)) - V(\tau_i, x) \le 0, \quad i \in \mathbb{N}.$$

If along any eventually vanishing solution of equations (1.1) and (1.2), at least one of the following conditions holds:
(i) the function dV/dt is not eventually vanishing,
(ii) the sequence $\{\Delta V_i\}$ is not eventually vanishing,
then the zero solution of system (1.1) and (1.2) is uniformly asymptotically stable.

Proof. The stability property of the zero solution can be proved just as in Theorem 2.1. Theorem 3.1 implies that the solution $x = 0$ is uniformly stable, i.e. for any $\varepsilon > 0$ there exists $\delta = \delta(\varepsilon) > 0$ such that for all $t_0 \in \mathbb{R}_+$ and $x_0 \in B_\delta$ the inequality $\|F(t, t_0, x_0)\| \le \varepsilon$ holds for all $t > t_0$. Let us prove that any solution $F(t, t_0, x_0)$ with such initial conditions possesses property (1.4).

Consider the function $v(t) = V(t, F(t, t_0, x_0))$. It is not increasing and is bounded below; hence the limit

$$\lim_{t\to\infty} v(t) = \eta \ge 0$$

exists. Let us prove that $\eta = 0$. Suppose that the opposite is true:

$$\eta = \lim_{t\to\infty} V(t, F(t, t_0, x_0)) > 0. \tag{3.8}$$

Consider the sequence of points $\{x_k\}$ where $x_k = F(t_0 + k\omega, t_0, x_0)$. Taking into account the fact that $\|x_k\| \le \varepsilon < H$, one can conclude that there exists a subsequence converging to the point $x_* \in B_\varepsilon$. Without loss of generality, we shall assume that the sequence $\{x_k\}$ itself converges to the point $x_* \ne 0$. Since the function V is continuous in x and periodic in t, the equality $V(t_0, x_*) = \eta$ must be satisfied. Consider the semitrajectory $F(t, t_0, x_*)$ for $t \ge t_0$ and the function $V_*(t) = V(t, F(t, t_0, x_*))$ along the trajectory. By the assumptions of the theorem, the function $V_*(t)$ is not increasing; moreover, there exists either an instant t_1 where the trajectory is continuous such that

$$dV(t_1, F(t_1, t_0, x_*))/dt < 0$$

or a discontinuity point τ_s such that

$$V(\tau_s^+, F(\tau_s^+, t_0, x_*)) - V(\tau_s, F(\tau_s, t_0, x_*)) < 0.$$

This means that there exists an instant $t_* > t_0$ such that

$$V(t_*, F(t_*, t_0, x_*)) = \eta_1 < \eta.$$

Since the sequence $\{x_k\}$ converges to the point x_*, the following inequality holds:

$$\|F(t_*, t_0, x_*) - F(t_*, t_0, x_k)\| < \gamma$$

for all $k > k_0(\gamma)$, for an arbitrary number $\gamma > 0$. Hence

$$\lim_{k\to\infty} V(t_*, F(t_*, t_0, x_k)) = \eta_1. \tag{3.9}$$

Taking into account the periodicity of system (1.1) and (1.2), we can write

$$F(t_*, t_0, x_k) = F(t_*, t_0, F(t_0 + k\omega, t_0, x_0)) = F(t_* + k\omega, t_0, x_0). \tag{3.10}$$

Indeed, the trajectories of system (1.1) and (1.2) starting at instants t_0 and $t_0 + k\omega$ at the point x_k, respectively, will move to the points $F(t_*, t_0, x_k)$ and $F(t_* + k\omega, t_0, x_0)$ respectively during time $\Delta t = t_* - t_0$; this proves (3.10). The periodicity of the function $V(t,x)$ in t yields the equality $V(t_*, x) = V(t_* + k\omega, x)$. Hence taking into account (3.10), condition (3.9) can be rewritten as follows:

$$\lim_{k\to\infty} V(t_* + k\omega, F(t_* + k\omega, t_0, x_0)) = \eta_1. \tag{3.11}$$

But since $\eta_1 < \eta$, relation (3.11) contradicts the inequality $V(t, x(t, t_0, x_0)) \geq \eta$. This contradiction proves that assumption (3.8) was incorrect, i.e. $\eta = 0$. Condition (3.7) justifies (1.4), which proves the asymptotic stability of the trivial solution. By using Theorem 3.2, we conclude that the zero solution of system (1.1) and (1.2) is uniformly asymptotically stable. □

Theorem 3.4. *Suppose that every solution of system (1.1) and (1.2) with initial data* $(t_0, x_0) \in \mathbb{R}_+ \times B_H$ *is bounded, there exists a function* $V(t,x)$ *which is periodic in t with the period* ω*, continuously differentiable on the domain* G*, and satisfies the conditions*

$$|V(t,x)| \leq b(\|x\|), \quad b \in \mathcal{K}; \tag{3.12}$$

$$\frac{dV}{dt} \geq 0 \ \text{ for } \ (t,x) \in G, \tag{3.13}$$

$$\Delta V_i(x) = V(\tau_i^+, x + J_i(x)) - V(\tau_i, x) \geq 0. \tag{3.14}$$

Moreover, suppose that, along any eventually vanishing solution of equations (1.1) and (1.2), at least one of the following conditions holds:
(i) the function dV/dt *is not eventually vanishing,*
(ii) the sequence $\{\Delta V_i\}$ *is not eventually vanishing.*
If any arbitrary small neighborhood of the origin for any $t > 0$ *contains a point* x *such that* $V(t,x) > 0$*, then the zero solution of system (1.1) and (1.2) is unstable.*

Proof. Suppose that $\varepsilon < H$ is a positive number. Let us choose an arbitrary $t_0 \in \mathbb{R}_+$ and an arbitrary small $\delta > 0$. We shall prove that there exist $x_0 \in B_\delta$ and $t > t_0$ such that $\|F(t, t_0, x_0)\| > \varepsilon$. To this end, take $x_0 \in B_\delta$ so that $V(t_0, x_0) = V_0 > 0$. Suppose the opposite:

$$\|F(t, t_0, x_0)\| \leq \varepsilon \tag{3.15}$$

for all $t > t_0$. Condition (3.12) implies

$$|V(t,x)| < V_0 \ \text{ for } \ \|x\| < b^{-1}(V_0) = \eta, \quad t \in \mathbb{R}_+.$$

Taking into account assumptions (3.13)-(3.15), we conclude that the semitrajectory $F(t, t_0, x_0)$ satisfies the conditions

$$\eta \leq \|F(t, t_0, x_0)\| \leq \varepsilon.$$

Consider the sequence of points $\{x_j\}$ given by $x_j = F(t_0 + j\omega, t_0, x_0)$ $(j = 1, 2, \ldots)$. Taking into account the fact that this sequence belongs to a compact set, we can choose a subsequence converging to the point x_* satisfying conditions $\eta \leq \|x_*\| \leq \varepsilon$. Without loss of generality, we assume that the sequence $\{x_j\}$ converges to the point x_*.

The function $v(t) = V(t, F(t, t_0, x_0))$ is monotone nondecreasing and bounded above by a constant $b(\varepsilon)$; hence the limit

$$\lim_{t\to\infty} v(t) = \lim_{t\to\infty} V(t, F(t, t_0, x_0)) = v_0 = V(t_0, x_*)$$

exists, and we have

$$V(t, F(t, t_0, x_0)) \leq v_0 \tag{3.16}$$

for $t \geq t_0$. Now, let us consider the semitrajectory $F(t, t_0, x_*)$ and $t > t_0$. By the assumption of the theorem, there exists a point t_1 such that

$$dV(t_1, F(t_1, t_0, x_*))/dt > 0$$

or a point τ_s such that

$$\Delta V_s = V(\tau_s^+, F(\tau_s, t_0, x_*) + J_s(F(\tau_s, t_0, x_*))) - V(\tau_s, F(\tau_s, t_0, x_*)) > 0.$$

This means that there exists an instant $t_* > t_0$ such that

$$V(t_*, F(t_*, t_0, x_*)) = v_1 > v_0.$$

Since the sequence $\{x_j\}$ converges to the point x_*, Lemma 1.1 implies the inequality

$$\|F(t_*, t_0, x_*) - F(t_*, t_0, x_j)\| < \gamma$$

for all $j > N(\gamma)$, for an arbitrary constant $\gamma > 0$. Hence

$$\lim_{j\to\infty} V(t_*, F(t_*, t_0, x_j)) = v_1. \tag{3.17}$$

Taking into account the periodicity of system (1.1) and (1.2), we can write

$$F(t_*, t_0, x_j) = F(t_* + j\omega, t_0, x_0). \tag{3.18}$$

The periodicity in t of the function $V(t, x)$ yields the equality $V(t_*, x) = V(t_*+j\omega, x)$, hence, by taking (3.18) into account, condition (3.17) can be rewritten as

$$\lim_{j\to\infty} V(t_* + j\omega, F(t_* + j\omega, t_0, x_0)) = v_1. \tag{3.19}$$

On the other hand, relation (3.19) contradicts inequality (3.16), because $v_1 > v_0$. This contradiction proves that asuumption (3.15) was incorrect, i.e. the trivial solution of system (1.1) and (1.2) is unstable; this proves the theorem. □

4. Stability and Instability Criteria by Means of Lyapunov Functions with Derivatives of High Orders

Suppose that $X(t,x)$ is a C^{j-1}-function: $X : G \to \mathbb{R}^n$, and V is a C^j-function $V : \mathbb{R}_+ \times \mathbb{R}^n \to \mathbb{R}$. Let us define $V_s : \mathbb{R}_+ \times B_H \to \mathbb{R}^n$:

$$V_s(t,x) := \frac{\partial V_{s-1}}{\partial t} + \sum_{i=1}^{n} \frac{\partial V_{s-1}}{\partial x_i} X_i(t,x), \quad s = 1, 2, \cdots, j,$$

for $t \neq \tau_k$ and $V_s(\tau_k, x) = V_s(\tau_k^-, x)$ where $V_0(t,x) := V(t,x)$. In particular

$$V_1(t,x) = \left.\frac{dV}{dx}\right|_{(1.1)} = \frac{\partial V}{\partial t} + \sum_{i=1}^{n} \frac{\partial V}{\partial x_i} X_i(t,x).$$

Definition 4.1. *We say that a function $V : \mathbb{R}_+ \times B_H \to \mathbb{R}$ belongs to the class $\mathcal{V}_m$ $(V \in \mathcal{V}_m)$, if V is m times continuously differentiable on any of the sets $G_k = (\tau_{k-1}, \tau_k) \times B_H$ and the finite limits*

$$\begin{aligned} V_r(t_0^-, x_0) &= \lim_{\substack{(t,x)\to(t_0,x_0)\\ (t,x)\in G_k}} V_r(t,x), \\ V_r(t_0^+, x_0) &= \lim_{\substack{(t,x)\to(t_0,x_0)\\ (t,x)\in G_{k+1}}} V_r(t,x), \quad r = 0, 1, \cdots, m. \end{aligned}$$

exist. Here $V_0 = V$.

It was shown in [22] that if system (1.1) and (1.2) is such that conditions of Theorem 1.1 hold, then solution (1.3) of system (1.1) and (1.2) is asymptotically stable. The goal of this section is to derive less stringent conditions for asymptotic stability in the case where the function V satisfies the condition $dV/dt \leq 0$ instead of (1.9).

In what follows, we need the following assertion.

Lemma 4.1. *Suppose that $h(t)$ is a scalar function having points of discontinuity of the first kind at $t = b_1$, $t = b_2, \ldots,$ where $0 < b_1 < b_2 < \ldots$, $\lim_{i\to\infty} b_i = \infty$, $h(b_i^-) = h(b_i)$, the function $h(t)$ is $j+1$ times continuously differentiable in each of the intervals (b_i, b_{i+1}), its derivatives $h'(t), h''(t), \cdots, h^{(j+1)}$* [1] *are bounded for $t \in \cup_{i=1}^{\infty}(b_i, b_{i+1}]$, the constants b_i satisfy property (P3), and*

$$\lim_{i\to\infty} [h(b_i^+) - h(b_i^-)] = 0, \quad \lim_{i\to\infty} [h^{(r)}(b_i^+) - h^{(r)}(b_i^-)] = 0 \quad r = 1, \ldots, j.$$

If

$$\lim_{t\to\infty} h(t) = 0, \tag{4.1}$$

then

$$\lim_{t\to\infty} h^{(r)}(t) = 0, \qquad r = 1, 2, \cdots, j.$$

Proof. Let us first show that

$$\lim_{t\to\infty} h'(t) = 0. \tag{4.2}$$

[1] By derivatives of any order at the points b_i we mean left derivatives.

Assume the contrary: there exist $\xi > 0$ and instants $T_m \in \mathbb{R}_+$, $m \in \mathbb{N}$ such that $T_m \to \infty$ as $m \to \infty$ and $|h'(T_m)| \geq 2\xi$. This implies that either $h'(T_m) \geq 2\xi$, or $h'(T_m) \leq -2\xi$. Since $h''(t)$ is bounded and the condition

$$\lim_{i\to\infty}[h'(b_i^+) - h'(b_i^-)] = 0$$

holds, it follows that there exist $\zeta > 0$ and $M_1 \in \mathbb{N}$ such that

$$h'(t) \geq \frac{3}{2}\xi \tag{4.3}$$

or

$$h'(t) \leq -\frac{3}{2}\xi \tag{4.4}$$

for $t \in [T_m - \zeta, T_m + \zeta]$, $m \geq M_1$. Hence, using the conditions of lemma we find that there exists $M_2 \in \mathbb{N}$ ($M_2 \geq M_1$) such that for $m \geq M_2$ the following inequalities are satisfied:

$$h(T_m + \zeta) \geq h(T_m) + \xi\zeta \tag{4.5}$$

in case (4.3), or

$$h(T_m + \zeta) \leq h(T_m) - \xi\zeta \tag{4.6}$$

in case (4.4). It follows from (4.1) that there exists $M \geq T_{M_2} > 0$ such that $|h(t)| < \frac{1}{2}\xi\zeta$ for $t \geq M$. On the other hand, (4.5) and (4.6) imply the following inequalities:

$$|h(T_m + \zeta)| \geq \frac{1}{2}\xi\zeta \quad \text{for} \quad m \in \mathbb{N},\ T_m \geq M.$$

But these inequalities contradict condition (4.1). The obtained contradiction proves (4.2).

Similarly, we can show that

$$\lim_{t\to\infty} h''(t) = 0, \cdots, \lim_{t\to\infty} h^{(j)}(t) = 0.$$

This completes the proof of the lemma. □

Theorem 4.1. *Suppose that system (1.1) and (1.2) is such that $X \in C^j$ for $(t,x) \in G$, properties (P1)-(P3) hold and there exist $V \in \mathcal{V}_{j+1}$, $w_1 \in \mathcal{K}$, and $w_2 \in \mathcal{K}$ such that*

(A) $V(t,x) \geq w_1(\|x\|)$ for $(t,x) \in \mathbb{R}_+ \times B_H$,

(B) $V_1(t,x) = \frac{dV}{dt} \leq 0$ for $(t,x) \in G$,

(C) $V_s(t,x)$ $(s = 1, \ldots, j+1)$ are bounded for $(t,x) \in G$,

(D) $\sum_{s=1}^{j} V_s^2(t,x) \geq w_2(\|x\|)$,

(E) $V(\tau_k^+, x + I_k(x)) - V(\tau_k, x) \leq 0$ *for* $x \in B_H$, $k \in \mathbb{N}$,

(F) *the constants τ_k satisfy property (P3) and*

$$\lim_{k\to\infty}[V_r(\tau_k^+, x + I_k(x)) - V_r(\tau_k, x)] = 0 \quad (r = 0, 1, \ldots, j)$$

uniformly with respect to $x \in B_H$;

Then solution (1.3) of system (1.1) and (1.2) is asymptotically stable.

Proof. To prove this theorem, let us use method of paper [32]. In view of Theorem 1.1 it follows from conditions (A), (B), and (E) that solution (1.3) of system (1.1) and (1.2) is stable. Take arbitrary $\varepsilon \in (0, H)$, $t_0 \in \mathbb{R}_+$ and $x_0 \in B_\delta$, where $\delta > 0$ is such that $\|F(t, t_0, x_0)\| < \varepsilon$ for $t > t_0$. Denote $v(t) = V(t, F(t, t_0, x_0))$. By $v(\tau_k)$ we mean, as before, $v(\tau_k - 0)$. The function $v(t)$ is positive and not increasing; therefore $\lim_{t\to\infty} v(t)$ exists and is nonnegative:

$$\lim_{t\to\infty} v(t) = \eta \geq 0. \tag{4.7}$$

Let us show that

$$\lim_{t\to\infty} v_1(t) = 0, \tag{4.8}$$

where $v_1(t) = V_1(t, F(t, t_0, x_0))$. Assume the contrary, i. e. assume that (4.8) does not hold. Then there exist $\beta > 0$ and a sequence $\{T_i\}_{i=1}^{\infty}$ $(T_i \in \mathbb{R}_+,\ i \in \mathbb{N},\ T_{i+1} > T_i,\ T_i \to \infty$ as $i \to \infty)$ such that $v_1(T_i) \leq -2\beta < 0$. By conditions (A), (B), (C), and (F) of the theorem, there exist $\gamma > 0$ and $M_1 \in \mathbb{N}$, such that $v_1(t) \leq -\beta$ for $t \in [T_i, T_i + \gamma]$, $i \geq M_1$. It follows from conditions (B) and (E) that

$$v(T_i + \gamma) \leq v(T_i) - \beta\gamma. \tag{4.9}$$

for $i \geq M_1$. Since

$$\lim_{i\to\infty} v(T_i) = \eta,$$

in view of (4.9), we have

$$\lim_{i\to\infty} v(T_i + \gamma) \leq \eta - \beta\gamma.$$

But since $T_i + \gamma \to \infty$ as $i \to \infty$, the resulting inequality contradicts (4.7). This contradiction proves (4.8).

Using (4.8) and Lemma 4.1 we find that

$$\lim_{t\to\infty} \sum_{s=1}^{j} v_s^2(t) = 0, \tag{4.10}$$

where $v_s(t) = V_s(t, F(t, t_0, x_0))$. Let us now show that the zero solution of system (1.1) and (1.2) is attractive. To do this, we have to show that

$$\lim_{t\to\infty} \|F(t, t_0, x_0)\| = 0. \tag{4.11}$$

Assume the contrary: suppose that there exist $\alpha > 0$ and a sequence $\{t_i\}$ $(i \in \mathbb{N},\ t_i \to \infty$ as $i \to \infty)$ such that $\|F(t_i, t_0, x_0)\| \geq \alpha$. It follows from condition (D) that

$$\sum_{s=1}^{j} v_s^2(t_i) \geq w_2(\|F(t_i, t_0, x_0)\|) \geq w_2(\alpha) > 0.$$

The resulting inequality contradicts (4.10), whence (4.11) holds. This completes the proof of the theorem. □

Remark 4.1. *If in this theorem we set* $j = 1,\ I_k(x) \equiv 0 \quad (k \in \mathbb{N},\ x \in B_H)$, *then we obtain the well-known Marachkov theorem [41].*

Denote

$$Q_h := \{(t, x) \in \mathbb{R}_+ \times B_h : V(t, x) > 0\}.$$

Theorem 4.2. *Suppose that system (1.1) and (1.2) is such that* $X \in C^j$ *for* $(t, x) \in G$, *properties (P1)-(P3) hold, and there exist functions* $V \in \mathcal{V}_{j+1}$ *and* $w \in \mathcal{K}$ *such that:*

(a) *the set* Q_ε *is not empty for all* $t \in \mathbb{R}_+,\ \varepsilon \in (0, H)$;

(b) $V(t, x), V_1(t, x), \ldots, V_{j+1}(t, x)$ *are bounded in* Q_H;

(c) $dV/dt = V_1(t, x) \geq 0$ *in* Q_H;

(d) $\sum_{s=1}^{j} V_s^2(t, x) \geq w(V(t, x))$;

(e) $\Delta V\Big|_{t=\tau_k} = V(\tau_k + 0, x + I_k(x)) - V(\tau_k, x) \geq 0$ *for* $x \in B_H,\ k \in \mathbb{N}$,

(f) *condition (F) of Theorem 4.1 holds.*

Then solution (1.3) of system (1.1) and (1.2) is unstable.

Proof. Choose arbitrary $\varepsilon \in (0, H),\ t_0 \in \mathbb{R}_+$, and suppose that δ is an arbitrary small positive number. Let us show that there exist an $x_0 \in B_\delta$ and a $t_1 > t_0$ such that $\|F(t_1, t_0, x_0)\| \geq \varepsilon$. We choose $x_0 \in B_\delta$ so that $V(t_0, x_0) > 0$. Such a choice is possible by condition (a).

Assume the contrary: for any $t > t_0$ the following inequality

$$\|x(t, t_0, x_0)\| < \varepsilon \tag{4.12}$$

holds, whence it follows that $(t, x) \in Q_\varepsilon$ and the function $V(t, F(t, t_0, x_0))$ is bounded. From conditions (c) and (e) we obtain

$$v(t) = V(t, F(t, t_0, x_0)) \geq V(t_0, x_0) = \beta > 0.$$

The function $v(t)$ is bounded and nondecreasing, therefore the limit

$$\lim_{t \to \infty} v(t) = \eta \geq \beta > 0.$$

exists. As in the proof of Theorem 4.1, we can show that

$$\lim_{t \to \infty} v_s(t) = 0, \quad s = 1, \ldots, j,$$

where $v_s = V_s(t, F(t, t_0, x_0))$, whence we see that (4.10) holds. Assumption (d) implies

$$\sum_{s=1}^{j} v_s^2(t) = \sum_{s=1}^{j} V_s^2(t, F(t, t_0, x_0)) \geq w(V(t, F(t, t_0, x_0)) \geq w(\beta) > 0 \tag{4.13}$$

for $t \geq t_0$. However, (4.10) and (4.13) contradict each other. This contradiction shows that assumption (4.12) is false, i.e. solution (1.3) of system (1.1) and (1.2) is unstable. □

Example 4.1. Consider the following system of differential equations with impulse effect

$$\begin{aligned} &dx/dt = -y - a(t)x, \quad dy/dt = x \qquad t \neq \tau_k \quad (k = 1, 2, \ldots), \\ &\Delta x\big|_{t=\tau_k} = -\tfrac{x}{k}, \qquad\qquad \Delta y\big|_{t=\tau_k} = 0, \end{aligned} \tag{4.14}$$

where $a(t) = 2 - \cos 2\pi t - \cos(2\pi \log_2(t+1)),\ \tau_k = 3k,\ k \in \mathbb{N}$. System (4.14) edmits the trivial solution

$$x = 0, \quad y = 0. \tag{4.15}$$

To study the stability of this solution, we use the Lyapunov function

$$V = \frac{1}{2}\left(x^2 + y^2\right).$$

We find

$$V_1 = \frac{dV}{dt} = -a(t)x^2, \qquad V_2 = \frac{dV_1}{dt} = (2a^2 - a')x^2 + 2axy,$$

$$V_3 = \frac{dV_2}{dt} = (6aa' - a'' - 4a^3 + 2a)x^2 + (4a' - 6a^2)xy - 2ay^2,$$

$$V_4 = \frac{dV_3}{dt} = (6a'^2 + 8aa'' - a''' - 24a^2a' + 6a' + 8a^4 - 10a^2)x^2 +$$

$$+(-28aa' + 6a'' + 14a^3 - 8a)xy + (6a^2 - 6a')y^2, \qquad V_5 = \frac{dV_4}{dt} =$$

$$= (20a'a'' + 10aa''' - a'''' - 60aa'^2 - 40a^2a'' + 12a'' + 80a^3a' - 60aa' - 16a^5 +$$

$$+34a^3 - 8a)x^2 + (-40a'^2 - 50aa'' + 8a''' + 118a^2a' - 32a' - 30a^4 + 40a^2)xy +$$

$$+(40aa' - 12a'' - 14a^3 + 8a)y^2.$$

The derivatives of $a(t)$ are of the form

$$a^{(s)}(t) = \alpha_s(t) + \beta_s \frac{\sin(2\pi \log_2(t+1))}{(t+1)^s} + \gamma_s \frac{\cos(2\pi \log_2(t+1))}{(t+1)^s},$$

where $\beta_s, \gamma_s \quad (s \in \mathbb{N})$ are constants, with $\beta_1 = 2\pi/\log 2,\ \gamma_1 = 0$,

$$\begin{aligned} \alpha_1 &= 2\pi \sin 2\pi t, & \alpha_2 &= (2\pi)^2 \cos 2\pi t, \\ \alpha_3 &= -(2\pi)^3 \sin 2\pi t, & \alpha_4 &= -(2\pi)^4 \cos 2\pi t. \end{aligned}$$

$a(t)$ is nonnegative, with $a(t) = 0$ for $t = t_n = 2^n - 1 \quad (n = 0, 1, 2, \ldots)$.

Consider the function $W = V_1^2 + V_2^2 + V_3^2 + V_4^2 + V_5^2$. Let us show that $W(t, x, y)$ is a positive definite function with respect to x and y. Let us first show that $W(t_n, x, y)$ is a positive definite function. Taking the relations $a(t_n) = 0,\ a'(t_n) = 0$ into account, we obtain:

$$V_1(t_n, x, y) = 0, \quad V_2(t_n, x, y) = 0, \quad V_3(t_n, x, y) = -a''(t_n)x^2,$$

$$V_4(t_n, x, y) = -a'''(t_n)x^2 + 6a''(t_n)xy,$$

$$V_5(t_n,x,y) = (-a''''(t_n) + 12a''(t_n))x^2 + 8a'''(t_n)xy - 12a''(t_n)y^2.$$

Note that $\gamma_2 = (2\pi)^2/\log^2 2 > 0$, whence we obtain

$$a''(t_n) > (2\pi)^2. \tag{4.16}$$

Using (4.16) we obtain

$$W(t_n,x,y) \geq V_3^2(t_n,x,y) + V_5^2(t_n,x,y) \geq (2\pi)^4x^4 + V_5^2(t_n,x,y).$$

Using the relation $V_5(t_n,0,y) = -12a''(t_n)y^2$ and inequality (4.16), we see that $W(t_n,x,y) > 2w(x^2+y^2)$, where $w \in \mathcal{K}$. The functions $a(t)$ and $a^{(s)}(t)$ $(s = 1,2,3,4)$ are uniformly continuous for $t \in \mathbb{R}_+$; therefore there exists $\varepsilon > 0$ such that

$$W(t,x,y) \geq w(x^2+y^2) \tag{4.17}$$

for $t_n - \varepsilon \leq t \leq t_n + \varepsilon$. Let us show that for

$$t \notin [t_n - \varepsilon, t_n + \varepsilon] \tag{4.18}$$

the function W also satisfies inequality (4.17) for any $n \in \mathbb{N}$. Note that, under condition (4.18), there exists $\omega > 0$ such that $a(t) > \omega$. Let us estimate W for such t:

$$W(t,x,y)\big|_{t\notin[t_n-\varepsilon,t_n+\varepsilon]} \geq (V_1^2+V_3^2)\big|_{t\notin[t_n-\varepsilon,t_n+\varepsilon]} \geq$$
$$\geq \psi(t,x,y) = \omega^2x^4 + [(6aa' - a'' - 4a^3 + 2a)x^2 + (4a' - 6a^2)xy - 2ay^2]^2.$$

Let us find $\psi(t,x,y)$ for $x = 0$:

$$\psi(t,0,y) = 4a^2(t)y^4 > 4\omega^2y^4.$$

Since $a(t), a'(t)$, and $a''(t)$ are bounded for $t \in \mathbb{R}_+$, we find that the function W is positive definite with respect to x, y for $t \notin [t_n - \varepsilon, t_n + \varepsilon]$. Therefore there exists a $w \in \mathcal{K}$ such that inequality (4.17) holds for $t \in \mathbb{R}_+$, $(x,y) \in \mathbb{R}^2$. Let us find

$$\Delta V\big|_{t=\tau_k} = \frac{1}{2}\Big[(x+\Delta x\big|_{t=\tau_k})^2 + (y+\Delta y\big|_{t=\tau_k})^2\Big] - \frac{1}{2}\Big(x^2+y^2\Big) = -\frac{x^2}{k}\Big(1-\frac{1}{2k}\Big) \leq 0.$$

Note that $V_1, V_2, V_3, V_4, V_5, V_6 := dV_5/dt$ can be expressed as

$$V_m = F_m(t)x^2 + P_m(t)xy + Q_m(t)y^2 \quad (m = 1,2,3,4,5,6),$$

where F_m, P_m, Q_m are bounded continuous functions of t. Therefore these functions are bounded for $x^2 + y^2 < H^2$, and

$$\Delta V_m\big|_{t=\tau_k} = F_m(\tau_k)\big[\big(x+\Delta x\big|_{t=\tau_k}\big)^2 - x^2\big] +$$
$$+P_m(\tau_k)\big[\big(x+\Delta x\big|_{t=\tau_k}\big)\big(y+\Delta y\big|_{t=\tau_k}\big) - xy\big] + Q_m(\tau_k)\big[\big(y+\Delta y\big|_{t=\tau_k}\big)^2 - y^2\big] =$$
$$= -F_m(\tau_k)\frac{x^2}{k}\Big(1-\frac{1}{2k}\Big) - P_m(\tau_k)\frac{xy}{k},$$

whence we obtain $\lim_{k\to\infty} \Delta V_m\big|_{t=\tau_k} = 0$ $(m = 1,\ldots,5)$.

Thus conditions (A)-(F) of Theorem 4.1 are satisfied, and we can conclude that solution (4.15) of system (4.14) is asymptotically stable.

5. Perturbed Systems with Impulse Effect

Along with system (1.1) and (1.2) we consider the system

$$\frac{dx}{dt} = X(t,x) + X_*(t,x), \;\; t \in \mathbb{R}_+, \; t \neq \tau_i, \tag{5.1}$$

$$x(\tau_i^+) - x(\tau_i) = J_i(x) + J_i^*(x) = I_i(x), \quad i \in \mathbb{N}, \tag{5.2}$$

where X_*, J_i^* satisfy properties (P1), (P2) respectively. Assume that both systems (1.1), (1.2) and (5.1), (5.2) have the trivial solution (1.3). For the cases when it is possible to find the explicit solution $F(t, t_0, x_0)$ of system (1.1) and (1.2), the criteria of asymptotic stability of the zero solution of system (5.1) and (5.2) were obtained in [10, 11]. We now demonstrate one of the possible applications of Theorem 2.2. when we have not such possibility (to find the explicit solution $F(t, t_0, x_0)$ of system (1.1) and (1.2)). Under the stated assumptions on the right-hand sides of system (1.1), (1.2) and (5.1), (5.2), the following theorem holds.

Theorem 5.1. *If solution (1.3) of system (1.1) and (1.2) is uniformly asymptotically stable,*

$$\lim_{t\to\infty} X_*(t,x) = 0 \tag{5.3}$$

holds uniformly with respect to $x \in B_H$ $(0 < H < \infty)$*, and the series* $\sum_{i=1}^{\infty} \|J_i^*(x)\|$ *converges uniformly with respect to* $x \in B_H$*, then the zero solution of system (5.1) and (5.2) is also uniformly asymptotically stable.*

Proof. Denote $x(t) = x(t, t_0, x_0)$ the solution of system (5.1) and (5.2) with initial data (t_0, x_0). Since solution $x = 0$ of system (1.1) and (1.2) is uniformly asymptotically stable, there exist a function $V(t,x)$ and Hahn functions g, b, c, satisfying the conditions of Theorem 2.2. Using Yoshizawa's theorem [53] we first estimate $D^+V(t,x)$ along a solution $x(t)$ of system (5.1) and (5.2) for $t \neq \tau_i$, $x \in B_{H_*}$ where $H_* < H$:

$$\begin{aligned} D^+V(t,x)\Big|_{(5.1),(5.2)} &= \limsup_{\xi\to 0^+} \frac{V(t+\xi, x+\xi X(t,x)+\xi X_*(t,x)) - V(t,x)}{\xi} \\ &\leq \limsup_{\xi\to 0^+} \frac{V(t+\xi, x+\xi X(t,x)+\xi X_*(t,x)) - V(t+\xi, x+\xi X(t,x))}{\xi} \\ &+ \limsup_{\xi\to 0^+} \frac{V(t+\xi, x+\xi X(t,x)) - V(t,x)}{\xi} \leq L_1\|X_*(t,x)\| + D^+V(t,x)\Big|_{(1.1),(1.2)}. \end{aligned} \tag{5.4}$$

In similar fashion we estimate the value of the jump of the function V along the trajectory $x(t)$ of system (5.1) and (5.2) at the instant τ_i:

$$\begin{aligned} \Delta V_i &= V(\tau_i^+, x + J_i(x) + J_i^*(x)) - V(\tau_i, x) \\ &= [V(\tau_i^+, x + J_i(x) + J_i^*(x)) - V(\tau_i^+, x + J_i(x))] + [V(\tau_i^+, x + J_i(x)) - V(\tau_i, x)] \\ &\leq V(\tau_i^+, x + J_i(x) + J_i^*(x)) - V(\tau_i^+, x + J_i(x)) \leq L_1\|J_i^*(x)\|. \end{aligned} \tag{5.5}$$

Recall that L_1 denotes the constant for the function V appearing in inequality (2.6).

Let us show that the zero solution of system (5.1) and (5.2) is uniformly stable. Pick an arbitrary $\varepsilon_1 > 0$ ($\varepsilon_1 < H_* < H$). Let $t_1 \in \mathbb{R}_+$ be sufficiently large. Let us show that there exists $\delta_1 = \delta_1(\varepsilon_1) > 0$ such that the solution $x(t) = x(t, t_1, x_1)$ of system (5.1) and (5.2) satisfies the condition $\|x(t)\| < \varepsilon_1$ for $t > t_1$ as soon as $x_1 \in B_{\delta_1}$. We set $\delta_1 = b^{-1}(\frac{1}{2}g(\varepsilon_1))$. We assume that the value t_1 satisfies the inequality $t_1 \geq T_1$ where T_1 is so large that $L_1\|X_*(t,x)\| < \gamma_1 = \frac{1}{2}c(\delta_1)$ for $t \geq T_1,\ x \in B_{\varepsilon_1}$ and $L_1 \sum\limits_i \|J_i^*(x)\| < \frac{1}{2}g(\varepsilon)$ for $x \in B_{\varepsilon_1}$ (here the summation is extended to those values of i, where $\tau_i \geq T_1$). Since (5.3) holds uniformly with respect to $x \in B_H$, the series $\sum\limits_{i=1}^{\infty} \|J_i(x)\|$ converges uniformly with respect x, and δ_1 depends only on ε_1, one can choose T_1 to be dependent only on ε_1. Using estimates (1.7) we obtain that

$$g(\|x(t)\|) \leq V(t, x(t)) \leq V(t_1, x_1) + \int_{t_1}^{t} D^+V(s, x(s))\Big|_{(5.1),(5.2)} ds + \sum_i \Delta V_i, \quad (5.6)$$

where the summation on the right-hand side of (5.6) is extended to those values of i, where $\tau_i \geq t_1$. Bearing in mind that $V(t_1, x_1) < \frac{1}{2}g(\varepsilon_1)$, $\sum\limits_i \Delta V_i < \frac{1}{2}g(\varepsilon_1)$ and $D^+V(t,x)\Big|_{(5.1),(5.2)} < -\frac{1}{2}c(\delta_1)$ for $\|x\| > \delta_1$ we obtain that at an arbitrary instant $t > t_1$ either the inequality $\|x(t)\| \leq \delta_1 < \varepsilon_1$ holds or

$$g(\|x(t)\|) \leq V(t, x(t)) \leq g(\varepsilon_1) - \frac{1}{2}c(\delta_1)(t - t_1) < g(\varepsilon_1),$$

whence $\|x(t)\| \leq \varepsilon_1$. Thus, it has been proved that for any $\varepsilon_1 > 0$ there exists a value $T_1 = T_1(\varepsilon_1) > 0$ such that for any $t_1 \geq T_1$ there is $\delta_1 = \delta_1(\varepsilon_1) > 0$ such that the inequality $\|x_1\| < \delta_1$ implies that $\|x(t)\| = \|x(t, t_1, x_1)\| < \varepsilon_1$ for $t > t_1$. From Lemma 1.1 and property (P3) we deduce that there exists $\delta > 0$ such that for any $t_0 \in [0, T_1]$ and $x_0 \in B_\delta$ the solution $x(t, t_0, x_0)$ satisfies the inequality $\|x(T_1, t_0, x_0)\| < \delta_1$ and therefore the inequality $\|x(t, t_0, x_0)\| < \varepsilon_1$ for $t > t_0$. Since δ_1 and T_1 depend only on ε_1, we conclude that δ also depends only on ε_1, which proves the uniform stability of the trivial solution of system (5.1) and (5.2).

We now show that the zero solution of system (5.1) and (5.2) is uniformly attractive. For that we choose an arbitrary $\varepsilon_1 > 0$ ($\varepsilon_1 < H_*$) and the corresponding value $\delta = \delta(\varepsilon_1) > 0$ in the definition of uniform stability. Let us show that for any $\varepsilon_2 > 0$ ($\varepsilon_2 < \varepsilon_1$) there exists a $\sigma = \sigma(\varepsilon_2) > 0$ such that $\|x(t, t_0, x_0)\| < \varepsilon_2$ for all $\|x_0\| < \delta,\ t_0 \in \mathbb{R}_+$ and $t \geq t_0 + \sigma$. For that we choose $\delta_2 = \delta_2(\varepsilon_2) > 0$ such that if the solution $x(t, t_0, x_0)$ of system (5.1) and (5.2), gets into B_{δ_2} at some instant, then it stays in B_{ε_2} for all t. It is possible to choose such δ_2 by the property of uniform stability of the zero solution of system (5.1) and (5.2) proved above. So we have the inequality $\|x(t, t_0, x_0)\| < \varepsilon_1$ for $t > t_0$. We estimate the period of time during which the solution $x(t, t_0, x_0)$ can belong to the set

$$\delta_2 \leq \|x\| \leq \varepsilon_1. \quad (5.7)$$

Let t_2 denotes an instant such that $L_1\|X_*(t,x)\| < \gamma_2 = \frac{1}{2}c(\delta_2)$ for $t \geq t_2,\ x \in B_{\varepsilon_1}$ and $L_1\|J_i^*(x)\| < \frac{1}{2}g(\varepsilon_1)$ for $\tau_i \geq t_2,\ x \in B_{\varepsilon_1}$. Since (5.3) holds uniformly with respect to $x \in B_H$, the series $\sum_{k=1}^{\infty} \|J_i^*(x)\|$ converges uniformly with respect to $x \in B_H$, and γ_2 depends only on ε_2, one can choose t_2 to be dependent only on ε_2. We show that the solution $x(t, t_0, x_0)$ leaves set (5.7) no later then at $t_3 = (b(\varepsilon_1) + \frac{1}{2}g(\varepsilon_1))/\gamma_2$. For that we estimate the value $V(t, x(t))$ for $t > t_2$:

$$0 < g(\delta_2) \leq V(t, x(t)) \leq V(t_2, x(t_2)) - \gamma_2(t - t_2) + \sum_i \Delta V_i \leq b(\varepsilon_1) + \frac{1}{2}g(\varepsilon_1) - \gamma_2(t - t_2). \tag{5.8}$$

We obtain the required estimate from inequalities (5.8). Thus, the value σ can be chosen in the form $\sigma = t_2 + t_3$, where t_2 and t_3 depend only on ε_2. This proves that the zero solution of system (5.1) and (5.2) is uniformly attractive and its domain of attraction contains the set B_δ. This completes the proof of the theorem. □

6. Conclusion

In this chapter, a system of ordinary differential equations with impulse effect at fixed instants is considered. The system is assumed to have the zero solution. It is shown that the existence of a corresponding Lyapunov function is a necessary and sufficient condition for the uniform asymptotic stability of the zero solution. Restrictions on perturbations of the right-hand sides of differential equations and impulse effect are obtained under which the uniform asymptotic stability of the zero solution of the "unperturbed" system implies the uniform asymptotic stability of the zero solution of the "perturbed" system. In the case of a periodic system with impulse effect, it is shown that if the trivial solution of the system is stable or asymptotically stable, then it is uniformly stable or uniformly asymptotically stable, respectively. By using the method of Lyapunov functions, the criteria of asymptotical stability and instability are obtained. The illustrative example is considered.

References

[1] N.U. Ahmed. Optimal impulse control for impulsive systems in Banach spaces. *International Journal of Differential Equations and Applications* , **1**:37–52, 2000.

[2] N.U. Ahmed. Some remarks on the dynamics of impulsive systems in Banach spaces. *Dynamics of Continuous, Discrete and Impulsive Systems, Series A* , **8**:261–274, 2001.

[3] N.U. Ahmed. *Dynamic Systems and Control with Applications* . World Scientific, Singapore, 2006.

[4] M.U. Akhmet. On the smoothness of solutions of impulsive autonomous systems. *Nonlinear Analysis*, **60**:311–324, 2005.

[5] M.U. Akhmetov and A. Zafer. Stability of the zero solution of impulsive differential equations by the Lyapunov second method. *Journal of Mathematical Analysis and Applications*, **248**:69–82, 2000.

[6] J. Angelova and A. Dishliev. Optimization problems for one-impulsive models from population dynamics. *Nonlinear Analysis*, **39**:483–497, 2000.

[7] J. Angelova, A. Dishliev, and S. Nenov. Comparison of zero-solutions of systems ODE via asymptotical stability. *Nonlinear Analysis*, **42**:339–350, 2000.

[8] D.D. Bainov and P.S. Simeonov. *Systems with impulse effect: stability, theory and applications*. Halsted Press, New York – Chichester – Brisbane – Toronto, 1989.

[9] S.R. Bernfeld, C. Corduneanu, and A.O. Ignatyev. On the stability of invariant sets of functional differential equations. *Nonlinear Analysis*, **55**:641–656, 2003.

[10] S.D. Borysenko, M. Ciarletta, and G. Iovane. Integro-sum inequalities and motion stability of systems with impulse perturbations. *Nonlinear Analysis*, **62**:417–428, 2005.

[11] S.D. Borysenko, M. Ciarletta, and G. Iovane. On the stability of motion of systems with impulse perturbations. *Dopov. Nats. Akad. Nauk Ukr. Mat. Prirodozh. Tekh. Nauki*, **1**:13–17, 2006.

[12] S.D. Borysenko, G. Iovane, and P. Giordano. Investigations of the properties motion for essential nonlinear systems perturbed by impulses on some hypersurfaces. *Nonlinear Analysis*, **62**(5):345–363, 2005.

[13] T.A. Burton. Uniform asymptotic stability in functional differential equations. *Proceedings of the American Mathematical Society*, **68**:195–199, 1978.

[14] T.A. Burton and L. Hatvani. On nonuniform asymptotic stability for nonautonomous functional-differential equations. *Differential and Integral Equations*, **3**:285–293, 1974.

[15] T.A. Burton and L. Hatvani. Stability theorems for nonautonomous functional-differential equations by Liapunov fuctionals. *Tõhoku Math. J.*, **41**:65–104, 1989.

[16] A. Cabada and E. Liz. Discontinuous impulsive differential equations with nonlinear boundary conditions. *Nonlinear Analysis*, **28**:1491 – 1497, 1997.

[17] V. Chellaboina, S.P. Bhat, and W.M. Haddad. An invariance principle for nonlinear hybrid and impulsive dynamical systems. *Nonlinear Analysis*, **53**:527–550, 2003.

[18] C. Corduneanu and A.O. Ignatyev. Stability of invariant sets of functional differential equations with delay. *Nonlinear Functional Analysis and Applications* , **10**(1):11–24, 2005.

[19] A. D'Onofrio. Pulse vaccination strategy in the SIR epidemic model: Global asymptotic stable eradication in presence of vaccine failures. *Mathematical and Computer Modelling*, **36**:473 – 489, 2002.

[20] A.K. Gelig and A.N. Churilov. *Stability and Oscillations of Nonlinear Pulse-Modulated Systems*. Birkhauser, Boston, 1998.

[21] Z.-H. Guan, J. Lam, and G. Chen. On impulsive autoassociative neural networks. *Neural Networks*, **13**:63–69, 2000.

[22] S.I. Gurgula and N.A. Perestyuk. On stability of the equilibrium in impulse systems. *Mat. Fiz*, **31**:9–14, 1982.

[23] W.M. Haddad, V. Chellaboina, and N.A. Kablar. Non-linear impulsive dynamical systems. Part I: stability and dissipativity. *Int. J. Control*, **74**(17):1631–1658, 2001.

[24] W.M. Haddad, V. Chellaboina, and S.G. Nersesov. Hybrid nonnegative and compartmental dynamical systems. *Mathematical Problems in Engineering*, **8**(6):493–515, 2002.

[25] W.M. Haddad, V. Chellaboina, and S.G. Nersesov. *Impulsive and hybrid dynamical systems: stability, dissipativity, and control*. Princeton University Press, Princeton, 2006.

[26] S. Hu and V. Lakshmikantham. Periodic boundary value problems for second order impulsive differential systems. *Nonlinear Analysis*, **13**:75–87, 1989.

[27] A.O. Ignatyev. On the stability in critical case of a pair of imaginary eigenvalues for nonautonomous systems. *Mat. Fiz.*, **31**:27–32, 1982.

[28] A.O. Ignatyev. On the asymptotic stability of nonautonomous systems. *Differentsial'nye Uravneniya*, **23**:2161–2163, 1987.

[29] A.O. Ignatyev. On the stability of equilibrium for almost periodic systems. *Nonlinear Analysis*, **29**(8):957–962, 1997.

[30] A.O. Ignatyev. On the asymptotic stability in functional-differential equations. *Proceedings of the Americal Mathematical Society*, **127**(6):1753–1760, 1999.

[31] A.O. Ignatyev and O.A. Ignatyev. On the stability in periodic and almost periodic difference systems. *Journal of Mathematical Analysis and Applications*, **313**(2):678–688, 2006.

[32] O.A. Ignatyev. On the partial asymptotic stability in nonautonomous differential equations. *Differential and Integral Equations*, **19**(7):831–839, 2006.

[33] S.K. Kaul and X. Liu. Generalized variation of parameters and stability of impulsive systems. *Nonlinear Analysis*, **40**:295–307, 2000.

[34] K.K. Kenzhebaev and A.N. Stanzhitskii. Invariant sets of impulsive systems and their stability. *Nonlinear Oscillations*, **7**:78–82, 2004.

[35] C. Kou, S. Zhang, and S. Wu. Stability analysis in terms of two measures for impulsive differential equations. *J. London Math. Soc.*, **(2)** 66:142–152, 2002.

[36] N.N. Krylov and N.N. Bogolyubov. *An Introduction to Nonlinear Mechanics*. Izd. Acad. Sci. Ukr. SSR, Kiev, 1937.

[37] V. Lakshmikantham, D.D. Bainov, and P.S. Simeonov. *Theory of impulsive differential equations*. World Scientific, Singapure – New Jersey – London, 1989.

[38] V. Lakshmikantham, S. Leela, and S. Kaul. Comparison principle for impulsive differential equations with variable times and stability theory. *Nonlinear Analysis TMA*, **22**:499–503, 1994.

[39] V. Lakshmikantham and X. Liu. On quasi stability for impulsive differential systems. *Nonlinear Analysis TMA*, **13**:819–829, 1989.

[40] I.G. Malkin. *Stability of Motion*. Nauka, Moscow, 1966.

[41] V.P. Marachkov. The theorem on the stability. *Bulletin of physics and mathematics society of Kazan University*, **12**:171–174, 1940.

[42] V.D. Milman and A.D. Myshkis. About stability of motion with pushes. *Sibirsk. Mat. Zh.*, **1**(2):233–237, 1960.

[43] A.D. Myshkis and A.M. Samoilenko. Systems with impulses in the present instants. *Mat. Sb.*, **74**(2):202–208, 1967.

[44] S.I. Nenov. Impulsive controllability and optimization problems in population dynamics. *Nonlinear Analysis*, **36**:881–890, 1999.

[45] N. Rouche, Habets. P., and M. Laloy. *Stability Theory by Liapunov's Direct Method*. Springer, New York, 1977.

[46] A.M. Samoilenko and N.A. Perestyuk. *Impulsive differential equations*. World Scientific, Singapure–New Jersey–London, 1995.

[47] P.S. Simeonov and D.D. Bainov. Stability with respect to part of the variables in systems with impulse effect. *J. of Math. Analysis and Applications*, **117**(1):247–263, 1986.

[48] R.J. Smith and L.M. Wahl. Distinct effects of protease and reverse transcriptase inhibition in an immunological model of HIV-1 infection with impulsive drug effects. *Bulletin of Mathematical Biology*, **66**:1259–1283, 2004.

[49] R.J. Smith and L.M. Wahl. Drug resistence in an immunological model of HIV - 1 infection with impulsive drug effects. *Bulletin of Mathematical Biology*, **67**:783–813, 2005.

[50] S. Tang and R.A. Cheke. State-dependent impulsive models of integrated pest management (IPM) strategies and their consequences. *Journal of Mathematical Biology*, **50**:257–292, 2005.

[51] S. Tang and L. Chen. Density-dependent birth rate, birth pulses and their population dynamic consequences. *Journal of Mathematical Biology*, **44**:185–199, 2002.

[52] Z. Wang and J. Sun. On impulsive dynamic systems. *Impulsive Dynamical Systems and Applications*, **3**:114–119, 2005.

[53] T. Yoshizawa. *Stability theory and the existence of periodic solutions and almost periodic solutions*. Springer, New York, 1975.

[54] X. Zhang, Z. Shuai, and K. Wang. Optimal impulsive harvesting policy for single population. *Nonlinear Analysis: Real World Applications*, **4**:639–651, 2003.

[55] Y. Zhang, B. Liu, and L. Chen. Extinction and permanence of a two-prey one-predator system with impulsive effect. *Mathematical Medicine and Biology*, **20**:309–325, 2003.

[56] Y. Zhou and H. Liu. Stability of periodic solutions for an SIS model with pulse vaccination. *Mathematical and Computer Modelling*, **38**:299 – 308, 2003.

Peer reviewers:
N.U. Ahmed, SITE and Department of Mathematics, 161 Louis Pasteur, University of Ottawa, Ottawa, Ontario, K1N6N5, Canada;
e-mail: ahmed@site.uottawa.ca

A.N.Stanzhitskii, Mechanical and Mathematical Department, Kiev National University, Vladimirskaya Street, 64, Kiev-01033, Ukraine;
e-mail: stom@mail.univ.kiev.ua

In: Progress in Nonlinear Analysis Research
Editor: Erik T. Hoffmann, pp. 391-431

ISBN: 978-1-60456-359-7

Chapter 17

EXAMPLES OF THE DISCRETE AGGLOMERATION MODEL WITH A TIME VARYING KERNEL

James L. Moseley
West Virginia Univ.

3.0. The Moment Problem

Next we review the development and solution of the Moment Problem. We also provide solutions of the Moment Problem for the examples. In addition, we provide the scaled times using the solution of the moment problem for the examples.

3.1. Development of the Moment Problem

We first review the development of the moment problem $Prob_M((I,t_0,M_0,A)$ = $Prob_{Mno}(I,t_0,\vec{n}_0,A)$ as a problem for the moment of a solution of $Prob_A(I,t_0,\vec{n}_0,A)$. That is, we show that the zeroth moment, $M_A(t)=\sum_{j=1}^{\infty} n_j(t)$, of a solution, $\vec{n}_A(t) = \{n_i^A(t)\}_{i=1}^{\infty} \in \vec{D}_{UBS}(I;t_0,\vec{n}_0)$, of $Prob_A(I,t_0,\vec{n}_0,A)$ will solve $Prob_M(I,t_0,M_0,A)$ = $Prob_{Mno}(I,t_0,\vec{n}_0,A)$. However, we do not yet claim that the solution to $Prob_M(I,t_0,M_0,A)$ = $Prob_{Mno}(I,t_0,\vec{n}_0,A)$ is the zeroth moment of a solution of $Prob_A(I,t_0,\vec{n}_0,A)$. This will be effected after we compute the solution of $Prob_A(I,t_0,\vec{n}_0,A)$ using the solution of $Prob_M(I,t_0,M_0,A)$ = $Prob_{Mno}(I,t_0,\vec{n}_0,A)$ and the initial condition $\vec{n}_0$. Thus there is a one-to-one correspondence between $Prob_A(I,t_0,\vec{n}_0,A)$ and $Prob_{Mno}(I,t_0,\vec{n}_0,A)\times IC_B$ where we consider problems with different initial conditions $\vec{n}_0$ to be different even when they generate the same M_0. Thus fixing t_0 and A, for $\vec{n}_0 = \{n_i^0\}_{i=1}^{\infty} \in IC_B$, we partition the solutions of $\mathrm{Prob}_A(I,t_0,\vec{n}_0,A)$ into sets which have the same $M_{no} = \sum_{j=1}^{\infty} n_j^{\,0}$ as $\mathrm{Prob}_M(I,t_0,M_0,A) = \mathrm{Prob}_{Mno}(I,t_0,\vec{n}_0,A)$ will have the same solution for all of these.

We begin by giving formal equations (since they involve infinite series that may not converge) for the zeroth moment of a solution to $\mathrm{Prob}_A(I,t_0,\vec{n}_0,A)$. We then provide sufficient conditions for these formal equations to be equations (i.e., for all of the infinite series to converge) and develop a problem for the zeroth moment of a solution to this problem. We need the following corollary to Theorem 2.3:

<u>COROLLARY 3.1</u>. Let $I\in \mathrm{Int}_o$ and $u_{i,j}\in A(I,R)$ for $i,j\in N$. If for all closed intervals $J\subseteq I$, there is a doubly infinite series of nonnegative constants $\left\{M_{umax}(i,j;J)\right\}_{i,j=1}^{\infty}$ such that for all $t\in J$ and $j\in N$, $|u_{i,j}(t)|\leq M_{umax}(i,j;J)$ and $\sum_{i=1}^{\infty}\sum_{j=1}^{\infty} M_{umax}(i,j;J) = \sum_{i=1}^{\infty} M_{umax}(i;J) = M_{umax}(J) < \infty$, then for all $i\in N$, the series $\sum_{j=1}^{\infty} u_{i,j}(t)$ converge pointwise to functions in $A(I,R)$ and $\sum_{i=1}^{\infty}\sum_{j=1}^{\infty} u_{i,j}(t)$ converges pointwise to a function in $A(I,R)$., the following corollary to Theorem 2.4 on rearrangement of summation terms:

<u>COROLLARY 3.2</u>. Assume $a_{i,j}\geq 0$ for $i,j\in N$. Then $\sum_{i=1}^{\infty}\sum_{j=1}^{i-1} a_{i,j}$ converges if and only if for all $j\in N$ we have that $a_i = \sum_{i=1}^{\infty} a_{i,j}$ converges and $\sum_{i=1}^{\infty} a_i$ converges. If either is the case, then

$$\sum_{i=1}^{\infty}\sum_{j=1}^{i-1} a_{i-j,j} = \sum_{i=2}^{\infty}\sum_{j=1}^{i-1} a_{i-j,j} = \sum_{i=1}^{\infty}\sum_{j=1}^{\infty} a_{i,j}\,. \qquad (3.1.1)$$

and (to consider the derivative of the zeroth moment) two additional corollaries to Theorem 2.3 on analytic convergent series (again see for example Wunsch, 1994):

COROLLARY 3.3. Let $I \in Int_o$ and $\vec{u}(t) = \{u_i(t)\}_{i=1}^{\infty} \in \vec{\mathcal{A}}(I, \mathbf{R}^{\infty})$. Suppose that for all $J \in Int_c(I)$, $u(t) = \sum_{j=1}^{\infty} u_j(t)$ converge uniformly on J. Then $\frac{du}{dt}$ is defined and analytic on I and $\frac{du}{dt} = \sum_{j=1}^{\infty} \frac{du_j}{dt} \in A(I,R)$.

COROLLARY 3.4. Let $I \in Int_o$ and $\vec{u}(t) = \{u_i(t)\}_{i=1}^{\infty} \in \vec{D}_{UBS}(I, \mathbf{R}^{\infty})$. Then $u(t) = \sum_{j=1}^{\infty} u_j(t)$ $\in A(I,R)$, $\frac{du}{dt} \in A(I,R)$ and $\frac{du}{dt} = \sum_{j=1}^{\infty} \frac{du_j}{dt}$.

To develop an equation for the derivative of the zeroth moment of a solution of $Prob_A(I,t_0,\vec{n}_0,A)$ we formally sum the terms in (2.4.1) from i = 1 to infinity to obtain the formal set of equations:

$$\sum_{i=1}^{\infty} \frac{dn_i}{dt} = \sum_{i=1}^{\infty} \frac{1}{2} \sum_{j=i}^{i-1} A(t)\, n_j n_{i-j} - \sum_{i=1}^{\infty} n_i \sum_{j=1}^{\infty} A(t)\, n_j\,, t \in I, \tag{3.1.2}$$

containing three formal series. We will show that if $(I,A,\vec{n}_0) \in Int(I_0,t_0) \times A(I_0,R) \times IC_B$ and $\vec{n}(t) = \{n_i(t)\}_{i=1}^{\infty} \in \vec{D}_{UBS}(I,t_0,\vec{n}_0)$, then the three formal infinite series converge absolutely at each $t \in I$ and uniformly on all closed intervals $J \subseteq I \in Int(I_0,t_0)$ to analytic functions and the formal set of equations (3.1.2) becomes an actual set of equations.

THEOREM 3.1.5. Let $(I,A,\vec{n}_0) \in Int(I_0,t_0) \times A(I_0,R) \times IC_B$. If $\vec{n} \in \vec{D}_{UBS}(I,t_0,\vec{n}_0)$, then the zeroth moment is $M(t) = \sum_{j=1}^{\infty} n_j(t)$, the depletion coefficient is $p(t) = A(t) \sum_{j=1}^{\infty} n_j(t)$,

$$\sum_{i=1}^{\infty} n_i \sum_{j=1}^{\infty} A(t)\, n_j = \sum_{i=1}^{\infty} \sum_{j=1}^{\infty} A(t)\, n_j\, n_i \quad , t \in I, \tag{3.1.3}$$

$$\sum_{i=2}^{\infty} \frac{1}{2} \sum_{j=1}^{i-1} A(t)\, n_{i-j}\, n_j = \frac{1}{2} \sum_{i=1}^{\infty} \sum_{j=1}^{\infty} A(t)\, n_i\, n_j\,, t \in I, \tag{3.1.4}$$

and the derivative of M(t) is

$$\frac{d}{dt}M(t)=\frac{d}{dt}\sum_{i=1}^{\infty}n_i=\sum_{i=1}^{\infty}\frac{dn_i}{dt},\quad t\in I. \qquad (3.1.5)$$

All exist in $\mathbb{A}$(I,R).

Proof. Let $(I,A,\vec{n}_0)\in Int(I_0,t_0)\times\mathbb{A}(I_0,R)\times IC_{M_0}$, $\vec{n}(t)=\{n_i(t)\}_{i=1}^{\infty}\in\vec{D}_{UBS}(I,t_0,\vec{n}_0)$, and $J\subseteq I_0$ be a finite closed interval. Since $A(t)\in\mathbb{A}(I_0,R)$, for any finite closed interval $J\subseteq I_0$ there exists a constant $M_{Amax}(J)$ such that $!A(t)!\leq M_{Amax}(J)$ for all $t\in J$. Since $\vec{n}\in\vec{D}_{UBS}(I,\mathbf{R}^{\infty})$, for any closed interval $J\subseteq I_0$ we have constants $M_{nmax}(i;J)$ such that for all $i\in N$, $!n_i(t)! < M_{nmax}(i;J)$ and $\sum_{i=1}^{\infty}M_{nmax}(i;J)=M_{nmax}(J)<\infty$. Hence $!A(t)\,n_i(t)\,n_j(t)! < M_{Amax}(J)\,M_{nmax}(i;J)\,M_{nmax}(j;J)$ and $\sum_{i=1}^{\infty}\sum_{j=1}^{\infty}M_{Amax}(J)\,M_{nmax}(i;J)\,M_{nmax}(j;J)=M_{Amax}(J)\,M_{nmax}(J)\,M_{nmax}(J)<\infty$. Thus the series on the right in (3.1.3) and (3.1.4) converge absolutely and uniformly on all J to analytic functions and we may rearrange terms. This yields the left hand sides of (3.1.3) and (using (3.1.1)), (3.1.4) so that (3.1.2) is an equation and $M(t)=\sum_{i=1}^{\infty}n_i(t)$ and $p(t)=A(t)\sum_{i=1}^{\infty}n_i(t)$ exist in $\mathbb{A}$(I,R). Since $\vec{n}\in\vec{D}_{UBS}(I,t_0,\vec{n}_0)$, Corollary 3.4 implies $dM(t)/dt\in$ exists in $\mathbb{A}$(I,R) and is given by (3.1.5). Q.E.D.

<u>THEOREM 3.6</u>. Let $(I,A,\vec{n}_0)\in Int(I_0,t_0)\times\mathbb{A}(I_0,R)\times IC_{M_0}$. If $\vec{n}_A(t)=\{n_i^A(t)\}_{i=1}^{\infty}\in\vec{D}_{UBS}(I,t_0,\vec{n}_0)$ is a solution to $Prob_A(\vec{D}_{UBS}(I,t_0,\vec{n}_0);A)$, then $M_0^A(t)=\sum_{j=1}^{\infty}n_j^A(t)\in R_{UBS}(I;t_0,\vec{n}_0)$ is a solution of $Prob_M(I,t_0,M_0,A)=Prob_{Mno}(I,t_0,\vec{n}_0,A)$ where $\vec{n}_0=\{n_i^0\}_{i=1}^{\infty}\in IC_B$ and $M_0=M_{no}=\sum_{j=1}^{\infty}n_j^0$.

Proof. Let $(I,A,\vec{n}_0)\in Int(I_0,t_0)\times\mathbb{A}(I_0,R)\times IC_B$ and $\vec{n}_A(t)=\{n_i^A(t)\}_{i=1}^{\infty}\in\vec{D}_{UBS}(I,t_0,\vec{n}_0)$ be a solution to $Prob_A(I,t_0,\vec{n}_0,A)$. By Corollary 3.5 we may sum (2.4.1) to obtain

$$\frac{dM_0^A}{dt}=\frac{1}{2}\sum_{i=1}^{\infty}\sum_{j=1}^{\infty}A(t)\,n_i^A\,n_j^A-\sum_{i=1}^{\infty}\sum_{j=1}^{\infty}A(t)\,n_j^A\,n_i^A$$

$$=-\frac{1}{2}\sum_{i=1}^{\infty}\sum_{j=1}^{\infty}A(t)\,n_i^A\,n_j^A=-\frac{1}{2}(M_0)^2,\ t\in I. \qquad (4.6)$$

Since $M_A(t_0)=\sum_{j=1}^{\infty} n_j^A(t_0) = \sum_{i=1}^{\infty} n_i^0 = M_{no}$, $M_A(t)$ solves $Prob_M(\mathbb{A}(I;t_0,M_0);A) = Prob_{Mno}(I,t_0,\vec{n}_0,A)$ where $\vec{n}_0 = \{n_i^0\}_{i=1}^{\infty} \in IC_B$ and $M_0 = \sum_{j=1}^{\infty} n_j^0$. Q.E.D.

COROLLARY 3.7. Let $(I,A,\vec{n}_0)\in Int(I_0,t_0)\times\mathbb{A}(I_0,R)\times IC_B$ and $I_{IV}^{Mno}(t_0,M_0,A)$ be the interval of validity for $Prob_M(I,t_0,M_0,A) = Prob_{Mno}(I,t_0,\vec{n}_0,A)$. If $Prob_A(I,t_0,\vec{n}_0,A)$ has a solution , then its interval of validity is no larger than $I_{IV}^{Mno}(t_0,\vec{n}_0,A)$.

Proof. Let $(I,A,\vec{n}_0)\in Int(I_0,t_0)\times\mathbb{A}(I_0,R)\times IC_B$. If $\vec{n}_A(t) = \{n_i^A(t)\}_{i=1}^{\infty} \in \vec{D}_{UBS}(I,t_0,\vec{n}_0)$ is a solution to $Prob_A(I,t_0,\vec{n}_0,A)$, then by Theorem 4.6 its zeroth moment $M_A(t) = \sum_{j=1}^{\infty} n_j^A(t)$ exists in $R_{UBS}(I,t_0,\vec{n}_0)$ and satisfies $Prob_M(I,t_0,M_0,A) = Prob_{Mno}(I,t_0,\vec{n}_0,A)$. Since $I_{IV}^{Mno}(t_0,\vec{n}_0,A)$ is the largest open interval on which the solution to $Prob_M(I,t_0,M_0,A) = Prob_{Mno}(I,t_0,\vec{n}_0,A)$ exists, the interval of validity for $Prob_A(I,t_0,\vec{n}_0,A)$ can be no larger than $I_{IV}^{Mno}(t_0,\vec{n}_0,A)$.

Q.E.D.

3.2. Solution of the Moment Problem

Suppose $(I,A,\vec{n}_0)\in Int(I_0,t_0)\times\mathbb{A}(I_0,R)\times IC_B$ and that $\vec{n}_A(t) = \{n_i^A(t)\}_{i=1}^{\infty} \in \vec{D}_{UBS}(I,t_0,\vec{n}_0)$ is a solution to $Prob_A(I,t_0,\vec{n}_0,A)$. Then by Theorem 3.6 , its zeroth moment, $M_A(t) = \sum_{i=1}^{\infty} n_i^A(t) \in R_{UBS}(I,t_0,\vec{n}_0)$, solves $Prob_M(I,t_0,M_{no},A) = Prob_{Mno}(I,t_0,\vec{n}_0,A)$:

$$\frac{dM}{dt} = -\frac{1}{2}A(t)\left(M_0\right)^2, \ \forall\ t\in I, \text{ where } M(t_0) = M_0 = M_{no} = \sum_{i=1}^{\infty} n_i^0 . \tag{3.2.1}$$

Although the ODE in (3.2.1) is nonlinear, and hence its interval of validity $I_{IV}^{M}(t_0,M_0,A)$ need not be I_0, it is separable and is easily solved. Solving we obtain

$$\int\frac{dM}{(M)^2} = -\int\frac{1}{2}A(t)dt . \tag{3.2.2}$$

Letting

$$\mathbb{A}(t) = \int_{s=t_0}^{s=t} A(s)\, ds \tag{3.2.3}$$

we obtain

$$-(M)^{-1} = -\frac{1}{2}A(t) - \frac{1}{2}C,\ \frac{2}{M} = A(t) + C,\ M = \frac{2}{A(t)+C} \tag{3.2.4}$$

Applying the initial condition in $M(t_0) = M_0$ we obtain

$$M_0 = \frac{2}{A(t_0)+C},\ M_0 = \frac{2}{C},\ C = \frac{2}{M_0}. \tag{3.2.5}$$

Hence, substituting (3.2.5) into (3.2.4) we obtain

$$M(t) = \frac{2}{A(t) + 2/M_0} = \frac{M_0}{1+(1/2)M_0 A(t)} = \frac{M_0}{D(t)}, \tag{3.2.6}$$

$$\frac{dM}{dt} = -\frac{(1/2)(M_0)^2 A(t)}{[1+(1/2)M_0 A(t)]^2} = -\frac{(1/2)(M_0)^2 A(t)}{[D(t)]^2},$$

$$\left.\frac{dM}{dt}\right|_{t=t_0} = -\frac{(M_0)^2 A(t_0)}{2} \tag{3.2.7}$$

$$R(t) = \frac{A(t)}{D(t)} = \frac{A(t)}{1+(1/2)M_0 A(t)} = \frac{2/M_0}{1+1/[M_0 A(t)/2]} \tag{3.2.8}$$

$$\begin{aligned}\frac{dR(t)}{dt} &= \frac{A(t)D(t) - A(t)(1/2)M_0 A(t)}{[D(t)]^2} \\ &= \frac{A(t)[1+(1/2)M_0 A(t)] - A(t)(1/2)M_0 A(t)}{[D(t)]^2} \\ &= \frac{A(t)}{[D(t)]^2}\end{aligned} \tag{3.2.9}$$

where

$$D(t) = 1 + (½)M_0 A(t). \tag{3.2.10}$$

By Corollary 3.7, we have that $I_{IV}^{A}(t_0, \vec{n}_0, A) \le I_{IV}^{Mno}(t_0, \vec{n}_0, A)$. Clearly, $I_{IV}^{M}(t_0, M_0, A)$ depends on the zeros of D(t). We use R(t) later in the solution of $Prob_A($ $\vec{D}_{UBS}(I, t_0, \vec{n}_0)$;A). Define $I_M = I_M(t_0,M_0,A)$, $I_A = I_A(t_0,M_0,A) \in Int_o(t_0,I_0)$ as follows. If $M_0 = 0$ or $A(t) \equiv 0$ (the trivial case), let $I_M = I_A = I_0$. If $M_0 \neq 0$ and $A(t) \not\models 0$, A(t) is not a constant and

D(t) has isolated zeros where A(t) = $-2/M_0$. Let I_M be the largest interval in $Int_o(t_0,I_0)$ where $M_0A(t) > -2$ and I_A be the largest interval in $Int_o(t_0,I_0)$ where $M_0A(t) > -1$. We have

<u>LEMMA 3.8</u>. I_A and I_M are connected, $I_A \subseteq I_M = I_{IV}^M(t_0, M_0, A) \subseteq I_N = \{t\in I_0: !M_0A(t)! > -2\}$, and $I_A \subseteq I_C = \{t\in I_0: !M_0 R(t)! < 2\} = \{t\in I_0: M_0A(t) > -1\}$.

Proof. I_A and I_M are connected, $I_A \subseteq I_M = I_{IV}^M$, and $I_A \subseteq \{t\in I_0: M_0A(t) > -1\}$ follow directly from their definitions. I_A may be smaller since it is required to be an interval. That $\{t\in I_0: !M_0 \Re(t)! < 2\} = \{t\in I_0: M_0A(t) > -1\}$ (which we then define as I_C) follows from the following set of equivalent inequalities:

$!M_0 R(t)! < 2$ if and only if $-2 < M_0 R(t) < 2$. Multiplying by $1 + (½) M_0A(t)$, we have two cases.

If $1 + (½) M_0A(t) > 0$.$-2 (1 + (½)M_0A(t)) < M_0 A(t) < 2 (1 + (½)M_0A(t))$
$(M_0A(t) > -2)$.

$$-2 - M_0A(t)) < M_0 A(t) < 2 + M_0A(t)$$
$$-2 - 2 M_0A(t)) < 0 < 2$$
$$-1 - M_0A(t)) < 0 < 1$$
$$- M_0A(t)) < 1 < 2$$
$$M_0A(t)) > -1 > -2$$

If $1 + (½) M_0A(t) < 0$.$-2 (1 + (½)M_0A(t)) > M_0 A(t) > 2 (1 + (½)M_0A(t))$
$(M_0\mathfrak{A}(t) < -2)$.

$$-2 - M_0A(t)) > M_0 A(t) > 2 + M_0A(t)$$
$$-2 - 2 M_0A(t)) > 0 > 2$$
$$-1 - M_0A(t)) > 0 > 1$$
$$- M_0A(t)) > 1 > 2$$
$$M_0A(t)) < -1 < -2$$

Thus in the second case, if $M_0A(t) < -2$, we must have $-1 < -2$ in order to have $M_0 R(t) < 2$. Since this is not true, the second case can not happen. Hence if $!M_0 R(t)! < 2$, we must have

$M_0A(t)) > -1$.

On the other hand, if $M_0A(t)) > -1$, then

$M_0A(t)) > -1 > -2$ $M_0A(t)) < 1 < 2$

$$-1 - M_0A(t)) < 0 < 1$$
$$-2 - 2 M_0A(t)) < 0 < 2$$
$$-2 - M_0A(t)) < M_0 A(t) < 2 + M_0A(t)$$
$$-2 (1 + (½)M_0A(t)) < M_0 A(t) < 2 (1 + (½)M_0A(t))$$

and since $M_0A(t)) > -1 > -2$ so that $1 + (½) M_0A(t) > 0$, we have

$$-2 < M_0 R(t) < 2$$

$$!M_0 \, \mathrm{R}(t)! < 2.$$

Hence $I_C = \{t \in I_0: \backslash ! M_0 \, \mathrm{R}(t)! < 2\} = \{t \in I_0: M_0 \mathrm{A}(t) > -1\}$. $I_A \subseteq I_C$ as I_A is just the largest interval containing t_0 where $M_0 \mathrm{A}(t) > -1$. Q.E.D.

When

$\vec{n}_0 = \{n_i^{\,0}\}_{i=1}^{\infty} \in IC_B$ and $M_0 = M_{no} = \sum_{i=1}^{\infty} n_i^{\,0}$, we use $I_{Mno} = I_{Mno}(t_0, \vec{n}_0, A)$, $I_{Ano} = I_{Ano}(t_0, \vec{n}_0, A)$, and $I_{Cno} = I_{Cno}(t_0, \vec{n}_0, A)$ for I_M, I_A, and I_C respectively. Later we show $I_{IV}^{A}(t_0, \vec{n}_0, A) = I_{Ano}(t_0, \vec{n}_0, A)$. For the physical case where $M_0 > 0$ and $A(t) > 0$ on I_0 we have $\mathrm{A}(t)$, $\mathrm{D}(t)$, and $\mathrm{R}(t)$ are increasing on $I_0 = (t_{0-}, t_0, t_{0+})$. Since $\mathrm{A}(t_0) = \mathrm{R}(t_0) = 0$ and $\mathrm{D}(t_0) = 1$, $\mathrm{A}(t)$, $\mathrm{D}(t)$, and $\mathrm{R}(t)$ are positive on (t_0, t_{0+}), $\mathrm{A}(t)$ and $\mathrm{R}(t)$ are negative on (t_{0-}, t_0), and $t_{M+} = t_{A+} = t_{C+} = t_{0+}$. Clearly $M(t)$ is positive and decreasing on $I_{pIV}^{M} = I_M = (t_{M-}, t_0, t_{0+})$, (coagulation does not occur). If, in addition, $[0,\infty) \subseteq I_0$ and $\lim_{t\to\infty} \mathrm{A}(t) = \infty$, then $I_{pIV}^{M} = I_M = (t_{M-}, t_0, \infty)$, $I_A = (t_{A-}, t_0, \infty)$, $\lim_{t\to\infty} M(t) = 0$ (coagulation occurs at infinity), and $\lim_{t\to\infty} \mathrm{R}(t) = 2/M_0^{\,0}$.

<u>THEOREM 3.9</u>. Let $(I, A, M_0) \in \mathrm{Int}(I_0, t_0) \times \mathcal{A}(I_0, R) \times R$ and $I_M = I_M(t_0, M_0, A)$, $I_A(t_0, M_0, A)$, and $\mathrm{R}(t)$ be as defined above. Then $\mathrm{Prob}_M(\mathcal{A}(I; t_0, M_0); A)$ has a unique solution $M(t)$ with interval of validity $I_{IV}^{M}(t_0, M_0, A) = I_M(t_0, M_0, A) \subseteq I_0$ and $\mathrm{R}(t_0) = 0$. If $M_0 = 0$, $M(t) = 0$ and $I_{IV}^{M}(t_0, M_0, A) = I_M(t_0, M_0, A) = I_0$. Otherwise, $M(t)$ is given by (3.2.6) and $I_{IV}^{M}(t_0, M_0, A)$ depends on the location of the zeros of $\mathrm{D}(t) = 1 + (½) M_0 \mathrm{A}(t)$ where $\mathrm{A}(t)$ is given by (3.2.3). If $M_0 > 0$ and $A(t) > 0$ on $I_0 = (t_{0-}, t_0, t_{0+})$, $\mathrm{R}(t) > 0$ on (t_0, t_{0+}) and $M(t)$ is positive and decreasing on $I_{pIV}^{M}(t_0, M_0, A) = (t_{M-}, t_0, t_{M+})$ where $t_{M+} = t_{0+}$, $t_{M-} = t_{0-}$ if $\mathrm{D}(t) > 0$ on I_0, and $t_{M-} = \sup\{t < t_0 : \mathrm{D}(t) = 0\}$ otherwise. If $[0, \infty) \subseteq I_0$ and $\lim_{t\to\infty} \mathrm{A}(t) = \infty$, then $\lim_{t\to\infty} M(t) = 0$ and $\lim_{t\to\infty} \mathrm{R}(t) = 2/M_0$.

We denote the (unique) solution of $\mathrm{Prob}_M(I, t_0, M_0, A)$ with interval of validity $I_{IV}^{M}(t_0, M_0, A)$ by $M_M(t)$ and the (unique) solution of $\mathrm{Prob}_M(I, t_0, M_0, A) = \mathrm{Prob}_{Mno}(I, t_0, \vec{n}_0 A)$ with interval of validity $I_{IV}^{Mno}(t_0, \vec{n}_0, M_0, A)$ by $M_{Mno}(t)$. We also let $\mathrm{D}_{no}(t) = 1 + (1/2) M_{no} \mathrm{A}(t) = \mathrm{D}(t; t_0, \vec{n}_0, A)$, and $\mathrm{R}_{no}(t) = \frac{\mathrm{A}(t)}{\mathrm{D}_{no}(t)} = \mathrm{R}(t; t_0, \vec{n}_0, A)$.

3.3 The Scaled Time Using the Solution of the Moment Problem

For future use, we wish the scaled time $\tau = h_p(t;t_0,p_g) = h_{AM}(t;t_0,A,M_g)$ when $p_g(t) = A_g(t)\, M_g(t)$ is given by $p_M(t) = A(t)\, M_M(t)$. Since $M_M(t)$, like I_M and I_A, depends only on t_0, M_0, and $A(t)$, we let $\tau = h_{AM}(t;A,M_M(t)) = h_{AM_M}(t;t_0,M_0,A)$ and use $I_{IV}^{M}(t_0,M_0,A) = I_M(t_0,M_0,A)=(t_{M-},t_0,t_{M+}) \supseteq I_A(t_0,M_0,A) = (t_{A-},t_0,t_{A+}) \leftrightarrow (\tau_{A-},0,\tau_{A+}) = I_A(t_0,M_0,A) \subseteq (\tau_{M-},0,\tau_{M+}) = I_M(t_0,M_0,A)$. We accent the physical context where $M_0 > 0$ and $A(t) > 0$ on I_0. For future use, when $\vec{n}_0 = \{n_i^{\,0}\}_{i=1}^{\infty} \in IC_B$ and $M_0 = M_{no} = \sum_{i=1}^{\infty} n_i^{\,0}$, we let $\tau = h_{AM_M}(t;t_0,A,M_{M_{no}}(t)) = h_{AM_{M_{no}}}(t;t_0,\vec{n}_0,A)$ and use $I_{IV}^{M_{no}}(t_0,\vec{n}_0,A) = I_{M_{no}}(t_0,\vec{n}_0,A) = (t_{Mno-},t_0,t_{Mno+}) \supseteq I_{A_{no}}(t_0,\vec{n}_0,A) = (t_{Ano-},t_0,t_{Ano+}) \leftrightarrow (\tau_{Ano-},0,\tau_{Ano+}) = I_{A_{no}}(A,M_{M_{no}}(\cdot;t_0,\vec{n}_0,A)) \subseteq (\tau_{Mno-},0,\tau_{Mno+}) = I_{M_{no}}(A,M_{M_{no}}(\cdot;t_0,\vec{n}_0,A))$. Again, To compute $\tau = h_{AM_M}(t;t_0,M_0,A)$ we use $p_M(t) = A(t)\, M_M(t)$ for $p_g(t) = A(t)\, M_g(t)$ in (2.4.11). For the trivial case, $\tau = t$. For the nontrivial case, where $M_M(t)$ is given by (3.2.6) we obtain

$$\int_{s=0}^{s=t} p_M(s)ds = \int_{s=t_0}^{s=t} A(s)\, M_M(s)\, ds = \int_{s=t_0}^{s=t} A(s)\frac{M_0}{D(s)} ds = \int_{s=t_0}^{s=t} A(s)\frac{M_0}{1+(1/2)M_0 A(s)} ds . \quad (3.3.1)$$

Letting

$$u(s) = D(s) = 1 + (1/2)M_0\, A(s),\ du = (1/2)M_0 A(s)\, ds \quad (3.3.2)$$

we have

$$\int_{s=t_0}^{s=t} p_M(s)ds = \frac{2}{M_0}\int_{u=1}^{u=1+(1/2)M_0 A(t)} \frac{M_0}{u} du = 2\ell n|1+(1/2)M_0 A(t)| - 2\,\ell n|1|$$

$$= \ell n(1+(1/2)M_0 A(t))^2 = -\ell n\left(\frac{1}{1+(1/2)M_0 A(t)}\right)^2 = -\ell n\left(\frac{1}{(D(t))^2}\right), \quad (3.3.3)$$

$$\frac{d\tau}{dt} = \exp\{-\int_{s=0}^{s=t} A(s)\, M_M(s)ds\} = \exp\left\{-\left[-\ell n\left(\frac{1}{(1+(1/2)M_0 A(s))^2}\right)\right]\right\}$$

$$= \frac{1}{(1+(1/2)M_0 A(t))^2} = \frac{1}{(D(t))^2}, \quad (3.3.4)$$

$$\tau = h_{AM_M}(t;t_0,M_0,A) = \int_{s=t_0}^{s=t} \frac{1}{(1+(1/2)M_0A(s))^2} ds = \int_{s=t_0}^{s=t} \frac{1}{(D(s))^2} ds, \tag{3.3.5}$$

$$\frac{d^2\tau}{dt^2} = -2$$

$$\frac{(1/2)M_0A(t)}{(D(t))^3} = -\frac{d\tau}{dt}\frac{M_0A(t)}{D(t)} = -\frac{d\tau}{dt}A(t)M_M(t), \tag{3.3.6}$$

We are particularly interested in the following time-varying analytic kernels:

$A(t) = A_0$ $\quad A_0 \neq 0$ $\quad$ Constant Kernel (3.3.7)

$A(t) = A_0 + b_0t$ $\quad A_0 \neq 0,\ b_0 \neq 0$ $\quad$ Linearly Time-Varying Kernel(3.3.8)

$A(t) = A_0 + b_0 \cos(\omega_0 t + \phi_0),\ A_0 \neq 0,\ b_0 > 0,\ \omega_0>0$ $\quad$ Oscillating Kernel. (3.3.9)

Hence we let $A(t) = A_0 + B(t)$ where $B(t)$ maybe chosen as 0, b_0 t, or $b_0 \cos(\omega_0 t + \phi_0)$ and $I_0 = R$ for the mathematical context. In the physical context we will assume $A_0 > 0$ as well as $A(t) > 0$ on I_0. Hence for $A(t) = A_0 + b_0t$ we have $I_0 = (-\gamma,\infty)$ if $b_0 > 0$ and $I_0 = (-\infty,\gamma)$ if $b_0 < 0$ where $\gamma = A_0/ b_0$ which places an additional constraint on t_0. For $A(t) = A_0 + b_0 \cos(\omega_0 t + \phi_0)$, we require $A_0 > !b_0!$ so that $I_0 = R$.

For the examples, along with τ we also wish to use (3.2.6) and (3.2.8) to compute

$$M(t) = \frac{2}{A(t)+2/M_0} = \frac{M_0}{1+(1/2)M_0A(t)} = \frac{M_0}{D(t)}, \tag{3.3.10}$$

$$R(t) = \frac{A(t)}{D(t)} = \frac{A(t)}{1+(1/2)M_0A(t)} = \frac{2/M_0}{1+1/[M_0A(t)/2]}. \tag{3.3.11}$$

For future use we let $R(t) = R(h^{-1}_{AM_M}(\tau;t_0,M_0,A)) = \tilde{R}(\tau;t_0,M_0,A)$,

$$\tau = h_{AM}(t;t_0,A,M_{M_{no}}) = h_{AM_M}(t;t_0,A,M_{no}) = h_{AM_{M_{no}}}(t;t_0,\vec{n}_0,A) \tag{3.3.12}$$

and $R_{no}(t) = R(h^{-1}_{AM_{M_{no}}}(\tau;t_0,\vec{n}_0,A)) = \tilde{R}_{no}(\tau;t_0,\vec{n}_0,A)$.

Summarizing, for each example we compute $A(t)$, $(1/2)M_0\, A(s)$, $D(s) = 1 + (1/2)M_0\, A(s)$, $M(t) = \frac{M_0}{D(t)}$, $R(t) = \frac{A(t)}{D(t)}$, and $\tau = h_{AM_M}(t;t_0,A,M_0)$. For $A(t) = A_0 + B(t)$ we let $Prob_M(A(I;t_0,M_0);A_0 + B(t)) = Prob_{MAB}(A(I;t_0,M_0);A_0,B(t))$ with solution $M(t) = M_{MA_0B}(t)$ and $\tau(t) = h_{AM_M}(t;t_0,M_0,A_0+B(t)) = h_{AM_{MA_0B}}(t;t_0,M_0,A_0,B)$ so that the change of variables yields $I_{IV}^{M}(t_0, M_0, A_0 + B(t)) = I_{IV}^{MA_0B}(t_0, M_0, A_0, B) = (t_{MAB-}, t_0, t_{MAB+}) \leftrightarrow (\tau_{MAB-}, 0, \tau_{MAB+}) = \mathsf{I}_{IV}^{MAB}(t_0, M_0, A_0, B)$. From (3.2.3) and (3.2.10), for $A(t) = A_0 + B(t)$ we have

$$\mathtt{A}(t) = \int_{s=t_0}^{s=t} (A_0 + B(s))ds = A_0(t - t_0) + \mathtt{B}(t) \text{ where } \mathtt{B}(t) = \int_{s=0}^{s=t} B(t)\, ds, \tag{3.3.13}$$

$$(1/2)M_0\, \mathtt{A}(t) = (1/2)M_0\, [A_0(t - t_0) + \mathtt{B}(t)] = (1/2)M_0A_0(t-t_0) + (½)M_0\, \mathtt{B}(t) \tag{3.3.14}$$

$$\begin{aligned} \mathtt{D}(t) &= 1 + (1/2)M_0\, \mathtt{A}(t) = 1 + (1/2)M_0A_0(t-t_0) + (½)M_0\, \mathtt{B}(t) \\ &= (1/2)M_0A_0[t + (\mathtt{B}(t)/A_0) - (t_0 - \alpha)] \\ &= (1/\alpha)[t - t_1 + (\mathtt{B}(t)/A_0)] \end{aligned} \tag{3.3.15}$$

where $\alpha = 2/(M_0\, A_0)$ and $t_1 = t_0 - \alpha$.

Using (3.3.10 and (3.3.12) in (3.2.6)-(3.2.9) and (3.3.4)-(3.3.5) we have

$$\begin{aligned} M_{MAB}(t) &= \frac{M_0}{\mathsf{D}(t)} = \frac{M_0}{1 + (1/2)M_0[A_0(t - t_0) + \mathsf{B}(t)]} = \frac{M_0}{1 + (1/\alpha)[t - t_0 + \mathsf{B}(t)/A_0]} \\ &= \frac{M_0\alpha}{\alpha + t - t_0 + \mathsf{B}(t)/A_0} = \frac{2/A_0}{t - (t_0 - \alpha) + \mathsf{B}(t)/A_0} = \frac{2/A_0}{t - t_1 + [\mathsf{B}(t)/A_0]} = \frac{2}{A_0(t - t_1) + \mathsf{B}(t)}, \end{aligned}$$

$$M_{MA_0B}(t_0) = \frac{2/A_0}{\alpha} = M_0, \tag{3.3.16}$$

$$\begin{aligned} \frac{dM_{MAB}}{dt} &= -\frac{(1/2)(M_0)^2\, \mathtt{A}(t)}{[\mathsf{D}(t)]^2} = \frac{(1/2)(M_0)^2[A_0 + B(t)]}{[1 + (1/2)M_0[A_0(t - t_0) + \mathsf{B}(t)]^2} \\ &= -\frac{2(M_0)^2[A_0 + B(t)]}{[M_0A_0(t - t_0) + M_0\mathsf{B}(t) + 2]^2} = -\frac{2[1 + B(t)/A_0]/A_0}{[t + [\mathsf{B}(t)/A_0] - (t_0 - \alpha)]^2} \\ &= -\frac{2[1 + B(t)/A_0]/A_0}{[t - t_1 + [\mathsf{B}(t)/A_0]]^2} = -\frac{2[A_0 + B(t)]}{[A_0(t - t_0) + \mathsf{B}(t) + 2/M_0]^2} \end{aligned} \tag{3.3.17}$$

$$\left.\frac{dM_{MAB}}{dt}\right|_{t=t_0} = -\frac{2[A_0 + B(t_0)]/(A_0)^2}{\alpha^2} = -\frac{(M_0{}^2[A_0 + B(t_0)]}{2} \tag{3.3.18}$$

$$\begin{aligned} \mathsf{R}^{AB}(t) &= \frac{\mathsf{A}(t)}{\mathsf{D}(t)} = \frac{2/M_0}{1 + 1/[M_0\mathsf{A}(t)/2]} = \frac{2/M_0}{1 + 1/[(M_0/2)(A_0(t - t_0) + \mathsf{B}(t))]} \\ &= \frac{2/M_0}{1 + 1/[(t - t_0 + \mathsf{B}(t)/A_0)/\alpha]} = \frac{2/M_0}{1 + \alpha/(t - t_0 + \mathsf{B}(t)/A_0)} \end{aligned}$$

$$= \frac{[2/M_0][t-t_0+B(t)/A_0]}{[t-t_0+B(t)/A_0]+\alpha} = \frac{[2/M_0][t-t_0+B(t)/A_0]}{t-t_1+B(t)/A_0} \quad (3.3.19)$$

$$\frac{d\tau}{dt} = \frac{1}{(D(t))^2} = \frac{4/(M_0)^2}{(A(t)+2/M_0)^2} = \frac{\alpha^2}{[t-t_1+(B(t)/A_0)]^2} \quad (3.3.20)$$

$$\tau = h_{ABM}(t;t_0,M_0,A_0,B(t)) = \int_{s=t_0}^{s=t} \frac{1}{(D(s))^2} ds = \int_{s=t_0}^{s=t} \frac{4/(M_0)^2}{(A(s)+2/M_0)^2} ds$$

$$= \int_{s=t_0}^{s=t} \frac{\alpha^2 ds}{[s-t_1+(B(s)/A_0)]^2}, \quad (3.3.21)$$

We wish to obtain $I_M = I_M(t_0,M_0,A) \subseteq I_N = \{t \in I_0 :! M_0 A(t)! > -2\}$ and $I_A = I_A(t_0,M_0,A) \subseteq I_C = \{t \in I_0 :! M_0 A(t)! > -1\}$. From the definitions of I_N and I_C, we wish to solve

$$M_0\, A(t) = M_0\, [\, A_0(t-t_0) + B(t)] > -r, \; r = 1(I_C), 2(I_N), \quad (3.3.22)$$

$$M_0\, A(t)/2 = (M_0/2)[\, A_0(t-t_0) + B(t)] > -r/2, \quad r = 1,2, \quad (3.3.23)$$

$$M_0\, A(t)/2 = (M_0 A_0/2)(t-t_0) + (M_0/2)\, B(t) > -r/2, \quad r = 1,2, \quad (3.3.24)$$

$$M_0\, A(t)/2 = (t-t_0)/\alpha + (M_0/2)\, B(t) > -r/2, \quad r = 1,2, \quad (3.3.25)$$

For the physical context we require $A_0 > 0$, $M_0 > 0$, (so that $\alpha > 0$) and $B(t) > -A_0$ (so that $A(t) = A_0 + B(t) > 0$) for $t \in I_0 = (t_{0-}, t_0, t_{0+})$. Let $t_{MAB+} = t_{0+}$. If $t - t_1 + (B(t)/A_0) > 0$ on I_0, let $t_{MAB-} = t_{0-}$. Otherwise, let $t_{MAB-} = \sup\{t < t_0 : t - t_1 + (B(t)/A_0) = 0\}$. Then $I_{pIV}^{MAB}(t_0,M_0,A_0,B) = (t_{MAB-},t_0,t_{MAB+}) \leftrightarrow (\tau_{MAB-},0,\tau_{MAB+}) = \mathcal{I}_{pIV}^{MAB}(t_0,M_0,A_0,B)$. Eqs. (3.3.22)-(3.3.25) confirm that $M(t)$ is positive and decreasing on $I_{pIV}^{MAB}(t_0,M_0,A)$ (coagulation does not occur) and $\mathcal{R}(t) > 0$ and that if $[0,\infty) \subseteq I_0$ (since $B(t) > -A_0$ we have $\lim_{t\to\infty} \mathcal{A}(t) = \lim_{t\to\infty}[\, A_0(t-t_0) + B(t)] = \infty$), then $\lim_{t\to\infty} M(t) = 0$ (coagulation occurs at infinity) and $\lim_{t\to\infty} R(t) = 2/M_0$.

For Example 1 where $B(t) = 0$, $I_0 = (t_{0-}, t_0, t_{0+}) = (-\infty, t_0, \infty)$. For Example 2 where $B(t) = b_0 t$, $I_0 = (t_{0-}, t_0, t_{0+}) = (-\infty, t_0, -\gamma)$ if $b_0 < 0$ and $I_0 = (t_{0-}, t_0, t_{0+}) = (-\gamma, t_0, \infty)$ if $b_0 > 0$.

For Example 3 where $B(t) = b_0 \cos(\omega_0 t + \phi_0)$, $I_0 = (t_{0-}, t_0, t_{0+}) = (-\infty, t_0, \infty)$.

3.3.1. The Scaled Time Using the Solution of the Moment Problem

EXAMPLE 1: THE CONSTANT KERNEL

Example 1. First consider the constant kernel where $B(t) = 0$ so that $A(t) = A_0 \in R$ is constant, $B(t) = 0$ and $I_0 = R$. We consider the nontrivial case where $M_0 \neq 0$ and $A_0 \neq 0$. The input parameters are t_0, M_0, and A_0. We may replace either M_0 or A_0 with $\alpha = 2/(M_0A_0)$. We may also replace t_0 with $t_1 = t_0 - \alpha$. From (3.3.13) and (3.3.14) we have

$$A(t) = A_0(t - t_0), \tag{3.3.1.1}$$

$$(\tfrac{1}{2})M_0\, A(t) = (\tfrac{1}{2})M_0\, A_0(t - t_0) = (t - t_0)/\alpha, \tag{3.3.1.2}$$

$$\begin{aligned} D(t) &= 1 + (1/2)M_0\, A(t) = 1 + (1/2)M_0A_0(t-t_0) \\ &= 1 + (t-t_0)/\alpha = [t - (t_0-\alpha)]/\alpha = (t - t_1)/\alpha, \end{aligned} \tag{3.3.1.3}$$

where $\alpha = 2/(M_0\, A_0)$, $t_1 = t_0 - \alpha$. $D(t)$ has a single zero at $t_1 = t_0 - \alpha$. Hence from (3.3.10), (3.3.11), (3.3.12) we have:

$$M_{MAB}(t) = M_{MAB1}(t) = \frac{M_0}{D(t)} = \frac{M_0}{(t-t_1)/\alpha} = \frac{M_0\alpha}{t-t_1} = \frac{2/A_0}{t-(t_0-\alpha)} = \frac{2/A_0}{t-t_1}, \tag{3.3.1.4}$$

$$\frac{d\tau}{dt} = \frac{1}{(D(t))^2} = \frac{\alpha^2}{(t-t_1)^2}, \tag{3.3.1.5}$$

$$\begin{aligned} R^{AB1}(t) = \frac{A(t)}{D(t)} &= \frac{2/M_0}{1+1/[M_0A(t)/2]} = \frac{2/M_0}{1+1/[(M_0/2)A_0(t-t_0)]} \\ &= \frac{2/M_0}{1+1/[(t-t_0)/\alpha]} = \frac{2/M_0}{1+\alpha/(t-t_0)} = \frac{[2/M_0](t-t_0)}{t-t_0+\alpha} \\ &= \frac{[2/M_0](t-t_0)}{t-(t_0-\alpha)} = \frac{\alpha A_0(t-t_0)}{t-t_1} \end{aligned} \tag{3.3.1.6}$$

$$\begin{aligned} \tau = h_{ABM}(t;t_0,M_0,A_0,0) &= h_{AB1}(t;t_0,\alpha) = \int_{s=t_0}^{s=t} \frac{\alpha^2 ds}{[s-t_1]^2} = \\ \alpha^2\left[-(t-t_1)^{-1} + (t_0-t_1)^{-1}\right] &= \left[-\frac{\alpha^2}{(t-t_1)} + \frac{\alpha^2}{\alpha}\right] = \left[\alpha - \frac{\alpha^2}{(t-t_1)}\right] \\ = \frac{\alpha[t-(t_0-\alpha)]-\alpha^2}{t-(t_0-\alpha)} &= \frac{\alpha t - \alpha t_0}{t-(t_0-\alpha)} = \frac{\alpha(t-t_0)}{t-(t_0-\alpha)} = \frac{\alpha(t-t_0)}{t-t_1}. \end{aligned} \tag{3.3.1.7}$$

From (3.3.1.6) we have

$\alpha(t - t_0) = \tau\, [t - t_1]$, $\alpha t - \alpha t_0 = \tau t - \tau t_1$, $\alpha t - \tau t = \alpha t_0 - \tau t_1$, $(\alpha - \tau)\, t = \alpha t_0 - \tau t_1$,

$$t = h_{AB1}^{-1}(t;t_0,\alpha) = \frac{\alpha t_0 - \tau t_1}{2\alpha - \tau} = \frac{\tau t_1 - \alpha t_0}{\tau - \alpha} \qquad (3.3.1.8)$$

so that from (3.3.1.5) we have

$$R^{AB1}(t) = \frac{A(t)}{D(t)} = \frac{2/M_0}{1+1/[M_0A(t)/2]} = \frac{2/M_0}{1+1/[(M_0/2)(A_0(t-t_0)+B(t))]}$$

$$= \frac{2/M_0}{1+1/[(t-t_0)/\alpha]} = \frac{2/M_0}{1+\alpha/(t-t_0)} = \frac{(2/M_0)(t-t_0)}{(t-t_0)+\alpha} = \frac{(2/(M_0A_0))A_0(t-t_0)}{t-(t_0-\alpha)}$$

$$= \frac{\alpha A_0(t-t_0)}{t-t_1} = \frac{\alpha A_0}{(t-t_1)/(t-t_0)} = \frac{2/M_0}{(t-t_1)/(t-t_0)} =$$

$$\frac{2/M_0}{(t-(t_0-\alpha))/(t-t_0)} = \frac{2/M_0}{(t-t_0+\alpha))/(t-t_0)} = \frac{2/M_0}{1+\alpha/(t-t_0)} = A_0\,\tau = \tilde{R}^{AB1}(\tau), \qquad (3.3.1.9)$$

Hence

$R^{AB1}(t_0) = 0 = \tilde{R}^{AB1}(0)$.

To obtain $I_N = \{t\in I_0:!M_0\mathcal{A}(t)! > -2\} = (t_{N-},t_0,t_{N+}) \supseteq I_M = (t_{M-},t_0,t_{M+}) \leftrightarrow (\tau_{M-},0,\tau_{M+}) = \mathrm{I}_M = \mathrm{I}_M(t_0,M_0,A) \subseteq \mathrm{I}_N(t_0,M_0,A)$ and $I_C = \{t\in I_0:!M_0A(t)! > -1\} = (t_{C-},t_0,t_{C+}) \supseteq I_A = (t_{A-},t_0,t_{A+}) \leftrightarrow (\tau_{A-},0,\tau_{A+}) = \mathrm{I}_A = \mathrm{I}_A(t_0,M_0^0,A) \subseteq \mathrm{I}_C(t_0,M_0^0,A)$, from (3.3.22) we wish to solve

$$M_0\,\mathcal{A}(t) = M_0\,[\,A_0(t-t_0)] > -r, \qquad r = 1,2$$
$$2\,[M_0\,A_0/2]\,(t-t_0) > -r, r = 1,2,$$
$$(t-t_0)/\alpha > -r/2, \qquad = 1,2, \qquad (3.3.1.10)$$

We consider two cases.

Example 1. $B(t) = 0$. Case 1. $\alpha = 2/(M_0^0A_0) < 0$ so that $-\alpha > 0$.

To obtain $I_N = \{t\in I_0:M_0A(t) > -2\}\supseteq I_M = (t_{M-},t_0,t_{M+})$ and $I_C = \{t\in I_0:M_0A(t) > -1\} \supseteq I_A = (t_{A-},t_0,t_{A+})$, we wish to solve

$$(t-t_0)/\alpha > -r/2, \qquad r = 1,2, \qquad (3.3.1.11)$$
$$t-t_0 < -r\,\alpha/2, \qquad r = 1,2, \qquad (3.3.1.12)$$
$$t < t_0 - r\,\alpha/2 = t_{1-}, \qquad r = 1,2, \qquad (3.3.1.13)$$

For $r = 2$, we have $t < t_0 - \alpha$, so that $I_N = I_{IV}^{MAB1a}(t_0,\alpha) = I_M = (-\infty,t_0,t_0 - \alpha)$. For $r = 1$, we have $t < t_0 - \alpha/2$, so that $I_C = I_A = (-\infty,t_0,t_0 - \alpha/2)$. To obtain $I_N \supseteq I_M = (t_{M-},t_0,t_{M+}) \leftrightarrow (\tau_{M-},0,\tau_{M+}) = \mathrm{I}_M = \mathrm{I}_M(t_0,M_0^0,A) \subseteq \mathrm{I}_N(t_0,M_0^0,A)$ and $I_C \supseteq I_A = (t_{A-},t_0,t_{A+}) \leftrightarrow (\tau_{A-},0,\tau_{A+}) = \mathrm{I}_A = \mathrm{I}_A(t_0,M_0^0,A) \subseteq \mathrm{I}_C(t_0,M_0^0,A)$, we need

$$h_{AB1}(\infty; t_0, \alpha) = \frac{\alpha(\infty - t_0)}{\infty - (t_0 - \alpha)} = \alpha.$$

$$h_{AB1}(t_0 - \alpha/2; t_0, \alpha) = \frac{\alpha(t_0 - \alpha/2 - t_0)}{t_0 - \alpha/2 - (t_0 - \alpha)} = \frac{\alpha(-\alpha/2)}{(-\alpha/2) + \alpha} = \frac{-\alpha^2/2}{\alpha/2} = -\alpha.$$

$$h_{AB1}(t_0 - \alpha; t_0, \alpha) = \frac{\alpha(t_0 - \alpha - t_0)}{t_0 - \alpha - (t_0 - \alpha)} = \infty.$$

Hence

$$I_{IV}^{MAB1a}(t_0, \alpha) = I_M = (-\infty, t_0, t_0 - \alpha) \supseteq I_A = (-\infty, t_0, t_0 - \alpha/2) \leftrightarrow (\alpha, 0, -\alpha) = \mathrm{I}_A \subseteq (\alpha, 0, \infty) = \mathrm{I}_M$$

Example 1. B(t) = 0. Case 2. $\alpha = 2/(M_0A_0) > 0$.

To obtain $I_N = \{t \in I_0: M_0\mathcal{A}(t) > -2\} \supseteq I_M = (t_{M-}, t_0, t_{M+})$ and $I_C = \{t \in I_0: M_0\mathcal{A}(t) > -1\} \supseteq I_A = (t_{A-}, t_0, t_{A+})$, we wish to solve

$(t - t_0)/\alpha > -r/2$, r = 1,2, (3.3.1.14)
$t - t_0 > -r\,\alpha/2$, r = 1,2, (3.3.1.15)
$t > t_0 - r\,\alpha/2 = t_{1-}$, r = 1,2, (3.3.1.16)

For r = 2, we have $t > t_0 - \alpha$, so that $I_N = I_{IV}^{MAB1a}(t_0, \alpha) = I_M = (t_0 - \alpha, t_0, \infty)$. For r = 1, we have $t > t_0 - \alpha/2$, so that $I_C = I_A = (t_0 - \alpha/2, t_0, \infty)$. To obtain $I_N \supseteq I_M = (t_{M-}, t_0, t_{M+}) \leftrightarrow (\tau_{M-}, 0, \tau_{M+}) = \mathrm{I}_M = \mathrm{I}_M(t_0, M_0, A) \subseteq \mathrm{I}_N(t_0, M_0, A)$ and $I_C \supseteq I_A = (t_{A-}, t_0, t_{A+}) \leftrightarrow (\tau_{A-}, 0, \tau_{A+}) = \mathrm{I}_A = \mathrm{I}_A(t_0, M_0, A) \subseteq \mathrm{I}_C(t_0, M_0, A)$, we use results from previous case to obtain

$$I_{IV}^{MAB1a}(t_0, \alpha) = I_M = (t_0 - \alpha, t_0, \infty) \supseteq I_A = (t_0 - \alpha/2, t_0, \infty) \leftrightarrow (-\alpha, 0, \alpha) = \mathrm{I}_A \subseteq (-\infty, 0, \alpha) = \mathrm{I}_M..$$

Example 2. $B(t) = b_0\,t$. **Case 3.** $\beta < (u_0)^2$ ($\delta < 1$) where $\delta = \beta/(u_0)^2 = 4/(M_0 b_0 (u_0)^2)$.

From (3.3.2.3)-(3.3.2.5) we have

$$\mathrm{A}(t) = (\tfrac{1}{2})\, b_0[(t + \gamma)^2 - (u_0)^2] \tag{3.3.2.65}$$

$$(\tfrac{1}{2})M_0\mathrm{A}(t) = [(t + \gamma)^2 - (u_0)^2]/\beta = [[(t + \gamma)/u_0]^2 - 1](u_0)^2/\beta = [[(t + \gamma)/u_0]^2 - 1]/\delta \tag{3.3.2.66}$$

$$\begin{aligned} \mathrm{D}(t) &= 1 + (\tfrac{1}{2})M_0\,\mathrm{A}(t) = [(t + \gamma)^2 + \beta - (u_0)^2]/\beta = [(t + \gamma)^2/(u_0)^2 + \beta/(u_0)^2 - 1]/(\beta/(u_0)^2) \\ &= [[(t + \gamma)/!u_0!]^2 + \delta - 1]/\delta = [[(t + \gamma)/!u_0!]^2 - (1 - \delta)]/\delta \end{aligned} \tag{3.3.2.67}$$

Hence $D(t)$ has zeros when $[(t+\gamma)/u_0]^2 = 1 - \delta$ (> 0 as $\delta < 1$). Hence when $|t+\gamma|/|u_0| = \sqrt{1-\delta}$, $|t+\gamma| = |u_0|\sqrt{1-\delta}$, $t+\gamma = \pm|u_0|\sqrt{1-\delta}$, or when $t = -\gamma \pm |u_0|\sqrt{1-\delta}$. Let

$$t_{3M\pm} = -\gamma \pm |u_0|\sqrt{1-\delta} \tag{3.3.2.68}$$

so that $t_{3M-} = -\gamma - |u_0|$

$\sqrt{1-\delta} < -\gamma < t_{3M+} = -\gamma + |u_0|\sqrt{1-\delta}$. If $u_0 = t_0 + \gamma < 0$ ($t_0 < -\gamma$), then $|u_0| = -u_0$ and $t_{3M-} = -\gamma - |u_0|\sqrt{1-\delta} = -\gamma + u_0\sqrt{1-\delta}$. If $u_0 = t_0 + \gamma > 0$ ($t_0 > -\gamma$), then $|u_0| = u_0$ and $t_{3M+} = -\gamma + |u_0|\sqrt{1-\delta} = -\gamma + u_0\sqrt{1-\delta}$.

More generally, to obtain $I_N = \{t \in I_0 : M_0 A(t) > -2\} \supseteq I_M = (t_{M-}, t_0, t_{M+})$ and $I_C = \{t \in I_0 : M_0 A(t) > -1\} \supseteq I_A = (t_{A-}, t_0, t_{A+})$, from (3.3.2.10) we wish to solve

$$M_0\, A(t) = 2M_0\, A(t)/2 = 2[[(t+\gamma)/|u_0|]^2 - 1]/\delta > -r,\ r = 1\ (I_C),\ 2\ (I_N) \tag{3.3.2.69}$$

We have equality when $2[[(t+\gamma)/|u_0|]^2 - 1]/\delta = -r$, or $[(t+\gamma)/|u_0|]^2 - 1 = -r\delta/2$, or $[(t+\gamma)/|u_0|]^2 = 1 - r\delta/2$. This is never true if $1 - r\delta/2 < 0$ or $1 < r\delta/2$, or $2/r < \delta$. For r = 2, we never have equality if $\delta > 1$ which is not this case. For r = 1, we never have equality if $\delta > 2$ which, again, is not this case. Hence for this case we always have $2/r > \delta$ and hence equality when $(t+\gamma)/|u_0| = \pm\sqrt{1 - r\delta/2}$ or when $t = -\gamma \pm |u_0|\sqrt{1 - r\delta/2}$. For r = 2, we will obtain I_N, the zeros $t_{3M}\pm$ of $D(t)$, and the end points of I_M. We have equality when

$$t = -\gamma \pm |u_0|\sqrt{1 - r\delta/2} = -\gamma \pm |u_0|\sqrt{1-\delta} = t_{2M\pm}\,. \tag{3.3.2.70}$$

For r = 1, we will obtain I_C and the end points of I_A. We have equality when

$$t = -\gamma \pm |u_0|\sqrt{1 - r\delta/2} = -\gamma \pm |u_0|\sqrt{1-\delta/2} = t_{2A\pm} \tag{3.3.2.71}$$

To obtain the endpoints of I_M and I_A, we need $\tau = h_{AB2}(t;\gamma,u_0,\beta) = h_{AB2,3}(t;\gamma,u_0,\delta)$. Recall $t_{3M-} = -\gamma - |u_0|\sqrt{1-\delta} < -\gamma < t_{3M+} = -\gamma + |u_0|\sqrt{1-\delta}$. We also have $t_{3A-} = -\gamma - |u_0|\sqrt{1-\delta/2} < -\gamma < t_{3M+} = -\gamma + |u_0|\sqrt{1-\delta/2}$. The relationship of $t_{3M\pm}$ and $t_{3A\pm}$ depends on the sign of δ.

There are four subcases, depending on u_0 and δ (or β).

Subcase 1: $t_0 < t_{2M-}$ which we show is equivalent to choosing

$$u_0 < 0\ (t_0 < -\gamma) \text{ and } 0 < \beta < (u_0)^2\ (0 < \delta < 1).$$

For the physical case, since $t_0 < -\gamma = -A_0/b_0$, we would have $I_0 = (-\infty,-\gamma)$ if $b_0 < 0$.

But this would imply $M_0 < 0$ as $\beta = 4/[M_0 b_0] > 0$. Since this is not physical, this is not a physical subcase.

Subcase 2: $t_{2M-} < t_0 < -\gamma$. which we show equivalent to choosing

$$u_0 < 0\ (t_0 < -\gamma) \text{ and } \beta < 0 < (u_0)^2\ (\delta < 0 < 1)$$

For the physical case, since $t_0 < -\gamma$, we have $I_0 = (-\infty,-\gamma)$ if $b_0 < 0$.

Subcase 3: $-\gamma < t_0 < t_{2M+}$ which we show is equivalent to choosing

$$u_0 > 0\ (t_0 > -\gamma) \text{ and } \beta < 0 < (u_0)^2\ (\delta < 0 < 1)$$

For the physical case, since $t_0 > -\gamma = -A_0/b_0$, we have $I_0 = (-\gamma, \infty))$ if $b_0 > 0$. But this would imply $M_0 < 0$ as $\beta = 4/[M_0 b_0] < 0$. Since this is not physical, this is not a physical subcase.

Subcase 4: $t_0 > t_{2M+}$, which we show is equivalent to choosing

$$u_0 > 0\ (t_0 > -\gamma) \text{ and } 0 < \beta < (u_0)^2\ (0 < \delta < 1)$$

For the physical case, since $t_0 > -\gamma = -A_0/b_0$, we have $I_0 = (-\gamma, \infty))$ if $b_0 > 0$.

We discuss each subcase and show these equivalences later. Note that t_{3M-} and t_{3M+} depend on t_0. Hence we can not choose t_{3N-} and t_{3N+} and then choose t_0 in the appropriate interval. However, we may choose A_0 and γ and then u_0 and δ. After choosing A_0 and γ, we may choose δ so that $0 < \delta < 1$ or $\delta < 0 < 1$. We may then choose u_0 to be positive or negative or ($t_0 < -\gamma$ or $t_0 > -\gamma$). We may then compute $t_{3M\pm} = -\gamma \pm ! u_0 ! \sqrt{1-\delta}$. Our choice of δ and u_0 determines the interval that t_0 falls in. This in turn determines the interval in which t lies and hence I_M.

We have physical cases when

Subcase 2: $t_{2M-} < t_0 < -\gamma$. or $u_0 < 0$ ($t_0 < -\gamma$) and $\beta < 0 < (u_0)^2$ ($\delta < 0 < 1$) when $b_0 < 0$ (decreasing agglomeration)

Subcase 4: $t_0 > t_{2M+}$. or $u_0 > 0$ ($t_0 > -\gamma$) and $0 < \beta < (u_0)^2$ ($0 < \delta < 1$) when $b_0 > 0$ (increasing agglomeration)

From (3.3.2.6)- (3.3.2.8) we have

$$M_{MAB2}(t) = \frac{M_0}{D(t)} = \frac{M_0\delta}{[(t+\gamma)/u_0]^2-(1-\delta)}$$

$$= \frac{4/[b_0(u_0)^2]}{[(t+\gamma)/u_0]^2-(1-\delta)} = \frac{4/[b_0(u_0)^2]}{(t-t_{3N-})(t-t_{3N+})} \tag{3.3.2.72}$$

$$\frac{d\tau}{dt} = \frac{1}{(D(t))^2} = \frac{\delta^2}{[(t+\gamma)/|u_0|]^2-(1-\delta)]^2}, \tag{3.3.2.73}$$

$$R^{AB2}(t) = \frac{A(t)}{D(t)} = \frac{2/M_0}{1+1/[M_0 A(t)/2]} = \frac{2/M_0}{1+\delta/[[(t+\gamma)/u_0]^2-1]}, \tag{3.3.2.74}$$

We now obtain $\tau = h_{AB2,3}(t;\gamma,u_0\delta)$ in terms of h₋(x).

$$\tau = h_{AB2}(t;\gamma,u_0,\beta) = h_{AB2,3}(t;\gamma,u_0,\delta) = \int_{s=t_0}^{s=t}\frac{1}{(D(s))^2}ds =$$

$$\int_{s=t_0}^{s=t}\frac{\delta^2}{[[(s+\gamma)/|u_0|]^2-(1-\delta)]^2}ds$$

$$= \delta^2\int_{s=t_0}^{s=t}\frac{1}{[[(s+\gamma)/|u_0|]^2-(1-\delta)]^2}ds \quad \delta<1 \text{ implies } 1-\delta>0$$

$$= \frac{\delta^2}{(1-\delta)^2}\int_{s=t_0}^{s=t}\frac{1}{[[(s+\gamma)/(|u_0|\sqrt{1-\delta})]^2-1]^2}ds =$$

$$\frac{\delta^2}{(1-\delta)^2}\int_{u=(t_0+\gamma)/(|u_0|\sqrt{(1-\delta)})}^{u=(t+\gamma)/(|u_0|\sqrt{(1-\delta)})}\frac{|u_0|\sqrt{(1-\delta)}}{[u^2-1]^2}du \tag{3.3.2.75}$$

Letting $u = (s+\gamma)/(|u_0|\sqrt{1-\delta})$ and (as $u_0 \neq 0$ or $t_0 \neq \gamma$) $\text{sgn}(u_0) = u_0/!u_0!$, we have du $= ds/(|u_0|\sqrt{1-\delta})$ so that

$$\tau = h_{AB2,3}(t;\gamma,u_0,\delta) = \frac{\delta^2|u_0|}{(1-\delta)^{3/2}}\int_{u=\text{sgn}(u_0)1/\sqrt{(1-\delta)}}^{u=(t+\gamma)/(|u_0|\sqrt{(1-\delta)})}\frac{1}{[u^2-1]^2}du$$

$$= \frac{\delta^2 |u_0|}{(1-\delta)^{3/2}} \left[h_-\left(\frac{(t+\gamma)}{|u_0|\sqrt{(1-\delta)}} \right) - h_-\left(\frac{\text{sgn}(u_0)}{\sqrt{(1-\delta)}} \right) \right] \tag{3.3.2.76}$$

Hence

$$\tau = \frac{\delta^2 |u_0|}{(1-\delta)^{3/2}} \left[h_-\left(\frac{t+\gamma}{|u_0|\sqrt{(1-\delta)}} \right) - h_-\left(\frac{-1}{\sqrt{(1-\delta)}} \right) \right] \text{ if } t_0 < -\gamma \tag{3.3.2.77}$$

$$\tau = \frac{\delta^2 |u_0|}{(1-\delta)^{3/2}} \left[h_-\left(\frac{(t+\gamma)}{|u_0|\sqrt{(1-\delta)}} \right) - h_-\left(\frac{1}{\sqrt{(1-\delta)}} \right) \right] \text{ if } t_0 > -\gamma \tag{3.3.2.78}$$

Note that the intervals ϑ_M and ϑ_A where τ lies are determined by the interval where t lies which is determined by where t_0 lies and hence by $\text{sgn}(u_0)$.

We now find $t = h_{AB2,3}^{-1}(t;\gamma,u_0,\delta)$ in terms of $h_-^{-1}(x)$. From (3.3.2.59) we have

$$\left[h_-\left(\frac{(t+\gamma)}{|u_0|\sqrt{(1-\delta)}} \right) - h_-\left(\text{sgn}(u_0)(1-\delta)^{-1/2} \right) \right] = \frac{(1-\delta)^{3/2}}{\delta^2 |u_0|} \tau$$

$$h_-\left(\frac{(t+\gamma)}{|u_0|\sqrt{(1-\delta)}} \right) = \frac{(1-\delta)^{3/2}}{\delta^2 |u_0|} \tau + h_-\left(\text{sgn}(u_0)(1-\delta)^{-1/2} \right)$$

$$\frac{(t+\gamma)}{|u_0|\sqrt{(1-\delta)}} = h_-^{-1}\left(\frac{(1-\delta)^{3/2}}{\delta^2 |u_0|} \tau + h_-\left(\frac{\text{sgn}(u_0)}{(1-\delta)^{1/2}} \right) \right)$$

$$t + \gamma = |u_0|(1-\delta)^{1/2} h_-^{-1}\left(\frac{(1-\delta)^{3/2}}{\delta^2 |u_0|} \tau + h_-\left(\frac{\text{sgn}(u_0)}{(1-\delta)^{1/2}} \right) \right)$$

$$t = -\gamma + |u_0|(1-\delta)^{1/2} h_-^{-1}\left(\frac{(1-\delta)^{3/2}}{\delta^2 |u_0|} \tau + h_-\left(\frac{\text{sgn}(u_0)}{(1-\delta)^{1/2}} \right) \right) \tag{3.3.2.79}$$

After we obtain I_N, I_M, I_C, and I_A, to obtain $I_M \supseteq I_A \leftrightarrow (\tau_{M-},0,\tau_{M+}) = \mathrm{I}_M = \mathrm{I}_M(t_0,M_0,A_0,b_0) \subseteq (\tau_{A-},0,\tau_{A+}) = \mathrm{I}_A = \mathrm{I}_A(t_0,M_0,A_0,b_0)$, we need the images of the points t_0, $\pm\infty$, $t_{3M\pm}$, $t_{3A\pm}$ and, for the physical case, $-\gamma$. From (3.3.2.59), we have

$$h_{AB2,3}(t_0;\gamma,u_0,\delta) =$$

$$=\delta^2 !u_0!(1-\delta)^{-3/2}[h_-((t_0+\gamma)(!u_0!)^{-1}(1-\delta)^{-1/2})-h_-(\text{sgn}(u_0)(1-\delta)^{-1/2})] = 0 \tag{3.3.2.80}$$

so that t_0 always maps to 0. Letting $u = \dfrac{t+\gamma}{|u_0|\sqrt{1-\delta}}$, we see that

$$\lim_{t\to\pm\infty}\frac{t+\gamma}{|u_0|\sqrt{1-\delta}} = \pm\infty,\quad \lim_{t\to t_{3M\pm}}\frac{t+\gamma}{|u_0|\sqrt{1-\delta}} = \frac{(-\gamma\pm|u_0|\sqrt{1-\delta})+\gamma}{|u_0|\sqrt{1-\delta}} = \pm 1, \qquad (3.3.2.81)$$

$$\lim_{t\to t_{3A\pm}}\frac{t+\gamma}{|u_0|\sqrt{1-\delta}} = \frac{(-\gamma\pm|u_0|\sqrt{1-\delta/2})+\gamma}{|u_0|\sqrt{1-\delta}} = \frac{\pm\sqrt{1-\delta/2})}{\sqrt{1-\delta}}. \qquad (3.3.2.82)$$

Hence using analysis

$$\lim_{t\to\pm\infty^-} h_-\left(\frac{t+\gamma}{|u_0|\sqrt{(1-\delta)}}\right) = \lim_{u\to\pm\infty} h_-(u) = 0, \qquad (3.3.2.83)$$

$$\lim_{t\to t_{3M-}^-} h_-\left(\frac{t+\gamma}{|u_0|\sqrt{(1-\delta)}}\right) = \lim_{u\to -1^-} h_-(u) = \infty, \qquad (3.3.2.84)$$

$$\lim_{t\to t_{3M-}^+} h_-\left(\frac{t+\gamma}{|u_0|\sqrt{(1-\delta)}}\right) = \lim_{u\to -1^+} h_-(u) = -\infty, \qquad (3.3.2.85)$$

$$\lim_{t\to t_{3M+}^-} h_-\left(\frac{t+\gamma}{|u_0|\sqrt{(1-\delta)}}\right) = \lim_{u\to 1^-} h_-(u) = \infty, \qquad (3.3.2.86)$$

$$\lim_{t\to t_{3M+}^+} h_-\left(\frac{t+\gamma}{|u_0|\sqrt{(1-\delta)}}\right) = \lim_{u\to 1^+} h_-(u) = -\infty, \qquad (3.3.2.87)$$

$$\lim_{t\to t_{3A\pm}} h_-\left(\frac{(t+\gamma)}{|u_0|\sqrt{1-\delta}}\right) = h_-\left(\frac{(-\gamma\pm|u_0|\sqrt{1-\delta/2}+\gamma)}{|u_0|\sqrt{1-\delta}}\right) = h_-\left(\frac{\pm\sqrt{1-\delta/2})}{\sqrt{1-\delta}}\right). \qquad (3.3.2.88)$$

Thus we let

$$\tau_{3\pm\infty} = \lim_{t\to\pm\infty^-} h_{AB2,3}(t;\gamma,u_0,\delta) = \lim_{t\to\pm\infty} \frac{\delta^2 |u_0|}{(1-\delta)^{3/2}}\left[h_-\left(\frac{t+\gamma}{u_0\sqrt{(1-\delta)}}\right) - h_-\left(\frac{\mathrm{sgn}(u_0)}{\sqrt{(1-\delta)}}\right)\right]$$

$$= \frac{\delta^2 |u_0|}{(1-\delta)^{3/2}}\left[\lim_{t\to\pm\infty} h_-\left(\frac{t+\gamma}{u_0\sqrt{(1-\delta)}}\right) - h_-\left(\frac{\mathrm{sgn}(u_0)}{\sqrt{(1-\delta)}}\right)\right] = \frac{\delta^2 u_0}{(1-\delta)^{3/2}}\left[0 - h_-\left(\frac{\mathrm{sgn}(u_0)}{\sqrt{(1-\delta)}}\right)\right]$$

$$= -\frac{\delta^2 |u_0|}{(1-\delta)^{3/2}}\left[h_-\left(\frac{\mathrm{sgn}(u_0)}{\sqrt{(1-\delta)}}\right)\right] = -\frac{\delta^2 |u_0|}{(1-\delta)^{3/2}}\left[h_-\left(\frac{\pm 1}{\sqrt{(1-\delta)}}\right)\right] \tag{3.3.2.89}$$

since if t is approaching $-\infty$, $t_0 < -\gamma$ and if t is approaching ∞, $t_0 > -\gamma$. We also let

$$\tau_{-1-} = \lim_{t\to t_{3M-}^-} h_{AB2,3}(t;\gamma,u_0,\delta) =$$

$$\lim_{t\to t_{3M-}^-} \frac{\delta^2 |u_0|}{(1-\delta)^{3/2}}\left[h_-\left(\frac{t+\gamma}{u_0\sqrt{(1-\delta)}}\right) - h_-\left(\frac{\mathrm{sgn}(u_0)}{\sqrt{(1-\delta)}}\right)\right] = \infty, \tag{3.3.2.90}$$

$$\tau_{-1+} = \lim_{t\to t_{3M-}^+} h_{AB2,3}(t;\gamma,u_0,\delta) =$$

$$\lim_{t\to t_{3M-}^+} \frac{\delta^2 |u_0|}{(1-\delta)^{3/2}}\left[h_-\left(\frac{t+\gamma}{u_0\sqrt{(1-\delta)}}\right) - h_-\left(\frac{\mathrm{sgn}(u_0)}{\sqrt{(1-\delta)}}\right)\right] = -\infty, \tag{3.3.2.91}$$

$$\tau_{1-} = \lim_{t\to t_{3M+}^-} h_{AB2,3}(t;\gamma,u_0,\delta) =$$

$$\lim_{t\to t_{3M-}^-} \frac{\delta^2 |u_0|}{(1-\delta)^{3/2}}\left[h_-\left(\frac{t+\gamma}{u_0\sqrt{(1-\delta)}}\right) - h_-\left(\frac{\mathrm{sgn}(u_0)}{\sqrt{(1-\delta)}}\right)\right] = \infty, \tag{3.3.2.92}$$

$$\tau_{1+} = \lim_{t\to t_{3M+}^+} h_{AB2,3}(t;\gamma,u_0,\delta) =$$

$$\lim_{t\to t_{3M+}^+} \frac{\delta^2 |u_0|}{(1-\delta)^{3/2}}\left[h_-\left(\frac{t+\gamma}{u_0\sqrt{(1-\delta)}}\right) - h_-\left(\frac{\mathrm{sgn}(u_0)}{\sqrt{(1-\delta)}}\right)\right] = -\infty. \tag{3.3.2.93}$$

We also let

$$\tau_{3\text{-}\gamma\pm} = h_{AB2}(-\gamma;u_0,\beta,\gamma) = \frac{\delta^2|u_0|}{(1-\delta)^{3/2}}\left[h_-\left(\frac{0}{u_0\sqrt{(1-\delta)}}\right)-h_-\left(\frac{\pm 1}{\sqrt{(1-\delta)}}\right)\right]$$

$$= -\frac{\delta^2|u_0|}{(1-\delta)^{3/2}}\left[h_-\left(\frac{\pm 1}{(1-\delta)^{1/2}}\right)\right] \qquad (3.3.2.94)$$

as where $-\gamma$ maps depends on whether $t_0 < -\gamma$ ($u_0 < 0$) or $t_0 > -\gamma$ ($u_0 > 0$). Now since $\tau_{3A-} < -\gamma < \tau_{3A+}$, we let

$$\tau_{3A\pm} = h_{AB2}(t_{3A\pm};u_0,\beta,\gamma) = \frac{\delta^2|u_0|}{(1-\delta)^{3/2}}\left[h_-\left(\frac{(t_{3A\pm}+\gamma)}{|u_0|\sqrt{(1-\delta)}}\right)-h_-\left(\frac{\pm 1}{\sqrt{(1-\delta)}}\right)\right]$$

$$= \frac{\delta^2|u_0|}{(1-\delta)^{3/2}}\left[h_-\left(\frac{-\gamma\pm|u_0|\sqrt{(1-\delta/2)}+\gamma}{|u_0|\sqrt{(1-\delta)}}\right)-h_-\left(\frac{\pm 1)}{\sqrt{(1-\delta)}}\right)\right]$$

$$= \frac{\delta^2|u_0|}{(1-\delta)^{3/2}}\left[h_-\left(\frac{\pm|u_0|\sqrt{(1-\delta/2)}}{|u_0|\sqrt{(1-\delta)}}\right)-h_-\left(\frac{\pm 1}{\sqrt{(1-\delta)}}\right)\right]$$

$$= \frac{\delta^2|u_0|}{(1-\delta)^{3/2}}\left[h_-\left(\frac{\pm\sqrt{(1-\delta/2)}}{\sqrt{(1-\delta)}}\right)-h_-\left(\frac{\pm 1}{\sqrt{(1-\delta)}}\right)\right]. \qquad (3.3.2.95)$$

Hence $(-\infty,t_0,t_{3M-})$ maps to $(\tau_{3-\infty}, 0, \infty)$, $(t_{3M-},t_0,-\gamma)$ to $(-\infty,0,\tau_{3-\gamma-})$, $(-\gamma, t_0,t_{3M+})$ to $(\tau_{3-\gamma+}, 0, \infty)$, and (t_{3M+}, t_0, ∞) to $(-\infty, 0, \tau_{3+\infty},)$.

Example 2. $B(t) = b_0\, t$. Case 3. $\beta < (u_0)^2$ ($\delta < 1$).

Subcase 1: $t_0 < t_{2M-} = -\gamma\ -!u_0!\ \sqrt{1-\delta}$ which is equivalent to $u_0 < 0$ ($t_0 < -\gamma$) and $0 < \beta < (u_0)^2$ $(0 < \delta < 1)$

We first show that these two characterizations of Subcase 1 are equivalent when $\delta < 1$. Suppose $t_0 < t_{2M-} = -\gamma\ -!u_0!\ \sqrt{1-\delta}$. then $t_0 + \gamma < -\ !u_0!\ \sqrt{1-\delta}$ or $u_0 < -!u_0!\ \sqrt{1-\delta} < 0$. Since $u_0 < 0$ we have $!u_0! = -u_0$ so that $u_0 < u_0\sqrt{1-\delta} < 0$. Hence $1 > \sqrt{1-\delta} > 0$ so that $1 > 1 - \delta > 0$. Hence $0 > -\delta > -1$ or $0 < \delta < 1$. Hence $0 < \beta < (u_0)^2$.

Now suppose that $u_0 < 0$ ($t_0 < -\gamma$) and $0 < \beta < (u_0)^2$ $(0 < \delta < 1)$. Then $!u_0! = -u_0$ and $0 > -\delta > -1$. Hence $1 > 1 - \delta > 0$ so that $1 > \sqrt{1-\delta} > 0$. Hence $!u_0! > !u_0!\ \sqrt{1-\delta} > 0$ so that $-!u_0! < -!u_0!\ \sqrt{1-\delta} < 0$. Since $!u_0! = -u_0$ we have $u_0 < -!u_0!$

$\sqrt{1-\delta} < 0$ or $t_0 + \gamma < -!u_0! \sqrt{1-\delta} < 0$. Hence $t_0 < -\gamma -!u_0! \sqrt{1-\delta} = t_{2M-} < -\gamma$.

Clearly $I_M = (-\infty, t_0, t_{2M-})$. Since $t_0 < t_{2M-} = -\gamma -!u_0! \sqrt{1-\delta} < -\gamma$ we have $t_0 + \gamma < 0$ so that $!u_0! = -u_0$ and hence $t_{2M-} = -\gamma -!u_0! \sqrt{1-\delta} = -\gamma + u_0\sqrt{1-\delta}$ Note that $t \in I_M = (-\infty, t_0, t_{3M-})$ iff $-\infty < t < t_{2M-} = -\gamma -!u_0! \sqrt{1-\delta} < -\gamma$, $-\infty < t + \gamma < -!u_0! \sqrt{1-\delta} < 0$, $-\infty < u = \frac{t+\gamma}{|u_0|\sqrt{1-\delta}} < -1 < 0$. Hence $u = \frac{t+\gamma}{|u_0|\sqrt{1-\delta}}$ maps $I_M = (-\infty, t_0, t_{3M-})$ to $\mathrm{I}_M = (-\infty, \frac{t_0+\gamma}{|u_0|\sqrt{1-\delta}}, -1)$. As $t_0 < -\gamma$, $u_0 = t_0 + \gamma < 0$ so that $!u_0! = -u_0$ and hence $\mathrm{I}_M = (-\infty, -\frac{1}{\sqrt{1-\delta}}, -1)$. Note that since $\delta > 0$, we have $-\delta < 0$, $1 - \delta < 1$, $\sqrt{1-\delta} < 1$, $-\sqrt{1-\delta} > -1$, $-\frac{1}{\sqrt{1-\delta}} < -1$ as expected.

Next we solve for I_N, I_M, I_C, and I_A. To obtain $I_N = \{t \in I_0 : M_0 \mathrm{A}(t) > -2\} \supseteq I_M = (t_{M-}, t_0, t_{M+})$ and $I_C = \{t \in I_0 : M_0 \mathrm{A}(t) > -1\} \supseteq I_A = (t_{A-}, t_0, t_{A+})$, we wish to solve

$$
\begin{array}{ll}
M_0^{\ 0}\,\mathrm{A}(t) = 2M_0^{\ 0}\,\mathrm{A}(t)/2 = 2[[(t+\gamma)/u_0]^2 - 1]/\delta > -r, & r = 1\ (I_C),\ 2(I_N) \\
[(t+\gamma)/u_0]^2 - 1 > -r\,\delta/2, & r = 1,2 \\
[(t+\gamma)/u_0]^2 > 1 - r\,\delta/2, & r = 1,2 \\
!(t+\gamma)/u_0! > \sqrt{1 - r\delta/2}, & r = 1,2 \\
!t+\gamma!/!u_0! > \sqrt{1 - r\delta/2}, & r = 1,2 \\
!t+\gamma! > !u_0! \sqrt{1 - r\delta/2}, & r = 1,2 \\
t+\gamma < -!u_0! \sqrt{1 - r\delta/2} \text{ or } t+\gamma > !u_0! \sqrt{1 - r\delta/2}, & r = 1,2 \\
t < -\gamma -!u_0! \sqrt{1 - r\delta/2} \text{ or } t > -\gamma +!u_0! \sqrt{1 - r\delta/2}. & r = 1,2
\end{array}
$$

Hence

$$I_C = \{t \in R : t < -\gamma -!u_0! \sqrt{1-\delta/2} \text{ or } t > -\gamma +!u_0! \sqrt{1-\delta/2}\}.$$

$$I_N = \{t \in R : t < -\gamma -!u_0! \sqrt{1-\delta} \text{ or } t > -\gamma +!u_0! \sqrt{1-\delta}\}.$$

Recall

$$t_{3M\pm} = -\gamma \pm !u_0! \sqrt{1-\delta} \tag{3.3.2.96}$$

$$t_{3A\pm} = -\gamma \pm !u_0! \sqrt{1-\delta/2}. \tag{3.3.2.97}$$

Hence

I_C ={t∈R: t < -γ -!u_0! $\sqrt{1-\delta/2}$ or t > -γ +!u_0! $\sqrt{1-\delta/2}$ }= (-∞, t_0, t_{3M-})∪(t_{3M+},t_0,∞)

.I_N ={t∈R: t < -γ -!u_0! $\sqrt{1-\delta}$ or t > -γ +!u_0! $\sqrt{1-\delta}$ }I_C = (-∞, t_0, t_{3A-})∪(t_{3A+},t_0,∞)

Hence

$$I_N = (-\infty, t_0, t_{3M-}) \cup (t_{3M+},t_0,\infty) \supseteq I_M = (-\infty, t_0, t_{3M-})$$

$$I_C = (-\infty, t_0, t_{3A-}) \cup (t_{3A+},t_0,\infty) \supseteq I_A = (-\infty, t_0, t_{3A-}).$$

Since 0 < δ < 1, we have 0 < δ/2 < δ < 1, 0 > -δ/2 > -δ > -1, 1 > 1-δ/2 >1 -δ > 0, 1 > $\sqrt{1-\delta/2}$ > $\sqrt{1-\delta}$ > 0, -1 < -$\sqrt{1-\delta/2}$ < -$\sqrt{1-\delta}$ < 0, and -1 < -$\sqrt{1-\delta/2}$ < -$\sqrt{1-\delta}$ < 0 < $\sqrt{1-\delta}$ < $\sqrt{1-\delta/2}$ < 1 -!u_0! < -!u_0! $\sqrt{1-\delta/2}$ < -!u_0! $\sqrt{1-\delta}$ < 0 < !u_0! $\sqrt{1-\delta}$ <!u_0! $\sqrt{1-\delta/2}$ < !u_0! u_0 < -!u_0! $\sqrt{1-\delta/2}$ < -!u_0! $\sqrt{1-\delta}$ < 0 < !u_0! $\sqrt{1-\delta}$ <!u_0! $\sqrt{1-\delta/2}$ < -u_0 t_0 +γ < -!u_0! $\sqrt{1-\delta/2}$ < -!u_0! $\sqrt{1-\delta}$ < 0 < !u_0! $\sqrt{1-\delta}$ <!u_0! $\sqrt{1-\delta/2}$ < -t_0 -γ t_0<-γ-!u_0! $\sqrt{1-\delta/2}$ <-γ-!u_0! $\sqrt{1-\delta}$ <-γ<-γ+!u_0! $\sqrt{1-\delta}$ <-γ+!u_0! $\sqrt{1-\delta/2}$ <-t_0-2γ

Hence $t_0 < t_{3A-} < t_{3M-} < t_{3M+} < t_{3A+}$ and I_M = (-∞, t_0, t_{3M-}) ⊇ I_A = (-∞, t_0, t_{3A-}).

To obtain I_M = (-∞, t_0, t_{3M-}) ⊇ I_A = (-∞, t_0, t_{3A-}) ↔ (τ_{A-},0,τ_{A+}) = $\mathcal{I}_A$ = $\mathcal{I}_A(t_0,M_0,A_0,b_0)$ ⊆ (τ_{M-},0,τ_{M+}) = $\mathcal{I}_M$ = $\mathcal{I}_M(t_0,M_0,A_0,b_0)$, we need the images of -∞, t_0, t_{3A-}, and t_{3M-} Recall

$$h_{AB2,3}(t_0;\gamma,u_0,\delta) = 0 \tag{3.3.2.98}$$

$$\tau_{3\pm\infty} = \lim_{t\to\pm\infty^-} h_{AB2,3}(t;\gamma,u_0,\delta) = -\frac{\delta^2|u_0|}{(1-\delta)^{3/2}}\left[h_-\left(\frac{\pm 1}{\sqrt{(1-\delta)}}\right)\right] \tag{3.3.2.99}$$

$$\tau_{-1-} = \lim_{t\to t_{3M-}^-} h_{AB2,3}(t;\gamma,u_0,\delta) = \infty \tag{3.3.2.100}$$

$$\tau_{3A\pm} = h_{AB2,3}(t_{3A\pm};u_0,\beta,\gamma) = \frac{\delta^2|u_0|}{(1-\delta)^{3/2}}\left[h_-\left(\frac{\pm\sqrt{(1-\delta/2)}}{\sqrt{(1-\delta)}}\right) - h_-\left(\frac{\pm 1}{(1-\delta)^{1/2}}\right)\right] \tag{3.3.2.101}$$

Hence I_M = (-∞, t_0,t_{3M-}) ⊇ I_A = (-∞,t_0,t_{3A-}) ↔ ($\tau_{3-\infty}$,0,τ_{3A-}) = $\mathcal{I}_A$ ⊆ ($\tau_{3-\infty}$,0,∞) = $\mathcal{I}_M$.

For the physical case, since t_0 < -γ = -A_0/b_0, we would have I_0 = (-∞,t_0,-γ) if b_0 < 0 . But this would imply M_0 < 0 as β = 4/[$M_0 b_0$] > 0. Since this is not physical, this is not a physical subcase.

Example 2. B(t) = b_0 t . Case 3. β < $(u_0)^2$ (δ < 1).

Subcase 2: $t_{2M-} < t_0 < -\gamma$. which is equivalent to $u_0 < 0$ ($t_0 < -\gamma$) and $\beta < 0 < (u_0)^2$ ($\delta < 0 < 1$)

We first show that these two characterizations of Subcase 2 are equivalent when $\delta < 1$. Suppose $t_{N-} = -\gamma\ -!u_0!\sqrt{1-\delta} < t_0 < -\gamma$. Then $-\,!u_0!\ \sqrt{1-\delta} < t_0 + \gamma < 0$ or $-!u_0!\sqrt{1-\delta} < u_0 < 0$. Since $u_0 < 0$ we have $!u_0! = -\,u_0$ so that $u_0\sqrt{1-\delta} < u_0 < 0$. Hence $\sqrt{1-\delta} > 1 > 0$ so that $1 - \delta > 1 > 0$. Hence $-\,\delta > 0 > -1$ or $\delta < 0 < 1$. Hence $\beta < 0 < (u_0)^2$.

Now suppose that $u_0 < 0$ ($t_0 < -\gamma$) and $\beta < 0 < (u_0)^2$ ($\delta < 0 < 1$). Then $!u_0! = -\,u_0$ and $-\,\delta > 0 >-1$. Hence $1 - \delta > 1 > 0$ so that $\sqrt{1-\delta} > 1 > 0$. Hence $!u_0!\sqrt{1-\delta} > !u_0! > 0$ so that .

$-!u_0!\sqrt{1-\delta} < -!u_0! < 0$. Since $!u_0! = -\,u_0$ we have $-!u_0!\sqrt{1-\delta} < u_0 < 0$ or

$-!u_0!\sqrt{1-\delta} < t_0 + \gamma < 0$. Hence $-\,\gamma\ -!u_0!\sqrt{1-\delta} = t_{3M-} < t_0 < -\gamma$.

Clearly $I_M = (t_{2M-}, t_0, -\gamma, t_{2M+})$. Since $t_{3M-} = -\gamma\ -!u_0!\sqrt{1-\delta} < t_0 < -\gamma$ we have $t_0 + \gamma < 0$ so that $!u_0! = -u_0$ and hence $t_{3M-} = -\gamma\ -!u_0!\sqrt{1-\delta} = -\gamma + u_0\sqrt{1-\delta}$. Note that $t \in I_M = (t_{2M-}, t_0, -\gamma, t_{2M+})$ iff $t_{2M-} = -\gamma\ -!u_0!\sqrt{1-\delta} < t < -\gamma < t_{2M+} = -\gamma + !u_0!\sqrt{1-\delta}$, $-!u_0!\sqrt{1-\delta} < t + \gamma < 0 < !u_0!\sqrt{1-\delta}$, $-1 < \frac{t+\gamma}{|u_0|\sqrt{1-\delta}} < 0 < 1$. Hence $u = \frac{t+\gamma}{|u_0|\sqrt{1-\delta}}$ maps $I_M = (t_{2M-}, t_0, -\gamma, t_{2M+})$ to $I_M = (-1, \frac{t_0+\gamma}{|u_0|\sqrt{1-\delta}}, 0,1)$. As $t_0 < -\,\gamma$, $u_0 = t_0 + \gamma < 0$ so that $!u_0! = -u_0$. Hence $I_M = (-1, -\frac{1}{\sqrt{1-\delta}}, 0,1)$.

Note that since $\delta < 0$, we have $-\,\delta > 0$, $1 - \delta > 1$, $\sqrt{1-\delta} > 1$, $-\sqrt{1-\delta} < -1$, $-\frac{1}{\sqrt{1-\delta}} > -1$ as expected.

Next we solve for I_N, I_M, I_C, and I_A. To obtain $I_N = \{t \in I_0 : M_0A(t) > -2\} \supseteq I_M = (t_{M-}, t_0, t_{M+})$ and $I_C = \{t \in I_0 : M_0A(t) > -1\} \supseteq I_A = (t_{A-}, t_0, t_{A+})$, we wish to solve

$$M_0\,A(t) = 2M_0\,A(t)/2 = 2[[(t+\gamma)/u_0]^2 - 1]/\delta > -r, \qquad r = 1,2$$

$$[(t+\gamma)/u_0]^2 - 1 < -r\,\delta/2, \qquad r = 1,2$$

$$[(t+\gamma)/u_0]^2 < 1 - r\,\delta/2, \qquad r = 1,2$$

$$!(t+\gamma)/u_0! < \sqrt{1-r\delta/2}, \qquad r = 1,2$$

$$!t+\gamma!/!u_0! < \sqrt{1-r\delta/2}, \qquad r = 1,2$$

$$!t+\gamma! < !u_0!\sqrt{1-r\delta/2}, \qquad r = 1,2$$

$$-!u_0!\sqrt{1-r\delta/2} < t+\gamma < !u_0!\sqrt{1-r\delta/2}, \qquad r = 1,2$$

$$-\gamma\ -!u_0!\sqrt{1-r\delta/2} < t < -\gamma + !u_0!\sqrt{1-r\delta/2}. \qquad r = 1,2$$

Hence

$$I_C = \{t \in R: -\gamma\ -!u_0!\sqrt{1-\delta/2} < t < -\gamma\ +!u_0!\sqrt{1-\delta/2}\}.$$

$$I_N = \{t \in R: -\gamma\ -!u_0!\sqrt{1-\delta} < t < -\gamma\ +!u_0!\sqrt{1-\delta}\}$$

.

Recall

$$t_{2M\pm} = -\gamma\ _{\pm}!u_0!\sqrt{1-\delta} \qquad (3.3.2.102)$$

$$t_{2A\pm} = -\gamma\ _{\pm}!u_0!\sqrt{1-\delta/2}\ . \qquad (3.3.2.103)$$

Hence

$$I_C = \{t \in R: -\gamma\ -!u_0!\sqrt{1-\delta/2} < t < -\gamma\ +!u_0!\sqrt{1-\delta/2}\} = (t_{2A-}, t_0, t_{2A+}) =.I_A$$

$$I_N = \{t \in R: -\gamma\ -!u_0!\sqrt{1-\delta} < t < -\gamma\ +!u_0!\sqrt{1-\delta}\} = (t_{2M-}, t_0, t_{2M+}) = I_M\ .$$

Since $\delta < 0 < 1$, we have $\delta < \delta/2 < 0 < 1$, $-\delta > -\delta/2 > 0 > -1$, $1-\delta > 1\ -\delta/2 > 1 > 0$, $\sqrt{1-\delta} > \sqrt{1-\delta/2} > 1 > 0$, $-\sqrt{1-\delta} < -\sqrt{1-\delta/2} < -1 < 0$, and $-\sqrt{1-\delta} < -\sqrt{1-\delta/2} < -1 < 0 < 1 < \sqrt{1-\delta/2} < \sqrt{1-\delta}$ $-!u_0!\sqrt{1-\delta} < -!u_0!\sqrt{1-\delta/2} < -!u_0! < 0 < !u_0! < !u_0!\sqrt{1-\delta/2} < !u_0!\sqrt{1-\delta}$ $-!u_0!\sqrt{1-\delta} < -!u_0!\sqrt{1-\delta/2} < u_0 < 0 < -u_0 < !u_0!\sqrt{1-\delta/2} < !u_0!\sqrt{1-\delta}$ $-!u_0!\sqrt{1-\delta} < -!u_0!\sqrt{1-\delta/2} < t_0+\ \gamma < 0 < -(t_0+\ \gamma) < !u_0!\sqrt{1-\delta/2} < !u_0!\sqrt{1-\delta}$ $-\gamma\ -!u_0!\sqrt{1-\delta} < -\gamma\ -!u_0!\sqrt{1-\delta/2} < t_0 < -\gamma < -t_0\ -2\gamma < -\gamma+!u_0!\sqrt{1-\delta/2} < -\gamma+!u_0!\sqrt{1-\delta}$

Hence $t_{2M-} < t_{2A-} < t_0 < t_{2A+} < t_{2M+}$ and $I_N = I_M = (t_{2M-}, t_0, t_{2M+}) \supseteq I_C = I_A = (t_{2A-}, t_0, t_{2A+})$.

To obtain $I_M = (t_{2M-},t_0,t_{2M+}) \supseteq I_A = (t_{2A-},t_0,t_{2A+}) \leftrightarrow (\tau_{A-},0,\tau_{A+}) = \mathrm{I}_A = \mathrm{I}_A(t_0,M_0^0,A) \subseteq (\tau_{M-},0,\tau_{M+}) = \mathrm{I}_M = \mathrm{I}_M(t_0,M_0^0,A)$, we need

$$h_{AB2,3}(t_0;\gamma,u_0,\delta) = 0 \qquad (3.3.2.104)$$

$$\tau_{-1+} = \lim_{t \to t_{3M-}^{+}} h_{AB2,3}(t;\gamma,u_0,\delta) = -\infty \qquad (3.3.2.105)$$

$$\tau_{1-} = \lim_{t \to t_{2M+}^{-}} h_{AB2,3}(t;\gamma,u_0,\delta) = \infty \qquad (3.3.2.106)$$

$$\tau_{3A\pm} = h_{AB2,3}(t_{3A\pm};u_0,\beta,\gamma) =$$

$$\frac{\delta^2 |u_0|}{(1-\delta)^{3/2}}\left[h_-\left(\frac{\pm\sqrt{(1-\delta/2)}}{\sqrt{(1-\delta)}}\right)-h_-\left(\frac{\pm 1}{(1-\delta)^{1/2}}\right)\right] \quad (3.3.2.107)$$

$$\tau_{3\text{-}\gamma\pm} = h_{AB2}(-\gamma;u_0,\beta,\gamma) = -\frac{\delta^2 |u_0|}{(1-\delta)^{3/2}}\left[h_-\left(\frac{\pm 1}{(1-\delta)^{1/2}}\right)\right] \quad (3.3.2.108)$$

Hence $I_M = (t_{3M-}, t_0, t_{3M+}) \supseteq I_A = (t_{3A-}, t_0, t_{3A+}) \leftrightarrow (\tau_{3A-},0,\tau_{3A+}) = \mathrm{I}_A \subseteq (-\infty,0,\infty) = \mathrm{I}_M$.

For the physical case, since $t_0 < -\gamma$, we have $I_0 = (-\infty,t_0,-\gamma)$ if $b_0 < 0$. Since $t_{2M-} < t_{2A-} < t_0 < -\gamma < t_{2A+} < t_{2M+}$, we have $I_{pM} = (t_{2M-}, t_0, -\gamma) \supseteq I_{pA} = (t_{2A-},t_0, -\gamma) \leftrightarrow (\tau_{3A-},0, \tau_{3\text{-}\gamma\text{-}}) = \mathrm{I}_{pA} \subseteq (-\infty,0, \tau_{3\text{-}\gamma\text{-}}) = \mathrm{I}_{pM}$.

Example 2. $B(t) = b_0\, t$. Case 3. $\beta < (u_0)^2$ ($\delta < 1$).

Subcase 3: $-\gamma < t_0 < t_{N+}$ which is equivalent to $u_0 > 0$ ($t_0 > -\gamma$) and $\beta < 0 < (u_0)^2$ ($\delta < 0 < 1$)

We first show that these two characterizations of Subcase 3 are equivalent when $\delta < 1$. Suppose $-\gamma.< t_0 < t_{N-} = -\gamma\ +!u_0!\ \sqrt{1-\delta}$ Then $0 < t_0 + \gamma < !u_0!\ \sqrt{1-\delta}$ or $0 < u_0 < !u_0!\ \sqrt{1-\delta}$. Since $u_0 > 0$ we have $!u_0! = u_0$ so that $0 < u_0 < u_0\sqrt{1-\delta}$. Hence $0 < 1 < \sqrt{1-\delta}$ so that $0 < 1 < 1\text{-}\ \delta$. Hence $-1 < 0 < -\delta$ or $1 > 0 > \delta$ or $\delta <0 < 1$. Hence $\beta < 0 < (u_0)^2$.

Now suppose that $u_0 > 0$ ($t_0 > -\gamma$) and $\beta < 0 < (u_0)^2$ ($\delta < 0 < 1$). Then $!u_0! = u_0$ and $-\ \delta > 0 >-1$. Hence $1 - \delta > 1 > 0$ so that $\sqrt{1-\delta} > 1 > 0$. Hence $!u_0!\ \sqrt{1-\delta} > !u_0! > 0$ so that . $!u_0!\ \sqrt{1-\delta} < !u_0! < 0$. Since $!u_0! = u_0$ we have $!u_0!\ \sqrt{1-\delta} < u_0 < 0$ or $!u_0!\ \sqrt{1-\delta} < t_0 + \gamma < 0$. Hence $-\gamma.< t_0 < -\ \gamma + !u_0!\ \sqrt{1-\delta} = t_{N-}$

Clearly $I_M = (t_{2M-},-\gamma, t_0, t_{2M+})$. Since $-\gamma < t_0 < t_{3M+} = -\gamma\ +!u_0!\ \sqrt{1-\delta}$, we have we have $t_0 + \gamma > 0$ so that $!u_0! = u_0$ and hence $t_{3M+} = -\gamma\ +!u_0!\ \sqrt{1-\delta} = -\gamma + u_0\sqrt{1-\delta}$. Note that $t\in I_M = (t_{2M-}, -\gamma, t_0, t_{2M+})$ iff $t_{3M-}= -\gamma\ -!u_0!\ \sqrt{1-\delta} < -\gamma < t < t_{2M+}= -\gamma+!u_0!\ \sqrt{1-\delta}$, $-!u_0!\ \sqrt{1-\delta} < 0 < t + \gamma < !u_0!\ \sqrt{1-\delta}$, $-1<0<\frac{t+\gamma}{|u_0|\sqrt{1-\delta}}<1$. Hence $u=\frac{t+\gamma}{|u_0|\sqrt{1-\delta}}$ maps $I_M = (t_{2M-}, -\gamma, t_0, t_{2M+})$ to

$\mathrm{I}_M = (-1, 0,\frac{t_0+\gamma}{|u_0|\sqrt{1-\delta}}\ ,1)$. As $t_0 < -\ \gamma$, $u_0 = t_0 + \gamma< 0$ so that $!u_0! = -u_0$. Hence $\mathrm{I}_M = (-1, 0, \frac{1}{\sqrt{1-\delta}}\ ,1)$. Note that since $\delta < 0$, we have $-\ \delta > 0$, $1 - \delta > 1$, $\sqrt{1-\delta}>1$, $\frac{1}{\sqrt{1-\delta}}<1$ as expected.

Next we solve for I_N, I_M, I_C, and I_A. To obtain $I_N = \{t\in I_0:M_0A(t) > -2\}\supseteq I_M = (t_{M-},t_0,t_{M+})$ and $I_C = \{t\in I_0:M_0A(t) > -1\}\supseteq I_A = (t_{A-},t_0,t_{A+})$, we wish to solve

$$M_0^0\, A(t) = 2M_0^0\, A(t)/2 = 2[[(t+\gamma)/u_0]^2 -1]/\delta > -r, \qquad r = 1,2$$
$$[(t+\gamma)/u_0]^2 -1 < -r\,\delta/2, r = 1,2$$
$$[(t+\gamma)/u_0]^2 < 1 -r\,\delta/2, r = 1,2$$
$$!(t+\gamma)/u_0! < \sqrt{1-r\delta/2}, \qquad r = 1,2$$
$$!t+\gamma!/!u_0! < \sqrt{1-r\delta/2}, \qquad r = 1,2$$
$$!t+\gamma! < !u_0!\sqrt{1-r\delta/2}, \qquad r = 1,2$$
$$-!u_0!\sqrt{1-r\delta/2} < t+\gamma < !u_0!\sqrt{1-r\delta/2}, \qquad r = 1,2$$
$$-\gamma -!u_0!\sqrt{1-r\delta/2} < t < -\gamma +!u_0!\sqrt{1-r\delta/2}. \qquad r = 1,2$$

Hence

$$I_C = \{t\in R: -\gamma -!u_0!\sqrt{1-\delta/2} < t < -\gamma +!u_0!\sqrt{1-\delta/2}\}.$$
$$I_N = \{t\in R: -\gamma -!u_0!\sqrt{1-\delta} < t < -\gamma +!u_0!\sqrt{1-\delta}\}.$$

Recall

$$t_{3M\pm} = -\gamma \pm!u_0!\sqrt{1-\delta} \qquad (3.3.2.109)$$

$$t_{3A\pm} = -\gamma \pm!u_0!\sqrt{1-\delta/2}. \qquad (3.3.2.110)$$

Hence

$$I_C = \{t\in R: -\gamma -!u_0!\sqrt{1-\delta/2} < t < -\gamma +!u_0!\sqrt{1-\delta/2}\} = (t_{3A-}, t_0, t_{3A+}) = .I_A$$
$$I_N = \{t\in R: -\gamma -!u_0!\sqrt{1-\delta} < t < -\gamma +!u_0!\sqrt{1-\delta}\} = (t_{3M-}, t_0, t_{3M+}) = I_M.$$

Since $\delta < 0 < 1$, we have $\delta < \delta/2 < 0 < 1$, $-\delta > -\delta/2 > 0 > -1$, $1-\delta > 1 -\delta/2 > 1 > 0$, $\sqrt{1-\delta} > \sqrt{1-\delta/2} > 1 > 0$, $-\sqrt{1-\delta} < -\sqrt{1-\delta/2} < -1 < 0$, and $-\sqrt{1-\delta} < -\sqrt{1-\delta/2} < -1 < 0 < 1 < \sqrt{1-\delta/2} < \sqrt{1-\delta}$ $-!u_0!\sqrt{1-\delta} < -!u_0!\sqrt{1-\delta/2} < -!u_0! < 0 < !u_0! < !u_0!\sqrt{1-\delta/2} < !u_0!\sqrt{1-\delta}$ $-!u_0!\sqrt{1-\delta} < -!u_0!\sqrt{1-\delta/2} < -u_0 < 0 < u_0 < !u_0!\sqrt{1-\delta/2} < !u_0!\sqrt{1-\delta}$ $-!u_0!\sqrt{1-\delta} < -!u_0!\sqrt{1-\delta/2} < -t_0 - \gamma < 0 < t_0 + \gamma < !u_0!\sqrt{1-\delta/2} < !u_0!\sqrt{1-\delta}$ $-\gamma -!u_0!\sqrt{1-\delta} < -\gamma -!u_0!\sqrt{1-\delta/2} < -t_0 -2\gamma < -\gamma < t_0 < -\gamma+!u_0!\sqrt{1-\delta/2} < -\gamma+!u_0!\sqrt{1-\delta}$

Hence $t_{3M-} < t_{3A-} < t_0 < t_{3A+} < t_{3M+}$ so that $I_N = I_M = (t_{3M-}, t_0, t_{3M+}) \supseteq I_C = I_A = (t_{3A-}, t_0, t_{3A+})$.

To obtain $I_M = (t_{3M-},t_0,t_{3M+}) \supseteq I_A = (t_{3A-},t_0,t_{3A+}) \leftrightarrow (\tau_{A-},0,\tau_{A+}) = \mathrm{I}_A = \mathrm{I}_A(t_0,M_0^0,A) \subseteq (\tau_{M-},0,\tau_{M+}) = \mathrm{I}_M = \mathrm{I}_M(t_0,M_0^0,A)$, we need

$$h_{AB2,3}(t_0;\gamma,u_0,\delta)=0 \tag{3.3.2.111}$$

$$\tau_{-1+}=\lim_{t\to t_{3M-}^{+}} h_{AB2,3}(t;\gamma,u_0,\delta)=-\infty \tag{3.3.2.112}$$

$$\tau_{1-}=\lim_{t\to t_{3M+}^{-}} h_{AB2,3}(t;\gamma,u_0,\delta)=\infty \tag{3.3.2.113}$$

$$\tau_{3A\pm}= h_{AB2,3}(t_{3A\pm};u_0,\beta,\gamma) = \frac{\delta^2|u_0|}{(1-\delta)^{3/2}}\left[h_-\left(\frac{\pm\sqrt{(1-\delta/2)}}{\sqrt{(1-\delta)}}\right)-h_-\left(\frac{\pm 1}{(1-\delta)^{1/2}}\right)\right] \tag{3.3.2.114}$$

Hence $I_M = (t_{3M-}, t_0,t_{3M+})\supseteq I_A = (t_{3A-},t_0,t_{3A+}) \leftrightarrow (\tau_{3A-},0,\tau_{3A+}) = \mathrm{I}_A \subseteq (-\infty,0,\infty) = \mathrm{I}_M$.

To consider the physical case, since $t_0 > -\gamma = -A_0/b_0$, we have $I_0 = (-\gamma, \infty))$ if $b_0 > 0$. But this would imply $M_0 < 0$ as $\beta = 4/[M_0b_0] < 0$. Since this is not physical, this is not a physical subcase.

Example 2. $B(t) = b_0\, t$. Case 3. $\beta < (u_0)^2$ ($\delta < 1$).

Subcase 4: $t_0 > t_{N+}$, which is equivalent to $u_0 > 0$ ($t_0 > -\gamma$) and $0 < \beta < (u_0)^2$ ($0 < \delta < 1$)

We first show that these two characterizations of Subcase 4 are equivalent when $\delta < 1$. Suppose $t_{2M-} = -\gamma + !u_0! \sqrt{1-\delta} < t_0$. Then $!u_0! \sqrt{1-\delta} < t_0 + \gamma$ or $!u_0! \sqrt{1-\delta} < u_0$. Since $u_0 > 0$ we have $!u_0! = u_0$ so that $u_0\sqrt{1-\delta} < u_0$. Hence $0 < \sqrt{1-\delta} < 1$ so that $0 < 1- \delta < 1$. Hence $-1 < -\delta < 0$ or $1 > \delta > 0$ or $0 < \delta < 1$. Hence $0 < \beta < (u_0)^2$.

Now suppose that $u_0 > 0$ ($t_0 > -\gamma$) and $0 < \beta < (u_0)^2$ ($0 < \delta < 1$). Then $!u_0! = u_0$ and $0 > -\delta > -1$. Hence $1 > 1 - \delta > 0$ so that $1 > \sqrt{1-\delta} > 0$ or $0 < \sqrt{1-\delta} < 1$. Hence $0 < !u_0! \sqrt{1-\delta} < !u_0!$. Since $!u_0! = u_0$ we have $0 < !u_0! \sqrt{1-\delta} < u_0$ or $0 < !u_0! \sqrt{1-\delta} < t_0 + \gamma$. Hence $-\gamma.<-\gamma + !u_0! \sqrt{1-\delta} = t_{2M-} < t_{0-}$

Clearly $I_M = (-\gamma, t_{2M+}\ t_0, \infty)$. Since $-\gamma < t_{2M+} = -\gamma + !u_0! \sqrt{1-\delta} < t_0$, we have $t_0 + \gamma > 0$ so that $!u_0! = u_0$ and hence $t_{3M+} = -\gamma + !u_0! \sqrt{1-\delta} = -\gamma + u_0\sqrt{1-\delta}$.

Note that $t\in I_M = (-\gamma, t_{3M+},t_0, \infty)$ iff $-\gamma < t_{3M+}= -\gamma+!u_0! \sqrt{1-\delta} < t < \infty$, $0 <!u_0! \sqrt{1-\delta} < t + \gamma < \infty$, $0<1<\frac{t+\gamma}{|u_0|\sqrt{1-\delta}}<\infty$. Hence $u=\frac{t+\gamma}{|u_0|\sqrt{1-\delta}}$ maps $I_M = (-\gamma, t_{3M+}\ t_0, \infty)$. to $\mathrm{I}_M = (0, 1,$

$\frac{t_0+\gamma}{|u_0|\sqrt{1-\delta}}$,∞). As $t_0 > -\gamma$, $u_0 = t_0 + \gamma > 0$ so that $!u_0! = u_0$. Hence $I_M = (0,1,\frac{1}{\sqrt{1-\delta}},\infty)$. Note that since $\delta < 0$, we have $-\delta > 0$, $1-\delta > 1$, $\sqrt{1-\delta} > 1$, $\frac{1}{\sqrt{1-\delta}} < 1$ as expected.

Next we solve for I_N, I_M, I_C, and I_A. To obtain $I_N = \{t\in I_0: M_0A(t) > -2\} \supseteq I_M = (t_{M-},t_0,t_{M+})$ and $I_C = \{t\in I_0: M_0A(t) > -1\} \supseteq I_A = (t_{A-},t_0,t_{A+})$, we wish to solve

$$
\begin{aligned}
M_0{}^0\,A(t) = 2M_0{}^0\,A(t)/2 = 2[[(t+\gamma)/u_0]^2 - 1]/\delta &> -r, && r = 1,2\\
[(t+\gamma)/u_0]^2 - 1 &> -r\,\delta/2, && r = 1,2\\
[(t+\gamma)/u_0]^2 &> 1 - r\,\delta/2, && r = 1,2\\
!(t+\gamma)/u_0! &> \sqrt{1-r\delta/2}, && r = 1,2\\
!t+\gamma!/!u_0! &> \sqrt{1-r\delta/2}, && r = 1,2\\
!t+\gamma! &> !u_0!\sqrt{1-r\delta/2}, && r = 1,2\\
t+\gamma < -!u_0!\sqrt{1-r\delta/2} \text{ or } t+\gamma &> !u_0!\sqrt{1-r\delta/2}, && r = 1,2\\
t < -\gamma - !u_0!\sqrt{1-r\delta/2} \text{ or } t &> -\gamma + !u_0!\sqrt{1-r\delta/2}. && r = 1,2
\end{aligned}
$$

Hence

$$I_C = \{t\in R: t < -\gamma - !u_0!\sqrt{1-\delta/2} \text{ or } t > -\gamma + !u_0!\sqrt{1-\delta/2}\}.$$

$$I_N = \{t\in R: t < -\gamma - !u_0!\sqrt{1-\delta} \text{ or } t > -\gamma + !u_0!\sqrt{1-\delta}\}.$$

Recall

$$t_{3M\pm} = -\gamma \pm !u_0!\sqrt{1-\delta} \qquad (3.3.2.115)$$

$$t_{3A\pm} = -\gamma \pm !u_0!\sqrt{1-\delta/2}. \qquad (3.3.2.116)$$

Hence

$$I_C = \{t\in R: t < -\gamma - !u_0!\sqrt{1-\delta/2} \text{ or } t > -\gamma + !u_0!\sqrt{1-\delta/2}\} = (-\infty, t_0, t_{3M-})\cup(t_{3M+},t_0,\infty).$$

$$I_N = \{t\in R: t < -\gamma - !u_0!\sqrt{1-\delta} \text{ or } t > -\gamma + !u_0!\sqrt{1-\delta}\} I_C = (-\infty, t_0, t_{3A-})\cup(t_{3A+},t_0,\infty)$$

Hence

$$I_N = (-\infty, t_0, t_{3M-})\cup(t_{3M+},t_0,\infty) \supseteq I_M = (t_{3M+}, t_0, \infty)$$
$$I_C = (-\infty, t_0, t_{3A-})\cup(t_{3A+},t_0,\infty) \supseteq I_A = (t_{3A+}, t_0, \infty).$$

Since $0 < \delta < 1$, we have $0 < \delta/2 < \delta < 1$, $0 > -\delta/2 > -\delta > -1$, $1 > 1-\delta/2 > 1-\delta > 0$, $1 > \sqrt{1-\delta/2} > \sqrt{1-\delta} > 0$, $-1 < -\sqrt{1-\delta/2} < -\sqrt{1-\delta} < 0$, and $-1 < -\sqrt{1-\delta/2} < -\sqrt{1-\delta} < 0 < \sqrt{1-\delta} < \sqrt{1-\delta/2} < 1$ $-!u_0! < -!u_0!\sqrt{1-\delta/2} < -!u_0!\sqrt{1-\delta} < 0 < !u_0!\sqrt{1-\delta} <!u_0!\sqrt{1-\delta/2} < !u_0!$ - $u_0 < -!u_0!\sqrt{1-\delta/2} < -!u_0!\sqrt{1-\delta} < 0 < !u_0!\sqrt{1-\delta} <!u_0!\sqrt{1-\delta/2} < u_0 - t_0 - \gamma < -!u_0!\sqrt{1-\delta/2} < -!u_0!\sqrt{1-\delta} < 0 < !u_0!\sqrt{1-\delta} <!u_0!\sqrt{1-\delta/2} < t_0 + \gamma$ $-t_0 - 2\gamma < -\gamma - !u_0!\sqrt{1-\delta/2} < -\gamma - !u_0!\sqrt{1-\delta} < -\gamma < -\gamma + !u_0!\sqrt{1-\delta} < -\gamma + !u_0!\sqrt{1-\delta/2} <$ t_0

Hence $t_{2A-} < t_{2M-} < t_{2M+} < t_{2A+} < t_0$ so that $I_M = (t_{3M+}, t_0, \infty) \supseteq I_A = (t_{3A+}, t_0, \infty)$.

To obtain $I_M = (t_{2M+}, t_0, \infty) \supseteq I_A = (t_{2A+}, t_0, \infty)$. $\leftrightarrow (\tau_{A-}, 0, \tau_{A+}) = \mathcal{I}_A = \mathcal{I}_A(t_0, M_0, A_0, b_0) \subseteq (\tau_{M-}, 0, \tau_{M+}) = \mathcal{I}_M = \mathcal{I}_M(t_0, M_0, A_0, b_0)$, we need the images of t_{3M+}, t_{3A-}, t_0, and ∞. Recall

$$h_{AB2,3}(t_0; \gamma, u_0, \delta) = 0 \tag{3.3.2.117}$$

$$\tau_{\pm\infty} = \lim_{t \to \pm\infty^-} h_{AB2,3}(t; \gamma, u_0, \delta) = -\frac{\delta^2 |u_0|}{(1-\delta)^{3/2}} \left[h_-\left(\frac{\pm 1}{\sqrt{(1-\delta)}} \right) \right] \tag{3.3.2.118}$$

$$\tau_{1+} = \lim_{t \to t_{2M+}^+} h_{AB2,3}(t; \gamma, u_0, \delta) = -\infty$$

$$\tau_{3A\pm} = h_{AB2,3}(t_{3A\pm}; u_0, \beta, \gamma) = \frac{\delta^2 |u_0|}{(1-\delta)^{3/2}} \left[h_-\left(\frac{\pm\sqrt{(1-\delta/2)}}{\sqrt{(1-\delta)}} \right) - h_-\left(\frac{\pm 1}{(1-\delta)^{1/2}} \right) \right] F \tag{3.3.2.119}$$

Hence $I_M = (t_{3M+}, t_0, \infty) \supseteq I_A = (t_{3A+}, t_0, \infty) \leftrightarrow (\tau_{3A+}, 0, \tau_{+\infty}) = \mathcal{I}_A \subseteq (-\infty, 0, \tau_{+\infty}) = \mathcal{I}_M$.

For the physical case, since $t_0 > -\gamma = -A_0/b_0$, we have $I_0 = (-\gamma, t_0, \infty))$ if $b_0 > 0$ .Since $t_{3M-} < t_{3A-} < -\gamma < t_0 < t_{3A+} < t_{3M+}$, we have $I_{pM} = (-\gamma, t_0, t_{3M+}) \supseteq I_{pA} = (-\gamma, t_0, t_{3A+}) \leftrightarrow (\tau_{3-\gamma-}, 0, \tau_{3A+}) = \mathcal{I}_{pA} \subseteq (\tau_{3-\gamma+}, 0, \tau_{3-\gamma+}) = \mathcal{I}_{pM}$.

Example 2. $B(t) = b_0 t$. **Case 4**. $\beta > (u_0)^2$ $(\delta > 1)$ where $\delta = \beta/(u_0)^2 = 4/(M_0 b_0 (u_0)^2)$.

From (3.3.2.3), (3.3.2.4), and (3.3.2.5), we have

$$\mathrm{A}(t) = (½)\, b_0[(t+\gamma)^2 - (u_0)^2] \tag{3.3.2.120}$$

$$(½)M_0\mathrm{A}(t) = [(t+\gamma)^2 - (u_0)^2]/\beta = [[(t+\gamma)/u_0]^2 - 1](u_0)^2/\beta = [[(t+\gamma)/u_0]^2 - 1]/\delta \tag{3.3.2.121}$$

$$\begin{aligned} \mathrm{D}(t) = 1 + (½)M_0\, \mathrm{A}(t) &= [(t+\gamma)^2 + \beta - (u_0)^2]/\beta = [(t+\gamma)^2/(u_0)^2 + \beta/(u_0)^2 - 1]/\beta/(u_0)^2 \\ &= [[(t+\gamma)/!u_0!]^2 + \delta - 1]/\delta = [[(t+\gamma)/!u_0!]^2 + (\delta - 1))]/\delta \end{aligned} \tag{3.3.2.122}$$

Since δ -1 > 0, D(t) has no real zeros so that $I_N = I_M = \mathbf{R}$.

More generally, to obtain $I_N = \{t\in I_0{:}M_0A(t) > -2\}\supseteq I_M = (t_{M-},t_0,t_{M+})$ and $I_C = \{t\in I_0{:}M_0A(t) > -1\}\supseteq I_A = (t_{A-},t_0,t_{A+})$, we wish to solve

$$M_0\,A(t) = 2M_0\,A(t)/2 = 2[[(t+\gamma)/!\,u_0!\,]^2 -1]/\delta > -r, \qquad r = 1,2$$

$$[(t+\gamma)/!\,u_0!\,]^2 -1 > -r\,\delta/2, \qquad r = 1,2$$

$$[(t+\gamma)/!\,u_0!\,]^2 > 1 -r\,\delta/2, \qquad r = 1,2$$

We have equality when $[(t+\gamma)/!\,u_0!\,]^2 = 1 -r\,\delta/2$, This is never true if $1 -r\,\delta/2 < 0$, or $1 < r\,\delta/2$. Since in this case, $\delta > 1$, there are no zeros for r = 2. For r = 1, this is never true when true when $1 < \delta/2$ or $2 < \delta$. Hence there are no zeros for $\delta > 2$. For r = 1 and $1 < \delta \leq 2$, we have zeros when $[(t+\gamma)/!\,u_0!\,]^2 = 1 - \delta/2$ or , $\dfrac{t+\gamma}{|u_0|} = \pm\sqrt{1-\delta/2}$, or $t+\gamma = \pm|u_0|\sqrt{1-\delta/2}$, or $t = -\gamma \pm |u_0|\sqrt{1-\delta/2}$. Let $t_{2A\pm} = -\gamma \pm |u_0|\sqrt{1-\delta/2}$.

There are three subcases, depending on signs of u_0 and $(u_0)^2 - \beta/2\,(1 - \delta/2)$.

Subcase 1. $u_0 < 0$ ($t_0 < -\gamma$) and $0 < (u_0)^2 < \beta \leq 2(u_0)^2$ $(0 < 1 \leq \delta < 2)$.

Subcase 2. $u_0 > 0$ ($t_0 < -\gamma$) and $0 < (u_0)^2 < \beta \leq 2(u_0)^2$ $(0 < 1 < \delta \leq 2)$.

Subcase 3. $0 < 2(u_0)^2 < \beta$ $(0 < 2 < \delta)$.

From (3.3.2.6)-(3.3.2.8) we have

$$M_{MAB2}(t) = \frac{M_0}{D(t)} = \frac{M_0\delta}{[(t+\gamma)/u_0]^2 + (\delta-1)} = \frac{4/[b_0(u_0)^2}{[(t+\gamma)/u_0]^2 + (\delta-1)}, \qquad (3.3.2.123)$$

$$\frac{d\tau}{dt} = \frac{1}{(D(t))^2} = \frac{\delta^2}{[(t+\gamma)/u_0]^2 + (\delta-1)]^2}, \qquad (3.3.2.124)$$

$$R^{AB2}(t) = \frac{A\,(t)}{D(t)} = \frac{2/M_0}{1 + 1/[M_0A\,(t)/2]} = \frac{2/M_0{}^0}{1+\delta/[[(t+\gamma)/u_0]^2 - 1]}, \qquad (3.3.2.125)$$

$$\tau = h_{AB2,4}(t;u_0,\beta,\gamma) = \int_{s=t_0}^{s=t} \frac{1}{(D(s))^2} ds = \int_{s=t_0}^{s=t} \frac{\delta^2}{[[(s+\gamma)/|u_0|]^2 + (\delta-1)]^2} ds$$

$$= \delta^2 \int_{s=t_0}^{s=t} \frac{1}{[[(s+\gamma)/|u_0|]^2 + (\delta-1)]^2} ds \quad \text{as } \delta -1 > 0$$

$$= \frac{\delta^2}{(\delta-1)^2}\int_{s=t_0}^{s=t}\frac{1}{[[(s+\gamma)/(|u_0|\sqrt{\delta-1})]^2+1]^2}\,ds =$$

$$\frac{\delta^2}{(\delta-1)^2}\int_{u=(t_0+\gamma)/(|u_0|\sqrt{(\delta-1)})}^{u=(t+\gamma)/(|u_0|\sqrt{(\delta-1)})}\frac{u_0\sqrt{(\delta-1)}}{[u^2+1]^2}\,du \qquad (3.3.2.126)$$

$$u=(s+\gamma)/(|u_0|\sqrt{\delta-1}),\ du = ds/(|u_0|\sqrt{\delta-1})$$

$$\tau = h_{AB2,4}(t;u_0,\beta,\gamma) = \frac{\delta^2|u_0|}{(\delta-1)^{3/2}}\int_{u=\mathrm{sgn}(u_0)/\sqrt{(\delta-1)}}^{u=(t+\gamma)/(|u_0|\sqrt{(\delta-1)})}\frac{1}{[u^2+1]^2}\,du$$

$$= \frac{\delta^2|u_0|}{(\delta-1)^{3/2}}\left[h_+\left(\frac{(t+\gamma)}{|u_0|\sqrt{(\delta-1)}}\right)-h_+\left(\frac{\mathrm{sgn}(u_0)}{\sqrt{(\delta-1)}}\right)\right] \qquad (3.3.2.127)$$

$$\tau = \frac{\delta^2|u_0|}{(1-\delta)^{3/2}}\left[h_+\left(\frac{t+\gamma}{|u_0|\sqrt{(\delta-1)}}\right)-h_+\left(\frac{-1}{\sqrt{(\delta-1)}}\right)\right] \text{ if } t_0 < -\gamma \qquad (3.3.2.128)$$

$$\tau = \frac{\delta^2|u_0|}{(1-\delta)^{3/2}}\left[h_+\left(\frac{(t+\gamma)}{|u_0|\sqrt{(\delta-1)}}\right)-h_+\left(\frac{1}{\sqrt{(\delta-1)}}\right)\right] \text{ if } t_0 > -\gamma \qquad (3.3.2.129)$$

Note that the intervals ϑ_M and ϑ_A where τ lies are determined by the interval where t lies which is determined by where t_0 lies and hence by $\mathrm{sgn}(u_0)$ and that $\tau =$

$$h_{AB2,4}(t_0;u_0,\beta,\gamma) = 0.$$

We now find t = $h^{-1}_{AB2,4}(t;\gamma,u_0,\delta)$ in terms of $h_+^{-1}(x)$. From (3.3.2.59) we have

$$\left[h_+\left(\frac{(t+\gamma)}{u_0\sqrt{(\delta-1)}}\right)-h_+\left((\delta-1)^{-1/2}\right)\right] = \frac{(\delta-1)^{3/2}}{\delta^2 u_0}\tau$$

$$h_+\left(\frac{(t+\gamma)}{u_0\sqrt{(\delta-1)}}\right) = \frac{(\delta-1)^{3/2}}{\delta^2 u_0}\tau + h_+\left((\delta-1)^{-1/2}\right)$$

$$\frac{(t+\gamma)}{u_0\sqrt{(\delta-1)}} = h_+^{-1}\left(\frac{(\delta-1)^{3/2}}{\delta^2 u_0}\tau+h_+\left((\delta-1)^{-1/2}\right)\right)$$

$$t+\gamma = u_0(\delta-1)^{1/2}h_+^{-1}\left(\frac{(\delta-1)^{3/2}}{\delta^2 u_0}\tau + h_+\left((\delta-1)^{-1/2}\right)\right)\sqrt{a^2+b^2}$$

$$t = -\gamma + u_0(\delta-1)^{1/2}h_+^{-1}\left(\frac{(\delta-1)^{3/2}}{\delta^2 u_0}\tau + h_+\left((\delta-1)^{-1/2}\right)\right) \qquad (3.3.2.130)$$

After we obtain I_N, I_M, I_C, and I_A, to obtain $I_M \supseteq I_A \leftrightarrow (\tau_{M-},0,\tau_{M+}) = \mathrm{I}_M = \mathrm{I}_M(t_0,M_0,A_0,b_0) \subseteq (\tau_{A-},0,\tau_{A+}) = \mathrm{I}_A = \mathrm{I}_A(t_0,M_0,A_0,b_0)$, we need the images of the points t_0, $\pm\infty$, $t_{4A\pm}$ and, for the physical case, $-\gamma^{\pm}$. From analysis we have

$\lim_{u\to\pm\infty} h_+(u) = \pm\pi/4$ and $h_+(0) = 0$.

Hence we let

$$\tau_{4\pm\infty\cdot} = \lim_{t\to\pm\infty} h_{AB2}(t;u_0,\beta,\gamma) = \lim_{t\to\pm\infty} \frac{\delta^2|u_0|}{(\delta-1)^{3/2}}\left[h_+\left(\frac{(t+\gamma)}{|u_0|\sqrt{(\delta-1)}}\right) - h_+\left(\frac{\mathrm{sgn}(u_0)}{\sqrt{(\delta-1)}}\right)\right]$$

$$= \frac{\delta^2|u_0|}{(\delta-1)^{3/2}}\left[\pm\frac{\pi}{4} - h_+\left(\frac{\pm 1}{\sqrt{(\delta-1)}}\right)\right], \qquad (3.3.2.131)$$

since if t is approaching $-\infty$, $t_0 < -\gamma$ and if t is approaching ∞, $t_0 > -\gamma$. For $1 < \delta \leq 2$, we let

$$\tau_{4A\pm\cdot} = \lim_{t\to 4A\pm} h_{AB2}(t;u_0,\beta,\gamma) = \lim_{t\to 4A\pm} \frac{\delta^2|u_0|}{(\delta-1)^{3/2}}\left[h_+\left(\frac{(t+\gamma)}{|u_0|\sqrt{(\delta-1)}}\right) - h_+\left(\frac{\mathrm{sgn}(u_0)}{\sqrt{(\delta-1)}}\right)\right]$$

$$= \frac{\delta^2|u_0|}{(1-\delta)^{3/2}}\left[h_+\left(\frac{-\gamma \pm |u_0|\sqrt{(1-\delta/2)}+\gamma}{|u_0|\sqrt{(1-\delta)}}\right) - h_+\left(\frac{\pm 1)}{\sqrt{(1-\delta)}}\right)\right]$$

$$= \frac{\delta^2|u_0|}{(1-\delta)^{3/2}}\left[h_+\left(\frac{\pm|u_0|\sqrt{(1-\delta/2)}}{|u_0|\sqrt{(1-\delta)}}\right) - h_+\left(\frac{\pm 1}{\sqrt{(1-\delta)}}\right)\right]$$

$$= \frac{\delta^2|u_0|}{(1-\delta)^{3/2}}\left[h_+\left(\frac{\pm\sqrt{(1-\delta/2)}}{\sqrt{(1-\delta)}}\right) - h_+\left(\frac{\pm 1}{\sqrt{(1-\delta)}}\right)\right]. \qquad (3.3.2.132)$$

Note that $\tau_{4A-} = \tau_{4A+}$ when $\delta = 2$.

$$\tau_{4\text{-}\gamma\pm\cdot} = \lim_{t\to-\gamma^{\pm}} h_{AB2,4}(t;u_0,\beta,\gamma) = \lim_{t\to-\gamma^{\pm}} \frac{\delta^2|u_0|}{(\delta-1)^{3/2}}\left[h_+\left(\frac{(t+\gamma)}{|u_0|\sqrt{(\delta-1)}}\right) - h_+\left(\frac{\mathrm{sgn}(u_0)}{\sqrt{(\delta-1)}}\right)\right]$$

$$= \lim_{t\to -\gamma^{\pm}} \frac{\delta^2 |u_0|}{(\delta-1)^{3/2}} \left[h_+(0) - h_+\left(\frac{\pm 1}{\sqrt{(\delta-1)}} \right) \right]$$

$$= -\frac{\delta^2 |u_0|}{(\delta-1)^{3/2}} \left[h_+\left(\frac{\pm 1}{\sqrt{(\delta-1)}} \right) \right] \tag{3.3.2.133}$$

Example 2. $B(t) = b_0\, t$. **Case 4.** $\beta > (u_0)^2$ $(\delta > 1)$ where $\delta = \beta/(u_0)^2 = 4/(M_0 b_0 (u_0)^2)$.

Subcase 1. $u_0 < 0$ ($t_0 < -\gamma$) and $0 < (u_0)^2 < \beta \le 2(u_0)^2$ $(0 < 1 < \delta \le 2)$.

To obtain $I_N = \{t \in I_0 : M_0 \mathbb{A}(t) > -2\} \supseteq I_M = (t_{M-}, t_0, t_{M+})$ and $I_C = \{t \in I_0 : M_0 \mathcal{A}(t) > -1\} \supseteq I_A = (t_{A-}, t_0, t_{A+})$, we wish to solve

$$M_0\, \mathbb{A}(t) = 2M_0\, \mathbb{A}(t)/2 = 2[[(t+\gamma)/u_0]^2 - 1]/\delta > -r, \qquad r = 1,2$$

$$[(t+\gamma)/u_0]^2 - 1 > -r\,\delta/2, \qquad r = 1,2$$

$$[(t+\gamma)/u_0]^2 > 1 - r\,\delta/2, \qquad r = 1,2$$

For $r = 2$, this is always true as $1 - \delta < 0$. Hence $I_N = I_M = \mathbf{R}$. For $r = 1$ we have $0 < 1 < \delta < 2 < 2 + \delta$, $0 < 1/2 < \delta/2 < 1 < 1 + \delta/2$, $-\delta/2 < 1/2 - \delta/2 < 0 < 1 - \delta/2 < 1$. Hence $[(t+\gamma)/u_0]^2 > 1 - \delta/2 > 0$,

$$!(t+\gamma)/u_0! > \sqrt{1-\delta/2}\,, \qquad r = 1,2$$

$$!t+\gamma!/!u_0! > \sqrt{1-\delta/2}\,, \qquad r = 1,2$$

$$!t+\gamma! > !u_0!\,\sqrt{1-\delta/2}\,, \qquad r = 1,2$$

$$t+\gamma < -!u_0!\,\sqrt{1-\delta/2} \text{ or } t+\gamma > !u_0!\,\sqrt{1-\delta/2}\,, \qquad r = 1,2$$

$$t < -\gamma - !u_0!\,\sqrt{1-\delta/2} \text{ or } t > -\gamma + !u_0!\,\sqrt{1-\delta/2}\,. \qquad r = 1,2$$

Since $t_{2A\pm} = -\gamma \pm !u_0!\,\sqrt{1-\delta/2}$, $I_C = (-\infty, t_{4A-}) \cup (t_{4A+}, \infty)$. Since $0 < 1 - \delta/2 < 1$, we have $0 < \sqrt{1-\delta/2} < 1$ and hence $0 > -\sqrt{1-\delta/2} > -1$, so that $-1 < -\sqrt{1-\delta/2} < 0 < \sqrt{1-\delta/2} < 1$. Hence we have $-!u_0! < -!u_0!\,\sqrt{1-\delta/2} < 0 < !u_0!\,\sqrt{1-\delta/2} < !u_0!$.

Now since $u_0 < 0$, $!u_0! = -u_0$, we have $u_0 < -!u_0!\,\sqrt{1-\delta/2} < 0 < !u_0!\,\sqrt{1-\delta/2} < -u_0$.

$$t_0 + \gamma < -!u_0!\,\sqrt{1-\delta/2} < 0 < !u_0!\,\sqrt{1-\delta/2} < -(t_0+\gamma).$$

$$t_0 < -\gamma - !u_0!\,\sqrt{1-\delta/2} < -\gamma < -\gamma + !u_0!\,\sqrt{1-\delta/2} < -\gamma - t_0 - \gamma.$$

$$t_0 < t_{2A-} < t_{4A+} < -\gamma - t_0 - \gamma.$$

Hence $I_A = (-\infty, t_0, t_{4A-})$

To obtain $I_M = (-\infty,t_0,\infty) \supseteq I_A = (-\infty,t_0,t_{4A-}) \leftrightarrow (\tau_{A-},0,\tau_{A+}) = \mathrm{I}_A = \mathrm{I}_A(t_0,M_0,A_0, b_0) \subseteq (\tau_{M-},0,\tau_{M+}) = \mathrm{I}_M = \mathrm{I}_M(t_0,M_0,A_0, b_0)$, we need

$$\tau_{4\pm\infty\cdot} = \lim_{t\to\pm\infty} h_{AB2}(t;u_0,\beta,\gamma) = \frac{\delta^2|u_0|}{(\delta-1)^{3/2}}\left[\pm\frac{\pi}{4} - h_+\left(\frac{\pm 1}{\sqrt{(\delta-1)}}\right)\right] \tag{3.3.2.134}$$

$$\tau_{4A\pm\cdot} = \lim_{t\to 4A\pm} h_{AB2}(t;u_0,\beta,\gamma) = \frac{\delta^2|u_0|}{(1-\delta)^{3/2}}\left[h_+\left(\frac{\pm\sqrt{(1-\delta/2)}}{\sqrt{(1-\delta)}}\right) - h_+\left(\frac{\pm 1}{\sqrt{(1-\delta)}}\right)\right]. \tag{3.3.2.135}$$

$$\tau_{4\text{-}\gamma\pm\cdot} = \lim_{t\to -\gamma^\pm} h_{AB2,4}(t;u_0,\beta,\gamma) = -\frac{\delta^2|u_0|}{(\delta-1)^{3/2}}\left[h_+\left(\frac{\pm 1}{\sqrt{(\delta-1)}}\right)\right]$$

Hence $I_M = \mathbf{R} = (-\infty, t_0, \infty) \supseteq I_A = (-\infty,t_0,t_{4A-}) \leftrightarrow (\tau_{4-\infty},0,\tau_{4A-}) = \mathrm{I}_A \subseteq (\tau_{4-\infty},0,\tau_{4+\infty}) = \mathrm{I}_M$.

For the physical case, since $t_0 < -\gamma = -A_0/b_0$, we have $I_0 = (-\infty,t_0,-\gamma)$ if $b_0 < 0$ .Since $t_0 < t_{2A-} < -\gamma$, we have $I_{pM} = (-\infty, t_0,-\gamma) \supseteq I_{pA} = (-\infty, t_0, t_{2A-}) \leftrightarrow (\tau_{4-\infty}, 0, \tau_{4A-}) = \mathrm{I}_A \subseteq (\tau_{4-\infty}, 0, \tau_{4-\gamma-}) = \mathrm{I}_M$. If $b_0 > 0$, we have no physical case.

Example 2. $B(t) = b_0\, t$.**Case 4**. $\beta > (u_0)^2$ $(\delta > 1)$ where $\delta = \beta/(u_0)^2 = 4/(M_0 b_0 (u_0)^2)$.

Subcase 2. $u_0 > 0$ ($t_0 < -\gamma$) and $0 < (u_0)^2 < \beta \leq 2(u_0)^2$ $(0 < 1 < \delta \leq 2)$.

To obtain $I_N = \{t\in I_0:M_0A(t) > -2\} \supseteq I_M = (t_{M-},t_0,t_{M+})$ and $I_C = \{t\in I_0:M_0A(t) > -1\} \supseteq I_A = (t_{A-},t_0,t_{A+})$, we wish to solve

$$M_0\, A(t) = 2M_0\, A(t)/2 = 2[[(t + \gamma)/u_0]^2 - 1]/\delta > -r, \qquad r = 1,2$$
$$[(t + \gamma)/u_0]^2 - 1 > -r\, \delta/2, \qquad r = 1,2$$
$$[(t + \gamma)/u_0]^2 > 1 - r\, \delta/2, \qquad r = 1,2$$

For r = 2, this is always true as $1 - \delta < 0$. Hence $I_N = I_M = \mathbf{R} = (-\infty,t_0,\infty)$. For r= 1 we have $0 < 1 < \delta < 2$, $0 < 1/2 < \delta/2 < 1$, $-\delta/2 < 1/2 - \delta/2 < 0 < 1 - \delta/2$. Hence

$$[(t + \gamma)/u_0]^2 > 1 - \delta/2 > 0,$$
$$!(t + \gamma)/u_0! > \sqrt{1-\delta/2}\,, \qquad r = 1,2$$
$$!t + \gamma!/!u_0! > \sqrt{1-\delta/2}\,, \qquad r = 1,2$$
$$!t + \gamma! > !u_0!\sqrt{1-\delta/2}\,, \qquad r = 1,2$$
$$t + \gamma < -!u_0!\sqrt{1-\delta/2} \text{ or } t + \gamma > !u_0!\sqrt{1-\delta/2}\,, \qquad r = 1,2$$
$$t < -\gamma - !u_0!\sqrt{1-\delta/2} \text{ or } t > -\gamma + !u_0!\sqrt{1-\delta/2}\,. \qquad r = 1,2$$

Since $t_{4A\pm} = -\gamma \pm !u_0! \sqrt{1-\delta/2}$, $I_C = (-\infty, t_{4A-}) \cup (t_{4A+}, \infty)$. Since $0 < 1 - \delta/2 < 1$, we have $0 < \sqrt{1-\delta/2} < 1$ and hence $0 > -\sqrt{1-\delta/2} > -1$, so that $-1 < -\sqrt{1-\delta/2} < 0 < \sqrt{1-\delta/2} < 1$. Hence we have $-!u_0! < -!u_0! \sqrt{1-\delta/2} < 0 < !u_0! \sqrt{1-\delta/2} < !u_0!$.

Now since $u_0 > 0$, $!u_0! = u_0$, we have

$$-u_0 < -!u_0! \sqrt{1-\delta/2} < 0 < !u_0! \sqrt{1-\delta/2} < u_0.$$

$$-(t_0 + \gamma) < -!u_0! \sqrt{1-\delta/2} < 0 < !u_0! \sqrt{1-\delta/2} < t_0 + \gamma.$$

$$-\gamma - t_0 - \gamma < -\gamma - !u_0! \sqrt{1-\delta/2} < -\gamma < -\gamma + !u_0! \sqrt{1-\delta/2} < t_0 .$$

$$-\gamma - t_0 - \gamma < t_{2A-} < t_{2A+} < t_0 .$$

Hence $I_A = (t_{2A+}, t_0, \infty)$.

To obtain $I_M = \mathbf{R} = (-\infty, t_0, \infty) \supseteq I_A = (t_{2A+}, t_0, \infty) \leftrightarrow (\tau_{A-}, 0, \tau_{A+}) = \mathrm{I}_A = \mathrm{I}_A(t_0, M_0, A) \subseteq (\tau_{M-}, 0, \tau_{M+}) = \mathrm{I}_M = \mathrm{I}_M(t_0, M_0, A)$, we need

$$\tau_{4\pm\infty} = \lim_{t\to\pm\infty} h_{AB2}(t;u_0,\beta,\gamma) = \frac{\delta^2 |u_0|}{(\delta-1)^{3/2}}\left[\pm\frac{\pi}{4} - h_+\left(\frac{\pm 1}{\sqrt{(\delta-1)}}\right)\right] \quad (3.3.2.136)$$

$$\tau_{4A\pm} = \lim_{t\to 2A\pm} h_{AB2}(t;u_0,\beta,\gamma) = \frac{\delta^2 |u_0|}{(1-\delta)^{3/2}}\left[h_+\left(\frac{\pm\sqrt{(1-\delta/2)}}{\sqrt{(1-\delta)}}\right) - h_+\left(\frac{\pm 1}{\sqrt{(1-\delta)}}\right)\right]. \quad (3.3.2.137)$$

$$\tau_{4-\gamma\pm} = \lim_{t\to -\gamma^{\pm}} h_{AB2,4}(t;u_0,\beta,\gamma) = -\frac{\delta^2 |u_0|}{(\delta-1)^{3/2}}\left[h_+\left(\frac{\pm 1}{\sqrt{(\delta-1)}}\right)\right] \quad (3.3.2.138)$$

Hence $I_M = \mathbf{R} = (-\infty, t_0, \infty) \supseteq I_A = (t_{2A+}, t_0, \infty) \leftrightarrow (\tau_{4A+}, 0, \tau_{4A+\infty}) = \mathrm{I}_A \subseteq (\tau_{4-\infty}, 0, \tau_{4+\infty}) = \mathrm{I}_M$.

For the physical case, since $t_0 > -\gamma = -A_0/b_0$, we have $I_0 = (-\gamma, t_0, \infty))$ if $b_0 > 0$. Since $t_{2A-} < -\gamma < t_{2A+} < t_0$, we have $I_{pM} = (-\gamma, t_0, \infty) \supseteq I_{pA} = (t_{2A+}, t_0, \infty) \leftrightarrow (\tau_{4A+}, 0, \tau_{4+\infty}) = \mathrm{I}_A \subseteq (\tau_{4-\gamma+}, 0, \tau_{4+\infty}) = \mathrm{I}_M$. Since $[t_0, \infty) \subseteq I_{pA} = (t_{2A+}, t_0, \infty)$, we have $\lim_{t\to\infty} M(t) = 0$ and $\lim_{t\to\infty} R(t) = 2/M_0$. If $b_0 < 0$, we have no physical case.

Example 2. $B(t) = b_0 t$.**Case 4.** $\beta > (u_0)^2$ $(\delta > 1)$ where $\delta = \beta/(u_0)^2 = 4/(M_0 b_0 (u_0)^2)$.

Subcase 3. $0 < 2(u_0)^2 < \beta$ $(0 < 2 < \delta)$.

To obtain $I_N = \{t \in I_0 : M_0 \mathbb{A}(t) > -2\} \supseteq I_M = (t_{M-}, t_0, t_{M+})$ and $I_C = \{t \in I_0 : M_0 \mathbb{A}(t) > -1\} \supseteq I_A = (t_{A-}, t_0, t_{A+})$, we wish to solve

$$M_0\, \mathbb{A}(t) = 2M_0\, \mathbb{A}(t)/2 = 2[[(t+\gamma)/u_0]^2 - 1]/\delta > -r, \qquad r = 1,2$$
$$[(t+\gamma)/u_0]^2 - 1 > -r\,\delta/2, \qquad r = 1,2$$
$$[(t+\gamma)/u_0]^2 > 1 - r\,\delta/2, \qquad r = 1,2$$

For r = 1 and 2, this is always true as $2 - \delta < 0$. Hence $I_N = I_M = I_C = I_A = \mathbf{R}$.

After we obtain I_N, I_M, I_C, and I_A, to obtain $I_M = \mathbf{R} = (-\infty, t_0, \infty) \supseteq I_A = \mathbf{R} = (-\infty, t_0, \infty) \leftrightarrow (\tau_{M-}, 0, \tau_{M+}) = \mathcal{I}_M = \mathcal{G}_M(t_0, M_0, A_0, b_0) \subseteq (\tau_{A-}, 0, \tau_{A+}) = \mathcal{I}_A = \mathcal{I}_A(t_0, M_0, A_0, b_0)$, we need the images of the points t_0, $\pm\infty$, $t_{4A\pm}$ and, for the physical case, $-\gamma^{\pm}$. Recall

$$\tau_{4\pm\infty} = \lim_{t\to\pm\infty} h_{AB2}(t; u_0, \beta, \gamma) = \frac{\delta^2 |u_0|}{(\delta-1)^{3/2}}\left[\pm\frac{\pi}{4} - h_+\left(\frac{\pm 1}{\sqrt{(\delta-1)}}\right)\right]. \qquad (3.3.2.139)$$

For $1 < \delta \le 2$,

$$\tau_{4A\pm} = \lim_{t\to 4A\pm} h_{AB2}(t; u_0, \beta, \gamma) = \frac{\delta^2 |u_0|}{(1-\delta)^{3/2}}\left[h_+\left(\frac{\pm\sqrt{(1-\delta/2)}}{\sqrt{(1-\delta)}}\right) - h_+\left(\frac{\pm 1}{\sqrt{(1-\delta)}}\right)\right]. \qquad (3.3.2.140)$$

$$\tau_{4-\gamma\pm} = \lim_{t\to -\gamma^{\pm}} h_{AB2,4}(t; u_0, \beta, \gamma) = -\frac{\delta^2 |u_0|}{(\delta-1)^{3/2}}\left[h_+\left(\frac{\pm 1}{\sqrt{(\delta-1)}}\right)\right] \qquad (3.3.2.141)$$

Hence $I_M = \mathbf{R} = (-\infty, t_0, \infty) \supseteq I_A = \mathbf{R} = (-\infty, t_0, \infty) \leftrightarrow (\tau_{4-\infty}, 0, \tau_{4+\infty}) = \mathcal{I}_A = (\tau_{4-\infty}, 0, \tau_{4+\infty}) = \mathcal{I}_M$.

For the physical case, we have two subsubcases. If $b_0 > 0$, we have $I_0 = (-\gamma, \infty)$). If $t_{2A-} < t_{2A-} < -\gamma < t_0 < t_{2A+} < t_{2M+}$, we have $I_{pM} = (-\gamma, t_0, \infty) = I_{pA} = (-\gamma, t_0, \infty) \leftrightarrow (\tau_{4-\gamma-}, 0, \tau_{4+\infty}) = \mathcal{I}_A = (\tau_{4-\gamma-}, 0, \tau_{4+\infty}) = \mathcal{I}_M$. Since $[t_0, \infty) \subseteq I_{pA} = (-\gamma, t_0, \infty)$, we have $\lim_{t\to\infty} M(t) = 0$ and $\lim_{t\to\infty} \mathbb{R}(t) = 2/M_0$. If $b_0 < 0$, we have $I_0 = (-\infty, -\gamma)$. If $t_{2M-} < t_{2A-} < t_0 < -\gamma < t_{2A+} < t_{2M+}$, we have $I_{pM} = (-\infty, t_0, -\gamma) = I_{pA} = (-\infty, t_0, -\gamma) \leftrightarrow (\tau_{4-\infty}, 0, \tau_{4-\gamma+}) = \mathcal{I}_A = (\tau_{4-\infty}, 0, \tau_{4-\gamma+}) = \mathcal{I}_M$.

CONCLUSIONS

The process of agglomeration of particles in a fluid environment is an integral part of many industrial processes and has been studied by the scientific community. The fundamental mathematical problem is the determination of the number of particles of each **particle-type** as a function of time for a system of particles that may agglutinate during two particle collisions. Almost no work has been done for systems where particle-type requires several variables. Efforts have been focused on a particle-type list with only one variable, size (or mass). This allows use of what is often referred to as the **coagulation equation** which has been well studied in aerosol research. Original work on this equation was done by Smoluchowski and it is also referred to as **Smoluchowski's equation**. The **agglomeration equation** is perhaps more descriptive since the term coagulation implies a process carried out until solidification whereas our focus is on the agglomeration process; that is, on the determination of a time-varying **particle-size distribution** even if coagulation is never reached.

Smoluchowski considered the agglomeration equation in a discrete form in his original work. Later, this equation was considered in a continuous form by Müller. In both cases, an **initial particle-size distribution** to specify the initial number of particles for each size is needed to complete the **initial value problem** (IVP) which we refer to as the **agglomeration model**. Since both forms have an infinite number of sizes, the **state (or phase) space** is infinite dimensional. The probability that any two particles will collide and agglutinate is taken into account by the **kernel**, $K_{i,j}(t)$. Solution of the agglomeration model yields an updated particle-size distribution giving **number densities** as time progresses. Existence and uniqueness results have been obtained for bounded kernels and for special unbounded kernels. For a general coagulation-fragmentation equation which allows a time-varying kernel, a local existence theorem for the physical case has been obtained. Formulas for solutions are known for specific examples of kernels and initial particle sizes for both the discrete and continuous models although the correspondence of results is not always clear. For a time-varying kernel that is independent of particle size, $K_{i,j}(t)= A(t)$, formulas for the solution of the discrete model were investigated by Moseley including the general (not necessarily physical) case.

In this paper we have extended and clarified Moseley's results for specific examples of interest to the scientific community. First we assumed that the kernel, A(t), is an analytic function of time and looked for analytic solutions. This assures uniqueness via Taylor series, but not local or global existence. We focused on obtaining a finite process for computing the particle-size distribution using algebraic and analytical operations. This is fundamentally different from solving via Taylor series or numerical techniques that yield approximations at grid points. Specifically, we focused on the development of solutions for the following examples of time-varying analytic kernels:

$A(t) = A_0$	A_0 ! 0	Constant Kernel
$A(t) = A_0 + b_0 t$	A_0 ! 0, b_0 ! 0	Linearly Time-Varying Kernel
$A(t) = A_0 + b_0 \cos(\omega_0 t + \varphi_0)$	A_0 ! 0, $b_0 > 0$, $\omega_0>0$	Oscillating Kernel.

For the linearly time-varying kernel, $b_0 > 0$ indicates increasing cohesion and $b_0 < 0$ indicates decreasing cohesion. Specifically, we have obtained formulas for all of these cases. That is, we have developed formulas for the number of particles of each particle size (or mass) as a function of time for each of these kernels.

REFERENCES

Drake, R.L., 1972, A General Mathematics Survey of the Coagulation Equation, in *Topics in Current Aerosol Research*, Hidy, G. M. And Brock (Eds), Pergamon Press, New York.

Goldberger, W. M., 1967, Collection of Fly Ash in a Self-agglomerating Fluidized Bed Coal Burner, in *Proceedings of the ASME Annual Meeting*, Pittsburg,PA, 67-WA/Fu-3.

Kaplan, W. 1991, *Advanced Calculus*, Fourth Ed. Addison-Wesley Publishing Company, New York.

Lu, B., 1987, The Evolution of the Cluster Size Distribution in a Coagulation System, *J. Stat. Phys*. 49, pp 669-684.

Marcus, A., 1965, Unpublished Notes, Rand Corporation, Santa Monica, California.

Marsden, J.E. and M.J. Hoffman, 1999, *Basic Complex Analysis*, Third Ed. W. H. Freeman, New York.

McLaughlin, D. J., W. Lamb, and A. C. McBride, 1997, An Existence and uniqueness Result for a Coagulation and Multi-fragmentation Equation, *SIAM J. Math. Anal.* 28, pp 1173-1190.

McLeod, J.B., 1962, On a Finite set of Nonlinear Differential Equations *II Quart. J. Math.* Oxford 13, pp 193-205.

Melzack, Z. A., 1957, A Scalar Transport Equation, *Trans. Am. Math Soc*. 85 pp 547-560.

Morganstern, D., 1955, Analytical Studies Related to the Maxwell-Boltzmann Equation, *J. Rational Mech. Anal.* 4, pp 533-555.

Moseley, J. L., 1998, The Moment Method for Solving an Agglomeration Model with Constant Kernel, in *Proceedings of the IASTED International Conference on Modelling and Simulation*, Pittsburg PA, May 13-16, pp 476-480.

Moseley, J. L., 1999, The Moment Method for Solving an Agglomeration Model with Linear Kernel, in *Proceedings of the IASTED International Conference on Modelling and Simulation*, Pittsburg PA, May 5-8, pp 200-204.

Moseley, J. L., 2000, The Moment Method for Solving an Agglomeration Model with Constant Kernel, in *Proceedings of the IASTED International Conference on Modelling and Simulation*, Pittsburg PA, May 15-17, pp 200-204.

Moseley, J. L., 2007, The Discrete Agglomeration Model with Time Varying Kernel, *Nonlinear Analysis: Real World Applications*, 8 pp 405-423.

Muller, H., 1928, Zur Allgemeinen Theorie Der Raschen Koagulation, *Kolloidchemische Beihefte* 27 pp 2123-2150.

Siegell, J. H., 1976, *Defluidization Phenomena in Fluidized Beds of Sticky Particles at High Temperatures*. Ph.D Thesis, City University of New York.

Smoluchowski, M. V., 1917, Versuch Einer Mathematichen Theorie der Koagulationskinetik Kollider Lossungen, *Z. Phys. Chem*. 92 pp 129-168.

Spouge, J. L., An Existence Theorem for the Discrete Coagulation-Fragmentation Equations, *Math Proc. Camb. Phil. Soc*. 96 pp 351-357.

Treat, R. P., 1990, An Exact Solution of the Discrete Smoluchowski Equation and Its Correspondence to the Solution in the Continuous Equation, *J. Phys.* A 23 pp 3003-3016.

White, W.H., 1980, A Global Existence Theorem for Smoluchowski's Coagulation Equations. Proc. Am. Math. Soc. 80 (2).

Wunsch, A. D., 1994, Complex Variables with Applications, Second Ed., Addison-Wesley Publsihing Company, New York.

Yu, H., Analysis of Algorithms for the Solutionof the Agglomeration Equation, Masters Thesis, West Virginia University, Morgantown, WV.

Ziff, R. M., 1980, Kinetics of Polymerization, *J. Stat. Phys.* 23 pp 241-263.

INDEX

A

B

H

I

N

O

P

Q

R

S

T

U

V

W

Y